Dear Sreeni

With my compliments

Thank for all your support

Regards

Mark

EDDY STRUCTURE IDENTIFICATION
IN FREE TURBULENT SHEAR FLOWS

FLUID MECHANICS AND ITS APPLICATIONS
Volume 21

Series Editor: R. MOREAU
MADYLAM
Ecole Nationale Supérieure d'Hydraulique de Grenoble
Boîte Postale 95
38402 Saint Martin d'Hères Cedex, France

Aims and Scope of the Series

The purpose of this series is to focus on subjects in which fluid mechanics plays a fundamental role.

As well as the more traditional applications of aeronautics, hydraulics, heat and mass transfer etc., books will be published dealing with topics which are currently in a state of rapid development, such as turbulence, suspensions and multiphase fluids, super and hypersonic flows and numerical modelling techniques.

It is a widely held view that it is the interdisciplinary subjects that will receive intense scientific attention, bringing them to the forefront of technological advancement. Fluids have the ability to transport matter and its properties as well as transmit force, therefore fluid mechanics is a subject that is particulary open to cross fertilisation with other sciences and disciplines of engineering. The subject of fluid mechanics will be highly relevant in domains such as chemical, metallurgical, biological and ecological engineering. This series is particularly open to such new multidisciplinary domains.

The median level of presentation is the first year graduate student. Some texts are monographs defining the current state of a field; others are accessible to final year undergraduates; but essentially the emphasis is on readability and clarity.

For a list of related mechanics titles, see final pages.

Eddy Structure Identification in Free Turbulent Shear Flows

Selected Papers from the IUTAM Symposium entitled:
"Eddy Structures Identification in Free Turbulent Shear Flows"
Poitiers, France, 12-14 October 1992

edited by

J. P. BONNET
CEAT/LEA,
Poitiers Cedex, France

and

M. N. GLAUSER
Clarkson University,
Potsdam, N.Y., U.S.A.

KLUWER ACADEMIC PUBLISHERS
DORDRECHT / BOSTON / LONDON

Library of Congress Cataloging-in-Publication Data

Eddy structure identification in free turbulence shear flows :
selected papers from the IUTAM symposium entitled: "Eddy structures
identification in free turbulent shear flows," Poitiers, 12-14
October 1992 / edited by J.P. Bonnet and M.N. Glauser.
p. cm. -- (Fluid mechanics and its applications ; v. 21)
Includes index.
ISBN 0-7923-2449-8
1. Turbulence--Congresses. 2. Shear flow--Congresses. 3. Eddies--Congresses. I. Bonnet, J. P. II. Glauser, M. N.
III. International Union of Theoretical and Applied Mechsanics.
IV. Series.
TA357.5.T87E33 1993
620.1'064--dc20 93-26851

ISBN 0-7923-2449-8

Published by Kluwer Academic Publishers,
P.O. Box 17, 3300 AA Dordrecht, The Netherlands.

Kluwer Academic Publishers incorporates
the publishing programmes of
D. Reidel, Martinus Nijhoff, Dr W. Junk and MTP Press.

Sold and distributed in the U.S.A. and Canada
by Kluwer Academic Publishers,
101 Philip Drive, Norwell, MA 02061, U.S.A.

In all other countries, sold and distributed
by Kluwer Academic Publishers Group,
P.O. Box 322, 3300 AH Dordrecht, The Netherlands.

Printed on acid-free paper

Printed in the Netherlands

Chairmen of the IUTAM Symposium:

J.P. Bonnet (CEAT/LEA) J.C.R. Hunt (Met. Office)

Scientific Committee:

R.J. Adrian (USA)
J.P. Bonnet (France)
M.N. Glauser (USA)
F. Hussain (USA)
Sir J. Lighthill (U.K.)
R. Moser (USA)
H. Sato (Japan)
R.A. Antonia (Australia)
H.E. Fiedler (Germany)
J.C.R. Hunt (UK)
M. Lesieur (France)
J.L. Lumley (USA)
A.A. Townsend (UK)

Organizing Committee:

J.P. Bonnet - J. Delville
CEAT/LEA

Sponsoring Organizations and Companies:

- Aérospatiale
- Alliant Computer Systems France
- B.P. Research
- Commissariat à l'Energie Atomique, Vaujours
- Centre National de la Recherche Scientifique (CNRS)
- Département de la Vienne
- European Community On Flow, Turbulence And Combustion (ERCOFTAC)
- International Union for Theoretical and Applied Mechanics (IUTAM)
- Ministère de la Défense (DRET/DGA)
- Ministère de la Recherche et de l'Espace
- National Science Fundation (NSF, USA)
- Région Poitou-Charentes
- Silicon Graphics, France
- Université de Poitiers
- UFR Centre d'Etudes Aérodynamiques et Thermiques
- Ville de Poitiers

Contents

PREFACE

The existence and crucial role played by large-scale, organized motions in turbulent flows are now recognized by industrial, applied and fundamental researchers alike. It has become increasingly evident that coherent structures influence mixing, noise, vibration, heat transfer, drag, etc... The acceleration of the development of both experimental and computational programs devoted to this topic has been evident at several recent international meetings.

One of the first questions which experimentalists or numerical analysts are faced with is: how can these structures be separated from the background turbulence? This is a nontrivial task because the coherent structures are generally embedded in a random field and the technique used to determine when and where certain structures are passing, or their averaged characteristics (in the more probable or dominant role sense) is directly related to the definition of the coherent structure. Several methods or approaches are available and the choice of a particular one is generally dependent on the desired information. This choice depends not only on the definition of the structure, but also on the experimental and numerical capabilities available to the researcher. For example, from an experimentalists point of view, some approaches require multiple sensors in order to obtain simultaneous instantaneous information for several spatial locations. Other approaches, however, only require knowledge at one or at most a few spatial locations. Moreover, it is necessary to choose the optimal experimental configuration in order to characterize the 3D structures in a given flow. The problem is equivalent to trying to find the simplest (minimum) probe arrangement that gains maximum significant information about the structures. Certainly one can state, without question, that the recent rapid advances in, for example, data collection, instrumentation, digital data processing, and increased computational, graphical and storage capabilities of computers have resulted in an exponential increase in the applications of the various methods.

It was clear to us that the different techniques available for the identification of eddy structures were in a mature enough state to justify a specialized meeting. This volume contains selected papers from such a meeting, the IUTAM symposium entitled "Eddy Structure Identification in Free Turbulent Shear Flows" which was held in Poitiers France, October 12-14, 1992. This Symposium was followed by an ERCOFTCAC[1] Workshop devoted to direct comparisons of several detection schemes and codes on common data bases. The workshop was a natural extension of the IUTAM meeting, and, it is our feeling, that it opened new avenues for the selection and implementation of detection procedures. A summary of this work shop has been presented in ERCOFTAC Bulletin # 17, June 1993. The papers presented at the IUTAM

[1]European Community On Flow, Turbulence And Combustion

meeting and in this volume focuss predominantly on Free Turbulent Shear Flows. This is mainly because of the availability of sufficiently resolved numerical and experimental data bases in free shear flows. It is important to note however that several of the detection schemes presented here are applicable in wall bounded shear flows as long as the unique problems associated with them are accounted for (e.g., probe interference, mesh requirements, etc.). The purpose of the IUTAM Symposium was to address some of the following questions:

- What is the state of the art of structure identification?
- How do these approaches work?
- How do the results from different techniques compare with one another?
- What kind of information can be obtained from the various methods (e.g., instantaneous results, vorticity, averages, statistics, etc...)?
- What are the new directions in detection methods and, correspondingly, what are their implications for turbulent flow analysis and prediction methods?

At the IUTAM symposium, 74 papers and posters were presented, 9 of which were invited lectures. The invited lectures were selected to present the different aspects of the various approaches and were given by R.J. Adrian, J.D. Chashechkin, J.A. Ferré, M. Hayakawa, F. Hussain, M. Lesieur, J.L. Lumley, A.E. Perry, and A.A. Townsend.
Sessions were devoted to the following topics:

- Nonconditional sampling such as proper orthogonal decomposition.
- Stochastic estimation.
- Pattern recognition.
- Conditional sampling.
- Eddy structure identification and detection.
- Coherent structures visualization and numerical computation.

From the 74 contributions, 7 papers, particularly devoted to theoretical or computational aspects, were selected to be published in a special issue of Theoretical and Computational Fluid Dynamics (Vol. 5, 1993). The titles of these contributions are given at the end of this preface. The 38 papers presented in this book were selected from the remaining papers by the Scientific Committee of the IUTAM Symposium. This volume begins with introductory remarks, followed by 11 sections focussed on the topics which are listed below:

- Large and Small Scale Dynamics
- Conditional Sampling 1: Vorticity-based Methods
- Conditional Sampling 2: Other Methods
- Wavelets
- Pattern Recognition Analysis
- Proper Orthogonal Decomposition
- Stochastic Estimation
- Combined or Comparative Approaches
- Topological Concepts
- Full Field Methods

• Compressible Flows

A note is perhaps in order here: The classification of papers into sections is always subjective, particularly when considering coherent structures, because of the overlap of the various approaches.

The papers included in this volume indicate that most of the detection techniques are now mature, well documented and generally easy to use. Normally, techniques have been viewed as separate entities. There are papers included in this volume, however, which suggest that different approaches can be implemented in tandem to further our understanding of the dynamics of coherent structures. This is a point that has been largely overlooked in the past. It is our hope that this book will contribute to the state of the art of coherent structure identification processes and will help guide future experimental, numerical and theoretical research in this area.

The Editors
Jean-Paul Bonnet
Mark N. Glauser

Contents of the **Special Issue of** *Theoretical and Computational Fluid Dynamics* Vol. 5, 1993:

"Affine Matching Pattern Recognition Analysis of Eddy Structures in a Numerically Simulated Boundary Layer," P. R. Voke.

"The Structure of Inhomogeneous Turbulence in Variable Density Nonpremixed Flames," K. K. Nomura and S. E. Elghobashi.

"Understanding Coherent Vortices through Computational Fluid Dynamics," M. Lesieur.

"On the Potential-Vortex Mode in Stably-Stratified Shear Flow," C. Staquet.

"Pressure Field, Vorticity Field and Coherent Structures in 2-D Incompressible Turbulent Flows," M. Larchevèque.

"A Spatiotemporal Decomposition of a Fully Inhomogeneous Turbulent Flow Field," M. Manhart and H. Wengle.

"Structure Extraction by Stochastic Estimation with Adaptive Events," S. Balachandar and R. J. Adrian.

COLOUR PLATES

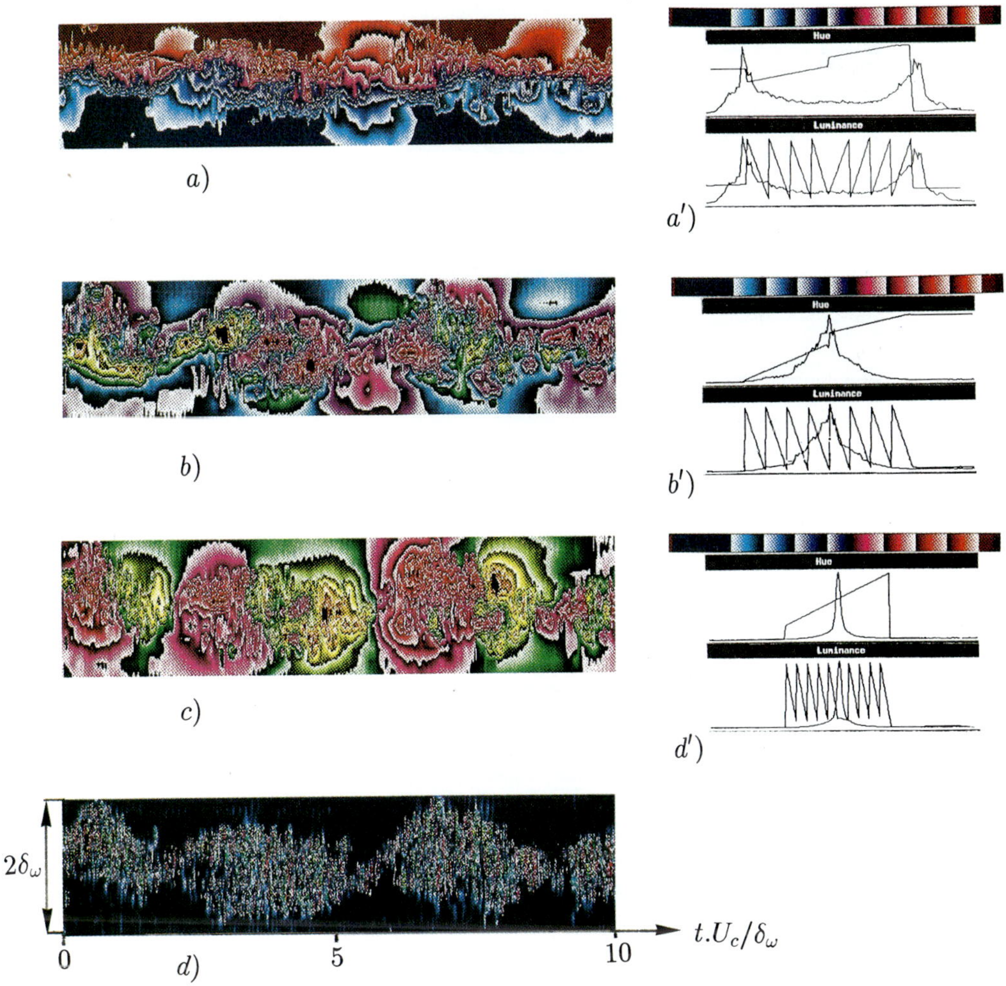

Figure 1: PFV of velocity rake being aligned across the mixing layer.

a) streamwise velocity: $U(y,t)$
b) fluctuating streamwise velocity: $u'(y,t)$
c) fluctuating lateral velocity: $v'(y,t)$
d) 2D vorticity : $\Omega_z(y,t)$

a') Color rule for Fig. 1a)
b') Color rule for Fig. 1b) and 1c)
d') Color rule for Fig. 1d)

From S. Bellin et al., p. 93.

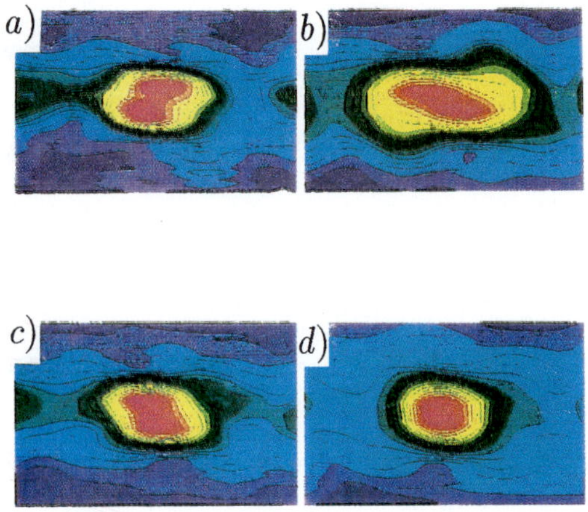

Figure 7: Iso-vorticity for $1.125\delta_\omega$ thickness, for several time shifts.
a) $-3\ 10^{-4}$; *b)* $-5\ 10^{-4}$; *c)* $3\ 10^{-4}$; *d)* $0s$.

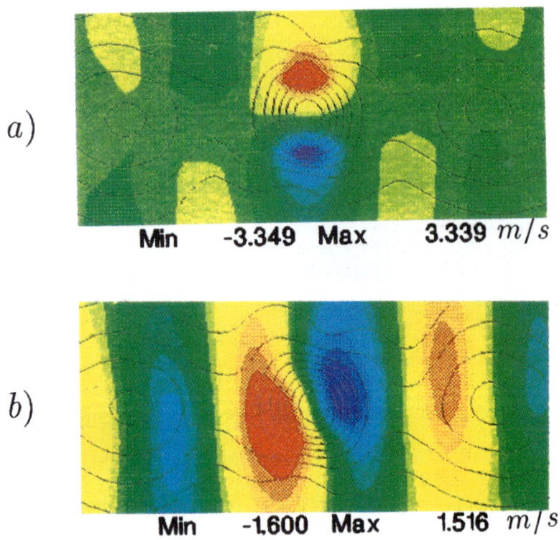

Figure 8: Triple Decomposition: Coherent Velocity, all events cumulated. *a)* u_c, *b)* v_c.

From S. Bellin et al., p. 99.

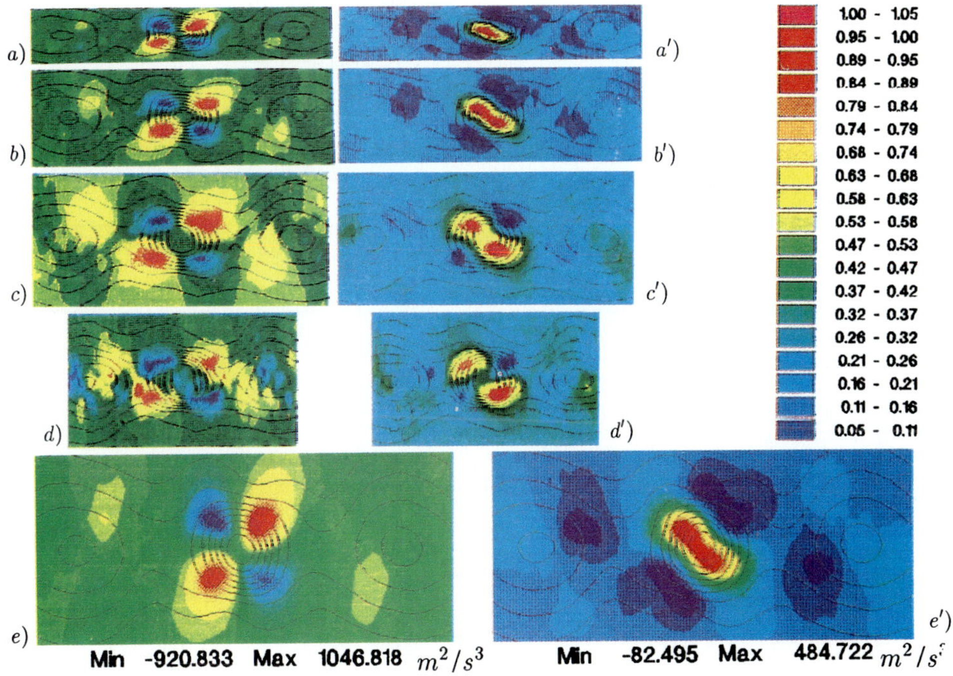

Figure 9: Triple Decomposition: Longitudinal and transversal productions for several thickness classes. *a*) $0.625\delta_\omega$, *b*) $0.875\delta_\omega$ *c*) $1.125\delta_\omega$ *d*) $1.375\delta_\omega$ *e*) all events.

From S. Bellin et al., p. 100.

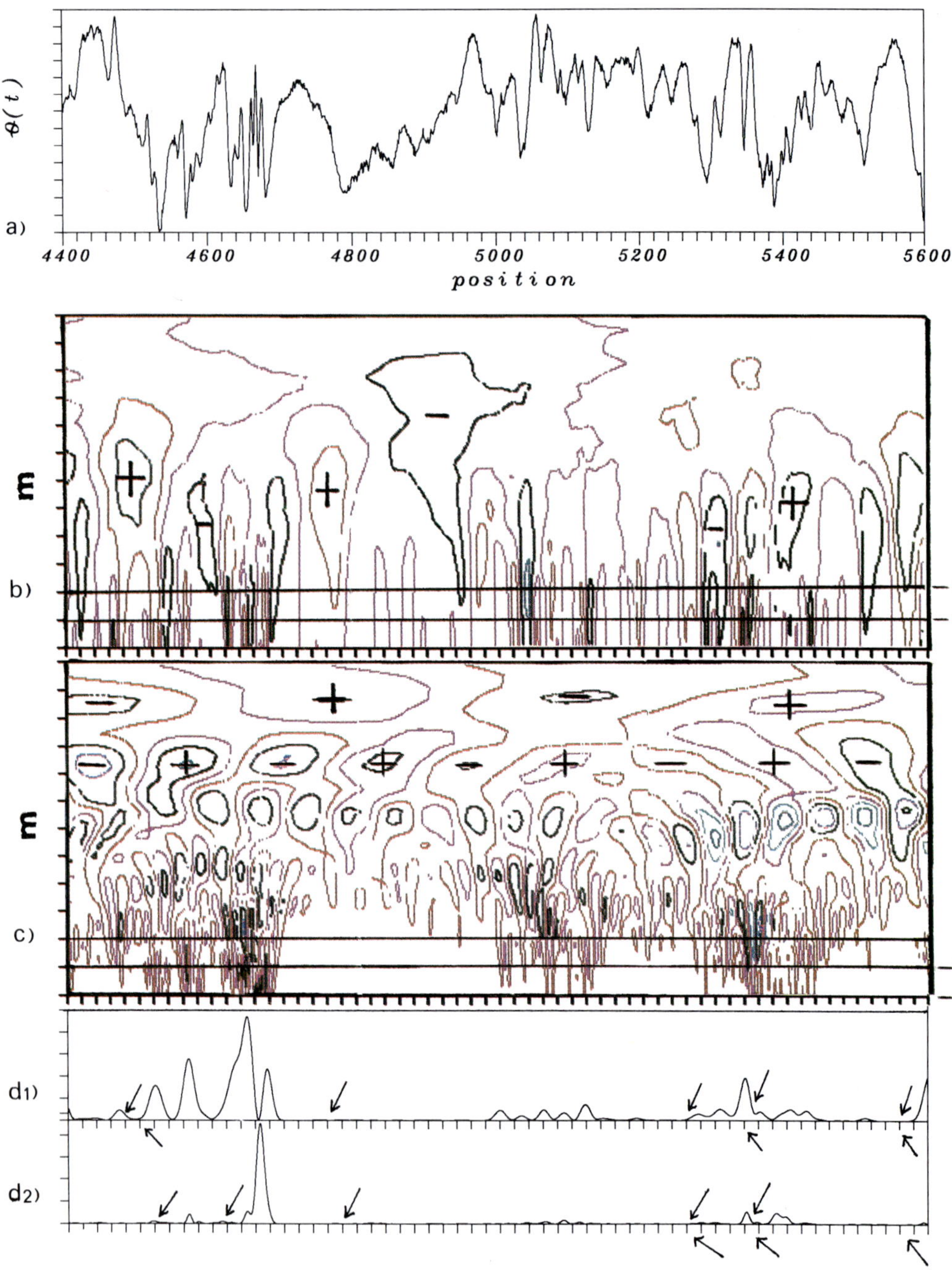

Fig. 3 Distributions of wavelet coefficients for a typical temperature signal obtained at $y^{+}=6$.
a). Temperature trace.; b). Haar wavelet.; c). Morlet wavelet.; d). Coefficients for one particular scale together with positions of Wag detected coolings using that window size: 1- Scale corresponding to 45 points; 2- 75 points.

From A. Benaissa et al., p. 164.

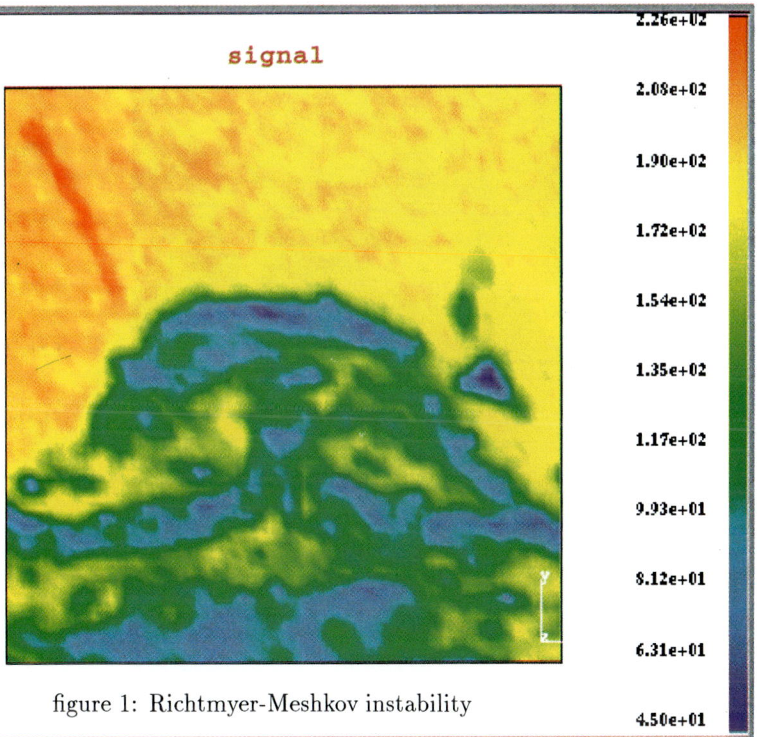

figure 1: Richtmyer-Meshkov instability

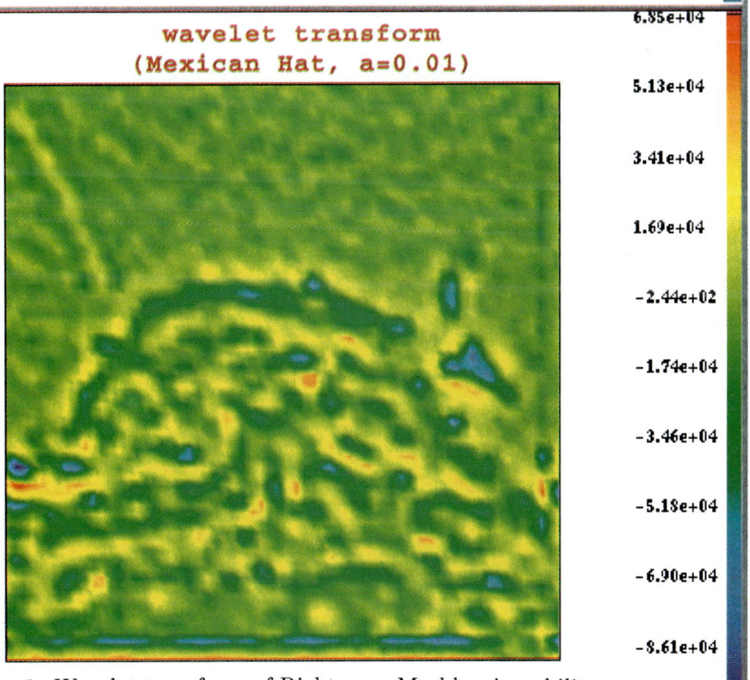

figure 2: Wavelet transform of Richtmyer-Meshkov instability, with the mexican hat wavelet

From Y. Lebret et al., p. 175.

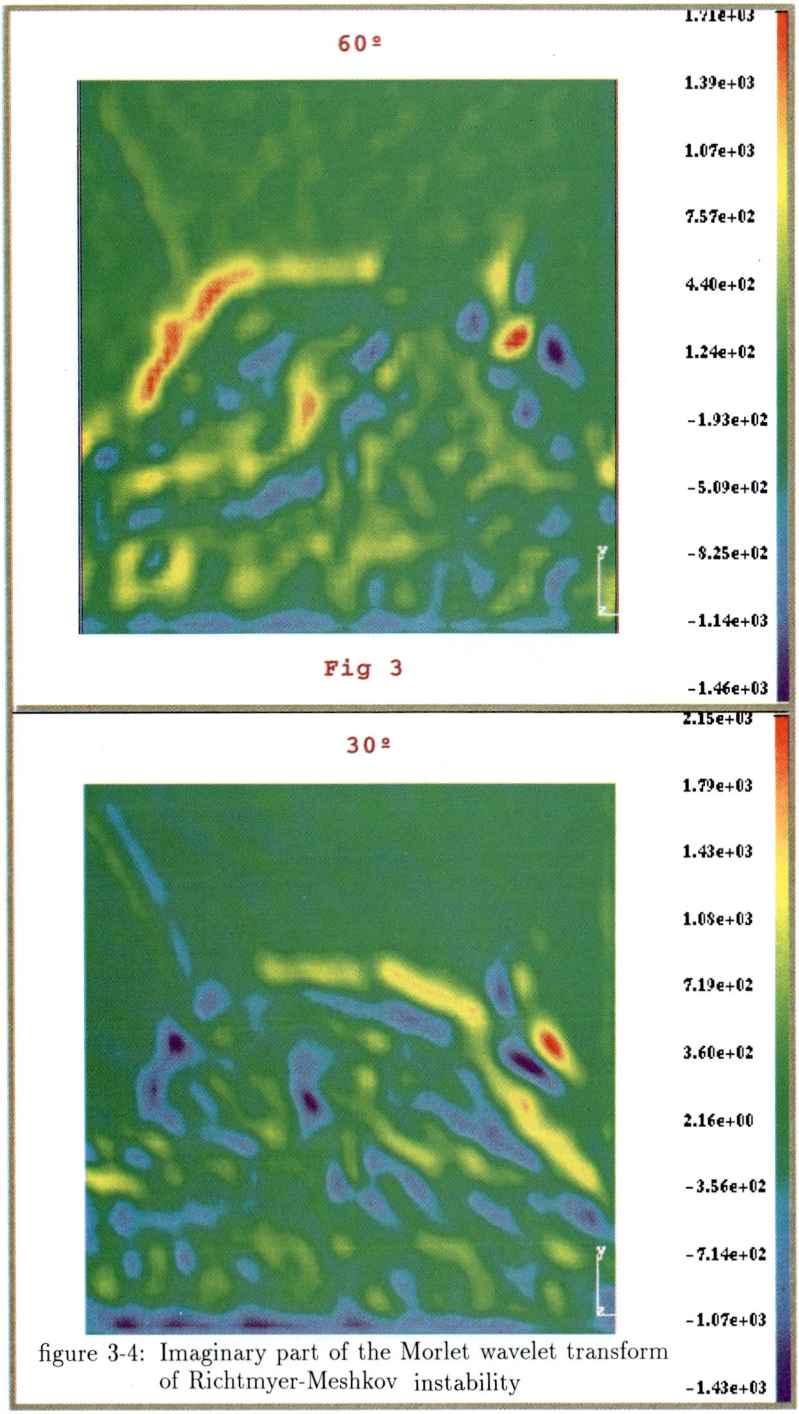

figure 3-4: Imaginary part of the Morlet wavelet transform of Richtmyer-Meshkov instability

From Y. Lebret et al., p. 176.

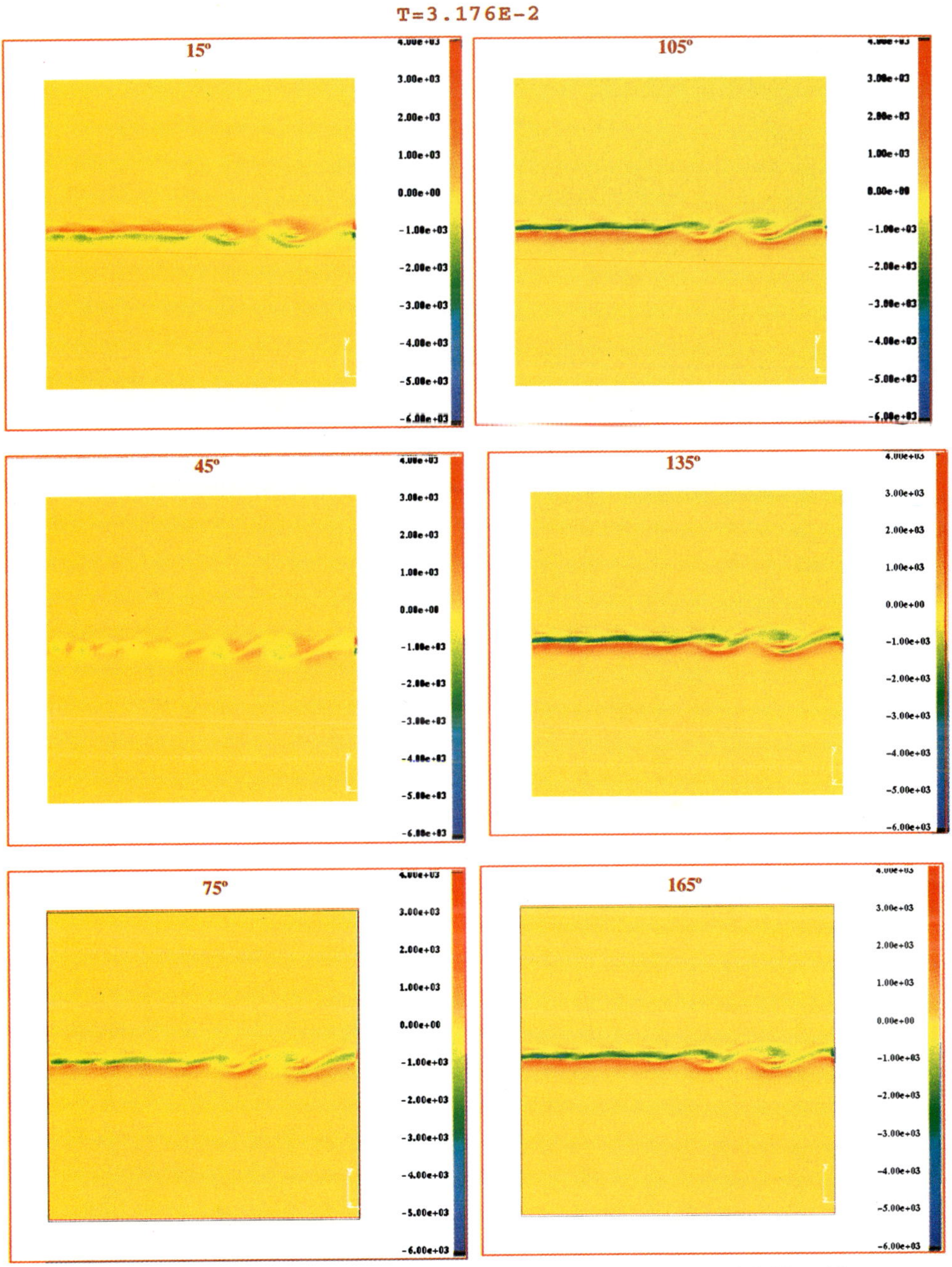

figure 5: Imaginary part of the Morlet wavelet transform of rotational fields, with different orientations, at $t = 3.176\ 10^{-2}\ s$

From Y. Lebret et al., p. 177.

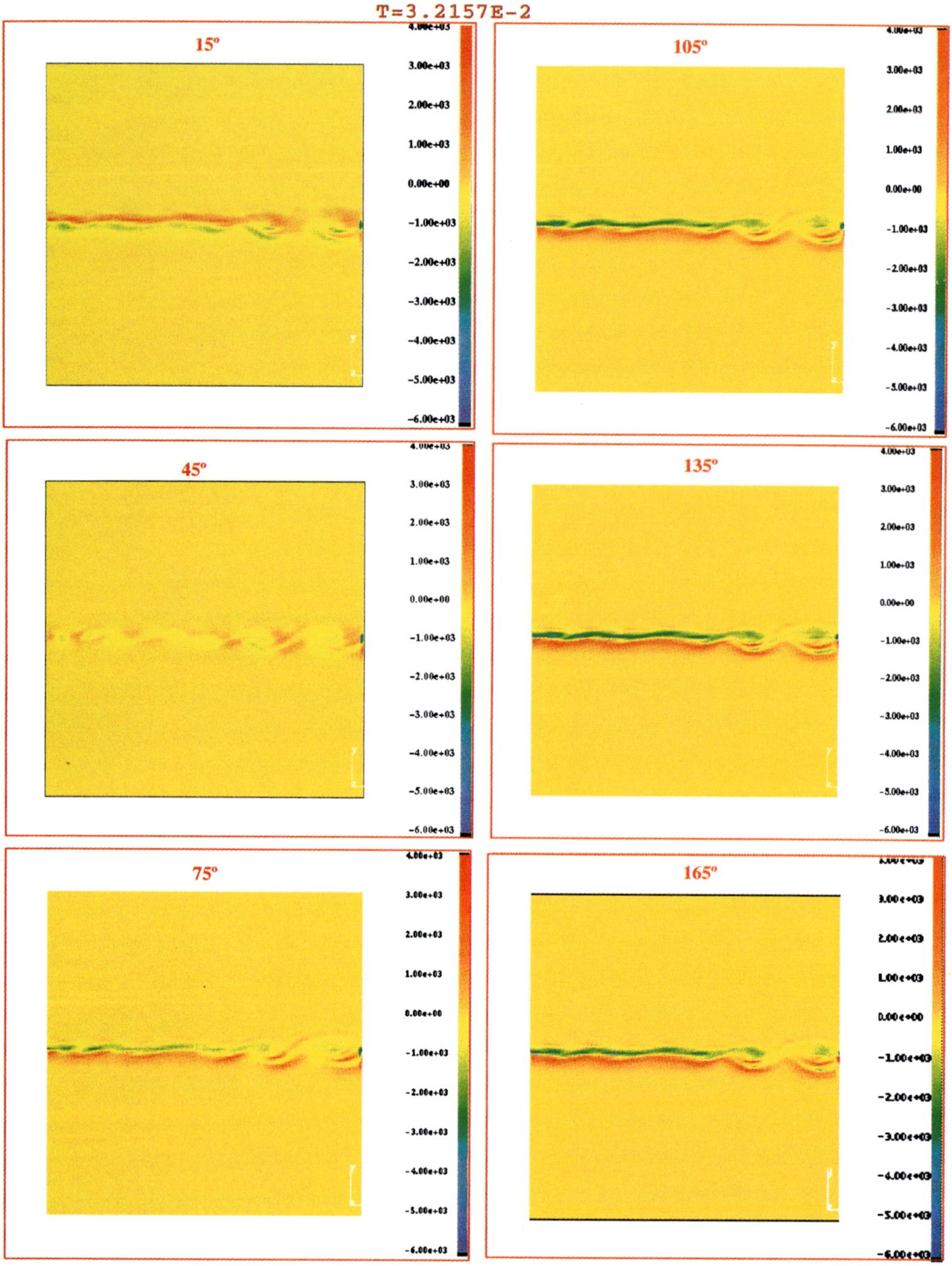

figure 6: Imaginary part of the Morlet wavelet transform of rotational fields, with different orientations, at $t = 3.2157\ 10^{-2}\ s$

From Y. Lebret et al., p. 178.

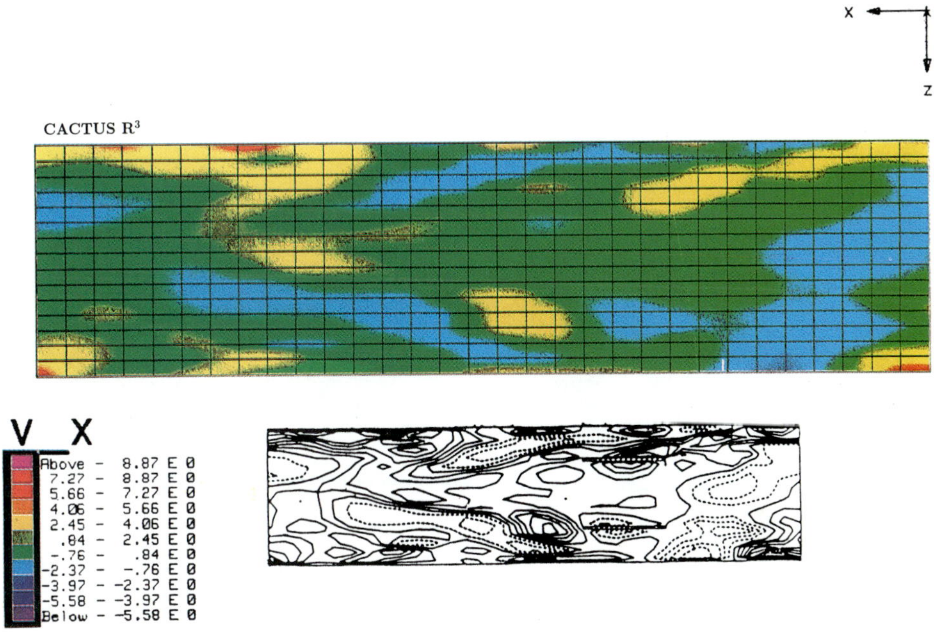

Fig.4 : Colour-coded u' intensity map, with original LES contour plot.

Fig.5 : Vertical section through the pressure field corresponding to Fig.4

From A.M. Savill et al., p. 200.

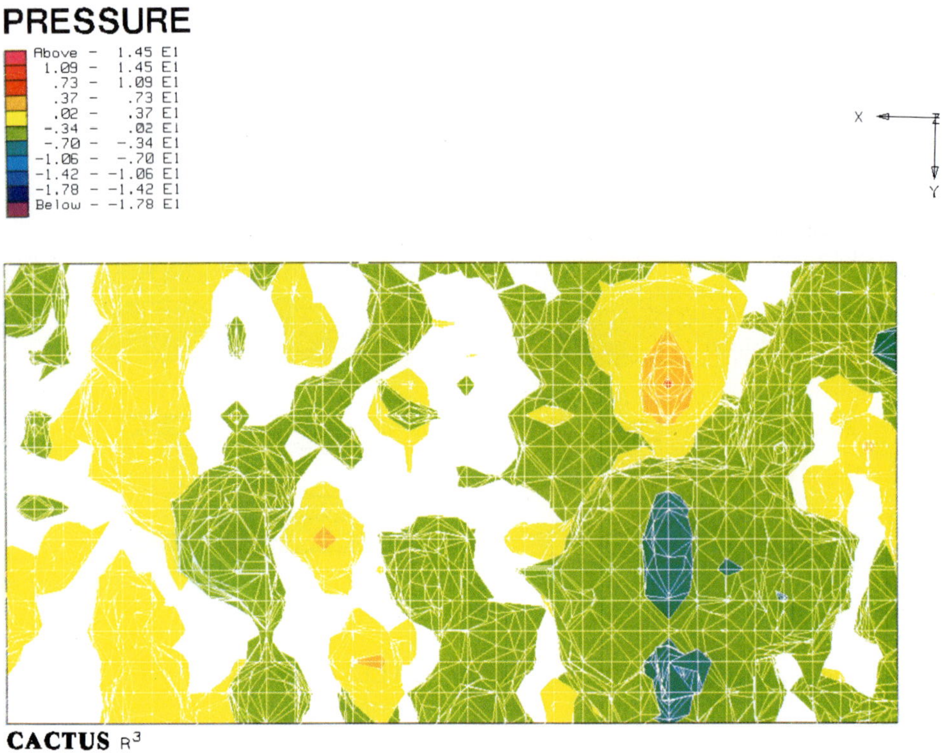

Fig.6 : Spanwise section through the pressure field between the lower manipulator and the bottom wall of the channel.

From A.M. Savill et al., p. 201.

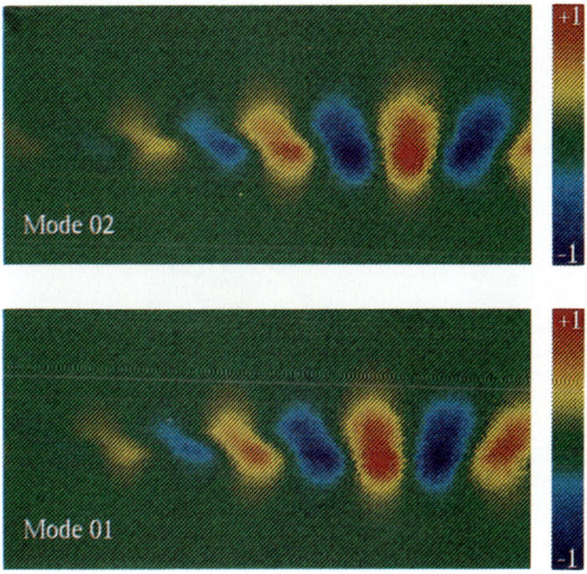

Figure 2: The mode 1 and 2 eigenfunctions, vertical component

From M. Rajaee et al., p. 261.

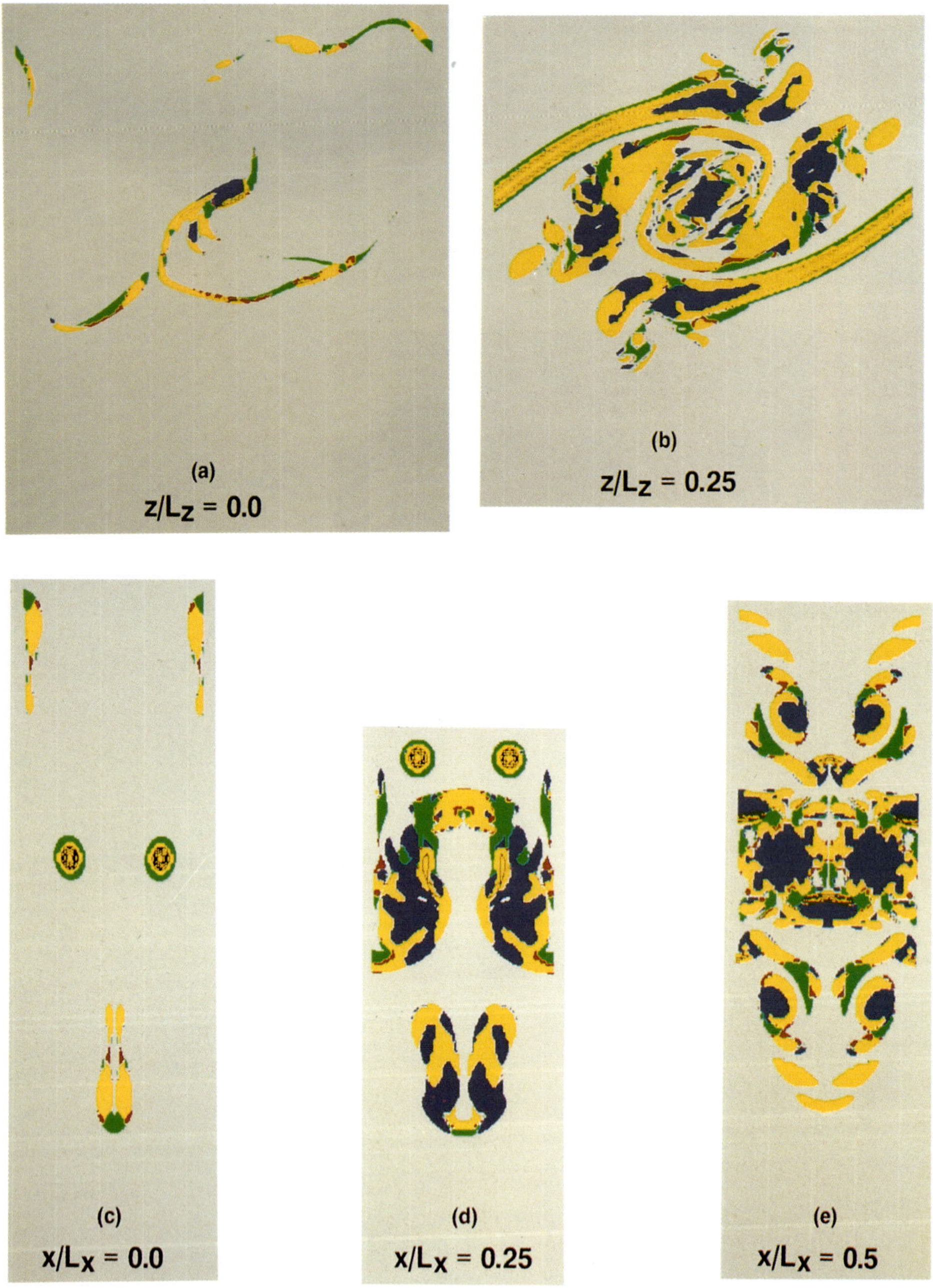

Figure 2. Topology map of xy-planes (a)-(b) and yz-planes (c)-(e) showing the local topology of the plane mixing layer for $|\omega| > 0.1|\omega|_{max}$. Black lines denote equally spaced $|\omega|$ contour lines for $|\omega| \geq 0.5|\omega|_{max}$.

From J. Soria and B.J. Cantwell, p. 388.

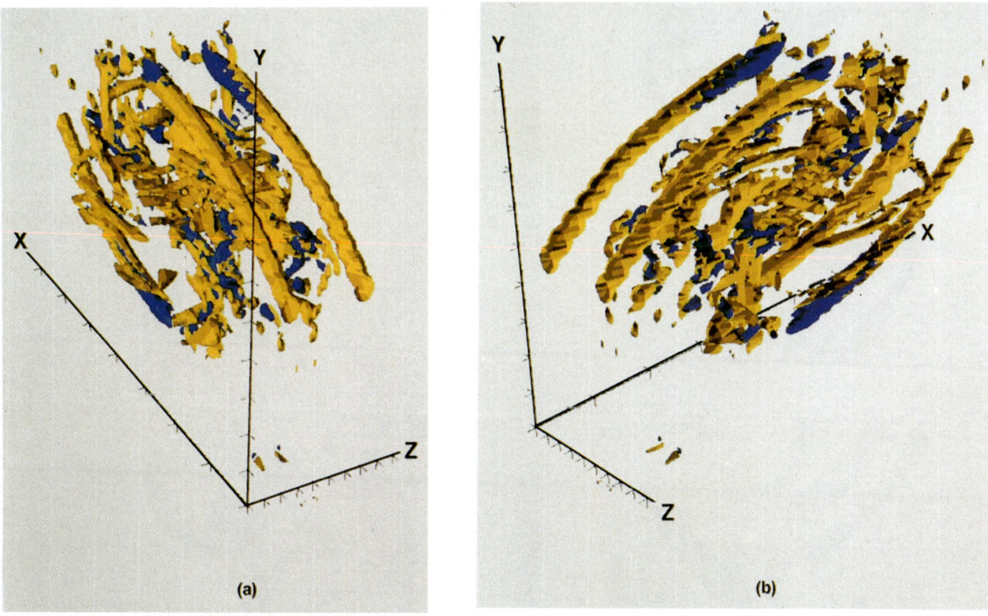

Figure 3. Visualisation of SF/S and UF/C focal (vortical) structures in the plane mixing layer shown as iso-surfaces with $|\omega| = 0.3|\omega|_{max}$.

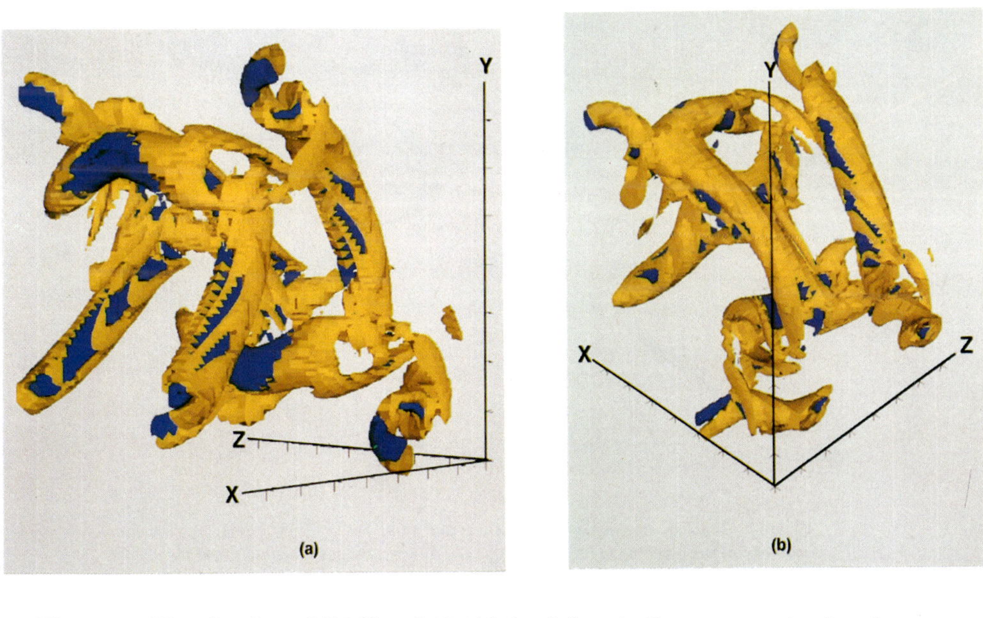

Figure 5. Visualisation of SF/S and UF/C focal (vortical) structures in the plane wake shown as iso-surfaces with $|\omega| = 0.4|\omega|_{max}$.

From J. Soria and B.J. Cantwell, p. 389.

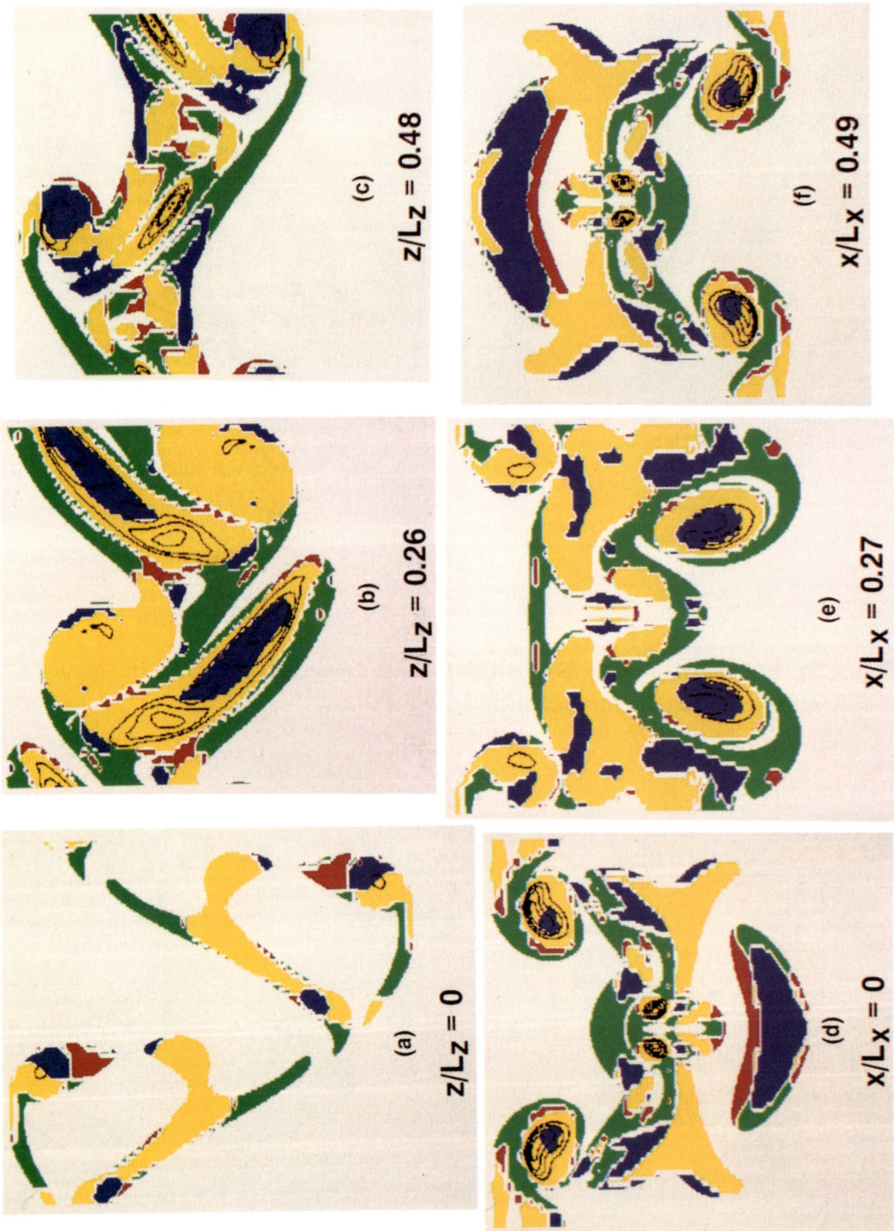

Figure 4 . Topology map of xy-planes (a)-(c) and yz-planes (d)-(f) showing the local topology of the plane wake for $|\omega| > 0.1|\omega|_{max}$. Black lines denote equally spaced $|\omega|$ contour lines for $|\omega| \geq 0.5|\omega|_{max}$.

From J. Soria and B.J. Cantwell, p. 390.

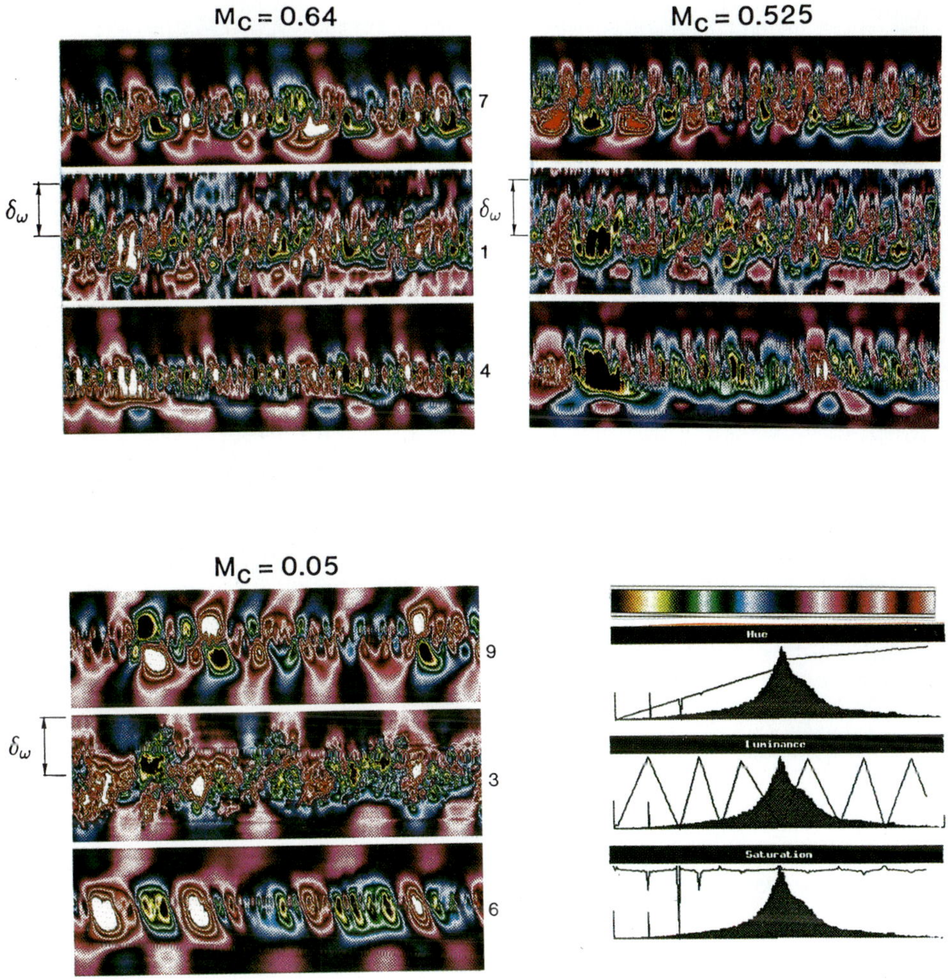

Figure 4: Pseudo Flow Visualizations of fluctuating instantaneous velocity:

M_c	0.64	0.525	0.05
direct meas.	1	2	3
first mode	4	5	6
second mode	7	8	9

From J.R. Debisschop et al., p. 462.

Introductory Remarks to the IUTAM conference and ERCOFTAC Workshop on "Eddy Structure Identification for free Shear Flows"

Hunt J.C.R.
U.K. Meteorological Office
Bracknell, U.K. RG12 2SZ

I would like to welcome you to this conference and explain why the study of the eddy structure of turbulence is, I believe, the most fruitful direction for turbulence research at present. This belief is based on the recent developments in the analysis understanding and description of turbulence, and the most appropriate techniques needed to study and calculate turbulent flows. These are summarised here as 6 statements (with some references to papers where these points have been explained).

1. Turbulent flows do not have (or even tend to) a universal structure; they all depend to a significant extent on initial and boundary conditions [Batchelor (1986) quoted by Hunt & Carruthers, 1990. Other evidence is reviewed by Hunt 1992].

2. At very high Reynolds number, the small scale eddy structure is approximately universal and self similar over a range of length scales (following Kolmogorov 1941), but some aspects appear to be non-universal (eg third, higher order, and non-isotropic moments) [Hunt & Vassilicos, 1991; Frisch, 1991].

3. Dynamics and kinematics of turbulence are defined as much by deterministic motions <u>within</u> characteristic eddies that persist over significant periods of time, and that have a complex internal structure with motions on many length scales (eg spirals, or turbulent "spots", "bursts"), as by random dynamical interactions <u>between</u> eddies, (Hunt & Vassilicos, 1991, Lundgren 1982, Hussain, 1986). The <u>distribution</u> of these characteristic eddies or eigensolm is influenced by the initial and boundary conditions if the degree of influence varies according to the flow.

4. For most kinds of shear flows, the energetic large scale eddy motions have common features in their structure as observed in individual realisations or coherent structures and in the statistics of the velocity fluctuations. (Townsend 1976; Hussain 1986; Hunt & Carruthers 1990. Antonia & Bisset 1991).

J. P. Bonnet and M. N. Glauser (eds.), Eddy Structure Identification in Free Turbulent Shear Flows, 1–4.

5. The strength of the evidence both for self similarity over a range of length scales, and for the existence of the common features of turbulence in different flows, depends on the measures used for studying turbulence. For example much higher values of Reynolds number are required to demonstrate small scale self-similar behaviour using spectra than using fractals [Vassilicos & Hunt, 1991], and measurements using conditional sampling, demonstrate the common eddy structure for shear flows more readily than measurements of spectra, correlations etc (eg Antonia & Bisset, 1991).

6. Different approximate methods of calculation are appropriate for different flows and different statistics (and therefore for different applied problems). For example in shear flows, where turbulence has an approximately general structure on a local scale, simple eddy viscosity methods are appropriate. On the other hand, for modelling flows where the local structure has no general form (such as turbulence driven by convection, or global atmospheric eddies) full computation of the unsteady motions is necessary, at least on the scale of the energetic eddies, (such as Large Eddy Simulation, or Numerical Weather Prediction [Hunt, 1992]).

It follows that the central activity of turbulence research should now be centred on the measurements, analyses and conceptual understanding of eddy structures, for different classes of turbulent flow. Researchers should in my opinion abandon the search for a completely general theory or model and also abandon the design of experiments with that goal in mind. Because eddy structure cannot be universal, this conference is mainly concerned with eddy structures in one class of turbulence, namely free shear flows, although there are many interesting papers on studies new analytical techniques (Aubry et al 1987) have been developed that may have application in other types of turbulent flow.

Finally, it should be remembered that research into eddy structure is providing the concepts that lead to modifications and new designs of turbulent flow to solve engineering and environmental problems. Some examples are given at this conference. Researchers should be encouraged to look for these applications, because they are sometimes better placed to see how the new concepts might be applied than the engineer, meteorologist and environmentalists dealing with practical problems on a daily basis. There should, in future, be more projects jointly conducted by researchers and the practitioners. Two important applications that interest me are the structure of the most intense gusts in atmospheric boundary layers (Hall et al 1993) and the use of bubbles to suppress vortex shedding and vibrations caused by ocean currents in the wake of circular cylinders in water currents, a problem of importance to the oil industry where drilling tubes are destroyed by such vibrations. These applications described at this conference Moros and Hunt (1993). It is also well known that devices for drag reduction and for improvements in combustion and noise suppression have been based on the results of recent research in this field (as mentioned in the paper of Hussain 1993).

I conclude by wishing the conference every success and by forecasting that the study of the eddy structure of turbulence will continue to be an essential component of both fundamental and applied research on turbulence.

References

[1] Antonia R.A. & Bisset D.K. 1991
"3-D aspects of the organised motion in the turbulent boundary layer"
In "Turbulence and coherent structures" (Ed. 0. Metais & M. Lesieur), pp.141-158, Kluwer

[2] Aubry N., Holmes P., Lumley J.L., Stone E. 1988
"The dynamics of coherent structures in the wall region of a turbulent boundary layer"
J. Fluid Mech., 192, pp.115-173

[3] Batchelor G.K. 1986
Opening remarks of First European Turbulence Conference, Lyon.

[4] Frisch U. 1991.
"From global scaling, à la Kolmogorov, to local multifractal scaling in fully developed turbulence"
Proc. Roy. Soc. A., 434, 89-100

[5] Hall C.D., Hunt J.C.R., Radford A.M., Carruthers D.J. & Weng W.S. 1993
"Forecasting tropical cyclones and near surface wind conditions"
Proc. ICSU/WMO Int. Symp. on Trop. Cyclone Disasters. Beijing Oct.1992 - In press

[6] Hunt J.C.R. & Carruters D.J. 1990
"Rapid distortion theory and the "problems" of turbulence"
J. Fluid Mech., 212, pp.497-532

[7] Hunt J.C.R. & Vassilicos J.C. 1991
"Kolmogorov's contributions to the physical and geometrical understanding of small scale turbulence and recent developments"
Proc. R. Soc. A (1991) 434, pp.183-240

[8] Hunt J.C.R. 1992
"Developments in computational modelling of turbulent flows"
In "ERCOFTAC workshop on numerical simulation of unsteady flows" (Ed. O. Pirroneau, et al) Cambridge Univ. Press

[9] Hussain F. 1986
"Coherent structures and turbulence"
J. Fluid Mech., 173, pp.303-356

[10] Hussain F. 1993
"Understanding coherent structures through vortex dynamics"
Proc. IUTAM Conf. on "Eddy structure identification", Kluwer

[11] Kolmogorov A.N. 1941
"The local structure of turbulence in incompressible fluid for very large Reynolds numbers"
Dokl. Akad Nauk SSSR 30, pp.301-305

[12] Lundgren T.S. 1982
"Strained spiral vortex model for turbulent fine structure"
Phys. Fluids 25, p.2193

[13] Moros A. & Hunt J.C.R. 1993
"Coherent structures in fluid-structure interaction phenomena"
Paper presented at the IUTAM Conf. on "Eddy structure identification", Poitiers, Oct. 92

[14] Townsend A.A. 1976
"Structure of turbulent shear flow"
Cambridge Univ. Press

[15] Vassilicos J.C. & Hunt J.C.R. 1991
"Fractal dimensions and spectra of interfaces with application to turbulence"
Proc. R. Soc. A 435, pp.505-534

Organised Eddies in Turbulent Shear Flows

Townsend A.A.
Emmanuel College
CB23AP Cambridge, U.K.

The subject of the conference might provoke a bystander to wonder why so much brain power and material resources are being applied to extracting recurrent patterns from chaotic turbulent flows and what we hope to do with the information if and when we have it. If we regard the scientific study of turbulent flow as beginning with the work of Osborne Reynolds and his mean value equations for mean velocity and mean turbulent kinetic energy, most work over the first fifty years took place in the absence of information about the mechanisms of turbulent transport of momentum and heat. Nevertheless, the early workers made important and enduring advances in basic understanding, in particular, by recognising that, although the transfer of mechanical energy to heat must be by working against viscous stresses, the magnitude of the transfer is nearly independent of the magnitude of the fluid viscosity. It was concluded that the transfer from motions on scales comparable with flow widths to the motions of very small scale that dissipate the energy depends on successive, quasi-independent transfers between "eddies" of increasingly smaller sizes. Today, that is expressed in the Kolmogorov theory of local isotropy which leads to the well-established $k^{-5/3}$ spectrum. Another was the deduction of the logarithmic distribution of velocity close to a solid boundary. These achievements depended mostly on consideration of equilibrium flows in which it was plausible to suppose that the motion was determined by simple boundary conditions and the use of dimensional analysis.

Any observation of water flows in rivers and tidal channels is sufficient to show that the rapid transport of momentum across the velocity gradient depends on the existence of organised movements with vertical and horizontal velocities being correlated over considerable distances but, looking back, there was a tendency to consider only the correlation without thought for the movements themselves. An important difference between motion of an incompressible fluid and the motion of molecules in a gas is that long-range interactions are not only present but may be dominant and so rule out use of theories using only local conditions. The change in attitude was slow but the beginning was two linked developments, the theory of isotropic turbulence of G.I. Taylor and the use of the hot-wire anemometer by Dryden, Simmonds and

J. P. Bonnet and M. N. Glauser (eds.), Eddy Structure Identification in Free Turbulent Shear Flows, 5–9.

many others. The work of G.I. Taylor first showed that the two-point velocity covariance could measure length scales of the turbulent motion and that the equations for the time evolution offered the possibility of constructing a theory of turbulence without making dubious assumptions. All of this led to many measurements of correlation functions and turbulent spectra and indeed to the local isotropy theory of Kolmogorov, but grid turbulence, the experimental approximation to isotropic turbulence, does not possess characteristic directions and the correlations can reveal little about eddy flow patterns. More interesting and informative functions are found in shear flows for which there are three characteristic directions, of the mean flow, of the mean velocity gradient and the transverse direction of near spatial homogeneity.

Early use of the term "large eddy" related to measurements of the energy balance for turbulence in the wake of a cylinder. The Reynolds mean value equation for the kinetic energy of the fluctuations at a point has terms for energy production by working against Reynolds shear stress, energy loss by viscous dissipation, net gain from advection by the mean velocity and the effect of lateral diffusive transport by velocity and pressure fluctuations. All of these terms could be measured with the simple anemometers of the day and lateral energy transport was found not only to be strong but also to be locally both up and down the gradient of turbulent energy. The conclusion was that important agents of transport were large circulations extending over at least half the width of the flow.

In 1958, H.L. Grant made measurements of correlations of the three velocity components with separations in the three co-ordinate directions, first in a cylinder wake and then in a boundary layer. From study of the wake measurements, he concluded that, if much of the turbulent energy were in the form of inclined double-roller eddies extending over the whole width of the flow, the complex form of the correlations could be explained. The conclusion was supported by application to the same data of the proper orthogonal decomposition analysis by J.L. Lumley. For the boundary layer, evidence for the double-roller structure was found in the outer part of the flow but the form of the correlation functions in the stream direction, particularly those of the streamwise component, changed very little with distance from the wall. An interesting point is that, if the rapid distortion approximation is used to calculate the effect of plane shear on initially isotropic turbulence, the calculated correlation functions for total strains of about two resemble very closely those measured in the wake. That could be if, during the development of the flow, there were periods of disorganised, quasi-isotropic motion, followed by "rapid distortion" growth and lastly by cessation of growth as the roller eddies become aligned with flow direction. While detailed evidence for such a growth-decay cycle has yet to be found, other growth-decay cycles have been observed.

Whatever the mechanism for control of average levels of turbulent intensities and stresses, the structures of simple free flows are broadly similar and their differences can be described by two velocity scales and a length scale. One velocity scale refers to the mean velocity profile and the other to turbulent intensities and rate of energy dissipation but the length scale is the flow width and the same for both. Using the three scales and the turbulent energy equation integrated over the flow width leads to a relation between non-dimensional rate of spread of the flow and form parameters that predicts, say, the rates of spread of jets and wakes. However, the entrainment velocities do not keep in constant ratio to either of the velocity scales and so the entrainment rates depend on the various flow geometries and so do not conform to the similarity of the turbulent structure.

For free flows, entrainment is essential for the supply of energy from the mean flow and S. Corrsin showed from oscilloscope traces of signals from a hot-wire anemometer that fluid in characteristically turbulent, highly rotational flow is separated from ambient fluid in potential flow by a remarkably sharp bounding surface. Every time the surface passes across the wire, the nature of the velocity fluctuation changes from one with very high frequency components to one with mostly much lower frequencies, and, at the time of transition, the sensor will lie on the bounding surface. By sampling other anemometers at the same time, L.S.G. Kovasznay and others have explored the flow fields near the surface. The simplest use of conditional sampling is to measure the intermittency ratio, the proportion of time that the wire is within the region of fully turbulent flow. I. Gartshore collected information about the distribution of the ratio with distance from the flow centre, and showed that the widths of the region of variation as ratios of the flow width are much larger for wakes than for jets and boundary layers. Taking into account the results of assuming structural similarity, it must be that the entrainment is forced to a rate set by the flow geometry, and the entrainment mechanism finds a suitable structure. For example, in a wake, the indentations of the surface are in the form of short, transverse and deep waves while, in boundary layers, they form shallow troughs extending in the direction of flow. An interpretation is that the deep indentations of the wake arise from Helmholtz instability of the surface whereas the shallower indentations of the boundary layer are forced by the double-roller eddies.

Conditional sampling has been very useful for the study of flow structures although it depends on an unambiguous specification of a condition. The feature is that it enables inspection at particular stages or eddy development, for example in the growth-decay cycle that exists in most developing flows. Perhaps the best-known use has been in the study of the bursting phenomenon in wall flows. On a smooth wall, break-away from the surface occurs in bursts separated by periods of comparatively quiescent flow, the “sweeps”, and the progress of the bursts into the outer flow can be followed. The origin of the bursts seems to be in the development of longitudinal roller eddies near the viscous sublayer of the kind revealed by another variable interval time averaging, in effect another form of conditional sampling. However, there must be some doubt about the necessity for bursts being set off by conditions in the sublayer since the turbulent motion over rough boundaries is, scaled with friction velocity and flow width, not distinguishable from that over smooth ones.

Just as the initial impetus to study details of flow structure came with the availability of hot-wire anemometers, the use of computers and data recording has increased greatly the possibilities of studying flow structures. A feature of much modern work is the use of arrays of anemometers with digital recording of the fluctuations for subsequent computer analysis, enabling the production of instantaneous velocity distributions over two-dimensional spaces with hundreds of data points. The main application has been for image recognition of flow structures, in particular using conditional biasing of the basic recognition routine to provide information about the flow structure to provide information about the flow structure at particular stages of the growth-decay cycle or of the entrainment process.

More recently, it has been possible to compute in detail both in space and time flow fields of the simpler free turbulent flows at flow Reynolds numbers sufficiently large to believe their general validity. With information of this kind, the possibility exists of identifying a flow pattern by seeking a match to one found by simple pattern recognition, tracing it back in time to its generation and forward in time to its eventual dissolution. Repeating the process with updated versions of the initial version of space-time history would lead to much better understanding of growth-decay cycles that would be very useful information for prediction schemes.

In my opinion, the best of the reasons for close investigation of eddy structures is that it leads to a basis for the understanding of the mechanics of the flow, and so to an informed attitude to the formulation of the generalised statements that are likely to be part of a prediction scheme. To some extent, the simple free turbulent flow - wakes, jets, mixing-layers and boundary layers - can be treated in terms of flow regions with motions determined by a small set of boundary conditions, and the application of overall conservation of momentum and kinetic energy fluxes. If the methods are extended to strongly curved flows or flows with considerable buoyancy fluxes, the predictions are very far out, essentially because the mechanism of energy transfer to the turbulent motion is very different. Even for the boundary layer, trouble arises if the motion in the wall layer is regarded as independent of that in the outer part of the flow. The argument is that most of energy production and dissipation occurs in the wall layer and consequently the boundary conditions for flow in the wall layer are the shear stress at its outer limit, the viscous no-slip condition at the wall and the shear stress at the wall. The outer flow is defined by the stress at its boundary with the wall layer and its depth. In a very elegant paper, Millikan showed that a consequence would be that the mean velocity distribution was logarithmic, but another consequence of the assumptions is that turbulent intensities and correlation functions scale with the wall stress and distance from the wall. Early measurements of intensities, correlations and pressure fluctuations all showed strong influence from the outer layer, and it is now thought that all the larger eddies of the flow - those contributing most to local Reynolds stress - have flow fields extending to the wall where all contribute considerably to fluctuations parallel to the wall. After all, the local scale in the wall layer is proportional to distance from the wall and the most natural way for this to happen is if all the larger eddies are constrained by a wall condition of zero normal velocity.

Besides the simple flows that develop smoothly, there are some interesting, possibly significant flows which undergo a rapid change of environment after development in steady conditions. Well-known examples are boundary layers moving from a surface of one roughness to a different surface or entering a region of strong adverse pressure gradient, both of which may be treated by using the attached eddies model. Another is the behaviour of a plane mixing layer as it passes around a 90° corner with a rapid diminution in the stress-intensity ratio, explicable as the effect of the strongly curved ambient flow on the initial double-roller eddies. Even more spectacular is the effect on a plane wake as it enters a region of ambient flow with strong extension parallel to the generating cylinder. The Karman-like eddies becomes much stronger and the entrainment rate much larger as a result of direct energy transfer from the ambient flow field. In any such case, a good knowledge of flow structures must be a starting point for understanding the flow response.

I. Large and Small Scales Dynamics

NEW APPROACHES TO VORTEX DYNAMICS: CORE DYNAMICS, HELICAL WAVES, AND INTERACTION WITH FINE SCALE TURBULENCE

FAZLE HUSSAIN

Department of Mechanical Engineering
University of Houston, Houston
TX 77204-4792, U.S.A.

Some new aspects of vortex dynamics, coherent structures and interaction with fine-scale turbulence are reviewed. In the absence of any appropriate measurement technology for the 3-D instantaneous vorticity field we rely on direct numerical simulation of the Navier-Stokes equations using spectral methods in an 128^3 periodic cube. Recognizing from our study of the vortex reconnection mechanism that vortex lines associated with coherent structures are frequently helical, we study first the evolution of a laminar vortex column with a nonuniform core at a Reynolds number $Re \equiv \Gamma/\nu = 665$, and then the vortex's evolution in the presence of fine-scale homogeneous, isotropic turbulence. We show that core dynamics, ignored so far, can be very significant in vortex dynamics and vortex interactions and consists of travelling vorticity wave packets along the vortex axis. We also show that core dynamics can be better explained by *complex helical wave decomposition* — a new mathematical tool — than in terms of coupling between swirl and meridional flow. Our results reveal a new mechanism of direct coupling between coherent structures and fine scale turbulence and hence the failure of the hypothesis of local isotropy. The fine scales are energized by the coherent vortex at the latter's boundary and evolve into larger scale spiral structures through pairing and become polarized. When these structures become sufficiently strong, they induce bending waves on the vortex, thus guaranteeing their continued survival.

1. Core Dynamics

Here we build on our long-standing proposal that coherent structures in fluid turbulence should be characterized by coherent vorticity — the underlying instantaneously space-correlated vorticity — and that vortex dynamics is a tractable avenue for understanding the evolutionary dynamics of coherent structures, their role in turbulent transport phenomena (namely entrainment, mixing, heat transfer, chemical reaction, and generation of drag and aerodynamics noise), and their interaction with fine-scale turbulence.

Investigation of vortex lines in the bridges during reconnection of two anti-parallel vortex tubes (Melander & Hussain 1988, referenced as MH) revealed that these lines are helical, hence inducing flow along the vortex axes; as a result of this, there is an inviscid 'vorticity smoothing mechanism' that causes enstrophy production opposite to that which would otherwise be expected and counters vorticity augmentation caused by vortex stretching in the same cross section. Such twisting of vortex lines and consequent

J. P. Bonnet and M. N. Glauser (eds.), Eddy Structure Identification in Free Turbulent Shear Flows, 13–25.

axial flow have also been observed in vortical structures in transitional flow of circular jets (Melander *et al.* 1991). The present study is also of fundamental interest in hydrodynamics as the lack of a clear understanding of the core dynamics in a 3D vortex is one of the principal obstacles in laminar vortex dynamics. The reason is that the motion of the axis of a curved vortex is very sensitive to local changes in the vortex core size (*e.g.* local induction approximation) — hence the importance of the core dynamics. Core dynamics is also of great interest for computations and modeling, because discretization of a vorticity field into a large number of vortex filaments constitutes one of the popular and attractive vortex methods (Leonard 1985). Usually the self-induced axial flow inside the filaments is ignored, either by using a fixed (time-independent) core size (Crow 1970), or by using a uniform core radius (although time-dependent) to conserve the volume of vortical fluid (*e.g.* Ashurst & Meiburg 1988). A noticeable exception is the filament method recently developed by Lundgren & Ashurst (1989), who employed self-consistent model equations for the evolution of the variable circular core. Our results here show that core dynamics is persistent in the inviscid limit and thus strongly suggest that core dynamics can not be ignored. We also find, not surprisingly, that core diameter variations can not disappear by inviscid effects (Moore & Saffman 1972). Moreover, our simulations of turbulent shear flows show strong polarization of the vortical structures (revealed by helical vortex lines) and thus suggest that it is perhaps also ill-advised to ignore the self-induced axial flow.

In order to focus on the effects of axial flow on the dynamics of vortical structures, we drastically idealized the problem to that of an axisymmetric vortex with a sinusoidal variation of core size. This way we eliminate the complications of curvature and the flow induced by the adjacent structure such as threads (MH). Because of the axial variation of the core diameter, there is also an axial variation of swirl and hence twisting of vortex lines soon after $t = 0$. Figure 1 shows a schematic of the vortex at a time $t > 0$ with vortex lines lying on the shaded surface denoting an axisymmetric vortex surface (not a vorticity level surface, but a surface on which rv = constant); a 90° cutaway is introduced to enhance comprehension of meridional streamlines, and swirl and axial velocity profiles. The vortex evolves as a result of the interaction between meridional flow and swirl. The velocity components u,v and w in the (r, θ, z) coordinates are governed by the following equations:

$$D_t\zeta = \text{viscous term}; \qquad \zeta \equiv rv; \tag{1.1}$$

$$D_t\eta = r^{-4}\partial\zeta^2/\partial z + \text{viscous term}; \quad \eta \equiv \omega_\theta/r \tag{1.2}$$

where $D_t = \partial_t + u\partial_r + w\partial_z$ denotes material derivative in the meridional flow, whose streamfunction ψ is related to u, w and ω_θ as follows:

$$u = \psi_z/r, \quad w = -\psi_r/r, \quad \text{and} \quad r\omega_\theta = \psi_{rr} - \psi_r/r + \psi_{zz}.$$

The axial variation of the swirl, ζ, produces a coiling of vortex lines and hence an axial pumping of fluid in a direction that will tend to smooth out core size variation and hence opposite to the axial variation of $|\omega|$ (*i.e.*, w is opposite to the vorticity flux). Swirl is large where core size is small. Also, the twisting is stronger near the axis than at the outer edge. Note that both swirl and meridional flow would have undergone uneventful decay were it not for the coupling term $r^{-4}\partial\zeta^2/\partial z$ between the two fields. Azimuthal vorticity, ω_θ, initially zero, is thus immediately created by axial expansion and contraction of the vortex tube via the coupling term. This is the basic mechanism for motion of vorticity waves along the vortex axis. It can be shown that the maximum twisting of vortex lines occurs at the inflexion points of $\zeta^2(z)$. Creation of ω_θ is necessary to start the meridional flow. Since the meridional flow first forms counter circulating cells (we call them primary cells), the vorticity peak (at the smallest initial core) bifurcates into two which travel in opposite directions. Figure 2 shows cross sections of isovorticity surfaces. The initial enstrophy peak bifurcates into two equal peaks of considerably smaller amplitude as seen by comparing figures 2a and 2b and recalling the periodic boundary conditions. The changes in the vortex's core size are linked to such changes in vorticity magnitude, essentially because the circulation of the vortex is conserved. The inviscid mechanism that alters the vorticity magnitude of a material particle is the enstrophy production, which represents the production (annihilation) of vorticity by local stretching (compression) of infinitesimal vortex tubes, thus causing both positive and negative production. The initial vorticity peak splits up because of a strong local axial compression (due to the negative production). Diffusion can also decrease (but can not increase) peak values of the vorticity magnitude; this viscous mechanism is, however, much slower and is not responsible for the rapid change between figures 2a and 2b.

The vorticity transport seen in figure 2 appears to take place in the form of "wave packets", because mass and vorticity transports are in opposite directions. This is caused by enstrophy production P_ω which allows variations in vorticity magnitude to propagate along vortex lines. We observe that P_ω changes sign near each wave packet; P_ω causes enstrophy to be produced in front of each packet and annihilated in the back of the packet, hence producing the packet's motion along the vortex axis. Note that for later frames, where the packets are in close proximity or overlap, the situation is less clear. Here we face a dilemma, which is roughly similar to that occurring for the wave motions governed by the linear 1D wave equation ($\partial^2\phi/\partial t^2=\partial^2\phi/\partial z^2$): namely, are the packets reflected off each other when they collide (figure 2g) or do we have a superposition of packets moving in opposite directions? For the linear 1D wave equation, the method of characteristics or d'Alembert formula elegantly resolves this problem. However, our present problem is nonlinear and the mathematical extraction of wave packets moving in opposite directions is a nontrivial matter. In sec. 2 we show how these wavepackets are readily separated (*i.e.* extracted) by the "complex helical wave decomposition".

The low enstrophy bubble in figure 2i forms in the region with a strong local compression. Since the compression is only significant near the axis, it follows that the vorticity magnitude will decrease rapidly near the axis, but not near the edge of the vortex core. Hence the vortex becomes locally hollow, as is clearly seen in figure 2i.

The bubble is not a permanent feature of the vorticity field (see figures 2i,l). Its disappearance is mainly due to a change in the enstrophy production, but is also due to "fill in" by viscous diffusion. One might therefore think that the bubble is no more than a temporary peculiarity of vorticity evolution. However, we shall see in sec. 2 that when the polarized wave packets are properly extracted from the vorticity field, a *permanent bubble is* associated with each packet.

In the interpretation of this phenomenon as wavepackets which bounce back and forth while undergoing viscous decay, the first rebound induces a meridional flow reversal within each primary cell, thereby generating a secondary cell; this soon dominates by pushing out the primary cell. The wavepacket behavior of the vorticity is associated with enstrophy production, P_ω, which can be related to ψ in inviscid flows as:

$$P_\omega \approx -\frac{4\omega_z^2}{r^2}\frac{\partial\psi}{\partial z} \tag{1.3}$$

Thus, P_ω can be inferred at any time from the meridional flow streamline pattern.

It is important to focus on the generation of meridional flow, hence $\eta = \omega_\theta/r$. We find that for inviscid flow,

$$D_t\eta = \text{(axial vorticity) x (differential rotation)},$$

the rotation being along an axisymmetric vortex surface; thus the instantaneous value of η does not affect instantaneous material generation of η, *i.e.* $D_t\eta$. Because of axial transport of vorticity as wavepackets, η and $D_t\eta$ also oscillate in time, η leading $D_t\eta$ by a phase shift of about 90°; this phase shift controls η-transport. The two η-transport mechanisms (convection of η by meridional flow and η-generation by differential rotation along axis) are in opposite directions (hence slower η-transport) in the primary cell, but in the same direction (hence faster η-transport) in the secondary cell.

By investigating this flow for increasingly higher Reynolds number we have found that the core size variations do not decay in the high Re-limit. Moreover, we find that the wave motion is more rapid at higher Reynolds numbers, albeit the wave speed remains finite as Re $\rightarrow \infty$. This result has significant implications for computational vortex filament models (*e.g.* Moore & Saffman 1972) where it is assumed that core variations are smoothed out by the above wave motion. This assumption becomes dubious in the light of our findings.

2. Helicity and Helical Wave Decomposition

The helical nature of flow has become the focus of considerable interest in recent years (Moffatt 1969). Of the three quantities: helicity density $h = \mathbf{u}\cdot\omega$, relative helicity $h_r = h/(|\mathbf{u}||\omega|)$, and helicity integral $H = \int \mathbf{u}\cdot\omega dV$, only the last is Galilean invariant and is a conserved quantity for inviscid flow. From $|\omega\wedge\mathbf{u}|^2 = \mathbf{u}^2\omega^2/(1- h_r^2)$ and the Navier-Stokes equations

$$\mathbf{u}_t + \omega\wedge\mathbf{u} = -\nabla(P/\rho + \mathbf{u}^2/2) + \nu\,\Delta\,\mathbf{u}, \tag{2.1}$$

it is common to claim that cascade is small if $h_r \cong \pm 1$. But such thinking is fraught with many pitfalls. Although $h_r \cong \pm 1$ does imply small $|\omega\wedge\mathbf{u}|$, it does not imply small $\nabla\wedge\omega\wedge\mathbf{u}$ as high frequency oscillations of $\omega\wedge\mathbf{u}$ are possible. In some frames, $\omega\wedge\mathbf{u}$ can be large but purely potential so that vorticity obeys purely diffusion equation and hence there is no cascade. A rectilinear vortex is a glaring example. Thus, for cascade suppression, Beltramization (*i.e.* $h_r = 1$) is sufficient, but not necessary. An example of cascade suppression for which $\nabla\wedge\omega\wedge\mathbf{u} = 0$ is not satisfied is a laminar vortex ring which travels at a constant speed.

Thus, while H is a topological property, h and h_r are of questionable use. But recognizing that local helical structure in a flow contains some essential physics, we propose that *complex helical wave decomposition* (Moses 1971; Lesieur 1990) is the relevant tool for understanding vortex dynamics and interactions. This decomposition in essence expands the field variables (*e.g.* **u**, ω) in terms of the eigenfunctions of the curl operator, which has only real eigenvalues in a periodic box or in infinite space. Moreover, the eigenfunctions can be chosen to form a complete set of orthonormal basis functions. In Fourier space, for each wavevector **k** there are two eigenmodes corresponding to the positive and negative eigenvalues $|\mathbf{k}|$ and $-|\mathbf{k}|$. All functions which are linear combinations of eigenfunctions of positive eigenvalues can be called *right-handed* as trajectories or vector lines locally form right-handed helices. Similarly, we call linear combinations of eigenfunctions corresponding to negative eigenvalues *left-handed.* Thus, for an incompressible velocity field $\mathbf{u} = \mathbf{u}_R + \mathbf{u}_L + \nabla\phi$. Since $\nabla\phi$ is the projection of **u** onto the eigenspace corresponding to the eigenvalue $|\mathbf{k}| = 0$ of the curl operator and since $\nabla\cdot\mathbf{u} = 0$, we have $\Delta\phi = 0$. If the potential part of the flow is constant at infinity, or for a periodic box, $\nabla\phi$ is a constant. Thus $\mathbf{u}_R$ and $\mathbf{u}_L$ are unique, and both translationally and rotationally invariant. Similarly, $\omega = \omega_R + \omega_L$ (assuming no rotation at infinity).

The decomposition of a flow field into polarized *(i.e.* right- and left-handed) components is rooted in the intrinsic physical nature of the Navier-Stokes equations: the eigenmodes of the curl operator are exact solutions of the Euler equation and constitute a special class of solutions called Beltrami flows with constant abnormality (*i.e.* ω = constant **u**). Because of orthogonality of the eigenfunctions of the curl operator, helicity takes the simple form

$$H = H_R + H_L \tag{2.2}$$

where

$$H_R = \int \mathbf{u}_R \cdot \omega_R dV \geq 0 \,; \quad H_L = \int \mathbf{u}_L \cdot \omega_L dV \leq 0 \tag{2.3}$$

The helical wave decomposition also provides a clearer insight into the flow physics: cascade is inhibited wherever $(\omega_R + \omega_L)\wedge(\mathbf{u}_R + \mathbf{u}_L)$ and its first spatial derivative are small. One can also easily show that cascade from interactions of modes is small when the modes are of the same parity *(i.e.* same handedness) but large when they are of opposite parity. One can thus infer that polarized vortices are persistent, hence "coherent", features of the flow.

We apply the decomposition to obtain some understanding of our axisymmetric vortex in the physical space. Remember that vector lines of $\omega_R(\omega_L)$ are right-handed (left-handed) helices. In general the physical space distributions of ω_R and ω_L overlap. The $|\omega_R|$ (above the axis) and $|\omega_L|$ (below the axis) distributions of our vortex are shown as functions of time in figure 3 (note identical distributions of $|\omega_R|$ and $|\omega_L|$ at t = 0 in figure 3a). This evolution is much more revealing than the evolution of the h_r field (not shown). The polarized components move in opposite directions. Aside from the nonlinear interaction between ω_R and ω_L there is an obvious resemblance to d'Alembert formula for the lD wave equation.

Note that $|\omega_R|$ is antisymmetric to $|\omega_L|$ in z with respect to $z = l$ at all times, coinciding in frame (a). The helical wave decomposition thus gives unique separation of ω into ω_R and ω_L components, a feature we think is essential for deeper understanding of coherent structures in turbulent flows. Note that $|\omega_R|$ peak moves to the left instead of to the right as one would expect; this is a consequence of nonlinear interaction (discussed below). In fact, in the case of an isolated vortex with only right-handed polarization, $|\omega_R|$ peak does indeed move to the right. In figure 3 we see that as they move, the wavepackets deform: they broaden by diffusion, but also elongate and form a bubble by nonlinear effects. The nonlinearity is also responsible for the breakup of the initial front-back symmetry, *i.e.* steepening at the front with a tail at the back.

To understand the nonlinear interactions between polarized components, we extract each component by the projection operators (P^+, P^-). Here we discuss only ω_R (by symmetry, ω_L equation is obvious)

$$\frac{\partial \omega_R}{\partial t} = \underset{\text{I}}{-\{\nabla\wedge(\omega_R\wedge\mathbf{u}_R)\}} + \underset{\text{II}}{\{P^-[\nabla\wedge(\omega_R\wedge\mathbf{u}_R)]\}} - \underset{\text{III}}{\{P^+[\mathbf{u}_L\cdot\nabla\omega_R]\}}$$
$$+\underset{\text{IV}}{\{P^+[\omega_R\cdot\nabla\mathbf{u}_L]\}} - \underset{\text{V}}{\{P^+[\nabla\wedge(\omega_L\wedge\mathbf{u})]\}} + \underset{\text{VI}}{\{\frac{1}{Re}(\Delta\omega_R)\}} \tag{2.4}$$

Term I is the inviscid self-evolution; term II is the generation of ω_L by evolution of ω_R; term III is the contribution to ω_R by its advection by $\mathbf{u}_L$; term IV is stretching of ω_R by $\mathbf{u}_L$; term V is the ω_R generation by ω_L, consisting of self-interaction of ω_L and the right-handed part of stretching and advection of ω_L by $\mathbf{u}_R$; term VI is the viscous diffusion term. Terms II-V represent coupling between the left- and right-handed components. We have evaluated these terms for the vortex and find that in general their dominance in decreasing order are as follows: III, I, V, IV, VI and II.

The evolution of ω_R can be largely understood in terms of two effects: its self-evolution, and advection by the $\mathbf{u}_L$ field. The self-evolution (term I) attempts to move the ω_R-packet to the right by adding vorticity at the front and annihilating it at the back. But because $\mathbf{u}_L$ moves the ω_R-packet to the left, the leftward motion dominates, although it is slower than $\mathbf{u}_L$. Also, because the velocity field has the largest axial component on the axis, it pushes the front of the packet backward at a faster rate near the axis (figure 3g), hence causing the formation of a low enstrophy bubble (figure 3i). The enhanced elongation is a combined effect of terms III and V.

The above gives a glimpse of how helical wave decomposition provides a new tool for analysis of core dynamics in terms of tractable interaction of polarized velocity and vorticity wave packets (see Melander & Hussain (1992) for details). Such physical insight into the core dynamics is impossible via the usual tools of vortex dynamics such as the Biot-Savart law and local induction approximation. We are hopeful that many other important vortex phenomena (two discussed briefly below) can be understood better by this approach. Vortex breakdown is another obvious candidate (subject of a separate study by us).

3. Interaction with Fine-scale Turbulence

In an attempt to understand the coupling between large-scale coherent structures and fine-scale turbulence, we have repeated the simulation reported earlier with the axisymmetric sinusoidal coherent vortex (called CV) immersed in a background of isotropic fine-scale incoherent turbulence. Care was taken to assure that the initial rms level was sufficiently high for the fine-scale turbulence to survive long enough to dynamically interact with CV. The fine-scale size was chosen to be the smallest permitted by the resolution, but still with a spectral gap between it and that of CV; see figure 4; spectrum labelled (a), a schematic of the initial vorticity distribution is shown as an insert. We find that the scales in the background incoherent vorticity grow progressively as evidenced by the elimination of the initial spectral gap in the spectra labelled (b-q) in figure 4. At the end of the simulation (nondimensional time t=8.0) the largest incoherent scales are comparable to CV; note that there is no separate peak at high wavenumbers in spectrum labelled (q) in figure 4. The spectral gap is gradually filled up not by diffusive mechanisms, but by interaction between incoherent turbulence and CV. While the size of the incoherent turbulence remains uniform in space, the strength is higher at the boundary of CV where it is wrapped around CV and energized by stretching induced by CV. This is not different from the growth of axisymmetric vortices on an impulsively rotated rod except that in our case the curved vortices around CV are intrinsically asymmetric, induced by the shear in the boundary of CV.

The spiral structures formed from organization of the incoherent turbulence are indeed quite different from the spiral sheet-like structures proposed by Lundgren (1982). The structures we find are rod-like spiral patterns wrapped around CV with senses aligned or opposed to the swirl of CV (figure 5). The helical wave decomposition is also helpful in understanding the coupling between CV and incoherent turbulence. The spiral structures are found to have a tendency to be highly polarized: they are either predominantly right-handed or predominantly left-handed. We indeed find a strong correspondence between peaks of incoherent turbulence in the boundary of CV and high values of the degree of polarization. By overlaying the isosurfaces of $|\omega_R|$ and $|\omega_L|$ we observed that the polarized components tend to occupy separate locations in the boundary layer around the vortex. This is a clear indication of small scale polarization and spatial separation of right- and left-handed components. Such spatial separation of polarized components has also been observed in the transition of a temporally evolving circular jet. The incoherent turbulence spirals undergo growth through merger by (truly inviscid) pairing mechanism and not by fusion through viscous decay and diffusion. In order for the spiral vortices to pair, their transport (axial with respect to CV) is essential. There are two distinct mechanisms for such transport: i) transport of nearby opposite signed vortices as dipoles, and ii) self-induced transport due to curvature of the azimuthally aligned vortices (not present in the 2D case). Because of the shear of CV, the small-scale vortex spiral undergoes differential stretching, which provides the spiral an intrinsic tendency to become polarized.

In virtually all turbulence theory, local isotropy, which also forms the cornerstone of Kolmogorov's hypothesis, is universally assumed and implies decoupling of large and fine scales. By extrapolating our results (figure 5), we conjecture, for the case of very high Reynolds number, a hierarchy of structures of successively smaller scales, *i.e.* a fractal cascade. This supports our long-held doubt regarding the validity of the concept of local isotropy. Of course local isotropy does not require individual fine scales to be isotropic, but merely the lack of any statistical preference for the fine-scale structure's orientation. We feel that the fine scales retain some preferred orientation with respect to the coherent structures.

The CV-in-turbulence of course decays, and these unforced flows will eventually laminarize, in a time scale depending on both Reynolds number and the initial rms turbulence level. If the level is too low, CV recovers axisymmetry; if too high, CV is completely disrupted. For intermediate values, interactions discussed above occur. However, in this case there is the interesting possibility of excitation of bending waves on CV. Deformation of CV, indicating the bending waves (stronger waves at higher Reynolds number), is clear in figure 6. Without bending waves, the spiral structures tend to be axisymmetric as they wrap around CV, thus diminishing their stretching by CV. When bending waves are excited, CV ceases to be axisymmetric and is then in a position to continue the stretching of the spiral incoherent turbulence. This is an example of feedback (or backscatter) from the smaller scales to CV. Thus we have here a mechanism of coherent structure/fine-scale interaction: the former organizes the latter by vortex stretching and the latter, if Reynolds number is high enough, induces bending waves in the former to generate the mechanism of its own survival.

In turbulent shear flows, the large-scale coherent structures would tend to have preferred configurations. We will now examine how the anisotropy of the largest scale CV cascades down to the smallest scales. The first step in this cascade is discussed above. Namely, that CV gives rise to the formation of secondary vortical structures in the boundary layer around the coherent vortex (figure 5) through alignment and subsequent amalgamation of small scale structures. These secondary structures may then, provided that they are sufficiently strong, dominate the strain rates in their immediate vicinity. If the Reynolds number is sufficiently large, so as to allow the existence of even smaller scales, we can then imagine that these smaller scales align and organize azimuthally around the secondary structures. In the limit of infinite Reynolds number, this line of reasoning leads to a hierarchy of increasingly smaller scale structures as shown schematically in figure 7. Thus, for very high Reynolds numbers we propose a fractal scenario of vorticity cascade from the largest scale to the smallest scale. Since the smallest scales are closely coupled with the CV, which will have preferred orientations in any turbulent shear flow, this study provides a direct challenge to the hypothesis of local isotropy, as well as provides a mechanism for small-scale anisotropy.

This lecture is based on joint research collaboration with Professor M.V. Melander of the Southern Methodist University in Dallas, Texas. This is a progress report of our continuing research activities on vortex dynamics and their connection to coherent structures in shear flow turbulence. This research is funded by the Air Force Office of Scientific Research under grant F49620-92-J-0200. The assistance of D. Virk in the preparation of this report is gratefully acknowledged.

References

ASHURST, W. T. & MEIBURG, E., 1988 Three-dimensional shear layers via vortex dynamics. *J. Fluid Mech* . **189**, 87.

CROW, S. C., 1970 Stability theory for a pair of trailing vortices. *AIAA J.* **8**, 2172.

KIDA, S., TAKAOKA, M. & HUSSAIN, F., 1989 Reconnection of two vortex rings. *Phys. Fluid* A **1**, 630.

LEONARD, A., 1985 Computing three-dimensional flows with vortical elen *Rev. Fluid Mech.* **17**, 523.

LESIEUR, M., 1990 Turbulence in Fluids (Kluwer Acad. Publ.).

LUNDGREN, T. S., 1982 Strained spiral vortex model for turbulent fine structure. *Phys. Fluids.* **25**, 2193.

LUNDGREN, T. S. & ASHURST, W. T., 1989 Area-varying waves on curved vortex tubes with application to vortex breakdown. *J. Fluid Mech.* **200**, 283.

MELANDER, M. V. & HUSSAIN, F., 1988 Cut-and-connect of two antiparallel vortex tubes CTR Report S88, Stanford U., pp.254-286.

MELANDER, M. V. & HUSSAIN, F., 1990 Topological Aspects of Vortex Reconnection. In *Topological Fluid Mechanics.* eds. H.K. Moffatt & A. Tsinober, pp.485-499 (Cambridge U. Press).

MELANDER, M. V. & HUSSAIN, F., 1992 Polarized vorticity dynamics on a vortex column. (submitted).

MELANDER, M. V., HUSSAIN, F. & BASU, A., 1991 Breakdown of a circular jet into turbulence.*Turb. Shear Flows 8,* Munich pp.15.5.1-5.

MOORE, D. W. & SAFFMAN, P. G., 1972 The motion of a vortex filament with axial flow. *Phil. Trans. Roy. Soc.* A **272**, pp.403-429.

MOSES, H. E., 1971 Eigenfunction of the curl operator, rotationally invariant Helmholtz theorem and application to electromagnetic theory and fluid mechanics. *SIAM J. Appl. Math.* **21**, pp.114-144.

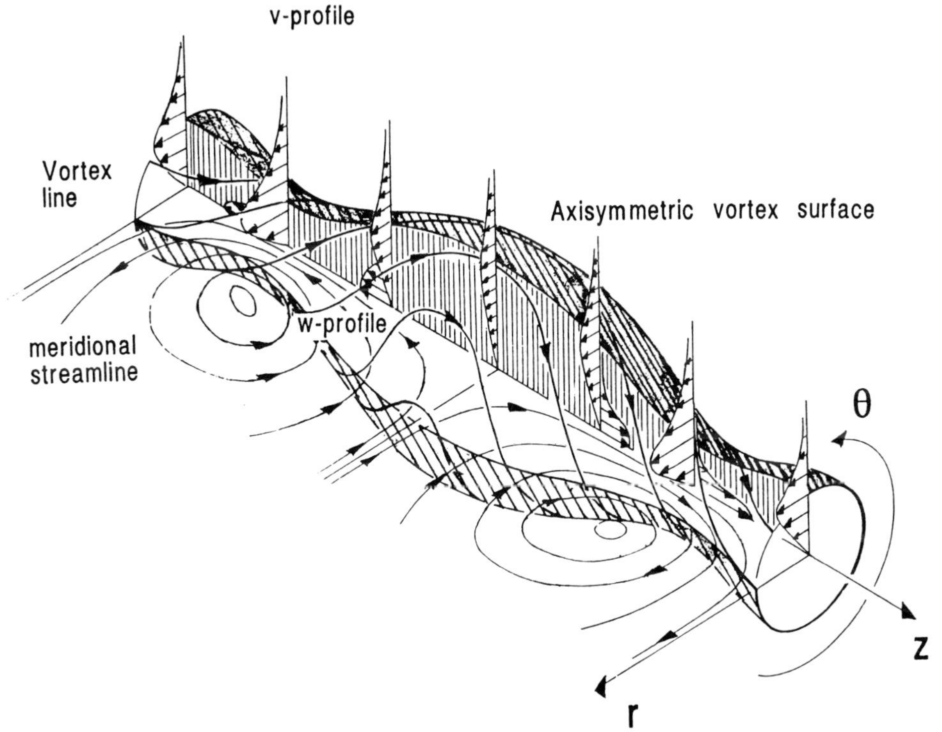

Fig. 1. Axisymmetric vortex surface with a 90° cutaway on which vortex lines lie. Also shown are streamlines in a meridional plane, and v(r) and w(r) profiles.

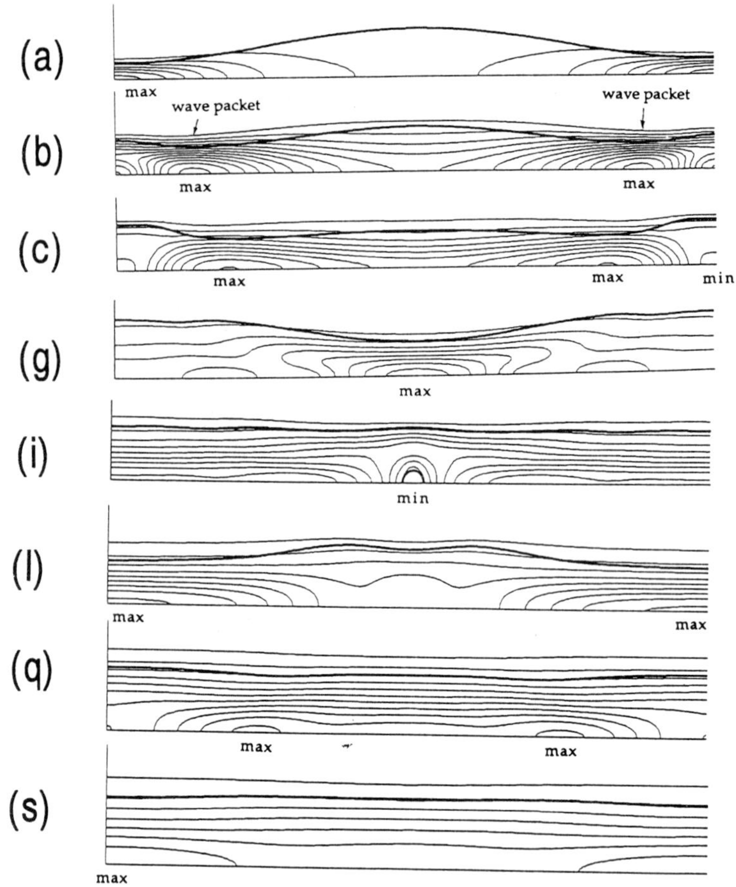

Fig. 2. Evolution of the instantaneous vorticity magnitude. In each frame the boundary of the vortex given by kinematic vorticity number N_k=1 is overlaid.

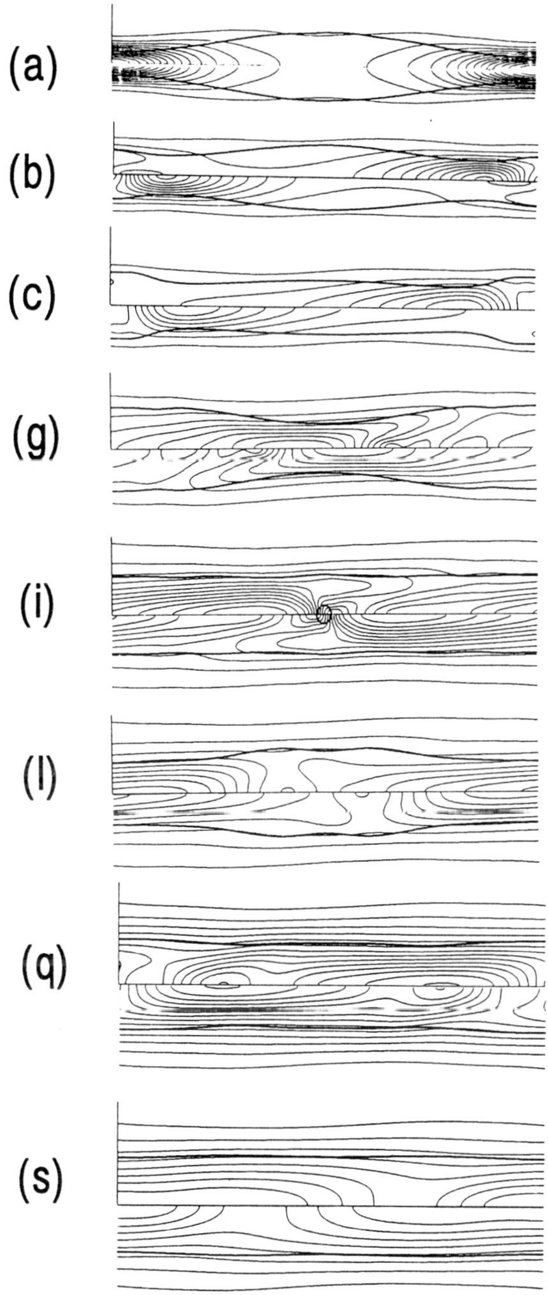

Fig. 3. Evolution of $|\omega_R|$ (above) and $|\omega_L|$ (below) the axis, with N_k=1 contour overlaid.

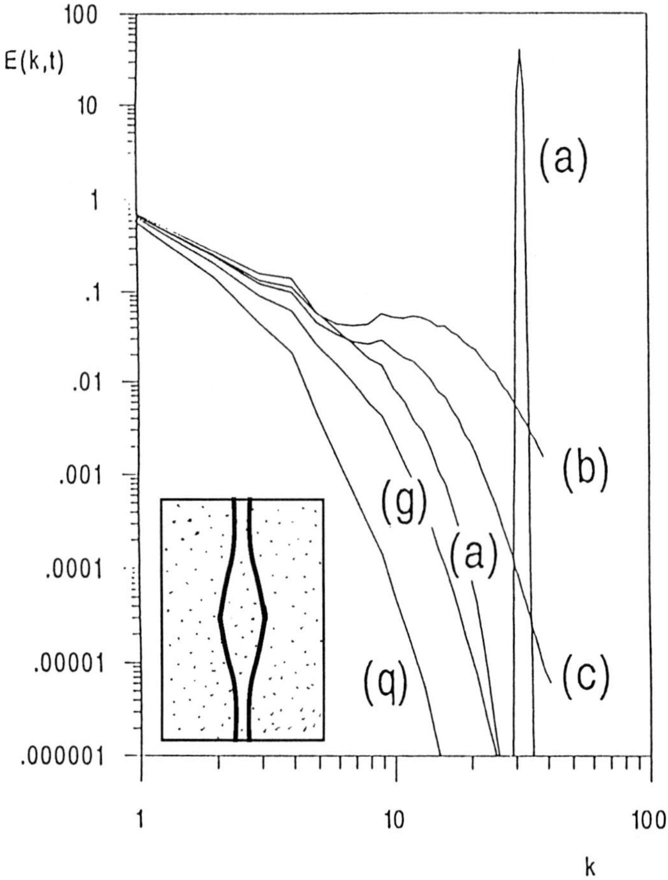

Fig. 4. Energy spectrum E(|**k**|, t) for Re=665 case with initially superposed turbulence. A schematic of the initial vorticity field is shown as an insert. Letters correspond to various nondimensional times: (a) t=0, (b) t= 0.5, (c) t= 1.0, (g) t=3.0, (q) t= 8.0.

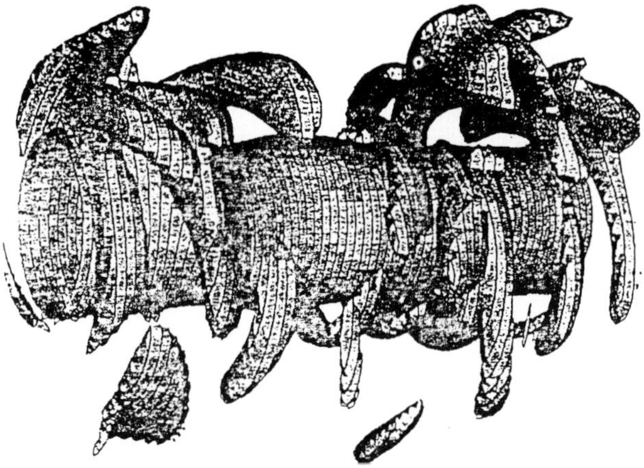

Fig. 5. |ω| surface at 10% of the peak value.

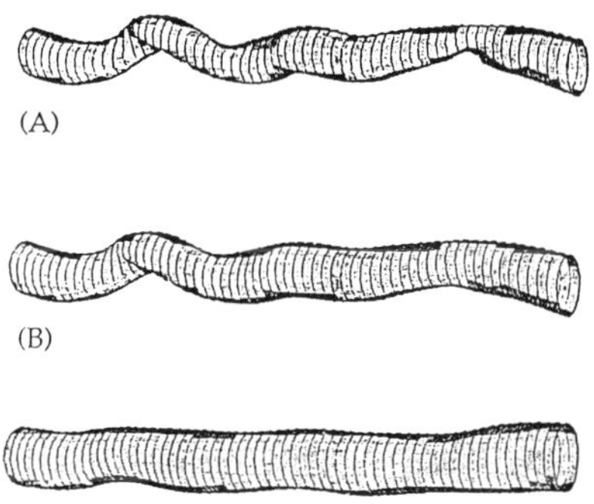

Fig. 6.
65% |ω| surface at the end of simulations with increasing Reynolds numbers: (A) Re=2500, (B) Artificial viscous term of the form $\nu_4 \Delta^2 \mathbf{u}$. (C) Re=665.2, the same as in figure 5.

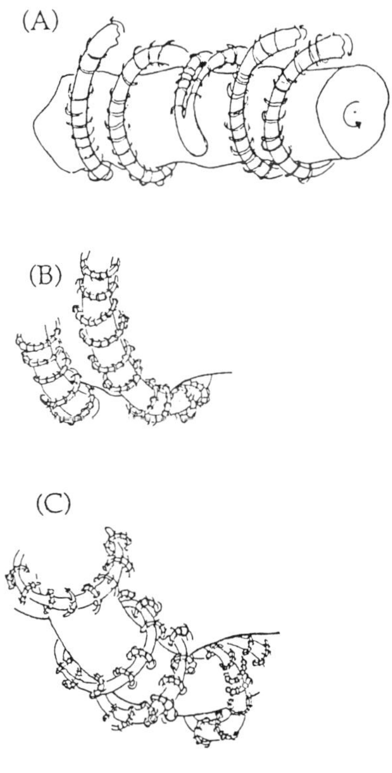

Fig. 7. Schematics of fractal scenario.

IMPULSE ANALYSIS OF FRACTAL AND MULTIFRACTAL STRUCTURE OF THE FIELD OF DERIVATIVES IN TURBULENT FLOWS

A. Bershadskii and A. Tsinober
Department of Fluid Mechanics, Faculty of Engineering,
Tel-Aviv University, Tel-Aviv 69978, Israel

Summary
A relation between power law spectrum of fluctuations of velocity derivatives and fractal dimensions of corresponding surfaces is obtained in the inertial range. It is shown that two values of surface fractal dimensions correspond to a single spectral power law. Good agreement with experimental data is obtained for Kolmogorov turbulence.

1. Introduction

There is an increasing interest in relations between fractal and multifractal properties of turbulent fields of derivatives and spectra of fluctuations of velocity and concentration of a passive scalar (see, for example [1-6]). Of particular interest is the relation(s) between the spectra of fluctuations of the rate of energy dissipation and the fractal dimension(s) of volume occupied by these fluctuations. However, the experimental difficulties in obtaining both of these quantities are well known. Therefore, it seems appropriate to look for relations between the fractal dimension of surfaces(s) and spectra of individual squared velocity derivatives. These characteristics are much easier for experiments.

2. A Simple Relation Between the PDF of Pulse Duration and the Energy Spectrum of a Random Signal $\xi(t)$

A signal of dissipation rate of energy is strongly intermittent or in other words exhibits a pulse structure even at rather moderate Reynolds numbers [7]. At larger Reynolds numbers the pulse structure of the signal becomes more pronounced as can be seen, for example, in [8,9]. Note that in [8,9] only measurements of an individual velocity derivative have been made.

Generally, the relation between the pulse structure of a signal of some quantity $\xi(t)$ and its spectral characteristics should depend (among other things) also on the details of the shape of the pulses. However, at least at large Reynolds numbers, the peculiarities of each individual pulse should have characteristic time scale much smaller than that of the pulse as a whole. It appears, therefore, that these peculiarities make a contribution to the proportionality coefficient in the power law spectrum of $\xi(t)$ (such as (8), see below).

We introduce first a deterministic function g(t) for the shape of pulses, such that it vanishes outside of the pulse and its Fourier transform

J. P. Bonnet and M. N. Glauser (eds.), Eddy Structure Identification in Free Turbulent Shear Flows, 27–34.

$$G(\omega) = \int_{-\infty}^{-\infty} g(t)e^{-i\omega t}\,dt \qquad (1)$$

In the following we will make use of the relation between the energy spectrum $E(\omega)$ and the probability density function of pulse duration $W(t)$ [10]

$$E_\xi(\omega) = \frac{2a^2}{T_0}\left\{[1 + (\sigma/a)^2]F_0(\omega) + 2\underline{\mathrm{Re}}\left[\frac{Q(\omega)Q_1(-\omega)\theta_1(\omega)}{1 - \theta_1(\omega)\theta_2(\omega)}\right] + \frac{G(0)\tau_0^2}{T_0}\,\delta(\omega)\right\} \qquad (2)$$

where

$$\langle\xi^2\rangle = \int_0^\infty E_\xi(\omega)d\omega\ , \quad F_0(\omega) = \int_0^\infty x^2|G(\omega x)|^2\, W(x)dx \qquad (3)$$

$$Q(\omega) = \int_0^\infty xG(\omega x)W(x)dx\ , \quad Q_1(\omega) = \int_0^\infty xG(-\omega x)e^{i\omega x}\,W(x)dx\ , \qquad (4),(5)$$

and $\theta_1(\omega)$ and $\theta_2(\omega)$ are the characteristic functions of two random variables: pulse duration and pauses between the pulses, respectively; $\delta(\omega)$ - is the delta function; a, T_0, σ, τ_0 - some constants and, finally $\underline{\mathrm{Re}}$ means the operation of taking the real part. Recall that a characteristic function of a random variable with probability density function $\omega(t)$ is defined as

$$\theta_1(\omega) = \int_0^\infty \omega(t)e^{i\omega t}\,dt \qquad (6)$$

The relation (2) is valid under the assumption that the pulse duration and the pauses between pulses are statistically independent.

In case of a power law

$$w(t) \sim t^{-(\mu+2)}\ , \quad \mu = \mathrm{const} \qquad (7)$$

and it follows from (2)-(7) that

$$E_\xi(\omega) \sim \omega^{-1+\mu} \qquad (8)$$

for large ω, provided that $|\omega^{\mu+1}\theta_2(\omega)| \to \infty$ with $\omega \to \infty$. This last property is valid for processes with PDF of pause duration of the type (7).

3. Estimates of effective surface of a turbulent field s and their fractal dimensions

In order to make such estimates we employ the Taylor hypothesis, i.e., replace x by $U_{mean}t$ [11]. to obtain

$$\langle x\rangle = \int_{\eta/L}^{1} xW(x)dx \sim (\eta/L)^{-\mu}\ , \qquad \langle x^2\rangle = \int_{\eta/L}^{1} x^2W(x)dx \sim 1 - (\eta/L)^{1-\mu}\ , \tag{9),(10}$$

where η - is the Kolmogorov scale and L - is the integral scale. The effective surface of a turbulent field can then be estimated in three ways

$$S \sim \langle x\rangle^2 \sim (\eta/L)^{-2\mu} \tag{11}$$
$$S \sim \langle x\rangle \cdot 1 \sim (\eta/L)^{-\mu} \tag{12}$$

(in dimensionless form 1 corresponds to L) and

$$S \sim \langle x^2\rangle \sim 1 - (\eta/L)^{1-\mu} \tag{13}$$

Since $1 - \mu > 0$ [11] (see also (8)) it follows that while the estimate (13) shows that the surface is regular, (i.e., S remains finite when $\eta/L \to 0$), the estimates (11) and (12) are characteristic of fractal surfaces. It is known [1] that for fractal surfaces

$$S \sim (\eta/L)^{2-D_\sigma} \tag{14}$$

where D_σ is the fractal dimension of the surface. Hence it follows from (11), (12) and (14) that

$$D_\sigma = 2 + 2\mu \tag{15}$$

$$D_\sigma = 2 + \mu \tag{16}$$

in the cases corresponding to (11) and (12) respectively.

4. The case of Kolmogorov turbulence

In single probe experiments the obtained signal is a one-dimensional section of 3-D active turbulent zones. Therefore as a crude estimate of the PDF of pulse duration W(t) the following expression can be adopted

$$W(x) \sim \int_{x}^{\infty} \rho(r)dr \tag{17}$$

where $\rho(r)$ - is the PDF of active zones with the characteristic scale r [3].
In the power law interval [3, 12]

$$\rho(r) \sim r^{-\beta} \tag{18}$$

and for $\beta > 1$

$$W(x) \sim x^{1-\beta} \tag{19}$$

Hence it follows from (7)

$$\beta = 3 + \mu \tag{20}$$

It has been shown in [3] that for Kolmogorov turbulence $\beta = 10/3$, so that from (20) it follows that $\mu = 1/3$ and from (15) and (16) $D_\sigma = 8/3$ and $D_\sigma = 7/3$ respectively. The value $D_\sigma = 7/3$ has been observed in a large number of experiments [1]. As for the value $D_\sigma = 8/3$ it is of particular interest in the experiments described in [2]. In these experiments the value $D_\sigma \sim 8/3$ was observed in the internal part of the jet, while the value $D_\sigma \sim 7/3$ was observed at the outer part of the jet [13]. Recalling that the values $D_\sigma = 8/3$ and $D_\sigma = 7/3$ correspond to estimates (11) and (12) of the fractal surface, the difference seems to become clear. Indeed, the estimate (11), which does not take into account the external scale, is more adequate in the internal regions of the jet, while the estimate (12) takes into account the external scale explicitely.

The experiments described in [14] deserve special attention for several reasons. From our point one of the important aspects are the detailed measurements of various properties of the derivatives of streamwise velocity fluctuations and in particular the spectra of fluctuations of squared derivatives. Observations of signals of fluctuations of squared derivatives of streamwise velocity exhibited a distinctly pulsed structure [15]. It is also noteworthy that in these experiments the $k^{-5/3}$ interval in the energy spectrum was observed over more than two decades.

Spectra of fluctuations of squared velocity derivatives adapted from [14] are shown in fig. 1 The straight line drawn in the inertial interval corresponds to the slope -2/3, i.e., to the value $\mu = 1/3$ (8).

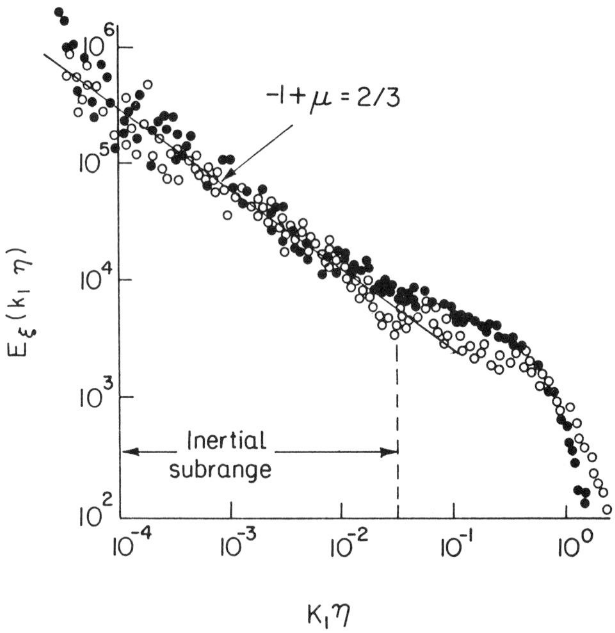

Fig. 1 Spectra of squared derivative of streamwise velocity fluctuations in the atmospheric boundary layer at 15m height from the ground [14]. • - $\xi_1 = (u^{-1}\partial u/\partial t)^2$; o - $\xi_2 = (U^{-1}\partial u/\partial t)^2$.

5. Multifractal asymptotics and surface fractal dimensions

Though in the experiments [14] no direct measurements of fractal dimensions of surfaces have been made, measurements of moments of the squared derivative have been performed, which together with the results in [3] allow to find the fractal dimensions of surfaces. Indeed, the intermittent nature of a signal $\xi^q(t)$ becomes more pronounced with increasing q (q > 1). It is convenient to use the relation

$$\langle \xi_r^q \rangle / \langle \xi \rangle^q \sim (r/L)^{-\mu_q} \tag{21}$$

for analysis of such signals, where

$$\mu_q = (d - D_q)(q - 1) \tag{22}$$

is the so-called intermittency exponent, d - is the topological dimension of space (usually 3 or 2). It has been shown in [3] that

$$D_q = \frac{D_\infty q - b}{q - 1}, \quad b = \text{const} \tag{23}$$

corresponds to pulsed signals at q large enough. The expression (23) was earlier proposed in [16]. However, as has been shown in [3] there exists a problem of relation of (23) to the behavior of D_q at small q. In any case we will use here (23) for large enough q only. It has been also shown in [3] that $D_\sigma = D_\infty$. Therefore it is straightforward to obtain D_σ from (22) using the results of measurements of moments. Indeed, it follows from (23)

$$\mu_q = (d - D_\sigma)q + b - d \tag{24}$$

and the slope μ_q as a function of q in (23) allows to find D_σ directly.

There is another important peculiarity of the experiments described in [14]. Namely, two types of normalization has been performed and respectively statistical properties of two quantities have been obtained: $\xi_1 = (u^{-1}\partial u/\partial t)^2$ and $\xi_2 = (U^{-1}\partial u/\partial t)^2$, where U is the mean velocity. It is natural to associate the estimate (11) with the measurements of ξ_1, while the estimate (12) - with the measurements of ξ_2. The dependence of μ_q versus q adapted from [14] is shown in fig 2 and appears to be linear for both types of measurements. Using (23) it follows that $D_\sigma = 2.56$ for the first case and $D_\sigma = 7/3$ for the second case. In the first case the difference from the theoretical value 8/3 is less than 4%. This corresponds to an exponent in the spectrum (8) equal to -0.71 which in fig. 1 is almost indistinguishable from -2/3. The deviation of D_σ from the value 8/3 is most probably due to use of local Taylor hypothesis ($\xi_1 = (u^{-1}\partial u/\partial t)^2$), which leads to errors in the region of large scales. Note that $D_\sigma = 8/3$ was obtained in [2] from direct measurements. In the second case the agreement, as can be expected, is better. The results obtained from the data in [14] are supported by the data obtained for a laboratory boundary layer [12] and jet flows [18]. These are also shown in fig. 2 and exhibit linear dependence of μ_q on q at large q with a slope $d - D_\sigma = 2/3$, i.e., $D_\sigma = 7/3$.

It is noteworthy that the value $D_\sigma = 8/3$ was obtained already in [17]u for a passive scalar in Kolmogorov turbulence.

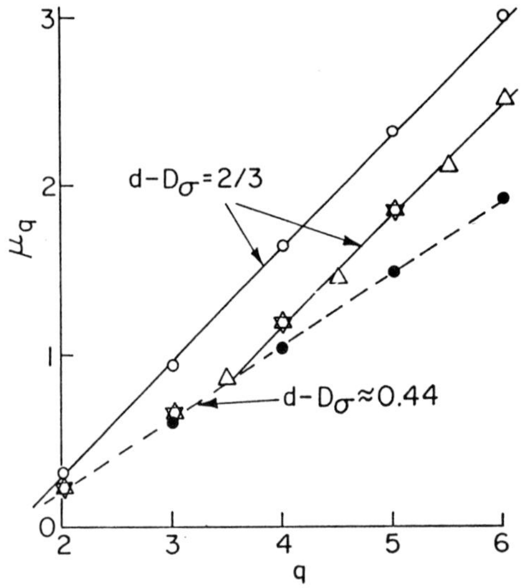

Fig. 2 Intermittency exponents versus order of the moments of squared derivative of streamwise velocity fluctuations. • - ξ_1 [14] , 0 +- ξ_2 [14] , Δ - [12] , ∇ - [18].

Concluding Remarks

We would like to emphasize that - as follows from the results of this note - the Kolmogorov spectrum $k^{-5/3}$ (without any correction to the exponent -5/3) is directly related to the spectrum of squared derivatives of streamwise velocity fluctuations $k^{-2/3}$ (i.e., $\mu = 1/3$). This is a rather nontrivial conclusion. Indeed, the intermittency theory of Kolmogorov-Obukhov [11] is based on the relation between μ ($\neq 0$) in the spectrum of dissipation fluctuations and the correction to the -5/3 power in the energy spectrum. Similar approach has been adopted in many subsequent models of intermittency. The results of this note point to the following alternative: either the above models are in error or the properties of a single squared derivative of velocity fluctuations (dissipation surrogate) are essentially different from those of the fluctuations of the rate of dissipation.

Recent experiments (experimental and numerical) [7, 19, 20] clearly indicate that there exist essential differences between statistical properties of the rate of dissipation fluctuations and it's surrogate.

References

[1] K.R. Sreenivasan, Fractals and multifractals in fluid turbulence, *Annu. Rev. Fluid Mech.* **23**, 539 (1991).

[2] P. Constantin, I. Procaccia and K.R. Sreenivasan, Fractal geometry of isoscalar surfaces in turbulence; theory and experiments, *Phys. Rev. Lett.*, **67**, 1739 (1991).

[3] A. Bershadskii and A. Tsinober, Asymptotic fractal and multifractal properties of

turbulent dissipative fields, *Phys. Lett.*, **A165**, 37 (1992).

[4] J.C. Vassilicos and J.C.R. Hunt, Fractal dimensions and spectra of interfaces with application to turbulence, *Proc. Royal Soc. Lond.* **A435**, 505 (1991).

[5] R. Kraichnan, Fractal geometry of isoscalar surfaces in turbulence, *Phys. Rev. Lett.*, **67**, 3634 (1991).

[6] A.J. Chorin, Equilibrium statistics of a vortex filament with applications, *Commun. Math. Phys.* **141**, 619 (1991).

[7] A. Tsinober, E. Kit and T. Dracos, Experimental investigation of the field of velocity gradients in turbulent flows, *J. Fluid Mech.*, in press (1992).

[8] K.R. Sreenivasan and C. Meneveau, The fractal facets of turbulence, J. Fluid Mech., 172, 357 (1986).

[9] C. Meneveau and K.R. Screenivasan, The multifractal spectrum of the dissipation field in turbulent flows, *Nuclear Phys.* **B**. (Proc. Suppl.) 2, 49 (1987).

[10] D. Middleton, *An introduction to statistical communication theory* (N.Y. McGraw-Hill, 1960).

[11] A.S. Monin and A.M. Yaglom, *Statistical Fluid Mechanics*, V.2 (MIT Press, Cambridge, MA, 1975).

[12] C. Meneveau and K.R. Sreenivasan, The multifractal nature of turbulent energy dissipation, *J. Fluid Mech.* **224**, 429 (1991).

[13] R.R. Prasad and K.R. Sreenivasan, The measurement and interpretation of fractal dimensions of surfaces in turbulent flows, *Phys. Fluids* **A2**, 792 (1990).

[14] M.Z. Kholmyansky, Measurments of turbulent microfluctuations of the wind-velocity derivative in the surface layer, *Atmosph. and Oceanic Phys.* **8**, 472 (1972).

[15] The authors are grateful to M.Z. Kholmyansky for this information and for fruitful discussions of his experiments.

[16] T. Nakano, Generalized dimensions of three-dimensional turbulence, Phys. Lett. **A140**, 395 (1989).

[17] B. Mandelbrot, On the geometry of homogeneous turbulence, with stress on the fractal dimension of the iso-surfaces of scalars, *J. Fluid Mech.*, 72, 401 (1975).

[18] R.A. Antonia, B.R. Satyaprakash and F. Hussain, Statistics of fine-scale velocity in turbulent plane and circular jets, *J. Fluid Mech.* **119**, 55 (1982).

[19] I. Hosokawa and K. Yamamoto, Evidence against the Kolmogorov refined similarity hypothesis, *Phys. Fluids* **A4**, 457 (1992).

[20] A. Tsinober, Some properties of velocity derivatives in turbulent flows as obtained from experiments (laboratory and numerical), to be published in the Proccedings of

the Les Houches School on *Turbulence in Spatially Extended Systems*, Les Houches, France, Jan. 20-31, 1992.

The genesis of longitudinal vortices in free and bounded shear layers

W.R.C. PHILLIPS

Department of Mechanical & Aeronautical Engineering
Clarkson University, Potsdam NY 13699-5725, U.S.A.

Mechanisms giving rise to longitudinal vortices in quasi-parallel turbulent shear layers are considered. The flow is described using the Generalized Lagrangian-mean equations of Andrews & McIntyre and is treated as a wave mean-flow interaction. Specifically, the wave field simultaneously acts both to distort the mean flow from its laminar form and account for turbulence fluctuations. Both spanwise-independent two-dimensional and three-dimensional wave fields are explored. In each case mechanisms exist to form longitudinal vortices; however the vortices grow exponentially fast only when excited by two-dimensional waves.

1. Introduction

Longitudinal vortices have long been observed in free (Miksad 1972) and bounded shear layers (Bakewell & Lumley 1967). The flow need not be turbulent, but when it is such vortices are a well defined example of coherent structures. Their role, at least in the case of bounded turbulent flows, is reasonably well understood but their origin is not: In either bounded or free shear flows that are spanwise independent (in the mean), the mean vorticity vector is spanwise and for longitudinal vortices to form a mechanism or mechanisms must exist to reorient and amalgamate that vorticity into concentrated streamwise cores. The object of the present work is to isolate such mechanisms.

To achieve this goal an oft suggested path is to apply linear stability theory to the mean turbulent profile. But while "the mean velocity profile doubtless plays a part...it is inconceivable...that strong, large-scale turbulent fluctuations should not also do so: [indeed] to restrict attention to linear instability of the mean profile seems naive" (Craik 1985). Here we attempt to account for large scale fluctuations in the form of a complex finite amplitude wave field.

Before doing so, however, we recall that vortex lines move with the fluid (in an inviscid flow) and that if the flow also contains an oscillatory component, the mean fluid velocity is not the Eulerian mean but rather the Lagrangian mean velocity, the difference being the Stokes drift. The latter quantities do not appear naturally in an Eulerian representation of the flow and for that reason we employ Andrews and McIntyre's (1978) Generalized Lagrangian-mean (GLM) equations. These equations were constructed with wave/mean flow interactions in mind and are outlined in §2; they are an exact mapping of Navier Stokes.

J. P. Bonnet and M. N. Glauser (eds.), Eddy Structure Identification in Free Turbulent Shear Flows, 35–41.

Phillips (1989, 1990, 1991) and Leibovich (1991) discuss the GLM approach with bounded flows in mind, but it is equally applicable to unbounded flows whose streamwise growth is sufficiently slow to invoke the quasi-parallel approximation.

We consider mechanisms that induce longitudinal vortices when the wave field is either two- (§3) or three-dimensional (§4). Longitudinal vortices are seen to grow exponentially fast only when the wave field is two dimensional. What is meant by the term 'wavefield' is discussed in §5.

2. Background

2.1 The Generalized Lagrangian-mean equations

Andrews & McIntyre's (1978) Generalized Lagrangian-mean (GLM) equations are an exact and very general Lagrangian-mean description of the back effect of oscillatory disturbances upon the mean state. The formulation is based upon an exact Lagrangian-mean operator $\overline{(\ )}^L$, corresponding to any given Eulerian-mean operator $\overline{(\ )}$, through an exact disturbance-associated particle displacement field $\xi(\mathbf{x}, t)$ and is valid provided the mapping $\mathbf{x} \mapsto \mathbf{x} + \xi$ is invertible. In consequence dependent variables are given an Eulerian description with position $\mathbf{x}$ and time t as independent variables. The Lagrangian-mean velocity $\bar{\mathbf{u}}^L$ is then the velocity field describing trajectories about which the fluctuating particle motions have zero mean, when *any* averaging process is applied.

Here we are interested in fluids of constant density and flows in which all mean quantities save the mean pressure $\mathcal{P}$ are independent of the streamwise $x = x_1$, direction; then for homentropic flows of constant density in a non-rotating reference frame

$$q_{i,t} + q_j q_{i,j} - p_j(q_{j,i} - q_{i,j}) + \mathcal{P}_{,i} = X_i \tag{1}$$

where repeated indices imply summation, commas denote partial differentiation, $(x_1, x_2, x_3) \equiv (x, y, z)$ and $q_i = \bar{u}_i^L - p_i$. The vector wave property p_i, the pseudo-momentum per unit mass, is

$$p_i = -\overline{\xi_{j,i} u_j^\ell}, \tag{2}$$

where $\bar{D}^L \xi_j = u_j^\ell$ and the operator $\bar{D}^L$ is defined as $\bar{D}^L = \partial/\partial t + \bar{u}_j^L \partial/\partial x_j$. X_i are disspative terms. We shall restrict attention to inviscid fluids, so X_i is identically zero, although viscosity may be included. Finally, in terms of the Eulerian mean velocity $\bar{\mathbf{u}}$, we have $\bar{\mathbf{u}}^L = \bar{\mathbf{u}} + \mathbf{d}$, where $\mathbf{d}$ is the Stokes drift.

2.2 Shear flows of variable order and $O(\epsilon)$ waves

We consider an $O(\epsilon^s)$ $(s \geq 0)$ primary shear flow subject to small spanwise perturbations whose strength relative to the primary shear is measured by a small parameter δ; δ is assumed sufficiently small that linearization with respect to it yields a good approximation to the equations governing the spanwise periodic disturbance. Then

$$\mathbf{q} = \epsilon^s\{[Q_1(z), 0, Q_3(z)] + \delta[q_1(y,z,t), \epsilon^n q_2(y,z,t), \epsilon^n q_3(y,z,t)]\} \quad (n \geq 0). \tag{3}$$

The disturbance is manifest by periodic waves that propagate in the x-direction.

At this point we shall assume the waves are three dimensional, although we shall also consider two-dimensional waves. The primary waves are of constant amplitude with slope characterized by the small parameter ϵ, and give rise to an $O(\epsilon^2)$ primary pseudomomentum field P_i. But distortion to the wave field may occur owing to the $O(\epsilon^s\delta)$ induced secondary velocity component, which in turn induces an $O(\epsilon^{s+2}\delta)$ component of pseudomomentum (Craik 1982). Thus

$$\mathbf{p} = \epsilon^2\{[P_1(y,z), 0, P_3(y,z)] + \epsilon^s\delta[p_1(y,z,t), p_2(y,z,t), p_3(y,z,t)]\}. \tag{4}$$

Then [from (1) with (3) and (4)] to $O(\epsilon^s\delta)$ the streamwise momentum equation takes the form

$$\frac{\partial q_1}{\partial t} + \epsilon^{s+n}\delta(q_2\frac{\partial q_1}{\partial y} + q_3\frac{\partial q_1}{\partial z}) + \frac{1}{\delta}(\epsilon^s Q_3 + \epsilon^2 P_3)\frac{dQ_1}{dz} + (\epsilon^s Q_3 + \epsilon^2 P_3)\frac{\partial q_1}{\partial z}$$

$$+\epsilon^{s+2}\delta(p_2\frac{\partial q_1}{\partial y} + p_3\frac{\partial q_1}{\partial z}) + \epsilon^{s+n}q_3\frac{dQ_1}{dz} + \epsilon^{s+2}p_3\frac{dQ_1}{dz} = 0. \tag{5}$$

And if we introduce the vorticity associated vector field $\mho = curl\mathbf{q}$, its streamwise component to $O(\epsilon^{s+n}\delta)$ is

$$\frac{\partial \mho_1}{\partial t} + \epsilon^{s+n}\delta(\frac{\partial \mho_1 q_2}{\partial y} + \frac{\partial \mho_1 q_3}{\partial z}) + \epsilon^{s+2}\delta(\frac{\partial \mho_1 p_2}{\partial y} + \frac{\partial \mho_1 p_3}{\partial z}) + \frac{\partial}{\partial z}(\mho_1(\epsilon^s Q_3 + \epsilon^2 P_3))$$

$$-\frac{\epsilon^{2-n}}{\delta}\frac{dQ_1}{dz}\frac{\partial P_1}{\partial y} + \epsilon^{2-n}\{\frac{\partial q_1}{\partial y}\frac{\partial P_1}{\partial z} - \frac{\partial q_1}{\partial z}\frac{\partial P_1}{\partial y}\}$$

$$-\epsilon^{s+2-n}\frac{dQ_1}{dz}\frac{\partial p_1}{\partial y} + \epsilon^{s+2-n}\delta\{\frac{\partial q_1}{\partial y}\frac{\partial p_1}{\partial z} - \frac{\partial q_1}{\partial z}\frac{\partial p_1}{\partial y}\} = 0. \tag{6}$$

Finally, in terms of more familiar variables, the streamwise velocity perturbation $u_1 = q_1 + p_1 - d_1$ and the streamwise component of vorticity,

$$\Omega_1 = \mho_1 + \frac{\partial(d_2 - p_2)}{\partial z} - \frac{\partial(d_3 - p_3)}{\partial y}.$$

So if wave field distortion is negligible we may identify q_1 with u_1 and $\mho_1$ with Ω_1.

3. Two dimensional waves

Our object is to elicit mechanisms which lead to the growth of $\mho_1$ with time and we shall begin by considering spanwise (y) independent two-dimensional waves. In our first example we set $\mathbf{P} = [P_1, 0, 0]$; then $Q_3 = 0$ and

$$\frac{\partial q_1}{\partial t} + \epsilon^{s+n}q_3\frac{dQ_1}{dz} = 0 \tag{7}$$

$$\frac{\partial \mho_1}{\partial t} + \epsilon^{2-n}\frac{\partial q_1}{\partial y}\frac{\partial P_1}{\partial z} - \epsilon^{s+2-n}\frac{dQ_1}{dz}\frac{\partial p_1}{\partial y} = 0. \tag{8}$$

The equations are coupled so we must set $n = (2-s)/2$ and rescale t. Then subject to boundary conditions (7) and (8) admit exponentially growing solutions. In physical terms, however, we note that velocity perturbations are not always of equal order: for example $n = 0$ for weak $O(\epsilon^2)$ shear flows, but is unity when the shear is $O(1)$, indicating that the transverse velocity components are $O(\epsilon)$ smaller than the axial perturbation! Likewise, wave field distortion may be ignored for shear of $O(\epsilon)$ or less [here the equations describe the Craik-Leibovich type 2 instability (Leibovich 1983)], but plays an important role at $O(1)$ (Craik 1982) where a further equation must enter to complete the set (7), (8). Of course longitudinal vortices may still form and grow exponentially fast, but wave field distortion acts (for all but the largest wavelengths) to inhibit that growth (Phillips & Wu 1993).

We now consider the case for which $\mathbf{P} = [0, 0, P_3]$. Here the equations can take more than one form, the most obvious being

$$\frac{\partial q_1}{\partial t} + \frac{1}{\delta}(\epsilon^s Q_3 + \epsilon^2 P_3)\frac{dQ_1}{dz} = 0, \tag{9}$$

and

$$\frac{\partial \mho_1}{\partial t} + \frac{\partial}{\partial z}(\mho_1(\epsilon^s Q_3 + \epsilon^2 P_3)) = 0. \tag{10}$$

This set is uncoupled. It would appear that the initial growth of q_1 is linear in t. In the case of bounded turbulent flows such growth gives rise to intense short-term local shear layers Phillips (1991). But over sufficiently large time, viscosity and $(\epsilon^s Q_3 + \epsilon^2 P_3)\frac{dQ_1}{dz}$ act to blur such layers. There is no amplification of $\mho_1$ but rather a convection and redistribution of it, assuming some exists; specifically

$$\mho_1 = \frac{f(t - \int \frac{dz}{Q_3})}{Q_3} \qquad (s > 2).$$

Less obvious are

$$\frac{\partial q_1}{\partial t} + \epsilon^{s+2} p_3 \frac{dQ_1}{dz} = 0, \tag{11}$$

$$p_i = p_i(\Phi, \phi, Q_1, q_1, \alpha, l) \tag{12}$$

and

$$\frac{\partial \mho_1}{\partial t} - \epsilon^{s+2-n}\frac{dQ_1}{dz}\frac{\partial p_1}{\partial y}. \tag{13}$$

Here the equations are coupled not through the dependent variables but rather through wave field distortion: that is the velocity field and wave field distort in concert. The distorted component of pseudomomentum is then given by (12), in which Φ and α denote the eigenfunction and wavenumber of the primary wavefield, which satisfy Rayleigh's equation, and ϕ and l denote the eigenfunction and wavenumber appropriate to the distorted wavefield which must satisfy a variant of Rayleigh's equation, $R(\Phi, \phi, Q_1, q_1, \alpha, l) = 0$ say. We shall not derive $p_i(\cdot)$ and $R(\cdot)$, but the approach is similar to that employed by Craik (1982). In this mechanism $n = 0$ for all s, so that velocity perturbations are the same order irrespective of shear. Moreover while exponential growth cannot be discounted, it is not mandatory and is unlikely to be sustained over large time. In short, spanwise distortion renders the wave field three dimensional, leading to growth more akin to that discussed in §4, which is algebraic.

To conclude our study of two dimensional waves we let $\mathbf{P} = [P_1, 0, P_3]$. Here we have the possibility of exponentially growing longitudinal vortices on an algebraically growing substrate. Relative to the evolving vortices, therefore, we consider the base flow to be quasi-steady and subtract it out. We are then left with (5) for q_1 and

$$\frac{\partial \mho_1}{\partial t} + \frac{\partial}{\partial z}(\mho_1(\epsilon^s Q_3 + \epsilon^2 P_3)) + \epsilon^{2-n}\frac{\partial q_1}{\partial y}\frac{\partial P_1}{\partial z} - \epsilon^{s+2-n}\frac{dQ_1}{dz}\frac{\partial p_1}{\partial y} = 0. \qquad (14)$$

The resulting mechanism is a further variant of Craik-Leibovich type 2.

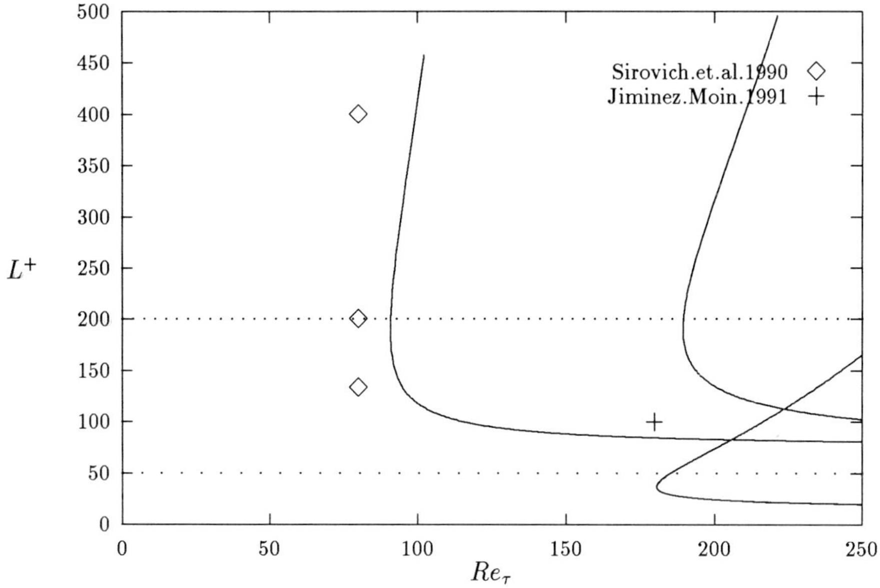

Figure 1. Neutral secondary curves for periodic Poiseuille flow (Phillips & Tu 1992).

Phillips (1991, 1993) and Phillips & Tu (1992) consider the development of secondary instabilities *via* this mechanism on a nonlinear equilibrium solution in periodic Poiseuille flow. They choose a nonlinear equilibrium solution whose streamwise and spanwise average (of velocity and kinetic energy) mimics that of fully developed turbulent channel flow. The secondary instability, which of course manifests as longitudinal shear-layer vortices, yields the neutral stability curve of figure 1, which depicts the spanwise periodicity of the vortices (L^+, in wall units) as a function of Reynolds number (Re_τ, based upon channel half width and friction velocity). Further, to compare with Direct numerical simulations, we include the $\mathbf{k} = (0,1,1), (0,2,1)$ and $(0,3,1)$ modes (implying streamwise vortices) calculated by Sirovich *et al* (1990), along with Jiminez and Moin's (1991) minimum width; that is the minimum periodic channel width necessary for turbulence to sustain itself at $Re_\tau = 180$. All lie close to the first neutral mode whose wavelength is *ci* that of streaks (*viz* 50-200 wall units). Moreover, in addition to the first mode, a higher-order mode is evident that is linearly most unstable at a Reynolds number higher than critical but with a much lower spanwise wavelength, about 40 wall units, a tad more than twice the empirical optimal spacing of riblets. Phillips & Tu (1992) conjecture that riblets excite this mode of instability. The vortices so formed are smaller in diameter than their flat plate counterparts and do not extend much beyond the buffer region. Detailed nonlinear calculations following their lifecycle are still in progress but early results suggest that their growth rate, while still exponential, is significantly less than their larger counterparts, possibly because of viscous damping. The net result is a reduced production of turbulent kinetic energy.

4. Three dimensional waves

We turn now to wave fields that contain an $O(\epsilon)$ oblique component, thereby requiring that the pseudomomentum be spanwise periodic. Then q_1 is apparently given by (9) with

$$\frac{\partial \mho_1}{\partial t} - \frac{\epsilon^{2-n}}{\delta}\frac{dQ_1}{dz}\frac{\partial P_1}{\partial y} = 0. \tag{15}$$

But since $\mho_1$, and thus q_3, grow as t, (9) is not the appropriate form for q_1; it should be

$$\frac{\partial q_1}{\partial t} + \epsilon^{s+n} q_3 \frac{dQ_1}{dz} + \epsilon^{s+2} p_3 \frac{dQ_1}{dz} = 0, \tag{16}$$

which indicates that the spanwise varying part of q_1 grows as t^2 and that $\epsilon^{2-2n-s}/\delta = O(1)$. Here we recover a family of flows, the first of which was considered in Eulerian form by Benny & Lin (1960) as an explanation for the presence of longitudinal vortices in transitory boundary layers. Further work on the $s = 0$ case was done by Benney (1964) and Craik (1970). Moreover when $s = 2$, and with P_1 replaced by the 1-component of Stokes drift, equations (15), (16) recover those of Craik & Leibovich (1976); such equations are thought to describe a particular case of Langmuir circulations which occur on the surface of lakes and oceans.

Algebraic growth will also occur when longitudinal vortices are deliberately introduced into the shear layer. In the absence of a mechanism to sustain the vortices, they will decay as do trailing vortices, like t^{-m} $(0 < m < 1)$, where m is determined by details of the vortex generating device (Phillips 1981). Then the spanwise varying part of q_1 grows as t^{1-m}, even though the vortex is decaying.

5. Discussion

To this point we have alluded to the presence of primary wave fields without qualification and should now like to say more about them. In the case of Poiseuille flow, for example, it has long been known (Noether 1921) that the viscous velocity profile may be distorted by the presence of two dimensional finite amplitude travelling waves. Indeed it happens that such waves are the attracting set for the periodic system (Zahn *et al* 1974). Moreover, the nonlinear primary instability they induce rapidly equilibrates, giving rise to an equilibrium surface in Reynolds number, wave number and disturbance energy space (Bayly *et al* 1988). But what is striking is that streamwise and spanwise averages of the (upper branch) equilibrated solution (arising from very few modes) depict velocity and fluctuating fields in close accord with those in turbulent channel flow, logarithmic law and all! (Rozhdestvensky & Simakin 1984). So in this instance the primary (finite amplitude) wave field is that supporting the equilibrium solution, and the formation of coherent structures the evolving secondary instability on that equilibria. In short the wave field provides a platform from which to study the aetiology of secondary instabilities; a platform that has the correct fluctuating energy distribution but none of the high frequency randomness that muddies interpretation of instantaneous measured or calculated flows.

The pertinent primary wave field in free shear layers will also be a nonlinear equilibrium solution, but whether it is sufficient to employ an equilibria based solely upon the dominant normal mode and its subharmonics in accord with Kelly's (1967) work on transitional flow, is unclear, although that would be an obvious starting point. Experiments indicate such equilibria exist, but only until the fundamental succumbs to three dimensional distortion. Since free shear flows grow spatially this may mean the equilibrium becomes increasingly unstable to three dimensional disturbances as streamwise Reynolds number increases. It could equally mean that spanwise perturbations in axial velocity are sufficiently strong to induce concurrent distortion to the primary wave field, giving rise to a scenario not unlike that discussed in §3.

This work was supported by the National Science Foundation, grant CTS-9008477.

REFERENCES

ANDREWS, D.G. & MCINTYRE, M.E. 1978 *J. Fluid Mech.*, **89**, 609-646.

BAYLY, B.J., ORSZAG, S.A. & HERBERT, T. 1988 *Ann. Rev. Fluid Mech.*, **20**, 359-391.

BAKEWELL, P & LUMLEY, J.L. 1967 *Phys. Fluids*, **10**, 1880-1889.

BENNEY, D.J. & LIN, C.C. 1960 *Phys. Fluids*, **3**, 656-657.

BENNEY, D.J. 1964 *Phys. Fluids*, **7**, 319-326.

CRAIK, A.D.D. 1970 *J. Fluid Mech.*, **41**, 801-821.

CRAIK, A.D.D. 1982 *J. Fluid Mech.*, **125**. 37-52.

CRAIK, A.D.D. 1985 *Wave Interactions and fluid flows,* Cambridge University Press

CRAIK, A.D.D., & LEIBOVICH, S. 1976 *J. Fluid Mech.*, **73**, 401-426.

JIMINEZ, J & MOIN, P. 1991 *J. Fluid Mech.*, **225**, 213-240.

KELLY, R.E. 1967 *J. Fluid Mech.*, **27**, 657-689.

LEIBOVICH, S. 1983 *Ann. Rev. Fluid Mech.*, **15**, 391-427.

LEIBOVICH, S. 1991 *Studies in Turbulence* (ed Gatski et al) Springer-Verlag 387-411.

MIKSAD, R.W. 1972 *J. Fluid Mech.*, **56** 695-719.

NOETHER, F 1921 *ZAMM 1*, **125**, 218-230.

PHILLIPS, W.R.C. 1981 *J. Fluid Mech.*, **105** 451-467.

PHILLIPS, W.R.C. 1989 *Zoran Zaric International Seminar 1988: Near wall turbulence* (ed S. Kline), Hemisphere. 736-747.

PHILLIPS, W.R.C. 1990 *Applied Mechanics Reviews*, **43**, S227-S231.

PHILLIPS, W.R.C. 1991 *Theoret. Comput. Fluid Dynamics*, **2**, 329-338.

PHILLIPS, W.R.C. 1993 (in preparation)

PHILLIPS, W.R.C. & TU, H.Y. 1992 *Bull. Am. Phys. Soc.*, **37**, 1794.

PHILLIPS, W.R.C. & WU, Z. 1993 *J. Fluid Mech.*, (submitted).

ROZHDESTVENSKY, B.L. & SIMAKIN, I.N. 1984 *J. Fluid Mech.*, **147** 261-289.

SIROVICH, L, BALL, K.S., & KEEFE, L.R. 1990 *Phys. Fluids A*, **2**, 2217-2226.

ZAHN, J.P., TOOMRE, J., SPIEGEL, E.A., GOUGH, D.O. 1974 *J. Fluid Mech.*, **64**, 319-345.

II. Conditional Sampling I: Vorticity-based Methods

VORTICITY-BASED CONDITIONAL SAMPLING FOR IDENTIFICATION OF LARGE-SCALE VORTICAL STRUCTURES IN TURBULENT SHEAR FLOWS

M.Hayakawa
Hokkaido University, N13 W8, Sapporo 060, Japan

This report is mainly concerned with a vorticity-based conditional sampling technique, which identifies large-scale vortical flow events in turbulent shear flows using multiple X-wire probes. Basic ideas and procedures of the technique are described, and several examples of the results are presented. Advantages and limitations of the technique are also discussed from an experimental point of view.

1 INTRODUCTION

1.1 Background

While the discovery of spatially coherent, recurrent, large-scale flow events (often called "coherent structures") in turbulent shear flows has been heavily owing to flow visualization studies, one has to rely upon a certain quantitative method in order to gain a further insight into the detailed characteristics and dynamical roles of those structures(e.g. [1], [2],[3],[4]). To date, various methods have been developed for this purpose, as exemplified in the present symposium[5].

The technique of conditional sampling and averaging has been one of the most widely used methods in laboratory experiments. The general description and historical overview of the technique are provided by Kaplan[6], Van Atta [7] and Antonia[8], among others. In recent years, the technique is also utilized by numerical analysists, since the increasing computer power has made it possible to regard the results of direct numerical simulation as "data" analogous to the data from laboratory experiments([9]).

What one obtains from the conditional sampling is a "conditional average", which is defined as an ensemble average taken over many events that satisfy a certain, prescribed condition[10]. Implicit working rules in obtaining the conditional average are: (1)the existence of identifiable, recurrent flow events, (2)the selection of similar events through conditioning, and (3)the decomposition of any sampling signal f into an ensemble-averaged(i.e. "coherent") part $\langle f \rangle$ and the remained part f_r through averaging; $f = \langle f \rangle + f_r$.

In spite of its broad use, the conditional sampling involves two major problems. One is concerned with "subjectivity", which enters in the process of conditioning, i.e., in the choice of proper conditioning signals and the decision of suitable conditions. In general, the ensemble average $\langle f \rangle$ more or less depends on the detection scheme used. Sato[11] mentioned that more patterns are hidden in more random field, and

J. P. Bonnet and M. N. Glauser (eds.), Eddy Structure Identification in Free Turbulent Shear Flows, 45–64.

in the limit we can find any pattern in a perfectly random field. This is of course an extreme case, but such a dangerous aspect always resides in conditional sampling techniques.

The other problem is the so-called "jitter", which occurs in the process of sampling/averaging. This is caused by the fact that a signal sampling point and a detection (or "trigger") point are usually different, both in space and time. Consequently various random factors of individual flow events enter into the averaging process and lead to the loss of phase information, eventually causing a large degradation of the ensemble-averaged result([10],[12],[13]). Since variations of individual events in their detail, movement and history arise from the inherent nature of turbulence, any conditional sampling technique cannot be entirely free from the jitter problem.

In a sense, the development of conditional sampling techniques has been a history of researchers' efforts in settling the two difficult problems.

1.2 Vorticity-Based Technique

To adapt a conditional sampling technique to the identification of organized structures, whose appearance is generally at best quasi-periodic, one has to choose a proper conditioning signal, which ought to be a characteristic flow property associated with the flow events under consideration.

In 1980, Hussain[14] proposed a definition of coherent structures: "a coherent structure is a connected turbulent fluid mass with 'instantaneously phase-correlated vorticity' ('coherent vorticity' in his terminology) over its spatial extent." This led him to believe that "vorticity" should be the best flow property to recognize coherent structures. While the usefulness of vorticity in the quantitative study of organized turbulence structures had been often suggested(e.g. [6],[15]), difficulty in high resolution measurements of vorticity even at a point, let alone at multiple points, in a flow prevented its practical use in the experimental studies. But the above operational definition by Hussain implied that coherent structures could be identified by measuring relatively low frequency vorticity components, rather than a sufficiently wide frequency range of vorticity fluctuations.

Motivated by Hussain's idea, Tso[16] employed, in his study of a far field round jet, a rake of seven X probes to measure the azimuthal component of (instantaneous) vorticity in a meridian plane, and extracted vortical structures with referring to the local peak of filtered vorticity concentration. His method appeared insensitive to the effects of jitter, because structures were captured at their (approximate) centers without use of any auxiliary reference signal for trigger. Since then, this "vorticity-based conditional sampling technique" (hereinafter, referred to as "vorticity-based technique" for brevity) has been applied to other free shear flows and found to be a useful method for the investigation of coherent structures[3,17].

In this paper, basic procedures involved in the vorticity-based technique are described first. Then, several results are presented in comparison with those obtained by other conditional sampling techniques. Discussion is mainly made in terms of instantaneous and ensemble-averaged vorticity fields, without entering into detailed turbulence mechanisms. Finally, the merit and demerit of the present technique and some related issues are discussed. Consideration is restricted to either two-dimensional or axisymmetric free shear flows.

2 BASIC PROCEDURES: EDUCTION SCHEME

2.1 Calculation of Vorticity

For an illustrative purpose, let us consider an array of N number of X probes with a relative separation Δs, aligned in the direction of the main shear of a plane shear layer (Fig.1). Coordinate axes are designated as x, y and z in the streamwise-, transverse-(main-shear-) and spanwise direction respectively, and the corresponding velocity components are u, v and w.

From a time series of simultaneous u and v signals at N y locations, the spanwise vorticity component ω_z is calculated with Taylor's hypothesis and a central difference approximation (Fig.2);

$$\omega_z = \frac{\partial v}{\partial x} - \frac{\partial u}{\partial y} \approx -\frac{1}{U_c}\frac{\Delta v}{\Delta t} - \frac{\Delta u}{\Delta s}$$

Here U_c is the average convection velocity of structure centers in the x direction, and Δt is the time step between successive data points. For instance, the vorticity at a point A in Fig.2 is estimated as follows, in an approximation to the instantaneous spanwise vorticity (in this section, ω will be used in place of ω_z, for brevity);

$$(\omega)_A \approx -\frac{1}{U_c\Delta t}\{\frac{v_{21}+v_{22}}{2} - \frac{v_{11}+v_{12}}{2}\} - \frac{1}{\Delta s}\{\frac{u_{12}+u_{22}}{2} - \frac{u_{11}+u_{21}}{2}\}$$

Thus, from the N number of X probes, we have ω values at (N − 1) y locations.

2.2 Recognition

Since our primary concern is in large-scale features, we take the moving, short-time average of vorticity signals to have simultaneous records of low-pass filtered vorticity $\tilde{\omega}$ at (N−1) y locations. The signal smoothing is not performed in the y direction, not only because otherwise, locally high but small-scale vorticity patches should always

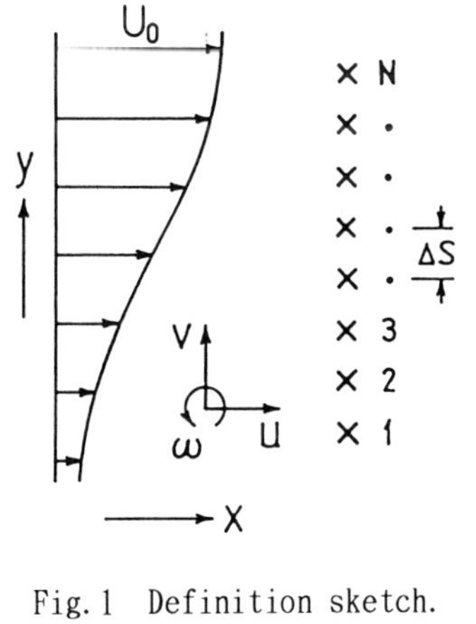

Fig. 1 Definition sketch.

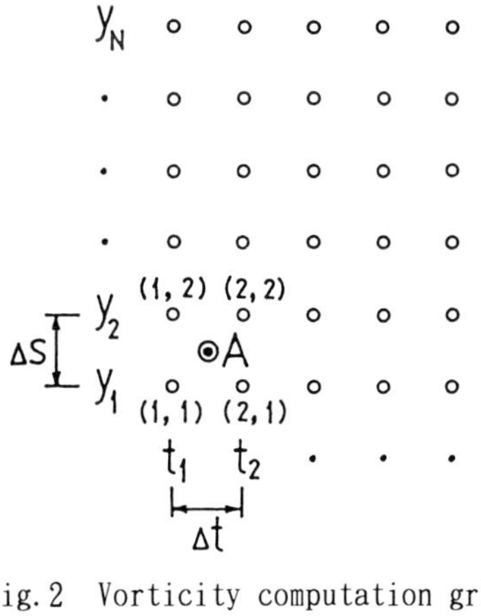

Fig. 2 Vorticity computation grid.

produce spurious bumps in $\tilde{\omega}$ values at the neighboring y locations, but also because the calculation of vorticity has already contained a spatial filtering effect due to the probe spacing Δs.

Figures 3(a)-(d) show four examples of the iso-contour map of smoothed vorticity, obtained in three different flow configurations: a circular cylinder wake(a,b), a plane mixing layer(c) and a round jet(d). In these figures, if time t is converted to streamwise distance x by $x = -U_c t$, the abscissa scale is compressed relative to the ordinate scale. The pseudo-instantaneous vorticity maps visualize reasonably well the occurrence of vorticity concentrations, and help us to recognize the presence of organized structures, more quantitatively than in flow visualization pictures.

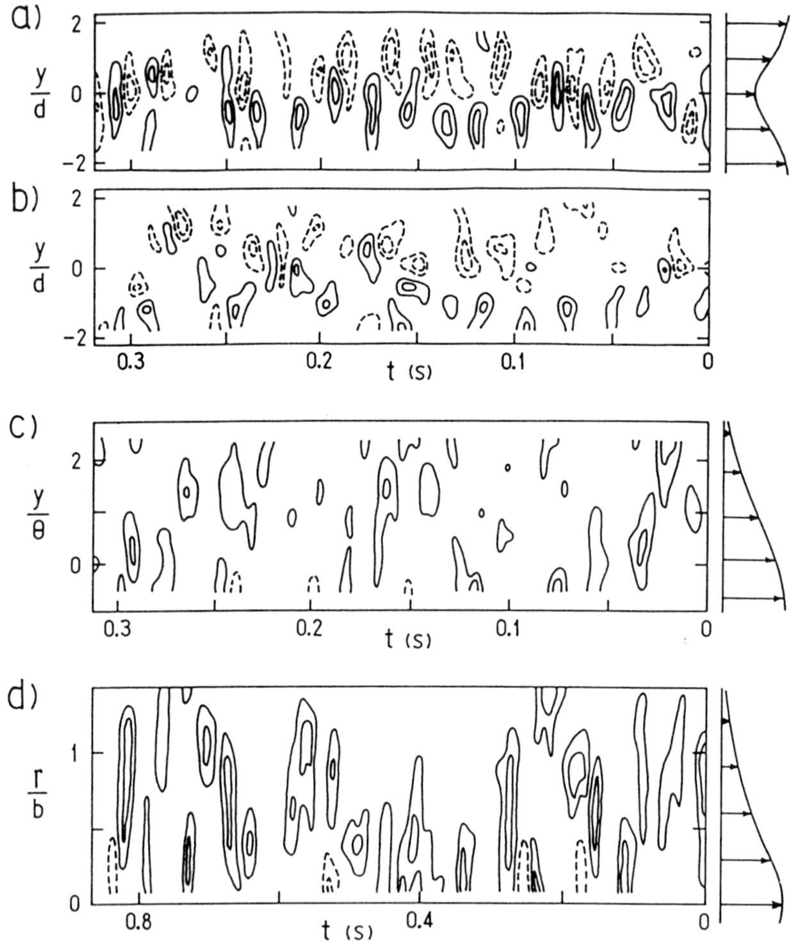

Fig. 3 Pseudo-instantaneous ω_z maps. Circular cylinder wake at a) $x/d=20$ and b) $x/d=40$ (Ref. [23]). c) Plane mixing layer at $x/\theta_e=1000$ (Ref. [32]). d) Circular jet at $x/D=50$ (Ref. [16]).

2.3 Transverse Detection Center

As the events of interest usually occur with wide dispersion across the shear layer, it is preferable to determine first a transverse detection center. For an ordinary purpose, the detection center is determined so as to coincide with the most probable y location of advecting structure centers. This is done by counting the number of $\tilde{\omega}$ peaks higher than a certain level, as a function of y (Fig.4). The detection center is assigned to the location y_c which gives the maximum probability. Then, the X-wire rake is relocated so that the middle location corresponds to y_c, as illustrated in Fig.4. This is the transverse detection center to be used in the following processes.

If the number of X probes is large enough for the probe spacing to be sufficiently narrow, this relocation is not essential. However, we have typically used eight X probes only, the relocation is necessary for efficient use of information from the limited number of the probes.

2.4 Detection: Conditioning

In the present technique, the detection conditions are imposed on the "strength" and "size" of (smoothed) vorticity concentrations. The strength is discriminated by applying a threshold Th_1 to $\tilde{\omega}$ at y_c;

$$\text{I.} \quad \tilde{\omega}(y_c) > Th_1 \ , \qquad Th_1 = k_1 S_M \ ; \qquad S_M = \left(\frac{\partial \bar{u}}{\partial y}\right)_{\max} .$$

Here the local maximum mean shear S_M is used for fixing the threshold level. The middle point during the interval of $\tilde{\omega}(y_c)$ being higher than Th_1 is assigned to a trigger instant t_c. Also, we check $\tilde{\omega}(y_c)$ to be transversely maximum at t_c;

$$\text{II.} \quad \tilde{\omega}(y_c, t_c) > \tilde{\omega}(y, t_c) \quad for \quad y \neq y_c .$$

The size criterion is applied to both the streamwise and transverse scales. The condition is that $\tilde{\omega}$ values around a point (y_c, t_c) are simultaneously greater than another threshold Th_2 (Fig.5);

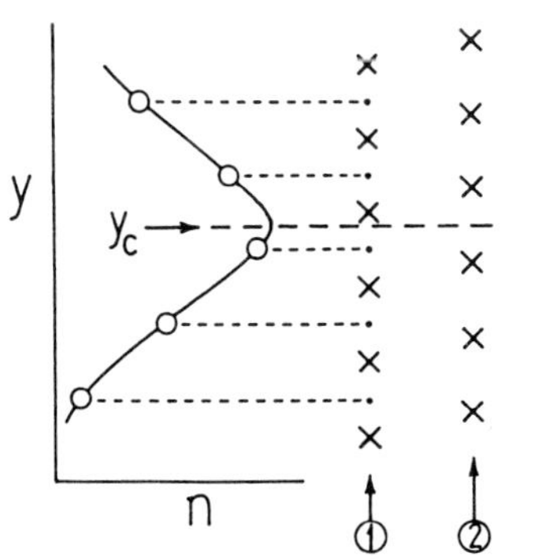

Fig. 4 Determination of the middle location of an X-wire rake; n is the number of vorticity peaks.

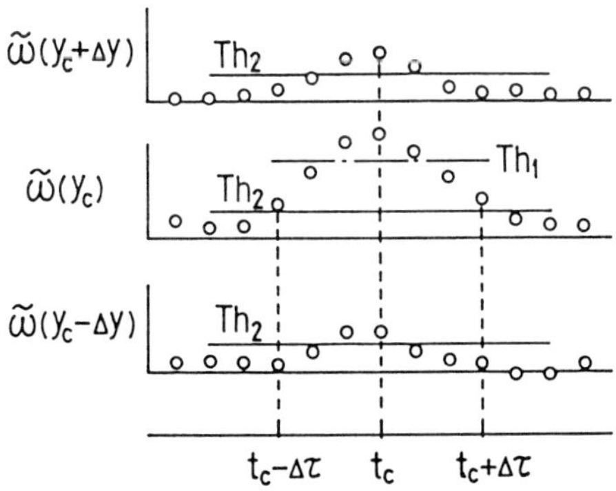

Fig. 5 Illustration showing the criteria on streamwise and transverse scales of vorticity concentrations.

III. $\tilde{\omega}(y_c \pm \Delta y, t_c) > Th_2$ (transverse size) , *and*

IV. $\tilde{\omega}(y_c, t_c \pm \Delta\tau) > Th_2$ (streamwise size) ; $Th_2 = k_2 S_M$.

Here $\Delta\tau$ is so chosen as $U_c\Delta\tau$ is nearly equal to Δy ; that is, a common condition is given to both the streamwise and transverse sizes. In other words, we assume the cross-section of structures to be nearly circular. However, if the selected events have a preferential (say, elliptical) shape, such a feature can survive through the detection process (see §3).

Large-scale events are accepted only when all the criteria above are satisfied, so that weaker, transversely-shifted, smaller-scale or highly distorted events are discarded.

2.5 Eduction: Ensemble Average

Once the trigger instants $t_{c,i} (i = 1, 2, --)$ are determined, the accepted realizations are relatively aligned with respect to each center (y_c, t_c), and ensemble-averages of velocity components, $\langle u \rangle$ and $\langle v \rangle$, are calculated. Then, the coherent vorticity $\langle \omega \rangle$ distribution is obtained by using again the Taylor hypothesis and the central difference approximation. It may be important to note that the ensemble averages are computed from original, unsmoothed velocity signals; the smoothed signals have been used only as a means of selecting similar large-scale events. Otherwise, ensemble-averaged results should suffer from excessive filtering effects, and the departure from the ensemble average, $f_r (= f - \langle f \rangle)$, would have no specific physical meaning.

2.6 Realignment: Signal Enhancement

While the essential procedure of the present technique has been outlined above, it is preferable to perform further alignment of individual structure centers, because smoothed vorticity peaks are usually not sharp enough to be identified clearly. For this purpose, we take the cross-correlation between the vorticity-signal segment $\tilde{\omega}(y_c)$ of each realization and the initial ensemble average $\langle \omega(y_c) \rangle$ (Fig.6). Each realization is then relatively shifted by the time delay of the peak correlation. In practice, the trigger instants $t_{c,i}$ which have been stored in a data file are revised to $t'_{c,i}$, as illustrated in Fig.6.

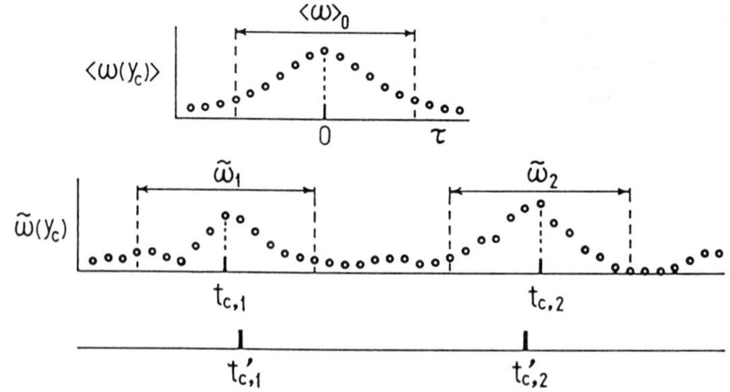

Fig. 6 Illustration showing the realignment process.

3 APPLICATIONS

The eduction scheme explained in the previous section has been applied to geometrically simple free shear flows by Hussain's group at Houston. Similar techniques based on the use of multiple X probes have been recently employed by Antonia's group at Newcastle and Bonnet's group at Poitiers. In this section, several results obtained by the present and similar vorticity-based techniques are presented, in comparison with the relevant results obtained by other techniques.

3.1 Circular Jet

In order to check the present eduction scheme, we first applied it to the near field of a circular free jet using a rake of seven X probes([18]). The jet flow was weakly excited at the jet preferred mode, $f_e D/U_0 \approx 0.3$. Here, f_e is the excitation frequency($= 44Hz$), D is the nozzle diameter($= 7.6cm$) and U_0 is the jet exit speed($= 11m/s$). Structures educed at x/D −3 and 8 are shown in Figs.7(a,b), where the locations of the probes are indicated by thick cross marks, and $\langle\omega\rangle$ values are made dimensionless with f_e. The number of finally accepted structures per second was $0.82 f_e$ at $x/D = 3$ and $0.40 f_e$ at $x/D = 8$.

For this particular case, detailed data by Hussain & Zaman[19] were available. They used local periodic u fluctuations on the jet axis as a trigger signal for phase-locked measurements. Figure 7(c) shows their result, when the trigger probe was placed at $x/D = 3$. Note that their data do not involve use of the Taylor hypothesis but essentially represent the actual vorticity distribution within the experimental uncertainty.

In comparing the two data sets, caution must be made that Figs.7(a,b) represent a selectively sampled, temporal picture at each x station, whereas Fig.7(c) shows a periodically sampled, spatial picture at a particular phase of excitation. Keeping this

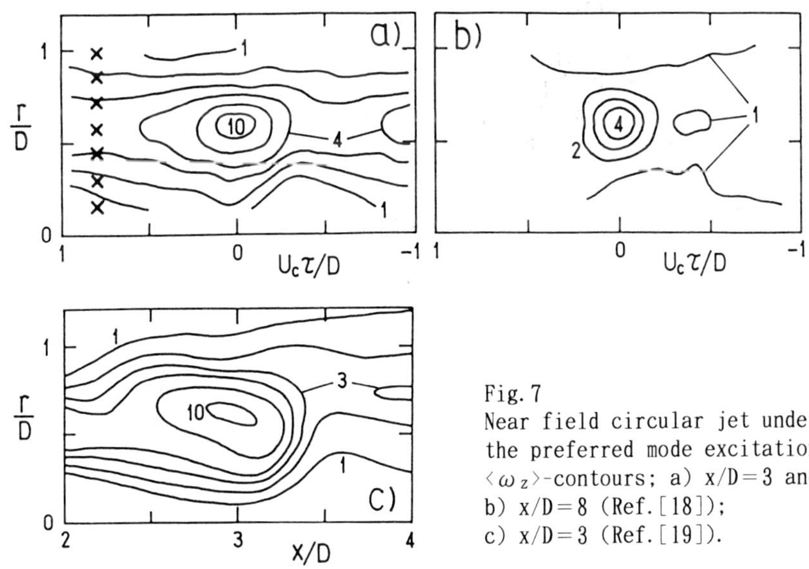

Fig. 7
Near field circular jet under the preferred mode excitation: $\langle\omega_z\rangle$-contours; a) x/D=3 and b) x/D=8 (Ref.[18]); c) x/D=3 (Ref.[19]).

difference in mind, the agreement between Figs.7(a) and 7(c) is fairly good. This indicates that the present eduction scheme works well and that the Taylor hypothesis based on U_c yields a good approximation in this case, where the preferred mode excitation prevents the occurrence of vortex pairing([19]).

Since the sharp signature of structure passage disappears beyond the potential core, the point trigger measurement did not reveal any clear structure for $x/D > 6$([19]). Figure 7(b) indicates, however, that at least 40 percent of the initially induced structures still retain their organization at $x/D = 8$. This example clearly shows a superior feature of our technique, which can detect structures without recourse to any external signal for trigger. Of course, because of the single radial plane measurement, it is not clear whether individual events are axisymmetric or spiral; the latter mode is known to be dominant in the region beyond the potential core([20]).

Recently, Chua & Antonia[21] adapted a similar X-wire rake technique to the jet near-field and identified the opposing and alternating modes even at $x/D = 4$, on the basis of filtered vorticity signals on a diametric plane.

In the fully developed state of a circular jet, Tso[16] employed a rake of 7 X probes, coupled with two single-wire probes which were displaced azimuthally by 90° on either sides of the rake and used for the mode discrimination. He found that the helical modes(m=±1) predominate over the axisymmetric(m=0) and double-helical modes(m=±2). Figures 8(a,b) show the ensemble averages of azimuthal vorticity and radial velocity component for a helical modes(m=+1) at $x/D = 50$. Since the helical structures intersect the jet diametric plane at an angle of roughly 60°([16]), the cross-section sliced by an x-r plane appears elongated more in the radial direction.

A comparable data set was reported by Komori & Ueda[22]. They made simultaneous measurements of the two velocity components(u *and* v) and temperature

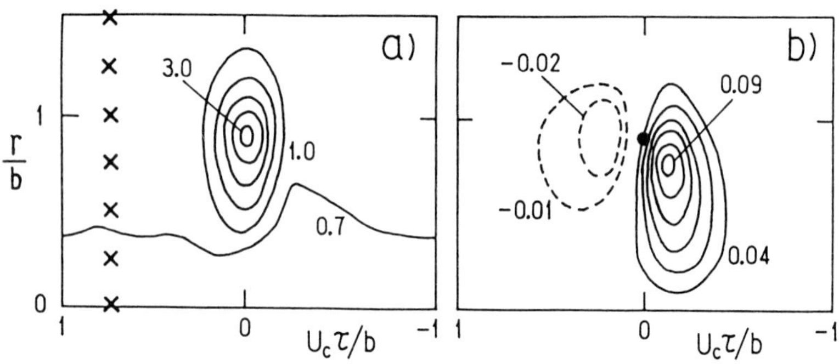

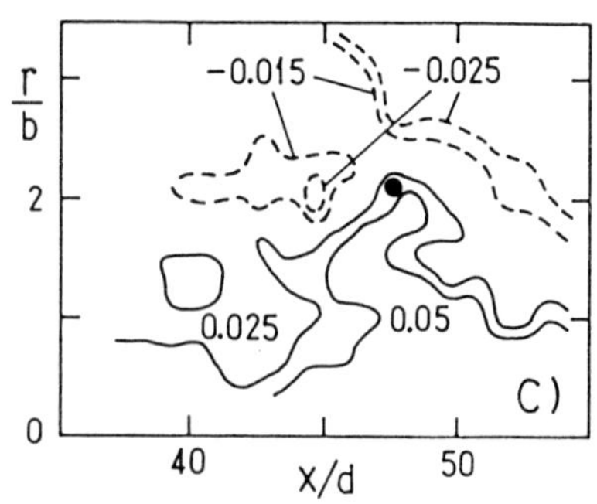

Fig. 8
Far field circular jets:
a) $\langle\omega_z\rangle/S_M$ and b) $\langle v\rangle/U_0$ (Ref. [16]);
c) $\langle v\rangle/(U_0-U_a)$ (Ref. [22]).

fluctuations in a slightly heated jet surrounded by a low speed (non-heated) co-flowing stream and adopted a pattern-averaging technique. The ensemble averages of the three quantities were obtained at 240 points around a trigger probe placed in the intermittent zone of the jet at $x/D = 47.5$. In spite of their elaborate technique(see [22] for detail), the iso-contours of $\langle v \rangle$ in Fig.8(c) look much more complicated than Fig.8(b). The reason may be that the events they captured contain structures of various sizes and different modes, and hence the ensemble-averaged result may have been smeared due to sizable effects of jitter.

3.2 Plane Wake

Taking advantage of the fairly periodic motion of shed vortices, many investigators have studied organized structures in the near region of bluff-body wakes using periodic sampling techniques. Since the periodicity is progressively lost with increasing the downstream distance from the body however, such techniques are effective at best up to $x/l = 10$; here, l is the lateral dimension of the body. Therefore, for farther downstream regions, the situation becomes quite similar to the excited jet case (for $x/D > 6$) discussed in the previous sub-section.

The intermediate region of a circular cylinder wake was studied in some detail by us[23] using a rake of 8 X-probes. The Reynolds number based on the cylinder diameter ($d = 2.7cm$) and the free-stream velocity ($U_0 = 7.0m/s$) was about 13000. Measurements were made at x/d =10, 20, 30 and 40, and most of the the data were taken by the rake covering mainly one side of the wake in order to analyse the details of structures to be educed.

Ensemble-averaged spanwise vorticity contours obtained at x/d =10 and 40 are compared in Figs.9(a,b). The detected and phase-aligned structures are depicted by solid lines and the opposite-signed vorticity contours corresponding to structures shed from the upper side of the cylinder are drawn by broken lines. Contour levels are non-dimensionalized by the local mean-flow variable, S_M, while coordinates are normalized by d (and U_c) for the sake of direct comparison of the structure scales between different x stations. The number of accepted events decreases from $0.37f_s$ to $0.13f_s$ between the two stations; here, f_s is the vortex shedding frequency corresponding to the Strouhal number of 0.21.

A comparison of the two structures at a same contour level (say, $\langle \omega_z \rangle / S_M = 1.0$)

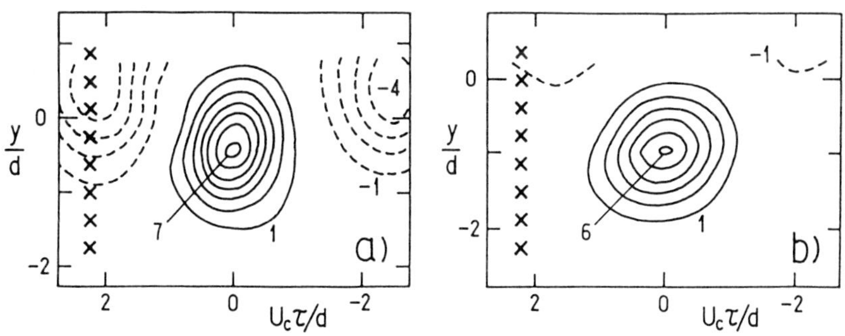

Fig. 9 Intermediate wake of a circular cylinder(Ref. [23]): $\langle \omega_z \rangle$-contours at a) x/d=10 and b) x/d=40.

suggests that typical structures do not grow in size with increasing x but their size relative to the local wake width decreases downstream. It is also apparent that the structure cross-section, which is initially somewhat elongated to the y direction(Fig.9a), becomes rounded and inclined as x increases(Fig.9b).

Since in the conditioning process, we have specified neither the upper bound scale of structures nor their sectional shape, the above observations should not be a consequence biased by the detection scheme used but reflect real features of the intermediate wake, where the (nominally) spanwise vortices shed from the cylinder are largely in a process of gradual decay. Such detailed features could not have been revealed, unless the detection method can capture the centers of individual events directly and align them adequately. If the detection point is distant away from the center of events, as is usually so in other conditional sampling techniques, then the ensemble average $\langle \omega_z \rangle$ would show a vaguer distribution.

In Figs.10(a)-(c), available data on bluff-body wakes obtained with different conditional sampling techniques([24],[25],[26],[27],[28],[29]) are compared, regarding to

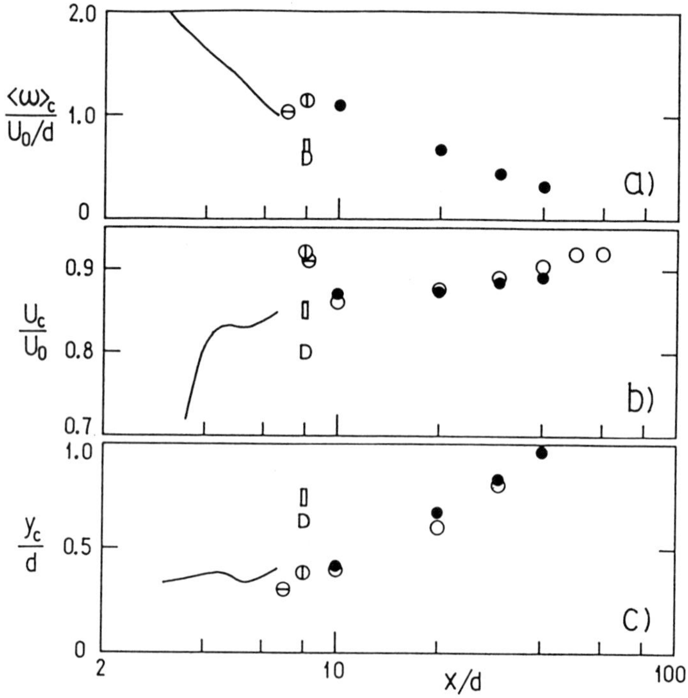

Fig. 10 Comparison of various bluff-body wake data. a) Peak coherent vorticity. b) Convection velocity. c) Average location of structure centers. Symbols are; —— : Cantwell & Coles (Re=140000), Φ : Kiya & Matsumura (16000), ⊖ : Armstrong & Barnes (21500), ○ : Zhuo & Antonia (5600), ● : Hussain & Hayakawa (13000), D : Davies(27000; D-shape cylinder), ▯ : Kiya & Matsumura (23000; normal plate).

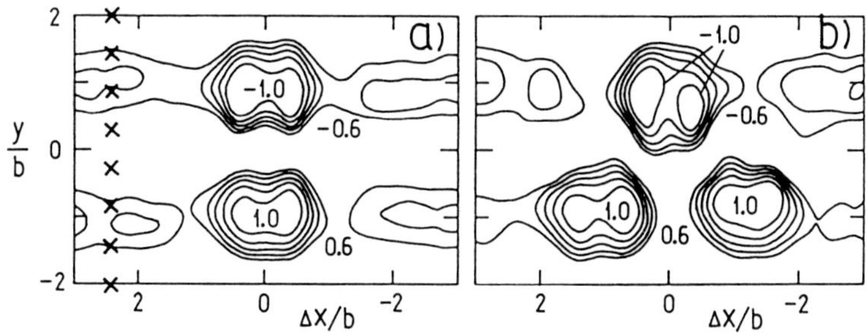

Fig. 11 Far wake of a circular cylinder(Ref. [30]): $\langle \omega_z \rangle$-contours for a)varicose mode and b)sinuous mode.

the streamwise variations of global structure properties: (a)the peak vorticity $\langle \omega_z \rangle_c$, (b)the convection velocity U_c, and (c)the average trajectory of structure centers y_c. The present data denoted by dark symbols are generally consistent with other data, except those of the D-shape cylinder([23]) and normal plate([27]) wakes, whose characteristics may be somewhat different from those of the circular cylinder wake. This comparison indicates that the present structure eduction, though being highly selective, properly identifies representative spanwise structures out of various events and provides a fairly objective result.

Bisset, Antonia & Browne[30] studied the far-field structure of a circular cylinder wake at $x/d = 420$ using a rake of 8 X probes, which covered both sides of the wake. They attempted at capturing the varicose and sinuous modes separately, the coexistence of which had been suggested by Wygnanski, Champagne & Marasli[31] on the basis of a linear stability analysis. Bisset et al. successfully educed the two modes by employing both the WAG (Window Averaged Gradient) detection applied to the filtered v fluctuations and the vorticity-based detection. Their result obtained by the latter technique is shown in Figs.11(a,b), where coordinates are made dimensionless with the local wake half-width b. It is noticed that the educed structures contain two vorticity maxima, the reason for which remains unclear.

3.3 Plane Mixing Layer

We applied the present technique to the self-preserving region of a single-stream plane mixing layer initiated from a turbulent boundary layer[32]. Measurements were made at two x stations: x/θ_e=1000 and 2000; here, θ_e is the initial boundary-layer momentum thickness. A rake of 8 X-probes covered the transverse extent of 60 percent of the (local) entire mixing-layer width B, because a preliminary test showed that large-scale vorticity-bearing events comparable in size with B were very rare.

The results are shown in Figs.12(a,b), where coordinates are made dimensionless with the local momentum thickness θ. The two $\langle \omega_z \rangle$ patterns are quite similar in both the geometry and the peak value, in accordance with the self-similar development of the mean velocity field. The structures have a streamwisely elongated shape with protrusion at the front and back portions. It is noted again that while the detection criteria were common to the two x stations (in terms of the local flow variables), the eduction scheme is capable to preserve different structure features if they were.

Under nearly the same experimental condition, Hussain & Zaman[33] educed coherent structures by using u fluctuations at the high-speed edge of the layer as a trigger signal. Their results are shown in Figs.12(c,d), which look apparently different from Figs.12(a,b). In particular, the $\langle\omega_z\rangle$ contours on the high-speed side show a bulgy appearance. The reason may be that the detection made at a distance from the center of event tends to yield ensemble-averaged flow properties exaggerated toward the trigger location and to distort their real distributions.

Another relevant information was provided by Metcalfe & Menon[32]. They applied an eduction scheme similar to the present one to their 3D direct numerical simulation data of a mixing layer, which evolves in time from an initial condition with random fluctuations. The Reynolds number of the simulated flow is roughly two-order in magnitude lower than that of the laboratory data discussed above. Because of the limited flow range and periodic boundary condition in the x direction, the ensemble average was obtained by taking data at successive spanwise planes as separate

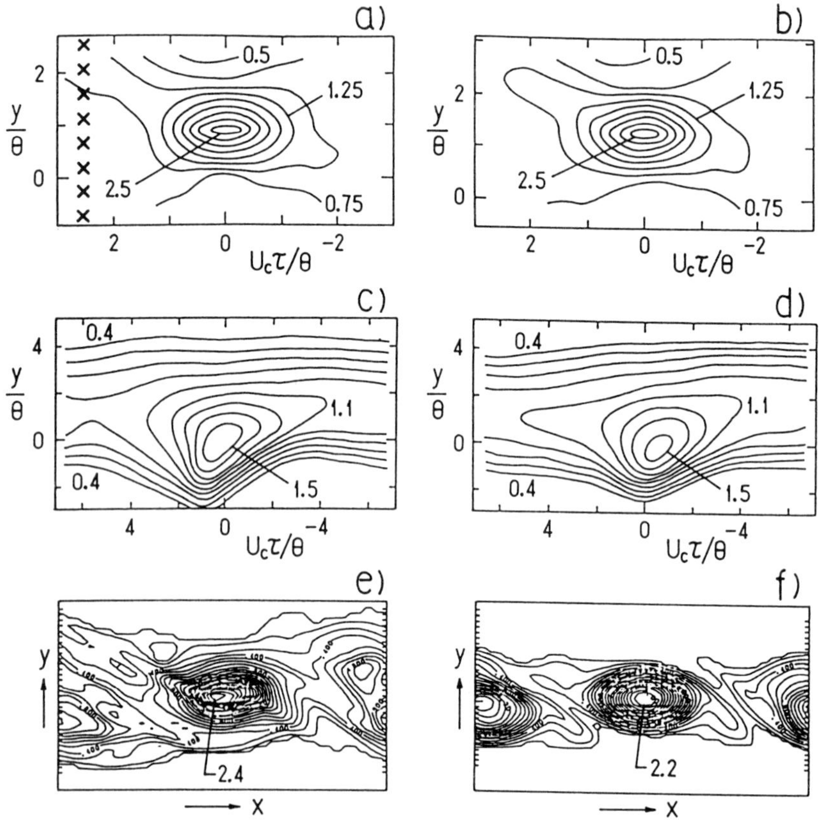

Fig. 12 Plane mixing layers: $\langle\omega_z\rangle$-contours; a) $x/\theta_e=1000$ and b) $x/\theta_e=2000$ (vorticity-based detection; Ref.[32]); c) $x/\theta_e=600$ and d) $x/\theta_e=4800$ (point-trigger detection; Ref.[33]); e)naturally evolving case and f)weakly forced case (direct numerical simulation data; Ref.[32]).

realizations. Results thus obtained are shown in Figs.12(e,f): one for a naturally evolving case(e) and the other for a weakly excited case(f).

In spite of several differences existing between the experimental and simulated flow conditions, the vorticity contour patterns in Figs.12(e,f) remarkably resemble those of our experimental data(Figs.12a,b) in their peak value and geometry, including the protrusions at the front and back of structures. This good correspondence suggests a possibility that the cross-sectional shape of typical structures in the turbulent plane mixing layer does not depend much on the local Reynolds number(see also Panides & Chevray[34]).

3.4 Other Applications

For plane shear flows, the X-wire rake technique can be readily extended to measurements of the transverse vorticity component ω_y, by rotating X probes by 90° and aligning them in the spanwise direction. This probe arrangement has been employed in order to study three-dimensional aspects of the intermediate wake by us[35] and of the far wake by Antonia & Bisset[36].

The experimental condition in the former case was identical with that mentioned in §3.2, and the number of X-probes used was again eight. A pseudo-instantaneous $\tilde{\omega}_y$ map at $x/d = 20$ is shown in Fig.13. There are frequent occurrences of $\tilde{\omega}_y$ concentration, whose scale and strength are comparable with those of spanwise vorticity concentrations(see Fig.3a). It is interesting to note that $\tilde{\omega}_y$ patches often exhibit a paired pattern with opposite-signed circulations, as marked by dotted lines in Fig.13. This vorticity map provides a quantitative evidence for the emergence of highly three-dimensional structures in the moderately near wake region. Similar distributions were also observed at $x/d = 40$([35]). In this connection, the pattern recognition analysis of Ferré & Giralt[37] revealed that "roller eddies", which are known for the far wake([38],[39],[40]), occur even at $x/d = 60$. This is qualitatively consistent with our observation.

From simultaneous records of ω_y at seven z locations, we educed three-dimensional structures sliced by an x-z plane. The eduction procedure was basically the same as used for the identification of spanwise structures. The result showed that when $\tilde{\omega}_y$ peaks of one sign were detected at a fixed z location, the ω_y concentration of the opposite sign preferentially appeared on one side of the detected (i.e., aligned) structure([35]), suggesting the presence of paired, counter-rotating vortices. However, the peak $\langle\omega_y\rangle$ value of the opposite signed structure was only about 20 percent of

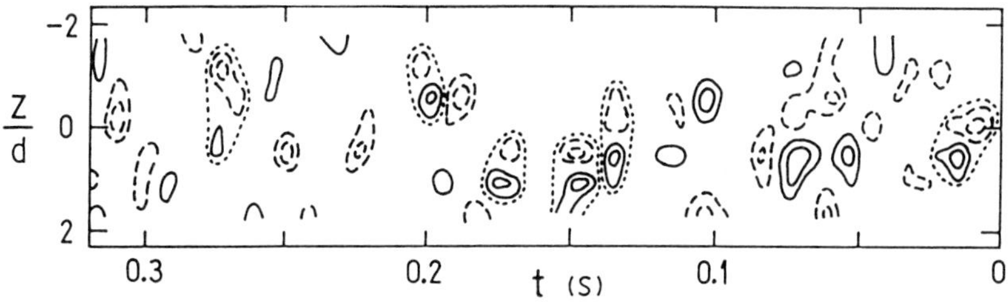

Fig. 13 Pseudo-instantaneous ω_y maps in the intermediate wake of a circular cylinder at x/d=20 (Ref.[35]).

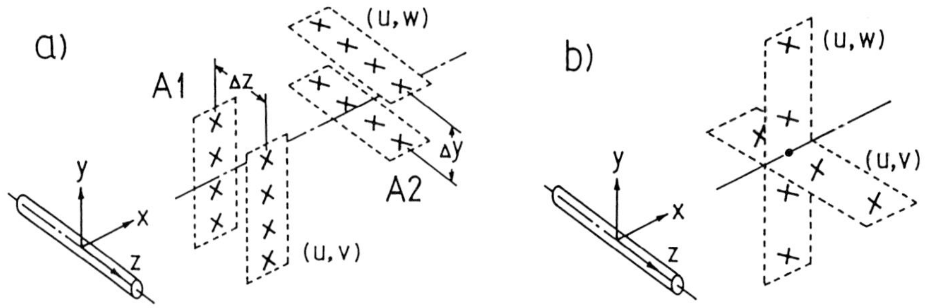

Fig. 14 Examples of X-wire rake arrangement, used in plane wake studies. a)Parallel rakes(Ref.[35]). b)Orthogonal rakes(Ref.[41]).

the corresponding value of the detected structure. This indicates that while paired $\tilde{\omega}_y$ patterns are often observed instantaneously, the relative spatial distance between such a pair significantly varies from one realization to another in both the x and z directions, due to various random factors(see [35] for detailed discussion).

In order to gain a further insight into the strongly three-dimensional structures, we attempted at simultaneous two-plane vorticity measurements [35], in which two parallel rakes were used for measurements of either spanwise or transverse vorticity component at separated two planes(Fig.14a). From two-plane ω_z measurements for several spanwise separations between the rakes (by a probe arrangement A1 in Fig.14a), we found that the typical spanwise extent of nominally two-dimensional vortices at $x/d = 20$ and 40 was roughly $b \sim 2b$; here, b is the wake half-width. Similarly, two-plane ω_y measurements were made with various combinations of two rakes separated in the y direction (by a probe arrangement A2 in Fig.14a). The result showed that the transverse coherence of ω_y concentrations was identifiable over a y range comparable with b but was mainly confined within one side of the wake, and that the most probable orientation of transverse structures with respect to the y axis was about 30°.

As to measurements of three-dimensional structures using multiple hot-wire probes, mention is to be made of two recent investigations for the far wake of a circular cylinder. Bisset, Antonia & Britz[41] employed two orthogonal rakes of X probes (coupled with a cold-wire probe in a slightly heated wake) and measured all three components of (large-scale) vorticity at a crossing point of the two rakes(Fig.14b). Ferré, Mumford, Savill & Giralt[42] applied the pattern-matched correlation technique, originally developed by Townsend[43], to velocity fluctuations taken by eight single-wire probes which were aligned either in the y or z direction. These studies shed more light on the relationship between double-roller and spanwise entrainment structures, an issue that has been a long-pending one since the early studies of Townsend[38], Grant[39] and Payne & Lumley[44].

4 ADVANTAGES AND LIMITATIONS

4.1 Advantages

Major advantageous features of the present technique are summarized below.

1) The whole procedure involved in the technique relies on an intrinsic flow property(i.e., "vorticity") of organized turbulence structures, and hence it is conceptually self-consistent and objective, provided that those events are presumed to be characterized by spatially correlated vorticity.
2) The presence of large-scale vortical events is recognized with filtered vorticity maps, which give a physical perception of instantaneous fields, more quantitatively than that from conventional flow visualization pictures.
3) The (approximate) center of each event is directly captured and is used as a phase reference for triggering ensemble averaging operation. This feature serves the following important merits.

a)Effects of jitter are minimal at the educed structure center, affording a sharper decomposition of the ensemble-averaged part and the remained part. This feature is particularly important in order to clarify the relative spatial distributions of large-scale and smaller-scale motions and to infer the mechanisms of entrainment, production and mixing.

b)Vortical events can be captured at any transverse/spanwise location, as far as they occur within the extent covered by the X-wire rake. This feature is similar to that of "delocalized conditional sampling" by Bellin, Delville, Vincendeau, Garem & Bonnet[45].

c)A condition on the sectional shape of structures can be included rather easily in the detection process (at the step III in §2.4), such as $\tilde{\omega}\,(y_c + \Delta y, t_c + \Delta\tau) > Th_2$ and $\tilde{\omega}\,(y_c - \Delta y, t_c - \Delta\tau) > Th_2$. Here, an optimum value of $\Delta\tau$ can be determined from the cross-correlation between $\tilde{\omega}\,(y_c)$ and $\tilde{\omega}\,(y_c \pm \Delta y, \tau)$. A similar procedure has been employed by Tso[16] for the mode discrimination in the far field jet, as mentioned earlier.

Some recent studies related to the advantages above need to be mentioned here. Instantaneous fields similar to our vorticity maps have been obtained by other techniques. Agui & Jimenez[46] obtained the instantaneous spatial distribution of spanwise vorticity in a circular cylinder wake with the aid of a particle-tracking method. Delville, Bellin, Garem & Bonnet[47] visualized instantaneous high shear regions in a plane mixing layer using a rake of 24 single-wire probes coupled with a color-coded image processing technique, called "pseudo flow visualization"(see also [45]).

Aside from direct numerical simulations, instantaneous 3D field data recently became measurable by novel experimental techniques. Let us mention here only two examples. Kasagi & Nishino[48] improved the 3D particle-tracking velocimetry system using three video cameras set at different viewing angles to a flow, and succeeded in fairly accurate measurements of three-dimensional velocity field in a turbulent channel flow. Meng & Hussain[49] developed a holographic particle velocimetry for measurements of velocity vector field in both 3D space and time, and demonstrated its promising capability in some vortical flows.

A good correspondence between low pressure regions and vortical structures has recently received attention both in theoretical and numerical studies(e.g. [50],[51]). Toyoda[52] developed a specially designed small pressure probe and obtained the ensemble average of static pressure in the near field of a circular jet excited at the vortex paring mode. Vortices undergoing pairing were well identified as the evolution of low pressure regions(see also [53]). Therefore, static pressure can be also a good flow property to characterizing vortical structures. In laboratory experiments, however, multi-point measurements of static pressure, which ought to be required to capture structure centers unambiguously, may not be an easy task.

4.2 Limitations

Limitations of the present technique are primarily due to the coarse resolution in measurements and the use of Taylor's hypothesis.

1) The spatial resolution depends on the number of X probes used and their separation relative to the thickness of a shear layer. .

a)Measurements of thin shear layers are difficult, unless specially designed miniature probes are available.

b)Vorticity concentrations smaller in size than the probe separation escape the measurement.

c)Transverse scale of identifiable events must be larger than $2\Delta s$, because otherwise, the center of events cannot be detected properly, except in the case that the transverse location of the centers is well fixed, say, via artificial excitation.

d)Calculated instantaneous vorticity values are crude approximation to the actual values. Vorticity estimated from a coarse resolution measurement(containing a spatial-filter function) may not be identical with filtered vorticity computed from the data of a finer resolution measurement.

e)Use of a larger number of probes requires a large capacity of laboratory computer along with a sufficiently high speed A/D conversion system for the acquisition and analyses of data, and hence it costs much.

2) The use of Taylor's hypothesis is unavoidable, since measurements are made at a fixed streamwise station. Note that the conversion of data from t to x, based on a constant convection velocity, is used in both the vorticity computation and the calculation of ensemble averages.

a)Any information about the spatial evolution of structures is limited within a short duration around a given instant.

b)Measurements in regions where structures are undergoing rapid evolution or interaction, such as vortex paring, may include significant error.

c)The applicability of Taylor's hypothesis, based on the structure celerity, to smaller-scale motions may be questionable, but is generally hard to be evaluated.

3) The process of conditioning involves subjectivity. Choice of thresholds in discriminating the strength and size of structures includes arbitrariness, and the number of selected events depends on the thresholds used.

With regard to the limitations listed above, some comments are necessary. Our experience, based on the use of a rake of 8 X-probes, has indicated that the spatial resolution is not necessarily a serious problem. Through a preliminary survey of instantaneous vorticity maps, we can choose a suitable probe arrangement to identify representative structures in the flow considered.

Vorticity fluctuations, whose scales are at least larger than the probe spacing, seem to be estimated reasonably well, as would be inferred from several examples presented in this paper. Moreover, velocity signals at N y locations contain spectral contents up to the Nyquist frequency of data sampling. Therefore the resolution of ensemble-averaged quantities through data interpolation is about a half the probe spacing, to the first-order approximation. For the time-to-space conversion of data, we have used the average streamwise celerity of structure centers, which was iteratively determined so as to be equal to the ensemble-averaged u component at the educed structure center. The pertinence of this choice for isolated structures was validated by Zaman & Hussain[54]. They warned, however, that when structure interactions like vortex pairing are involved, no convection velocity can be found with which the Taylor hypothesis works well. In such a case, phase-locked spatial measurements without use of the hypothesis appear to be a better approach, as was done by many investigators(e.g. [55],[56],[57],[58]).

Arbitrariness involved in the process of conditioning is a common problem in any conditional sampling techniques. In the present case, the arbitrariness corresponds to our judgement in selecting the events of 'similar' characteristics. Such subjectivity is unavoidable to some extent, as far as we try to extract dominant structures in turbulent shear flows.

Finally, it may be worth mentioning that the generic usefulness of the present eduction scheme has been confirmed in the numerical simulations of a plane mixing layer([32]) and a thick flat plate wake([59]), where there is no concern about the limitations due to the spatial resolution and Taylor hypothesis.

5 CLOSING REMARKS

The vorticity-based conditional sampling technique using a rake of multiple X-wire probes has been outlined, and its effectiveness has been shown through several examples.

Since recent progress in the laboratory technology has made it easier to employ a larger number of the probes simultaneously([60],[61],[62]), it is not currently difficult to perform the rake measurement which is more resolvable and accurate than the one presented in this paper. However the present approach may not be easily extended to the measurement of all three components of vorticity at multiple points in a flow, which is obviously desired to unveil the topology and dynamics of essentially three-dimensional organized structures in turbulence. While accurate measurements of three vorticity components at a point has become possible owing to the development of miniature vorticity probes(e.g. [63],[64]), the simultaneous use of the probes at many points in a flow is still beyond the current laboratory technology.

These problems in hot-wire probe measurements will be conquered by newly developing methods, such as 3D particle-tracking velocimetry([49]), holographic particle velocimetry([50]), NMR imaging velocimetry([65]) and scalar imaging velocimetry([66]); all these methods have an excellent capability in acquiring instantaneous 3D vector field data. However, even if a massive amount of 3D data are taken through the novel experimental techniques as well as direct numerical simulations, it is crucial how to focus on dynamically significant flow events in the flow considered and how to extract them with some least subjective means. It seems to us that "vorticity" should be a primarily important flow property to observe for that purpose. This has been a basic concept in the vorticity-based conditional sampling technique.

ACKNOWLEDGEMENT

I wish to express my cordial thanks to Professor F.Hussain for invaluable and stimulating discussions through long-term collaborative works, which provided the basis of this report. I would like to thank Professors Y.Kobashi, S.Iida, M.Kiya and H.S.Husain and Dr M.Ichijo for helpful discussions and encouragement. Thanks are also due to Professors R.A.Antonia, K.Toyoda and M.Matsumura for providing me with useful information regarding their experimental results.

REFERENCES

1. Roshko,A. 1976 Structure of turbulent shear flows: a new look. *AIAA J.* **14**, 1349-1357.

2. Cantwell,B.J. 1981 Organized motion in turbulent flow. *Ann.Rev.Fluid Mech.* **13**, 457-515.
3. Hussain,A.K.M.F. 1986 Coherent structures and turbulence. *J.Fluid Mech.* 173, 303-356.
4. Fiedler,H.E. 1987 Coherent structures. *Advances in Turbulence* **I**(ed. G.Comte-Bellot & j.Mathieu), 320-336.
5. Proc.IUTAM Symp.:"Eddy structure identification in free turbulent shear flows". Poitiers (1992).
6. Kaplan,R.E. 1973 Conditioned sampling technique. *Turbulence in Liquids,*, 274-283. Univ.Missouri-Rolla.
7. Van Atta,C.W. 1974 Sampling techniques in turbulence measurements. *Ann.Rev.Flu Mech.* **6**, 75-91.
8. Antonia,R.A. 1981 Conditional sampling in turbulence measurement. *Ann.Rev.Fluia Mech.* **13**, 131-156.
9. Hunt,J.C.R. 1988 Studying turbulence using direct numerical simulation: 1987 Center for Turbulence Research NASA Ames/Stanford Summer Programme. *J.Fluic Mech.* **190**, 375-392.
10. Blackwelder,R.F. 1977 On the role of phase information in conditional sampling. *Phys.Fluids* **20**, S232-S242.
11. Sato,H. 1983 Cognition and description of patterns in turbulent flows. *Proc.2nd Asian Congr.Fluid Mech.*. Sicence Press, Beijing.
12. Yule,A.J. 1979 Phase scrambling effects and turbulence data analysis. *Turb.Shear Flows* **2**, 263-281.
13. Hussain,A.K.M.F. 1983 Coherent structures — reality and myth. *Phys.Fluids* **26**, 2816-2850.
14. Hussain,A.K.M.F. 1980 Coherent structures and studies of perturbed and unperturbed jets. *Lecture Notes in Physics* **136**, 252-291.
15. Laufer,J. 1975 New trends in experimental turbulence research. *Ann.Rev. Fluid Mech.* **7**, 307-326.
16. Tso,J. 1983 Organized structures in the far field of a circular jet. Ph.D Thesis. Johns Hopkins Univ., Baltimore. See also, Tso,J. & Hussain,F. 1989 Organized motions in a fully developed turbulent axisymmetric jet. *J.Fluid Mech.* **203**, 425-448.
17. Hayakawa,M.& Hussain,A.K.M.F. 1985 Eduction of coherent structures in the turbulent plane wake. *Proc.5th Symp.Turb.Shear Flows,* Cornell Univ., 4.33-4.37.
18. Hayakawa,M.& Hussain,A.K.M.F. 1984 (unpublished data).
19. Hussain,A.K.M.F.& Zaman,K.B.M.Q. 1981 The 'preferred mode' of the axisymmetric jet. *J.Fluid Mech.* **110**, 39-71.
20. Batchelor,G.K.& Gill,A.E. 1962 Analysis of the stability of axisymmetric jets. *J.Fluid Mech.* **14**, 529-551.
21. Chua,L.P.& Antonia,R.A. 1992 Spatial organization of large structures in the near-field of a circular jet. *Fluid Dyn.Res.* **9**, 59-71.
22. Komori,S.& Ueda,H. 1985 The large-scale coherent stuctures in the intermittent region of the self-preserving round jet. *J.Fluid Mech.* **152**, 337-359.
23. Hussain,A.K.M.F.& Hayakawa,M. 1987 Eduction of large-scale structures in a turbulent plane wake. *J.Fluid Mech.* **180**, 193-229.
24. Davies,M.E. 1976 A comparison of the wake structure of a stationary and oscillating bluff body, using a conditional sampling technique. *J.Fluid Mech.* **75**, 209-231.
25. Cantwell,B.& Coles,D. 1983 An experimental study of entrainment and transport in the turbulent near wake of a circular cylinder. *J.Fluid Mech.* **136**, 321-374.

26.Kiya,M.& Matsumura,M. 1985 Turbulence structure in the intermediate wake of a circular cylinder. *Bull.JSME* **28**, 2617-2624.
27.Armstrong,B.J.& Barnes,F.H. 1987 A comparison of the structure of the wake behind a circular cylinder in a steady flow with that in a perturbed flow. *Phys.Fluids* **30**, 19-26.
28.Kiya,M.& Matsumura,M. 1988 Incoherent turbulence structure in the near wake of a normal plate. *J.Fluid Mech.* **190**, 343-356.
29.Zhou,Y.& Antonia,R.A. 1992 Convection velocity measurements in a cylinder wake. *Exps.Fluids* **13**, 63-70.
30.Bisset,D.K.,Antonia,R.A.& Browne,L.W. 1990 Spatial organization of large structures in the turbulent far wake of a cylinder. *J.Fluid Mech.* **218**, 439-461.
31.Wygnanski,I.,Champagne,F.& Marasli,B. 1986 On the large-scale structures in two-dimensional, small-deficit, turbulent wakes. *J.Fluid Mech.* **168**, 31-71.
32.Metcalfe,R.W.,Hussain,A.K.M.F.,Menon,S.& Hayakawa,M. 1987 Coherent structures in a turbulent plane mixing layer: a comparison between direct numerical simulations and experiments. *Turb.Shear Flows* **5**, 110-123.
33.Hussain,A.K.M.F.& Zaman,K.B.M.Q. 1985 An experimental study of organized motions in the turbulent plane mixing layer. *J.Fluid Mech.* **159**, 85-104.
34.Panides,E.& Chevray,R. 1990 Vortex dynamics in a plane, moderate-Reynolds-number shear layers. *J.Fluid Mech.* **214**, 414-435.
35.Hayakawa,M.& Hussain,F. 1989 Three-dimensionality of organized structures in a plane turbulent wake. *J.Fluid Mech.* **206**, 375-404.
36.Antonia,R.A.& Bisset,D.K. 1989 Three-dimensional aspects of the organized motion in a turbulent boundary layer. *Turbulence and Coherent Structures* (eds. M.Lesieur & O.Metais), 1-17.
37.Ferre,J.A.& Giralt,F. 1989 Pattern-recognition analysis of the velocity field in plane turbulent wakes. *J.Fluid Mech.* **198**, 27-64.
38.Townsend,A.A. 1956 The Structure of Turbulent Shear Flow. Cambridge Univ. Press.
39.Grant,H.L. 1958 The large eddies of turbulent motion. *J.Fluid Mech.* **4**, 149-190.
40.Mumford,J.C. 1983 The structure of the large eddies in fully developed turbulent shear flows: Part 2 The plane wake. *J.Fluid Mech.* **137**, 447-190.
41.Bisset,D.K.,Antonia,R.A.& Britz,D. 1990 Structure of large-scale vorticity in a turbulent far wake. *J.Fluid Mech.* **218**, 463-482.
42.Ferre,J.A.,Mumford,J.C.,Savill,A.M.& Giralt,F. 1990 Three-dimensional large-eddy motions and fine-scale activity in a plane turbulent wake. *J.Fluid Mech.* **210**, 371-414.
43.Townsend,A.A. 1979 Flow patterns of large eddies in a wake and in a boundary layer. *J.Fluid Mech.* **95**, 515-537.
44.Payne,F.R.& Lumley,J.L. 1967 Large eddy structure of the turbulent wake behind a circular cylinder. *Phys.Fluids* **10**, S194-S196.
45.Bellin,S.,Delville,J.,Vincendeau,E.,Garem,J.H.& Bonnet,J.P. 1992 Large scale structure characterization in a 2D mixing layer by pseudo-flow visualization and delocalized conditional sampling. In Ref.[5], V.2.1-2.8.
46.Agui,J.C.& Jimenez,J. 1987 On the performance of particle tracking. *J. Fluid Mech.* **185**, 447-468.
47.Delville,J.,Bellin,S.,Garem,J.H.& Bonnet,J.P. 1989 Analysis of structures in a turbulent, plane mixing layer by use of a pseudo flow visualization method based on hot-wire anemometry. *Advances in Turbulence* **2**(eds. H.H.Fernholz & H.E.Fiedler), 251-256.
48.Kasagi,N.& Nishino,K. 1991 Probing turbulence with three-dimensional particle-

tracking velocimetry. *Exps.Thermal Fluid Sci.* **4**, 601-612.
49.Meng,H& Hussain,F. 1991 Holographic particle velocimetry: a 3D measurement technique for vortex interactions, coherent structures and turbulence. *Fluid Dyn.Res* **8**, 33-52.
50.Larcheveque,M. 1992 Velocity field, vorticity field, pressure field and coherent structures. In Ref.[5], VII.11.1-11.5.
51.Sandham,N.D.& Reynolds,W.C. 1991 Three-dimensional simulations of large eddies in the compressible mixing layer. *J.Fluid Mech.* **224**, 133-158.
52.Toyoda,K.& Hussain,F. 1992 Eduction of vortical structures in a circular jet by pressure measurements. *Proc.5th Asian Congr.Fluid Mech.* Vol.**1**, 587-591.
53.Toyoda,K.& Shirahama,Y. 1992 Eduction of vortical structures by pressure measurements in noncircular jets. In Ref.[5], VII.2.1-2.5.
54.Zaman,K.B.M.Q.& Hussain,A.K.M.F. 1981 Taylor hypothesis and large-scale coherent structures. *J.Fluid Mech.* **112**, 379-396.
55.Browand,F.K.& Weidman,P.D. 1976 Large scales in the developing mixing layer. *J.Fluid Mech.* **76**, 127-144.
56.Zaman,K.B.M.Q.& Hussain,A.K.M.F. 1980 Vortex pairing in a circular jet under controlled excitation. Part 1. General jet response. *J.Fluid Mech.* **101**, 449-491.
57.Wygnanski,I.& Weisbrot,I. 1988 On the pairing process in an excited plane turbulent mixing layer. *J.Fluid Mech.* **195**, 161-173.
58.Husain,H.S.& Hussain,F. 1991 Elliptic jets. Part 2. Dynamics of coherent structure: pairing. *J.Fluid Mech.* **233**, 439-482.
59.Jeong,J.& Grinstein,F.F. 1992 Eduction of coherent structures in a numerically simulated plane wake. In Ref.[5], V.4.1-4.5.
60.Delville,J. 1992 Characterization of the organization in shear layers via proper orthogonal decomposition. In Ref.[5], VI.1.1-1.8.
61.Zhou,Y.& Antonia,R.A. 1992 A study of flow properties near critical points. In Ref.[5], IX.2.1-2.6.
62.Albanis,N.,Husain,H.S.& Hussain,F. 1992 (private communication).
63.Vukoslavcevic,P.,Wallace,J.M.& Balint,J.L. 1991 The velocity and vorticity vector fields of a turbulent boundary layer. Part 1. Simultaneous measurement by hot-wire anemometry. *J.Fluid Mech.* **228**, 25-51.
64.Tsinober,A.,Kit,E.& Dracos,T. 1992 Experimental investigation of the field of velocity gradients in turbulent flows. *J.Fluid Mech.* **242**, 169-192.
65.Kose,K. 1991 Instantaneous flow distribution measurements of the equilibrium puff in a circular pipe using NMR imaging. *Phys.Rev.A* **44**, 2495-2504.
66.Dahm,W.J.A.,Su,L.K.& Southerland,K.B. 1992 A scalar imaging velocimetry technique for fully resolved four-dimensional vector velocity field measurements in turbulent flows. *Phys.Fluids A* **4**, 2191-2206.

Eduction of Coherent Structures in a Numerically Simulated Plane Wake

J. Jeong*, F. F. Grinstein+, F. Hussain*, & N. Albanis*

* Dept. of Mech. Eng., U. of Houston, TX 77204-4792 USA

+ Naval Research Lab., Code 6410, Washington, D.C. USA

Abstract

Coherent structures (CS) are educed using a conditional sampling technique involving alignment of vorticity patches of largest size and strength; hence we educe dominant CS. A numerically simulated spatially evolving wake of a flat plate is used as the database, and the inflow condition for the simulated wake includes random velocity perturbations which emulate turbulent conditions at a plate exit in the laboratory. In addition to previously educed properties such as coherent vorticity and production, and incoherent Reynolds stress and turbulence intensity, other measures such as coherent pressure and passive scalar distributions are also studied. In spite of the geometry difference, the near-wake dynamics of the plate seem quite similar to that of a cylinder. For example, turbulence is mostly produced by vortex stretching of the ribs at the saddle and then advected to the structure center, where it accumulates, and is balanced by incoherent dissipation. The distribution of coherent passive scalar indicates that mixing occurs in the saddle regions and that the mixed fluid is advected into the structure center.

1. Introduction

CS in a circular jet and the wake of a circular cylinder were educed previously in laboratory flows using a linear rake of X-wires, which provided spanwise vorticity distribution in a spanwise plane (see [1], [2], the latter referenced as HH). Despite the unavoidable crude transverse resolution of the experimental data, these studies provided significant insights into the geometry and dynamics of CS. In this paper, a database of a spatially evolving flow reported previously by Grinstein *et al.* [3] (referenced as GHB) is used to educe CS and evaluate CS properties in more detail as well as other measures, which are difficult to obtain experimentally.

J. P. Bonnet and M. N. Glauser (eds.), Eddy Structure Identification in Free Turbulent Shear Flows, 65–75.

The numerical model is based on the solution of the unsteady inviscid conservation equations for mass, momentum, and energy of an ideal gas. The free-stream Mach number is $M_\infty = 0.6$. A passive scalar c with unity Schmidt number S_c, is advected with the velocity field in order to explore mixing in the flow. It is initialized by setting its free-stream values as $c = 1$ and $c = -1$ on the two sides of the wake. Based on U_∞, d, and effective numerical viscosity ν_{eff}, the characteristic Reynolds numbers for the wake flows simulated in this work are in the range $10^3 \lesssim Re \lesssim 4 \times 10^4$.

2. Eduction of CS

2.1 Eduction philosophy

Although our eduction approach in principle applies to the 3D vorticity field, limitation of experimental data forced us so far to use one vorticity component in a 2D plane, namely ω_z in the xy plane. Numerically, eduction of 3D structure based on the easily available 3D vorticity data is quite feasible except that it is computationally prohibitive to obtain 3D instantaneous data for long times. Therefore, in the present case, we educe the 2D structure in the xy plane considering statistical homogeneity in the z direction. The eduction process, in principle, consists of the following steps: (i) choose smoothed spanwise vorticity as a feature that denotes passage of large-scale structures, (ii) select the most dominant ones based on their vorticity magnitude and size, (iii) phase align the accepted structures and obtain their ensemble average, (iv) refine the phase average by an iterative procedure that is based on the cross-correlation of individual realizations (rejecting 'freak' structures) and their ensemble average, and then (v) compute by ensemble averaging relevant coherent and incoherent quantities that are important to the dynamics of the flow from the *unsmoothed* data. Thus, smoothing, used only for identification of structures in the random vorticity field, does not affect any of the educed quantities: coherent or incoherent.

One advantage of our eduction scheme over other eductions of CS, say using POD (proper orthogonal decomposition) or LSE (linear stochastic estimation) is the ability to address the dynamics of physically occurring CS in any flow. This is possible through a conditionally averaged equation derived from the Navier-Stokes equation after decomposing the instantaneous field into coherent and incoherent parts (see [4]). Another advantage is that the educed incoherent quantities provide information about the dynamics of the small-scale motions, not totally possible in POD or LSE.

2.2 Eduction scheme

Compared to laboratory experiments, the instantaneous flow field in the simulation is free from limitations such as the effects of probe interference and the highly restrictive Taylor's hypothesis. The computations also have the advantage of being free from errors of hot wire measurements due to high turbulence levels and flow reversal (specially in the near-wake). On the other hand, numerical simulations have limitations as well. The relatively large number of realizations required for the statistical studies, the extremely long times needed for phase-space analysis of the flow variables, and computations of 3D flows at practical Reynolds numbers can become prohibitive [4]. An unavoidable compromise that we were forced to make here was to regard each spanwise section (xy plane) of a roll near the location $x/d = 10$ as a separate realization. This was also done in the eduction of CS in numerically simulated mixing layers by Metcalfe *et al.* [5].

Since our primary interest is in large-scale events, the identification of structures was performed on the smoothed instantaneous spanwise vorticity data which were obtained by local spatial smoothing. In the present case, a nine point averaging (3 $\times$ 3 in an xy plane) was used; this amounts to 8 % of the average interval between successive structures. This operation smooths out the fine-scale fluctuations, but retains the underlying large-scale vorticity. Once again, note that all structure properties are finally educed from the unsmoothed data.

The first condition employed by the present eduction scheme is for the structure strength. Based on that, realizations (*i.e.* vorticity maps in the streamwise-transverse plane) with maximum smoothed ω_z exceeding $2S_M$ (where S_M is the maximum mean shear at $x/d = 10$) are detected at each spanwise plane. The choice of the local S_M for specifying the threshold is justified by the fact that both S_M and the structure strength evolve in x. The second condition is for the structure size. In order to specify this, ω_z is required to be higher than $0.3S_M$ within a distance of $0.25d$ from its peak in the xy plane.

Each realization is then aligned with respect to the peak of ω_z and averaged to obtain the zeroth-order ensemble average. Each realization is then shifted in both x and y directions to maximize its cross-correlation with the zeroth-order ensemble average in order to obtain sharper structure boundaries. This process is iterated until convergence in structure shape is achieved. The purpose of the iteration is to find the underlying, true center of each realization necessary for its optimal alignment before its inclusion for the ensemble averaging. Note that the location of maximum spanwise vorticity can only give a local peak and not the center of the overall structure, which can be obtained from the location of maximum cross-correlation in this study. The domain selected for the cross-correlation is shown in figure 5c. Realizations that produce weak cross-correlation coefficient, here below 0.8, or require shifts more than 0.5d were discarded because they are considered to be 'freak' *i.e.* significantly different from the ensemble-averaged structure. Note that this iteration process is similar to the pattern recognition technique, provided that the zeroth iteration ensemble-average is considered as the trial pattern. Here, the trial pattern is not guessed, but derived from signals themselves. After a new alignment of the realizations through the modified peak locations, the smoothed data are discarded and only the unsmoothed data are used to obtain the final ensemble average. Whatever survives the ensemble average is the coherent structure; the departure of each accepted realization from the average is the incoherent turbulence in that realization. Incoherent turbulence is then rms ensemble averaged to extract its statistical measures over the structure cross-section.

The data presented in this paper are spatial distributions of various phase-averaged quantities; velocities are normalized by the free-stream velocity U_∞, and vorticity and strain rate are normalized by the peak value of $\partial U/\partial y$ at the corresponding x station.

3. Instantaneous flow field

Figure 1a shows superimposed isosurfaces of spanwise (grid) and streamwise (solid) vorticity at a level 10% of the peak vorticity magnitude ($\Omega = (\omega_x^2 + \omega_y^2 + \omega_z^2)^{\frac{1}{2}}$) in the computational box; figure 1b shows isosurfaces of the vorticity magnitude at the same level as figure 1a, but as seen from the opposite side of the wake. The instantaneous visualization in figure 2 corresponds to a perspective view of figure 1, and shows isosurfaces for the passive scalar, c, associated with the particular values $c = 0$ (solid) and $c = \pm 0.5$ (grid), characterizing regions of large and moderate mixing between the two sides of the wake, respectively.

The vorticity introduced by the spanwise excitation in the transverse and streamwise directions is deformed, stretched and concentrated in longitudinal ribs as a result of the high strain-rates between rolls. The ribs appear in counter-rotating pairs distinctly superimposed on the rolls and can be initially interpreted as legs of horseshoe vortices. As modeled by GHB and indicated schematically in figure 3, the horseshoe structures initially develop together with the undulation of the newly shed roll 3 in the saddle region I. Wrapping the roll 2 of opposite sign shed on the other side of the wake, a horseshoe vortex is likewise formed with the opposite circulation in the saddle region III.

Eduction analysis of the simulated wake shows that when one mode of sinusoidal excitation is used at the inflow, one can find very good agreement with experimental results on the coherent quantities. However, basic incoherent properties reported in HH, such as occurrence of peak incoherent turbulence intensity at the spanwise roll center, are not observed. In order to emulate the inflow background turbulence, we imposed random perturbations on the streamwise and cross-stream velocities with a rms-level of 2.5 % of the free-stream velocity.

A particularly relevant issue is that of the dependence of the flow properties on outflow boundary conditions in the simulation. These conditions can trigger formation of rolls through feedback at the inflow boundary. They can also affect the evolution of the rolls directly, when the rolls are close to the outflow boundary. For the flows simulated in the present study, the feedback amplitude level is small compared to the imposed random perturbation level at the inflow boundary. Direct effects of the outflow boundary conditions on the structure properties are minimized by choosing the detection location sufficiently far away from the outflow boundary.

Figures 4a,b show isosurfaces of the vorticity magnitude and the passive scalar corresponding to the levels $0.1\Omega_{\mathbf{peak}}$ and $c = 0$ for the flow with the inflow conditions discussed above. From these, it is clear that rolls and ribs become much more disorganized as a result of the inflow perturbations.

4. Educed flow field

To increase the number of realizations available in a typical roll (indicated by an arrow in figure 4a), we consider spanwise planes of five successive structures captured near $x/d = 10$. This yields about six hundred realizations. Of these, half were rejected during iterative selection process, and the remaining half were optimally aligned before final ensemble-averaging. Figure 5 shows $\langle u - U_c \rangle$, $\langle v \rangle$, $\langle \omega_z \rangle$, $\langle S_{xy} \rangle$ in the xy plane. In all the figures, the circle denoting the contour level at 10 % of the peak $\langle \omega_z \rangle$ (figure 5c) serves as the common reference. The corresponding results obtained using one mode of sinusoidal spanwise excitation (not shown), have distributions quite similar to those in figure 5, indicating that the selection of the perturbation has little effect on these quantities. The distributions of $\langle u - U_c \rangle$, $\langle v \rangle$, $\langle \omega_z \rangle$ and $\langle S_{xy} \rangle$ shown here are quite similar to those obtained by HH in a cylinder wake. For reference, we marked the vortex center and the saddle by '+' and '×', respectively.

Due to the ribs, peaks of $\langle \omega_x^2 \rangle$ (figure 6a) occur in the saddle region and coincide with those of incoherent Reynolds stress (figure 6b). There is also significant $\langle \omega_x^2 \rangle$ at the center of the CS due to the undulation of the rolls. It can be argued that the incoherent Reynolds stress is produced by the induced motion of the ribs (intermediate-scale CS), and that ribs connect CS with opposite signs of ω_z (figures 6a, 5c). The coherent shear production also shows peaks near the saddle because of the large incoherent Reynolds stress and shear strain in those regions (figure 6c). Thus, the mechanism of turbulence production in the wake of the flat plate is the same as that in the cylinder wake (HH), namely vortex stretching of ribs by the shear strain caused by adjacent rolls. The incoherent turbulence along the ribs is apparent in figure 6d; however most of the incoherent turbulence occurs at the structure center, implying that incoherent turbulence produced at the saddle is transported to the center of the structure (see [2], [6]) and is accumulated there.

Coherent pressure $\langle p - p_o \rangle$, where p_o is the free-stream pressure, has a negative peak at the structure center and has positive value at the saddle, as expected (figure 6e). This suggests that the pressure signal can be used for the detection of the roll. The advantage of using the pressure signal is that it is invariant under any coordinate transformation, and that the negative peak pressure at the structure center is proportional to the circulation of the structure, so that one can directly detect the large-scale structure from it [7]. However, the pressure is high in the saddle region and locally drops in the ribs. Therefore, if the circulation of the ribs is not high enough, the pressure drop is not sufficient for their detection [8]. Note that the ribs can be easily detected using $(\omega_x^2 + \omega_y^2)$ even when they do not have circulation comparable to that of the rolls. Therefore, vorticity magnitude allows the identification of both CS and dynamically relevant ribs.

The coherent passive scalar level $\langle c \rangle$ (figure 7a) has large gradients in the neighborhood of the saddle region, reflecting that scalar mixing is the largest there. This is expected because passive scalars from the two sides of the wake are advected into the saddle region, and the interface between them is increased by rib-induced flow. Clearly, $|\langle c \rangle| = 0$ on the interface separating the two fluids on the two sides of the wake. However, a region of $|c| \ll 1$ would also indicate mixing by cross diffusion; while a region of $|\langle c \rangle| \ll 1$ may indicate both stirring and mixing, only $\langle c_r^2 \rangle$ would indicate mixing. Regions of local minimum of $|\langle c \rangle|$ are found to occur at the structure centers. Contours of incoherent passive scalar $\langle c_r^2 \rangle$ (figure 7b) have minimum values in the structure centers, showing that the core fluid is indeed mixed. Minimum incoherent passive scalar occurs at the structure center essentially due to turbulent diffusion.

The above discussion on the passive scalar and turbulence properties suggests a conceptual model for turbulent mixing as shown in figure 8. Roll-induced motion advects the unmixed passive scalar from two sides of the wake into the saddle region as well as toward the rib-roll contact region. In the saddle region, mixing starts due to further stretching of the scalar interface by rib-induced motion; at the rib-roll interface, mixing starts due to interactions between ribs and rolls. The partially mixed scalar is then advected to the structure center, where mixing is completed by the high turbulence there.

5. Conclusions

Random velocity fluctuations at the inflow boundary in the numerical simulation are necessary for the incoherent properties to be consistent with laboratory experimental data. The mechanism of turbulence production and the behavior of incoherent turbulence in a flat-plate wake is essentially the same as in a cylinder wake; *i.e.* turbulence is produced at the saddle and is advected into the structure center where it accumulates. However, the incoherent passive scalar shows a minimum at the structure center. Mixing of the passive scalar starts in the saddle region and the rib-roll contact region and is completed at the structure center by high level turbulence.

Acknowledgements

This research was supported by the Office of Naval Research and the Naval Research Laboratory. The calculations were performed at the computing facilities of NAS at NASA Ames Research Center.

References

[1] Tso, J. & Hussain, F., *J. Fluid Mech.* **203**, 425 (1989).
[2] Hayakawa, M. & Hussain, F., *J. Fluid Mech.*, **180**, 193 (1987).
[3] Grinstein, F. F., Hussain, F. & Boris, J. P., in *Advances in Turbulence 3*, eds. A.V. Johansson & P.H. Alfredsson, pp. 34–41, Springer-Verlag, Berlin (1991).
[4] Hussain, F. *Phys. Fluids*, **26** (10), 2816 (1983).
[5] Metcalfe, R. W., Hussain, F. S., Menon, S. & Hayakawa, M., *Turbulent Shear Flows 5*, ed. F. Durst et al., Springer-Verlag, Berlin, pp. 110–123 (1985).
[6] Hussain, F. & Zaman, K. B. M. Q., *J. Fluid Mech.*, **110**, 39 (1981).
[7] Robinson, S. K., *Ph. D. Dissertation* (1991).
[8] Jeong, J., *Ph. D. Dissertation* (1993).

a) b)

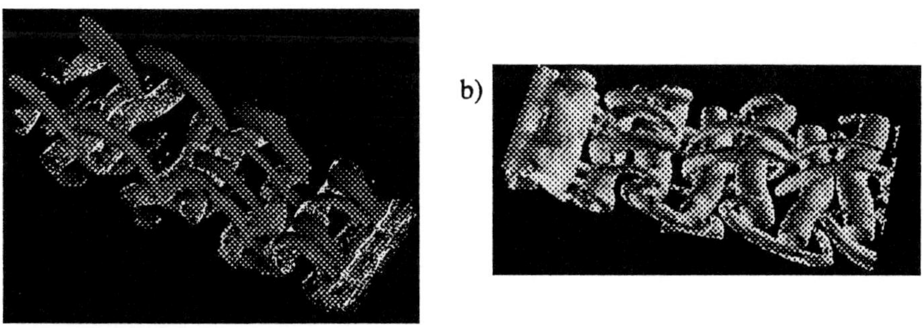

Figure 1. Instantaneous visualization of the plane wake perturbed with coherent (sinusoidal) spanwise-excitation; a) superimposed isosurfaces of ω_z (grid) and ω_x (solid), b) isosurfaces of Ω.

Figure 2. Isosurfaces of the passive scalar c for the structure shown in figure 1.

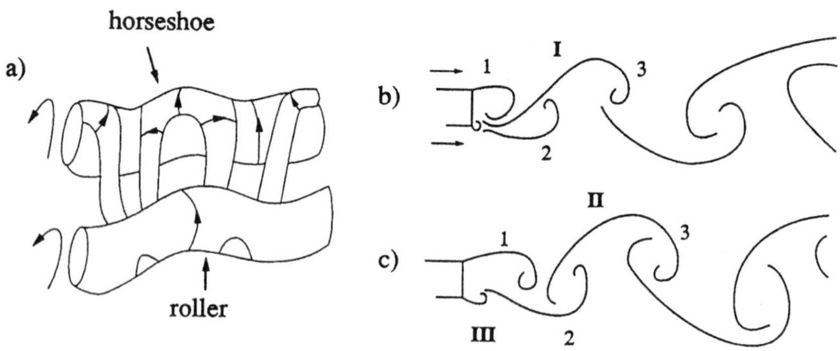

Figure 3. Topological model for the interaction between rolls and ribs in the near wake (from GHB [3]).

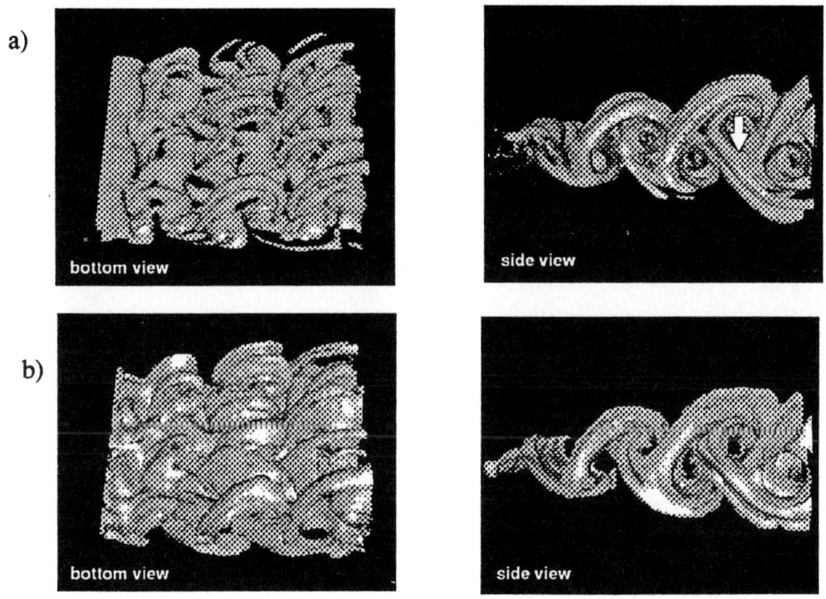

Figure 4. Instantaneous visualization of the plane wake perturbed with (incoherent) spanwise-excitation; a) isosurfaces of Ω, b) isosurfaces of the passive scalar c.

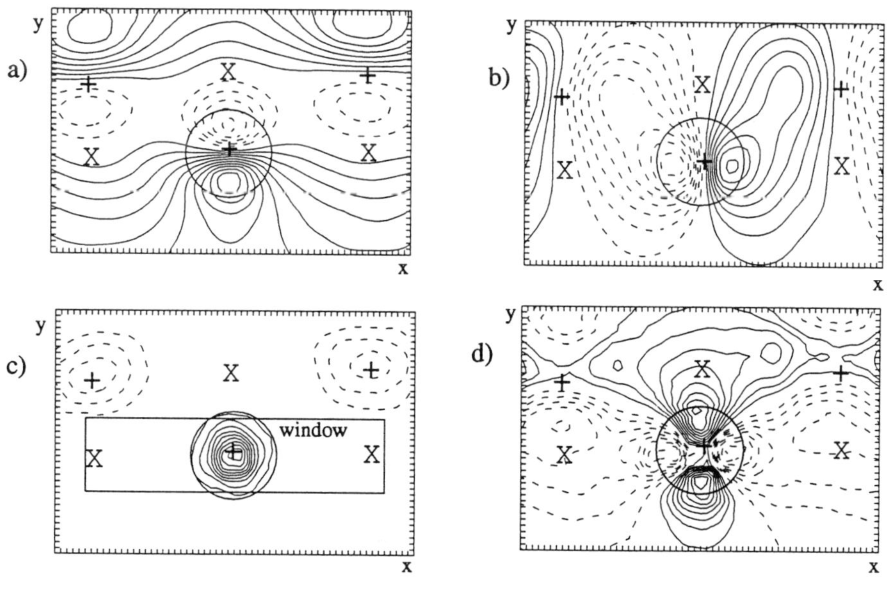

Figure 5. Educed structure properties; the structure boundary is indicated by a circle; a) $\langle u - U_c \rangle$, b) $\langle v \rangle$, c) $\langle \omega_z \rangle$, d) $\langle S_{xy} \rangle$.

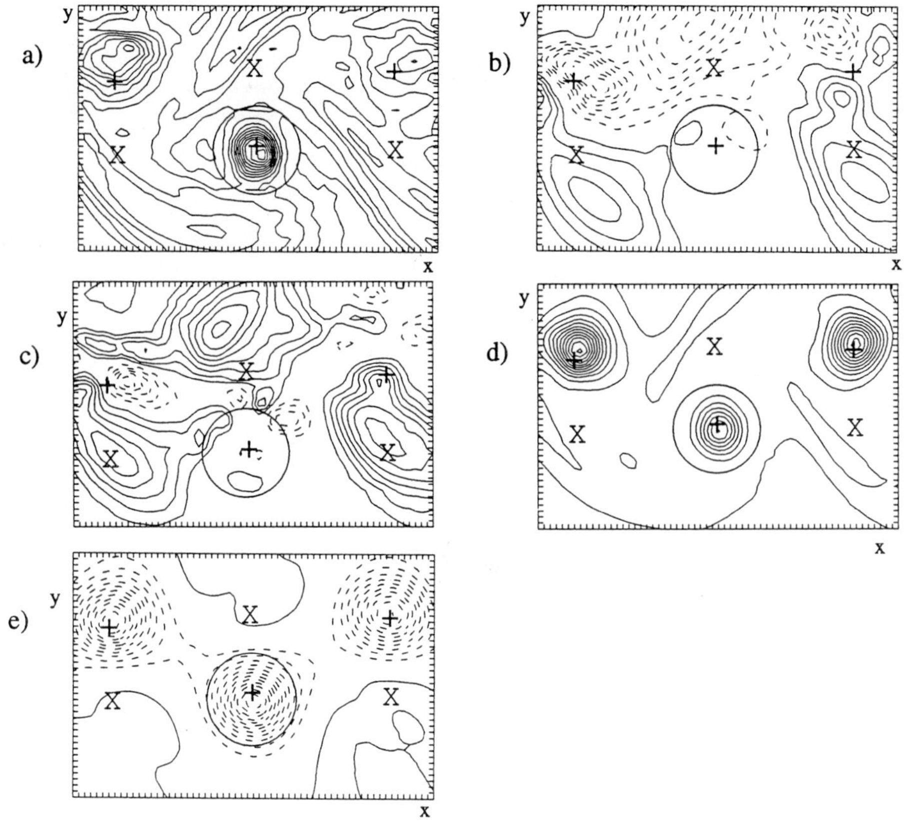

Figure 6. Educed structure properties; the structure boundary is indicated by a circle; a) $\langle \omega_x^2 \rangle$, b) Incoherent Reynolds stress $\langle u_r v_r \rangle$, c) Coherent shear production $P_s = -2\langle u_r v_r \rangle \langle S_{xy} \rangle$, d) Incoherent turbulent kinetic energy $\langle u_r^2 + v_r^2 \rangle$, e) Coherent pressure $\langle p - p_\circ \rangle$.

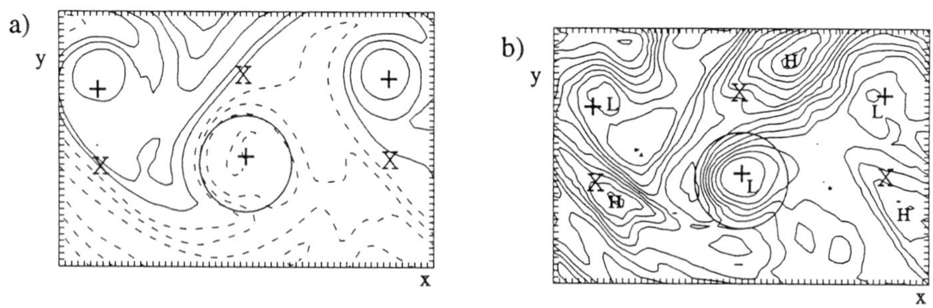

Figure 7. Educed structure properties; the structure boundary is indicated by a circle; a) Coherent passive scalar $\langle c \rangle$, b) Incoherent passive scalar $\langle c_r^2 \rangle$; 'L' and 'H' denote local minima and maxima respectively.

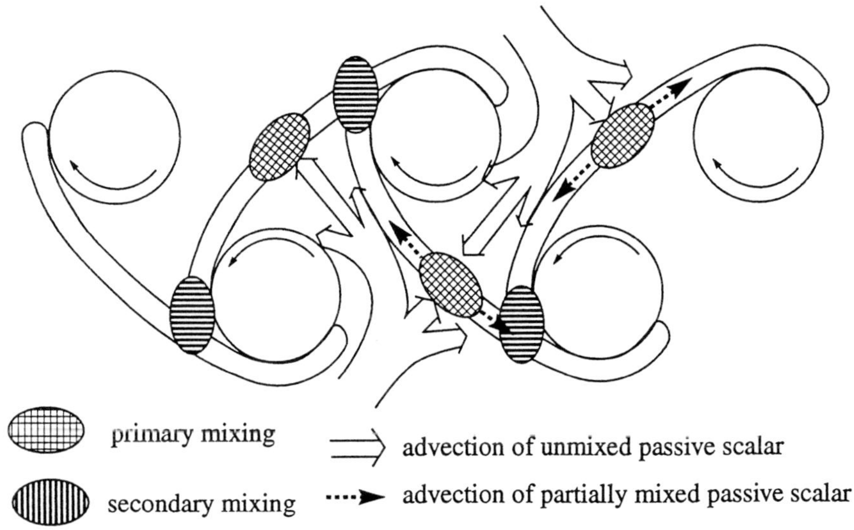

Figure 8. Conceptual model of turbulent mixing.

Acoustic Control of Vortical Structure in a Plane Jet

H. MAKITA and T. HASEGAWA
Toyohashi Univ. of Tech., 1-1 Tenpaku-cho Toyohashi 441, Japan.

Abstract

Characteristic features of coherent motions were experimentally analyzed in a transition region of a two-dimensional jet with parabolic velocity distribution. The symmetrical or the anti-symmetrical vortex arrangement was weakly emphasized in the jet through an acoustic method. Hotwire measurements through a conditional sampling method were conducted point by point throughout the entire transition region. The resultant figures of the flow fields were compared with the pictures given by flow visualization by a conditional multi-smokewire method. The velocity vector maps demonstrated the characteristic features of the streamwise development of the jet's macroscopic structure; the intensity of the vortical fluid motion, the streamwise growth of the jet's width, and the meandering motion of the jet. The vorticity contour maps illustrated the differences in behaviour between the two modes of coherent vortices in the transition region with respect to the streamwise changes in the scale, configuration, peak intensity, and convection velocity of the coherent vortices. The present results suggested good possibility of controlling the jet structure by the acoustic method.

Introduction

Coherent structures are known to exert decisive influences on the transition process of jets and it may be most important to know how these coherent structures behave in the transition region in order to clarify the jets' structure. In a plane jet with parabolic velocity distribution, coherent vortices usually anti-symmetrically arranged with respect to the jet axis are observed to develop downstream in its transition region. But, as Sato (1960) pointed out through hotwire measurements and spectral analyses, the other type of the coherent structure, i.e. vortices arranged symmetrically, was still latent in it. The present experiment aims to elucidate the differences in aspects of the streamwise development of the coherent vortices in the transition region of the parabolic plane jet between these two basic modes of the vortical structure. In a natural (unexcited) jet, it is not so easy to remove mutual contaminations from the other mode and to detect characteristic features inherent in each mode of the vortex arrangement independently. This is because the natural jet is presumed to have structure as a complicated mixture of the two modes.

In the present experiment, each mode of the vortex arrangement in the jet was weakly excited by injecting monotonic sound with characteristic frequency corresponding to each of the modes. The acoustic excitation of coherent motions in jets

J. P. Bonnet and M. N. Glauser (eds.), Eddy Structure Identification in Free Turbulent Shear Flows, 77–88.

has been examined by a number of experimental researchers, e.g., Sato (1960) for a parabolic plane jet, Zaman & Hussain (1980) for an axi-symmeteric jet and Hussain & Thompson (1980) and Thomas & Goldschmidt (1986) for plane jets with top-hat velocity distribution. The results of these experiments have surely given us plenty of important information on the mechanism of the coherent motions in the jets, but most of the information given by these experiments was rather concerned with phenomena in restricted areas of the jets.

The authors thought it nevertheless important to offer a general and detailed image how the coherent vortices vary their structure as they travel downstream in the transition regions. In the present measurement, conditional sampling was made throughout the entire transition region with respect to each two modes for the natural and the excited jets. Resultant ensemble averaged velocity vector maps and vorticity contour maps were compared with instantaneous pictures given by flow visualization through a multi-smokewire method.

Experimental setup and procedure

Figure 1 illustrates a schematic of the present experimental apparatus. A plane jet with parabolic velocity distribution was ejected into a test section from a slit exhaust of 10 mm × 400 mm in cross section at the end of a 1860 mm run-up section. The top and the bottom of the test section are covered by a pair of flat plates of 840 mm × 1165 mm to keep the two-dimensionality of the flow field. The jet had the center velocity of U_0 =6.0 m/s and the Reynolds number based on U_0 and the width of the exhaust, $2h$, was about 3850. Ahead of the measurement, the symmetry and the two-dimensionality of the flow field were checked by measuring the distributions of the mean velocity, U, and the turbulence intensities, u' and v'. Monotonic sounds of f_s =150 Hz for the symmetrical mode excitation and f_a =95 Hz for the anti-symmetrical mode excitation were injected to the jet from loudspeakers (a) and (b), respectively. A curved sound reflector on the ceiling, a carpet on the floor, and thick curtains around the test area effectively avoided direct reflection of the injected sound into the flow field. In both cases of the excitation, the sound intensity was about 60 dB on the jet axis and reference quasi-sinusoidal signals

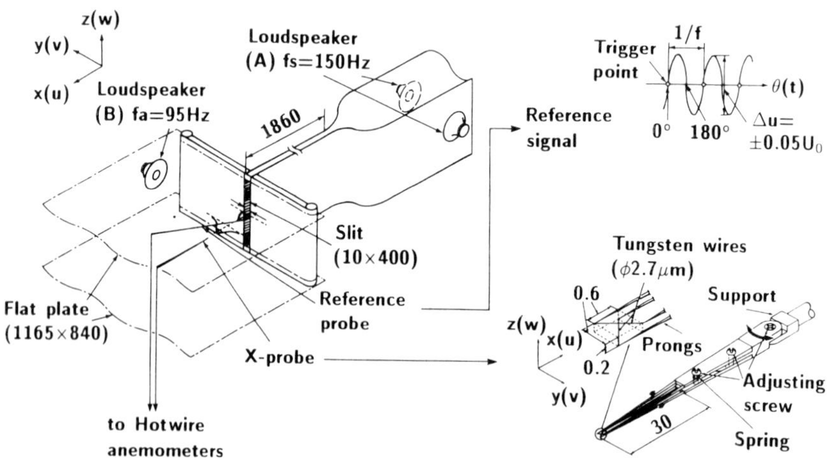

Fig.1 Experimental setup and details of X probe (unit: mm)

having the corresponding frequencies, 150 and 95 Hz, and the amplitude of $\pm 0.05 U_0$ ($= \pm 0.3$ m/s) were induced at the reference probe installed at $x/2h = 3$, $y/2h = 0$ for the symmetrical mode and at $x/2h = 3$, $y/2h = 0.4$ for the anti-symmetrical mode. The Strouhal number was 0.24 and 0.16, respectively.

Measurements were conducted by an adjustable X-wire probe illustrated on the right-hand side of Fig. 1. The probe has a measuring volume of about $0.6 \times 0.6 \times 0.2$ mm^3. The probe configuration, especially with respect to the right angle and the separation between the wires, was precisely adjusted under a microscope before each measurement. The conditional measurement was conducted point by point across the entire transition region of the parabolic plane jet in the four cases of the natural symmetrical mode, the natural anti-symmetrical mode, the symmetrically excited, and the anti-symmetrically excited jets. The conditioning was made based on a triggering electric pulse standing at $\theta = 0°$ as shown in Fig. 1, where θ is the phase angle of the reference signals. Therefore, the images of the flow fields produced as the results of the present measurements are almost free from the distortion caused by the transformation of timewise signals through the Taylor's frozen pattern hypothesis. In each sampling, the data were recorded for $0 \leq \theta \leq 360$, i.e. for one cycle of the vortex motion, to get θ-dependence of the vortical motions.

Flow visualization by a conditional multi-smokewire system was also made to get detailed instantaneous images of the coherent structures in the jet against the ensemble averaged figures obtained by the hotwire measurements. In the test section, a number of thin nichrome wires of 40 μm in diameter were placed parallel at every 15 mm across the jet. Photographs were taken through a glass window on the flat plate by flushing strobe lights from the outside of the test section. The multi-wire system made it possible to remove the distortion of the flow patterns caused by the hysteresis effect of smoke streaks due to their long stay in the flow field, as suggested by Wei & Sato (1984). We also employed diluted oil of paraffin liquid and benzene of 5 : 3 in volume ratio. It could have reduced the electric power required to heat up the wires so greatly that unnecessary fluctuations of the smoke streaks due to the buoyancy effect were scarcely observed even outside of the jet. In the present flow visualization, the timings of heating the wires and flushing the strobe lights were also triggered by reference pulses raised at $\theta = 0°$ with proper delay times.

Results and discussions

Pictures shown in Figs. 2a-d illustrate instantaneous flow patterns of the jet for (**a**) the natural symmmetrical mode, (**b**) the natural anti-symmetrical mode, (**c**) the symmetrically excited mode and (**d**) the anti-symmetrically excited mode. Owing to the reduction in the wire's heating power, buoyancy-caused fluctuations of the smoke streaks are scarcely observed along the wires in the fluid almost at rest outside of the jets. A staggered vortex street is mostly observed for the natural jet, sometimes even for the natural jet referred to the symmetrical mode shown in Fig. 2a, which suggests that the anti-symmetrical mode is more stable in the parabolic plane jet without the excitation. When the jet is acoustically excited, the coherent vortices are observed to appear from the more upstream location than the case of the natural jet. For the symmetrical mode excitation, small coherent vortices begin to develop from about $x/2h = 2.5$, symmetrically drawn up in two lines on both sides of the jet axis, see Fig. 2b. But, their vortex arrangement seems to change into somehow anti-symmetrical one around $x/2h = 7$ and then the coherent vortices with rigid structure begin to decay into many small random eddies. The scale of the vortex does not increase downstream so rapidly and the jet width is reduced from that of the natural jet. In the case of the anti-symmetrical mode excitation shown in Fig. 2c, the vortex

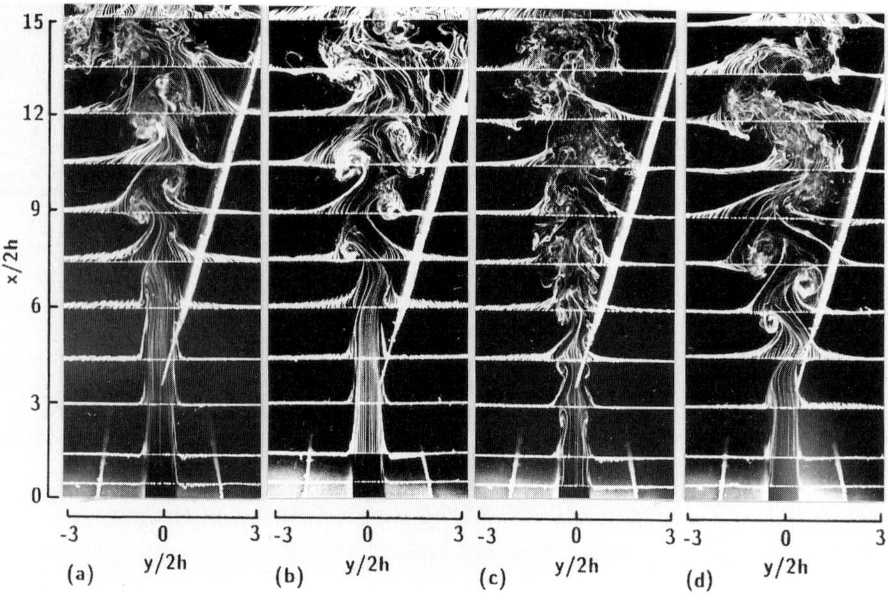

Fig.2 Flow patterns of the natural and excited jets at θ =0°, (a) natural (symmetrical mode), (b) natural (anti-symmetrical mode), (c) symmetrically excited, (d) anti-symmetrically excited.

Tab.1 Subregions in transition region of the natural and the excited jets.

	Liniear interaction region	Nonlinear interaction region	Randomization region
Natural	$x/2h \leq 5$	$5 < x/2h \leq 9$	$x/2h > 9$
Symmetrically excited	$x/2h \leq 3$	$3 < x/2h \leq 11$	$x/2h > 11$
Anti-symmetrically excited	$x/2h \leq 3$	$3 < x/2h \leq 9$	$x/2h > 9$

begins to develop at about $x/2h$ ~4. Its scale grows downstream most rapidly and the jet's width is apparently increased from that of the natural jet. As the results of spectrum analyses in addition to those of the flow visualization, the transition region of the plane jet is divided into three regions of a linear interaction region, a nonlinear interaction region, and a randomization region as tabulated in Tab. 1; see Makita, et. al., (1985).

Velocity vector maps of the natural and the excited jets at θ =0° are shown in Figs. 3a-c. As supposed by the results of the flow visualization in which the flow patterns of the the natural jets resembled closely to each other, the difference in the features of the flow fields was not apparent between the velocity vector maps in the two cases of the natural jets. Therefore, only the results for the symmetrical mode

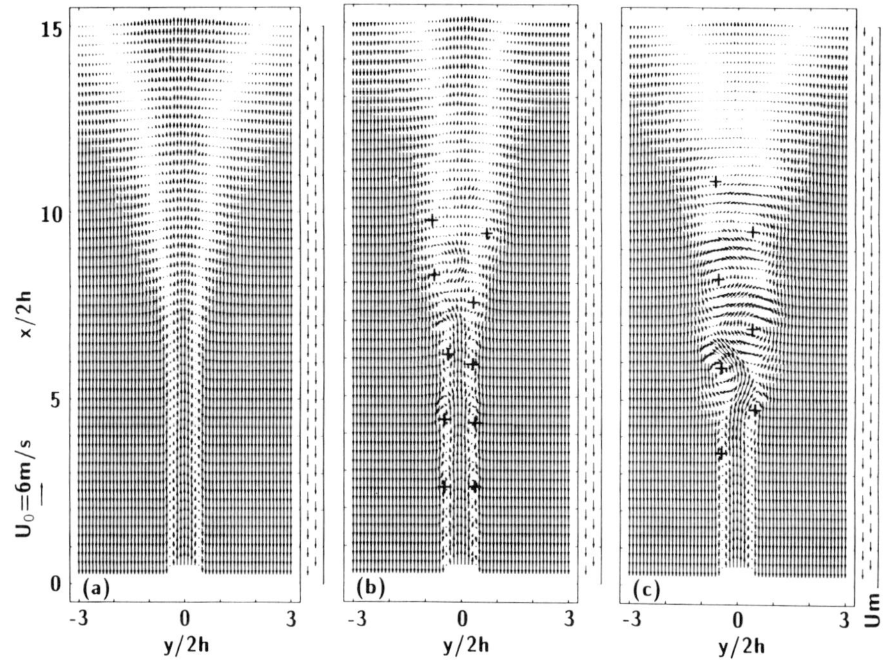

Fig.3 Velocity vector maps ($\langle u\rangle - U_m$ vs. $\langle v\rangle$) at θ =0°, (a) natural (symmetrical mode), (b) symmetrically excited, (c) anti-symmetrically excited, arrows at the right-hand side of each figure denote the local mean velocity U_m at the vortex center, where the vortex center is shown by the symbol, +.

are described with respect to the natural jet hereafter in the present paper. In these maps, the velocity vectors are illustrated in the form of $(\langle u\rangle - U_m, \langle v\rangle)$, where $\langle u\rangle$ and $\langle v\rangle$ are the phase-averaged velocity components and U_m is the mean velocity at each vortex center. The magnitude of U_m is illustrated on the right-hand side of each figure. The authors have examined the other ways of the expression such as $(\langle u\rangle, \langle v\rangle)$ and $(\langle u\rangle - U_c, \langle v\rangle)$, where U_c is the convection velocity of the coherent vortex. The former was not appropriate to express the vortical motions in the jets. The velocity vector maps in the form of $(\langle u\rangle - U_c, \langle v\rangle)$ raised large difference in impression from the visualized pictures with respect to the jet's macroscopic structure, such as the width of the jet, the mean velocity profile and the position of the vortex center. In this way of the expression, the vortical motion seemed apparently distorted, because U_c was almost constant throughout the transition region as stated later and the magnitude of the velocity vector does not become nearly zero at the center of each vortex. Images of the flow fields given by the maps in Fig. 3 agree quite well with those expressed by the flow visualization. Fluid dynamical motions in the jet's flow field, e.g., vortical motions of the fluid flow and the meandering motion of the jet, are expressed only ambiguously for the natural jet. This is because the vortices in the natural jets have weak coherency and the velocity fields of the natural jets expressed by the present method of the conditional sampling could not have eliminated the effects of mutual contamination between the two basic modes even if the samplings

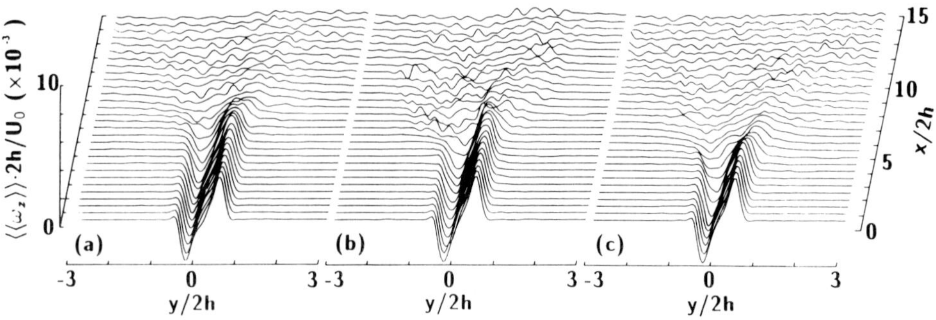

Fig.4 Bird's eye views of the distribution of $\langle\langle\omega_z\rangle\rangle$ in the transition region of the natural and the excited jets, (a) natural (symmetrical mode), (b) symmetrically excited, (c) anti-symmetrically excited, where $\langle\langle\omega_z\rangle\rangle$ is integrated phase averaged vorticity.

were strictly made based on the reference triggers. When the jet is excited, these motions became apparently observed and the characteristic differences in aspects of the streamwise evolution of the jet structure by the excitation mode are clearly expressed with respect to the intensity and the scale of the vortical motions, the magnitude of the meandering motion and the streamwise growth of the jet width.

It may be most appropriate to draw contour maps of vorticity in order to observe how the coherent vortices behave, or deform themselves through the mutual interactions with the adjacent vortices and/or affected by the mean velocity gradient, in the transition region. The z-component of the phase-averaged vorticity, $\langle\omega_z\rangle$, its coherent component, $\langle\omega_{zc}\rangle$, and total vorticity integrated for a period of the reference signal, $\langle\langle\omega_z\rangle\rangle$, were calculated from the velocity vectors as follows,

$$\langle\omega_z\rangle = \frac{\partial\langle v\rangle}{\partial x} - \frac{\partial\langle u\rangle}{\partial y}, \tag{1}$$

$$\langle\langle\omega_z\rangle\rangle = \frac{\partial\langle\langle v\rangle\rangle}{\partial x} - \frac{\partial\langle\langle u\rangle\rangle}{\partial y}, \tag{2}$$

and

$$\langle\omega_{zc}\rangle = \frac{\partial v_c}{\partial x} - \frac{\partial u_c}{\partial y} = \langle\omega_z\rangle - \langle\langle\omega_z\rangle\rangle \tag{3}$$

where u_c and v_c are the coherent velocity components in the triple decomposition analysis (Hussain, 1983) and $\langle\langle u\rangle\rangle$, $\langle\langle v\rangle\rangle$ are given by integrating $\langle u\rangle$, $\langle v\rangle$ for $0° \leq \theta \leq 360°$, being almost equal to the time-averaged velocity components, $\bar{u}$ and $\bar{v}$.

Figures 4a-c show the streamwise change of the spanwise distribution of $\langle\langle\omega_z\rangle\rangle$ for the natural (symmetrical mode) and the two modes of the excited jets. The dependence on the mode of the streamwise growth rate of the jet's width in the randomization region is clearly shown the figures. In the case of the symmetrical mode excitation shown in Fig. 4b, vorticity peaks begin to be rapidly collapsed into many small random peaks at the end of the nonlinear interaction region of $x/2h$ ~6-8, where the change in the vortex arrangement occurs. In the case of the anti-symmetrical mode excitation shown in Fig. 4c, the peaks decay a little more gradually from the more upstream region of about $x/2h$ ~5-6, than the other two

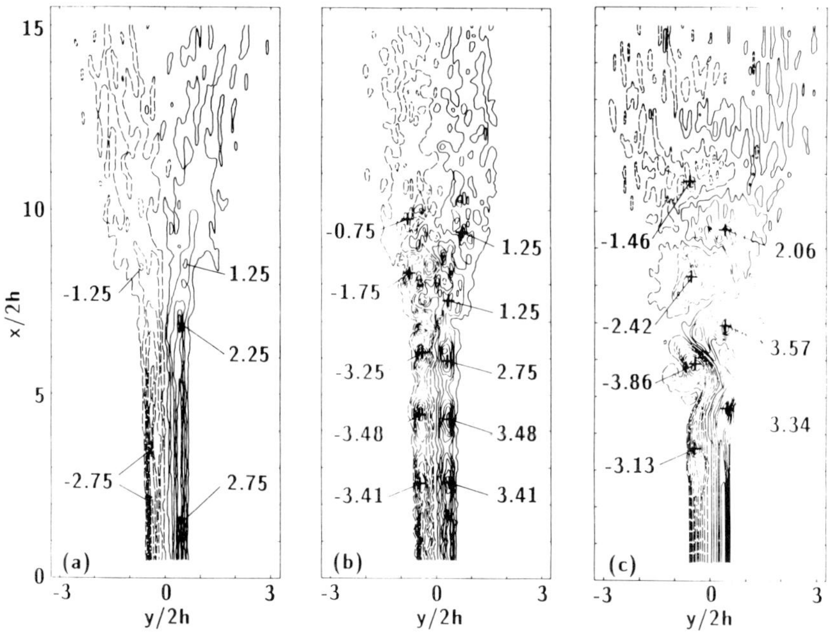

Fig.5 Contour maps of $\langle\omega_z\rangle \cdot 2h/U_0$ at θ =0° of the transition region for the natural and excited jets, (a) natural (symmetrical mode), (b) symmetrically excited, (c) anti-symmetrically excited, the lowest level: $\pm 0.25\times10^{-3}$, the other levels: 0.5×10^{-3} steps, —: positive value, $\cdots$: negative value, +: vortex center.

cases. Good axi-symmetry of the distribution at each streamwise location confirms the reliability of the present measurement.

The contour maps of $\langle\omega_z\rangle$ at θ =0° in Figs. 5a-c clearly demonstrate how differently the coherent vortices behave by the excitation modes. In the case of the symmetrical mode excitation, a pair of the coherent vortices with vorticity of opposite sign start to develop from $x/2h$ ~2.5, i.e., near the end of the linear interaction region. Their scale and peak vorticity grow larger slowly as it flows downstream, keeping their symmetrical arrangement until $x/2h$ ~5 in the nonlinear interaction region. At $x/2h$ ~6-7, the vortex arrangement brings about slipping off and becomes to be a little anti-symmetrical, though the symmetry of the arrangement of the large scale structure seems recovered in the randomization region. More carefully observed, the coherent vortex begins to be broken up into two or more pieces around here. For the anti-symmetrical mode excitation, the coherent vortex is observed to start developing on one side of the jet at $x/2h$ ~3.5 and the next vortex appears on the opposite side of the jet at $x/2h$ ~4.7. Such staggered arrangement is kept throughout the transition region. The vortex scale increases quite rapidly downstream in the nonlinear interaction region. The peak vorticity increases through the nonlinear interaction region until $x/2h$ ~6, but decreases rapidly in spite of the continuous increase in the vortex scale in the more downstream region. The grown-up large-scale coherent vortices

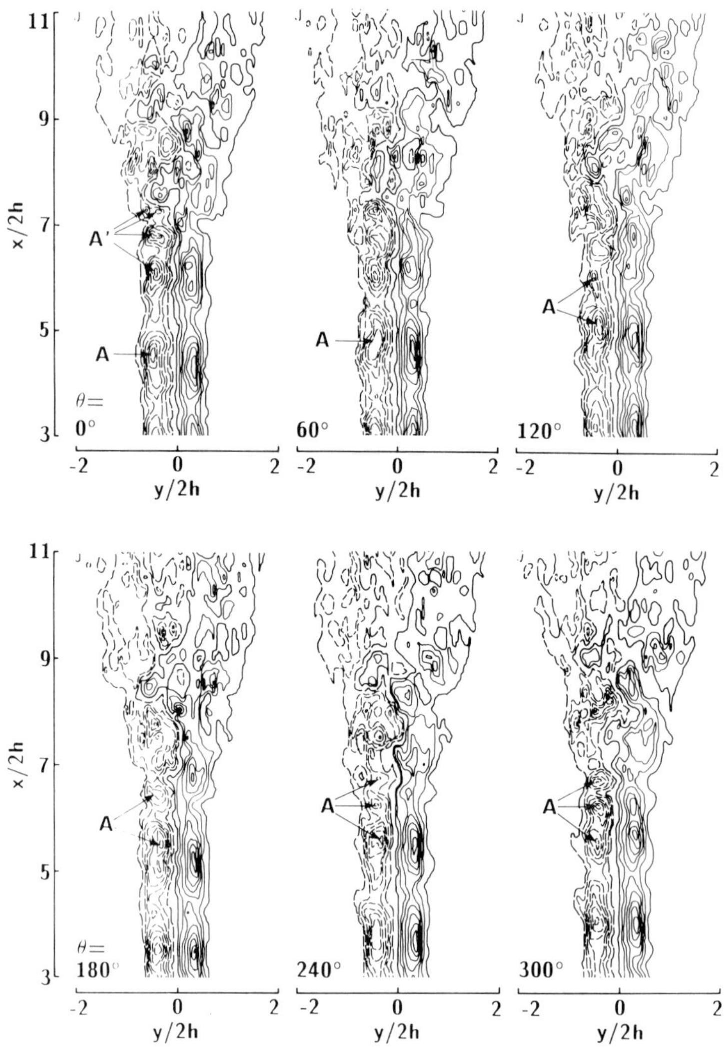

Fig.6 Contour maps of $\langle\omega_z\rangle \cdot 2h/U_0$ at various θ for the nonlinear region of the symmetrically excited jet, the lowest level: $\pm 0.25\times10^{-3}$, the other levels: 0.5×10^{-3} steps, —: positive value, $\cdots$: negative value.

with rigid structure are observed to decay into many small random eddies for about $x/2h \geq 11$ in the randomization region.

Figures 6 and 7 show the contour maps of $\langle\omega_z\rangle$ at various angles of θ between 0° and 360° around the nonlinear interaction regions for the symmetrically and the antisymmetrically excited jet, respectively. Complicated situations of the deformation of the coherent vortices are illustrated in detail in these figures. Now, let's start by

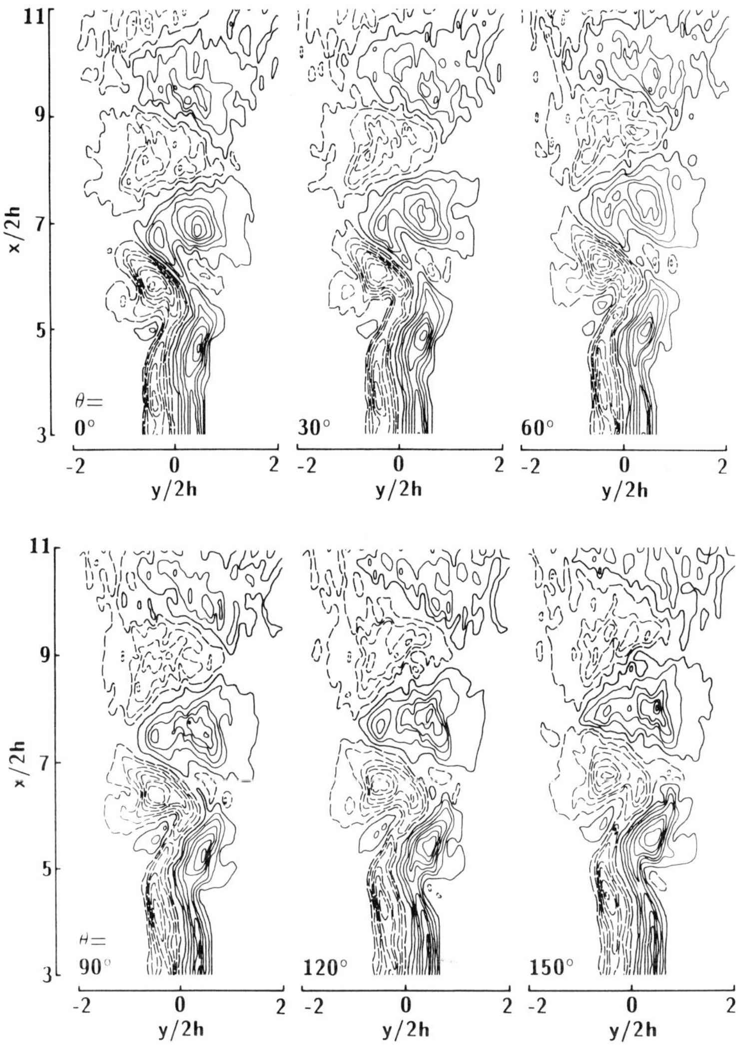

Fig.7 Contour maps of $\langle\omega_z\rangle \cdot 2h/U_0$ at various θ for the nonlinear region of the anti-symmetrically excited jet, the lowest level: $\pm 0.25\times10^{-3}$, the other levels: 0.5×10^{-3} steps, —: positive value, $\cdots$: negative value.

paying attention to a vortex pointed by an arrow (**A**) for the symmetrically excited jet at θ =0° shown in Fig. 6. The vortex is oval in shape keeping its apse line almost parallel to the jet axis and has about the same scale and peak intensity as the one on the opposite side. The vortex seems slightly elongated in the streamwise direction at θ =60° and becomes to have two-peak structure at θ =120°. At θ =180°, the vortex is broken into two pieces, and it forms three islands at θ =240°.

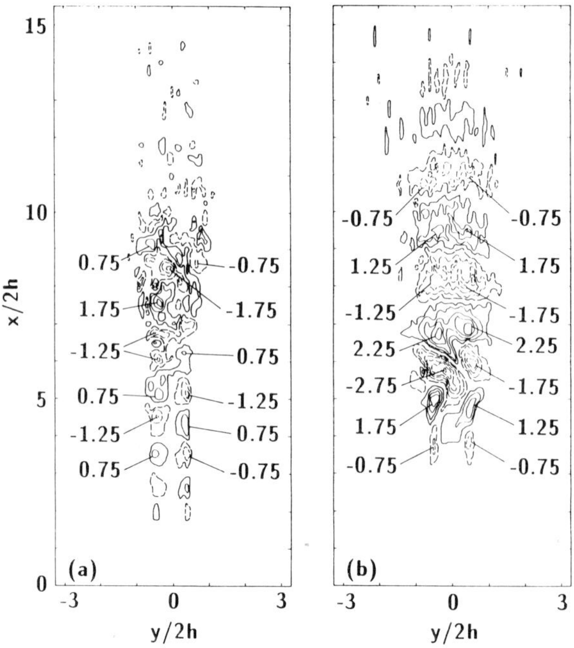

Fig.8 Contour maps of $\langle \omega_{zc} \rangle \cdot 2h/U_0$ at $\theta = 0°$ of the transition region for the natural and excited jets, (a) symmetrically excited, (c) anti-symmetrically excited, the lowest level: $\pm 0.25 \times 10^{-3}$, the other levels: 0.5×10^{-3} steps, —: positive value, $\cdots$: negative value.

Then, let's observe a vortex on the most downsream side among the islands. At $\theta = 300°$, it grows larger rapidly as it is pushed inward through the interactions with the vortices nearby. Finally, the vortex (**A**) is broken into several small vortices randomly scattered in the jet, as pointed by the arrows (**A'**) in the figure of $\theta = 360°$.

Figure 7 demonstrates the rapid streamwise growth of the coherent vortex in the nonlinear interaction region for the anti-symmetrical mode excitation. Different from the case of the symmetrical mode excitation, we could have obtained almost complete repeatability of the flow patterns in this case as observed in these contour maps. The vortex enlarges its scale rapidly in the streamwise direction to have structure apparently far more rigid than that of the symmetrically excited jet, though the difference in the maximum peak intensity is not so large between the two cases. These facts show that the anti-symmetrical arrangement is far more stable than the symmetrical one. It must be noted that the vortex pairing, observed by Hussain & Zaman (1980) and Thomas & Chu (1989) for the jets with top-hat velocity distribution, did not occur in the present experiment, as shown in Figs. 6 and 7. This may be because $\partial U/\partial y$ is gentler in the parabolic jet than in the top-hat jet and/or the present levels of the sound excitations could not have induced the coherent vortices strong enough to raise the pairing through their mutual interactions. The cases with more strong excitation are now under the examination in our laboratory.

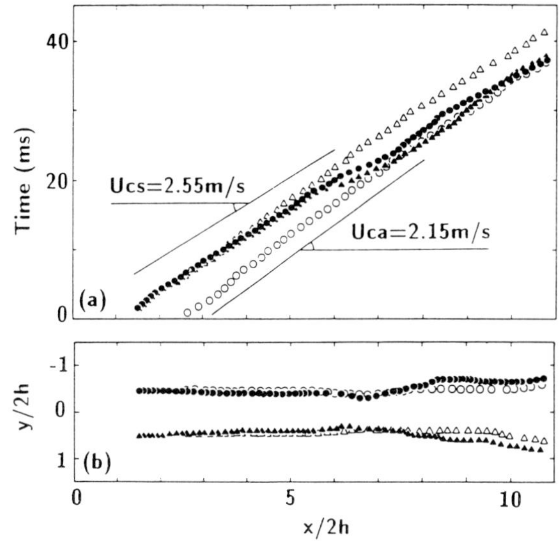

Fig.9 Trajectories of the vorticity peaks, (a) symbols denote locations of the vortices in the x–t plane, (b) the x–Y plane, the slopes in Fig.(a) give the convection velocities of the vortex, (U_{cs}; the symmetrically excited, U_{ca}; anti-symmetrically excited), ●: $+\langle\omega_z\rangle$, ▲: $\langle\omega_z\rangle$; symmetrically excited, ○: $+\langle\omega_z\rangle$, △: $\langle\omega_z\rangle$; anti-symmetrically excited.

Figures 8a-b illustrate the contour maps for the coherent component of the vorticity, $\langle\omega_{zc}\rangle$. As described by Eq. (2), the vorticity induced by the mean velocity gradient, $\langle\langle\omega_z\rangle\rangle$, is subtracted from $\langle\omega_z\rangle$ in calculating $\langle\omega_{zc}\rangle$. It must be noted that the coherent vortex starts to develop from the more upstream region in the case of the symmetrical vortex arrangement. In both of the cases, the center of the coherent vortex, or the vorticity peak, seems to be convected almost parallel to the jet's axis and slightly shift outwards as it flows downstream. The trajectories of the vortex center are estimated by drawing the contour maps of $\langle\omega_{zc}\rangle$ at every $\Delta\theta = 15°$ as plotted in Figs. 9a-b. As shown in Fig. 9b, the spanwise separation between the vortices on opposite sides of the jet seems roughly constant throughout the transition region. The convection velocity U_c of the vortex was determined by the timewise trajectories of the vortex center as shown in Fig. 9a. Different from the case of Goldschmidt et al.(1981), U_c was independent of U_m as stated in the preceding section and has almost the constant values of $U_{cs} = 2.55$ m/s for the symmetrical mode and $U_{ca} = 2.15$ m/s for the anti-symmetrical mode throughout the transition region. The values of U_c, however, could not be determined so correctly in the randomization region.

Conclusion

Two basic modes of the vortex arrangements of a plane jet with a parabolic velocity distribution were experimentally clarified through the acoustic excitation. The unexcited natural jet has the structure far more complicated than ever expected, because its flow field is a mixture of these two modes having quite different features. The present experiment also confirmed the possibility of the acoustic control of the jet structure.

References

1. Goldschmidt,V.W., Young,M.F. & Ott,E.S. 1981 *J. Fluid Mech.* **105** 327–345.
2. Hussain,A.K.M.F. 1983 *Phys. Fluids* **26** 2816–2850.
3. Hussain,A.K.M.F. & Thompson,C.A. 1980 *J. Fluid Mech.* **100** 397–431.
4. Makita,H., Ohtani,H. & Ishizumi,K. 1988 *Trans. Jpn. Soc. Mech. Eng.* **Ser.B 54-504** 1946–1952 (in Japanese).
5. Sato,H. 1960 *J. Fluid Mech.* **7** 53–80.
6. Thomas,F.O. & Chu,H.C. 1989 *Phys. Fluids* **A 1** 1566–1687.
7. Thomas,F.O. & Goldschmidt,V.W. 1986 *Trans. ASME J. Fluid Eng.* **108** 331-337.
8. Wei,Q.D., & Sato,H. 1980 *J. Fluid Mech.* **143** 153-172.
9. Zaman,K.B.M.Q. & Hussain,A.K.M.F. 1980 *J. Fluid Mech.* **101** 449-491.

III. Conditional Sampling II: Other Methods

LARGE SCALE STRUCTURE CHARACTERIZATION IN A 2D MIXING LAYER BY PSEUDO-FLOW-VISUALIZATION AND DELOCALIZED CONDITIONAL SAMPLING.

Bellin S.[†], Delville J., Vincendeau E., Garem J.H. & Bonnet J.P.
C.E.A.T Laboratoire d'Etudes Aérodynamiques 43 route de l'Aérodrome
F-86036 Poitiers cedex
[†]HORWELL, 53 av. Gabriel Péri F-92500 Rueil Malmaison.

Introduction.

The main purpose of this paper is to illustrate two particular techniques for Structure Eduction that can be used when the information on flows (velocity, pressure, passive contaminants, etc...) are simultaneously available in several locations. This situation particularly occurs when rakes of sensors are used, or for DNS or LES simulations.

• The first approach, the Pseudo-Flow-Visualization, allows one to simply handle tremendous amounts of data. This approach simulates conventional experimental visualizations and avoids problems linked to seeding effects. Qualitative and raw quantitative information concerning the organization of the flow can be obtained by this way; however, a great care has to be taken with this approach: a wrong use can lead to misinterpretation when one quantity is not properly considered.

• The second approach is an extension of the conditional average technique where the detector(s) locations are self adjustable. Due to the small level of rejected events, the Delocalized-Conditional-Sampling has a better efficiency than the conventional one. Moreover, it leads to new information on the structures in the flows: not only statistics on the size and position of the structures but also phase-shift within the structures may be retrieved.

Experimental apparatus and Flow characteristics.

This experimental investigation of the plane mixing layer is based on a 12 X hot-wires rake, for the present study, in the mean-gradient direction (Y). The sub-miniature probes are home made, with a square cross section of 1mm×1mm and W-Pt wires (dia. 2.5μm, length 0.5 mm). The probe holder (1.6 mm thick) is also developed in our Lab from a printed circuit board. In this experiment, the separation between probes is 6mm and the rake extent is about twice the local vorticity thickness δ_ω. By comparing spectra, space time correlations and higher order moments with the cor-

J. P. Bonnet and M. N. Glauser (eds.), Eddy Structure Identification in Free Turbulent Shear Flows, 91–101.

responding results obtained with conventional one or two single probes, it has been shown that the Hot-Wire-Rake (HWR) does not significantly perturb the flow, at least for the abovementioned characteristics [1]. Constant Temperature Anemometers built from TSI 1750 are used; the entire HWR is automatically calibrated in velocity, yaw and temperature. The experimental procedure is fully automated and the signals are simultaneously sampled at rates up to 100 kHz per channel.
The subsonic turbulent plane mixing layer under investigation has the following characteristics: external velocities of U_a=42.8 and U_b=25.2 m.s^{-1}, fully turbulent co-flowing boundary layers; the test section is 30x30 cm^2. All the experiments are performed 600mm downstream of the trailing edge of the splitting plate, in the asymptotic part of the mixing layer. The vorticity thickness is there 27.6 mm. More details on the flow characteristics and on the experimental apparatus can be found in [5, 1].

Pseudo-Flow Visualizations (PFV).

The use of an "instantaneous reconstruction" to analyze the behavior of the flow is not a new idea: Townsend [12], Browand & Troutt [3] or Keffer [9] used the instantaneous velocity profiles or time histories as an aid in the analysis of the behavior of turbulent shear flows. Delville et al [5] used HWR sampled at high frequency together with specific coloration rules, in order to precise the morphology of the subsonic plane mixing layer. The general purpose is then to create "pictures" which can lead to simple and direct analysis of the flow. This technique has been more recently used by Glauser and George [6]. Different characteristics of the organization of the flow can be evidenced according the selected coloration rule and the selected quantity.

PFV allows multiple sensor results to be analyzed from an Eulerian point of view. Then a very large amount of information can be easily handled with relatively small computers: the analysis being based mainly on manipulations of colors through the use of Look Up Tables (LUT).
In the present study, the PFV is used to visualize the longitudinal organization when the rake is perpendicular to the mixing layer axis (see figure 1)[1].
In the cases presented here, the coloration rule is based on the HLS (hue-luminance-saturation) color scheme. The saturation will be kept constant (its maximum value), only hue and luminance will then be used as parameters for the definition of colors. These parameters are directly related to the Probability Density Function (PDF) of the entire image minimizing the operator's influence.
The figure 1-a presents the PFV based on the instantaneous local velocities $u(y,t)$ (the corresponding coloration rule is given on fig. 1-a′)). In this case, the PDF exhibits

[1]The rake can be oriented parallel to the splitting plate; then the evolution of the three dimensional spanwise organization can be observed. In this last case, 2D Fourier transforms (in time and span) allow the transverse wavelengthes to be determined unambiguously. However, this particular application will not be developed here. Detailed results of this spanwise approach can be found in [1].

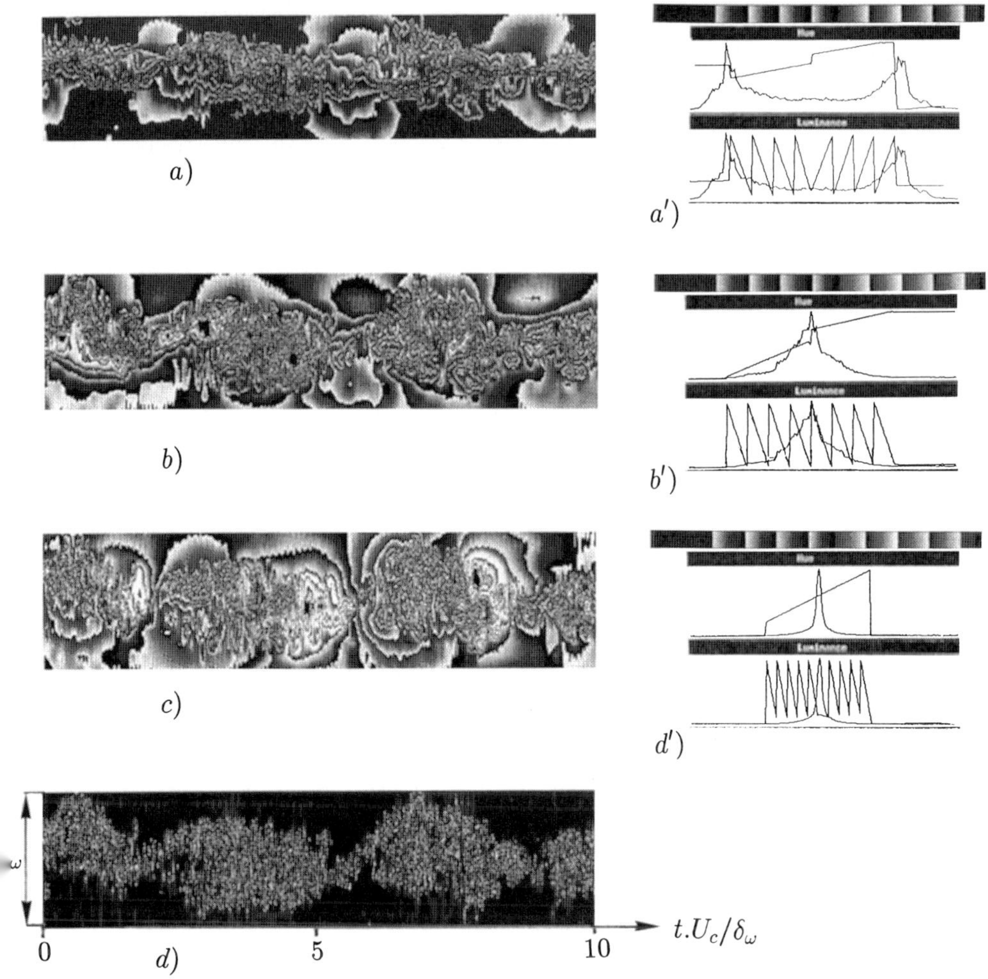

Figure 1: PFV of velocity rake being aligned across the mixing layer.

- *a*) streamwise velocity: $U(y,t)$
- *b*) fluctuating streamwise velocity: $u'(y,t)$
- *c*) fluctuating lateral velocity: $v'(y,t)$
- *d*) 2D vorticity : $\Omega_z(y,t)$
- *a'*) Color rule for Fig. 1a)
- *b'*) Color rule for Fig. 1b) and 1c)
- *d'*) Color rule for Fig. 1d)

Colour version in Colour Plates section, p. xvii.

two peaks corresponding to the external mean velocities. The assignment of colors corresponds to domains based on these 2 modes: Velocities less than the mode of the low velocity side are kept dark blue, while velocities greater than the mode of the high velocity side are kept dark red. Between these two modes the color bands corresponds to the difference between the actual velocity and the convection velocity $U_c = 0.5(U_a + U_b)$. A typical quasi-periodic organization can be observed. However, the puffs appearing here *do not correspond* to physical structures in the conventional sense. This can be proved by looking at the *fluctuations*, as plotted on fig. 1-b for the longitudinal component and fig. 1-c for the transversal one. Here, the same coloration rule is applied to PDF with single maximum (fig. 1-b′). The usual definition of Coherent Structures for mixing layers corresponds to positive fluctuations in the rapid side (velocity excess) and to negative values for the low speed side (velocity defect) [4]. This corresponds to the association of red parts in the upper side of the figure and blue (or green) parts in the lower part. It is then clear when comparing figures 1-a and 1-b that the CS are present *between* the puffs apparent on the PFV based on the local velocity. This feature is definitively confirmed by the PFV of the transversal component of the *vorticity*, Ω_z, fig. 1-d (the Taylor hypothesis with constant convection velocity U_c is used). The corresponding HLS rule is given on fig. 1-d′. In this case, the most probable range of vorticity is darkened in order to suppress the average vorticity linked to the mean gradient of velocity. Only the values with marginal probability are shown. With this coloration rule, the structure of the flow is quite evident and closely resembles to physical visualizations (for example [11]).
The visual detection of CS based on the PVF of the vorticity is in very good agreement with the application of the concept of phase correspondence of the fluctuations in the two sides of the layer.

It then has been evidenced that the observation of PFV based on instantaneous local velocity *for the direct conventional detection of CS*. However, this kind of PFV allows to easily point out the events occurring *between CS*. These events can be associated to the large spatial extend of the trace of the braids, particularly within the external region. This particular characteristic will then be used for improving the conditional process described in the next section.

Delocalized Conditional Sampling (DCS).

The approach of conditional sampling based on one or two points as detectors was developed by Zaman and Hussain [14], Browand and Weidman [4] and is extended in this study to the case of a rake of sensors. A special algorithm has been built [1] in order to automatically select the two optimal probes for the detection of the structure. This algorithm does not involve any relevant threshold conditions. It is then possible to detect structures of any size occurring at any transverse position. With this technique each probe plays a role in the detection procedure and is used

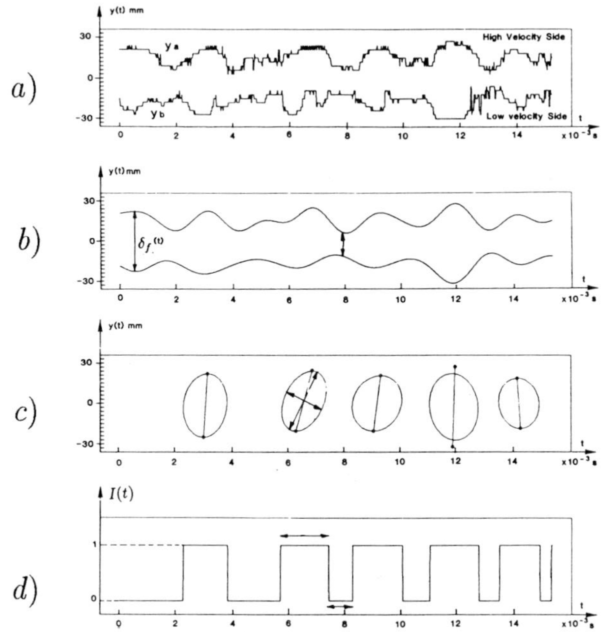

Figure 2: Events partition.

a) Edges extraction *c*) Extrema linking and quadratic fitting of the edges
b) Low pass filtering *d*) Introduction of a function of detection $I(t)$

to calculate ensemble averages. The DCS technique follows a two-step procedure: In the first step we determine a "pre-partition" of the flow allowing to split the time of observation into two parts, respectively corresponding to the times where CS are present or not. The second step will serve to precise the location and the characteristics of the CS within the pre-determined time periods.

First step: Pre-partition of the flow.

The simplest and, in some sense less ambiguous approach to select the different phases of the flow is to pay attention to the instantaneous velocity field. This is performed with the procedure described on fig. 2. By adopting a threshold value on the velocity, it is possible to extract the instantaneous edges corresponding to the PFV given fig. 1-a. The edges are then smoothed (fig. 2-b). A quadratic fit (with ellipses, fig. 2-c) which links the transversal extrema, allows to built a detector function, $I(t)$ (fig. 2-d). This function is 0 when CS can be present. During these periods, the Delocalized Sampling itself can be applied. This procedure is rather insensitive to the threshold

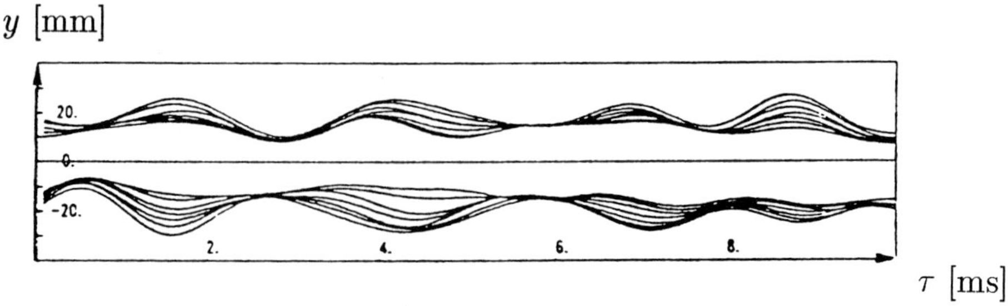

Figure 3: Influence of threshold level on the function of detection $I(t)$.
The threshold varies from 4 % (inner trace) to 10 % (external trace) by 1 % steps.

selected for the partition. Fig. 3 shows that a large variation of this threshold (from 4 to 10 % of the external velocities) does not strongly modify the location of the edges of the puffs and therefore the detector function $I(t)$. The main consequence of a threshold too small is essentially to bypass some $I = 1$ events, then to increase the time period of the application of the DCS.

Conventional ensemble averages can be directly obtained from this detector function. Figure 4 shows the iso-contours of the vorticity obtained when the reference time is either centered on $I = 0$ (figure 4-a) or on $I = 1$ (figure 4-b). It is obvious that the periods when $I = 0$ corresponds to the presence of CS , in agreement with the previous discussion.

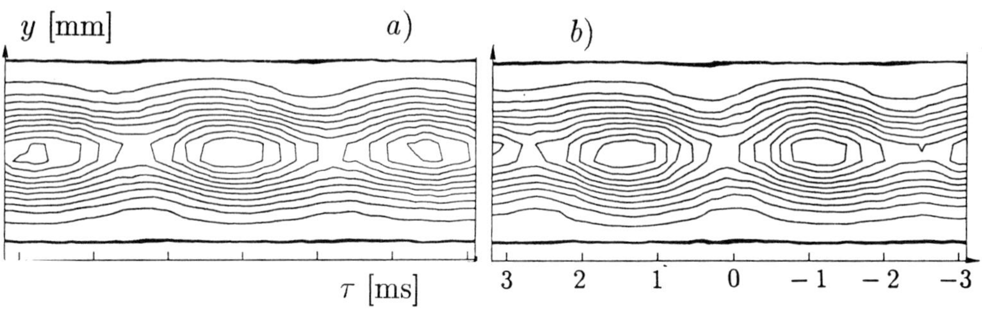

Figure 4: Signature of vorticity Ω_z based on the predetection function.
$a)$ centered on $I(t) = 0$, $b)$ centered on $I(t) = 1$.

Second step: DCS.

The pre-determined periods of time that correspond to high probable occurrence of CS having been selected by the first step, the specific DCS can now be applied. The principle is schematically described on fig. 5. For each period concerned, the algorithm can detect in space and time the two locations where the fluctuating streamwise component of velocity reaches its maximum value in the rapid side (D_1 in figure 5) and where it reaches its minimum in the low velocity side (D_2). Then, this technique allows to detect different scales of CS, that can be centered anywhere relative to the mixing layer axis. In addition, the time shift between the two detections gives an approximation of the stage of the evolution of the CS. This can be related to the different phases of the life of the structures.

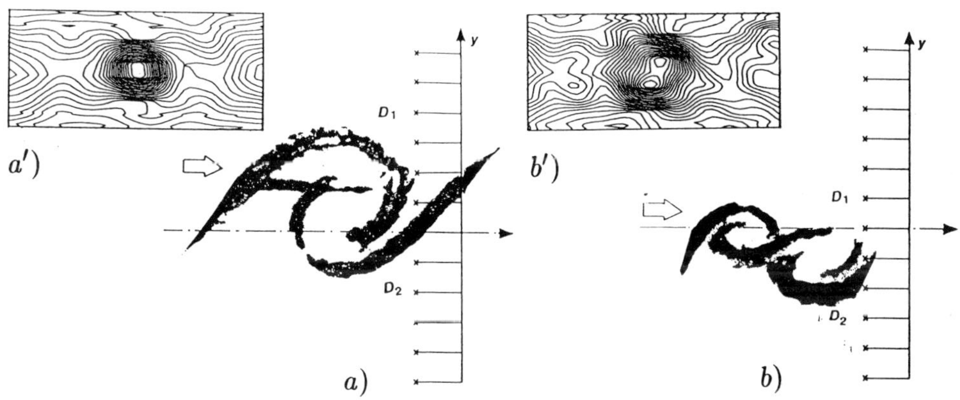

Figure 5: Delocalized Conditional Sampling. Schematic representation of "typical events" and corresponding coherent vorticity: *a*) Single core, *b*) Two cores. Corresponding measured typical iso-vorticity: *a'*) zero time shift, *b'*) non-zero time shift.

The DCS allows a lower rejection rate than conventional techniques and leads to built new statistics for several criteria. A detailed study of the statistics has been performed by Bellin [1] in order to analyze the efficiency of the technique. Figure 6 shows typical statistics which can be obtained.

The time interval between two successive structures is given on fig. 6-a). With a Taylor hypothesis, the more probable distance between structures can be evaluated to the order of $3\delta_\omega$. This value is in agreement with the results of other authors [2]

A characteristic scale of the CS can be obtained in taking into account both the distance between the reference probes and the time delay between the two detections (fig. 6-b). It can be observed that this scale is far from being normally distributed. The distribution is closer to a Poisson one than to Gaussian; the probability to encounter sizes larger than roughly 0.5 δ_ω is greater than for lower sizes. This low value for the mode has to be related to the definition of the scale itself (distance between the velocities extrema) which is lower than the physical spatial extend of the structures. However this scale remains roughly proportional to the size of the structure itself.

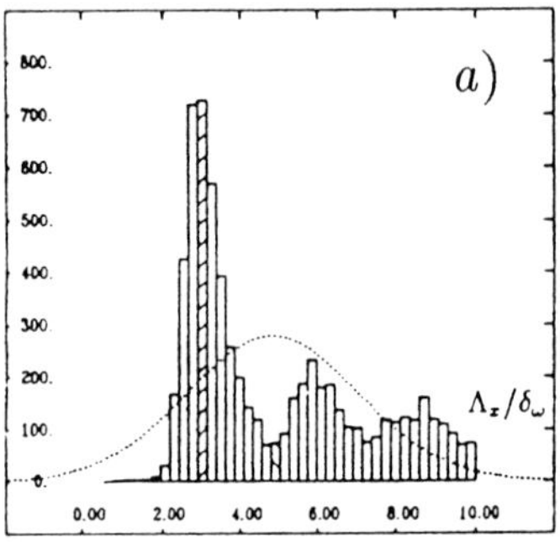

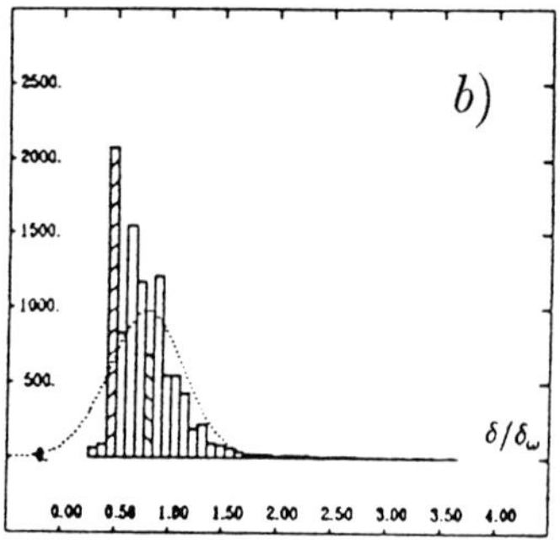

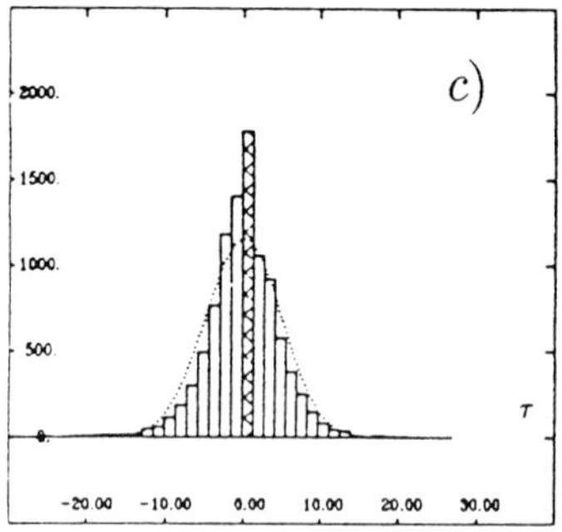

Figure 6: DCS. Typical histograms:
a) Distance between structures
b) Characteristic size of the structure
c) Time shift within structures.

Lastly, the time shift within structures is given on fig. 6-c. This characteristic looks to be quasi-normally distributed. With this parameter, the phase inside each structure can be estimated. It can be noticed that, from a single streamwise HWR location, several types of events can be individualized according their stage of evolution.

It is then possible to select a given class of size and to built the conditional averages of vorticity given in fig. 7 for different time shifts. The resulting image is then the average characteristics of CS at different ages of its evolution. This procedure can be systematically used for any available turbulent quantity. Moreover, marginal statistics can be more convenient for the analysis of these characteristics.

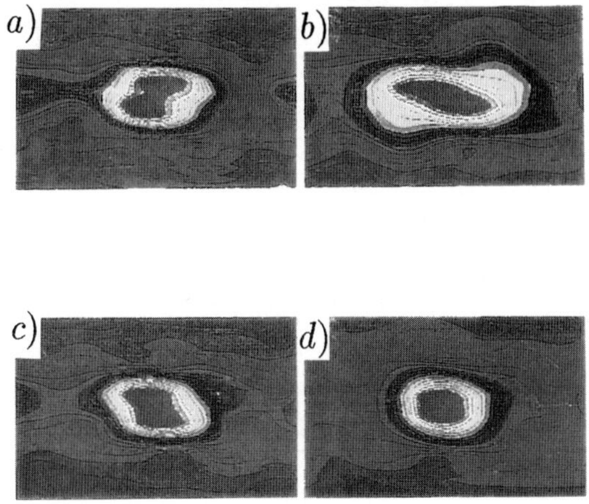

Figure 7: Iso-vorticity for $1.125\delta_\omega$ thickness, for several time shifts.
$a)$ $-3\ 10^{-4}$; $b)$ $-5\ 10^{-4}$; $c)$ $3\ 10^{-4}$; $d)$ $0s$.

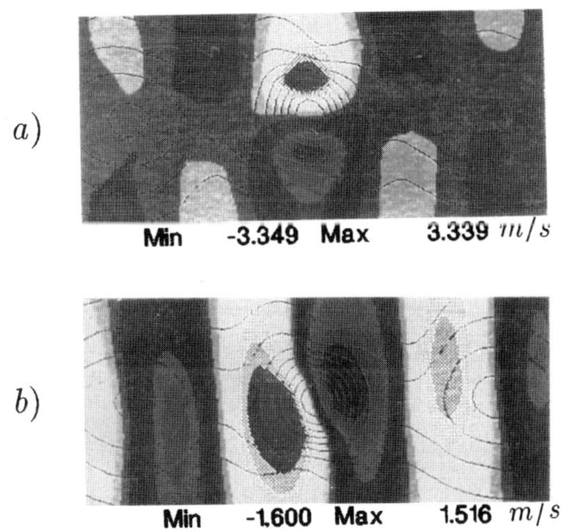

Figure 8: Triple Decomposition: Coherent Velocity, all events cumulated. $a)$ u_c, $b)$ v_c.

Colour version in Colour Plates section, p. xviii.

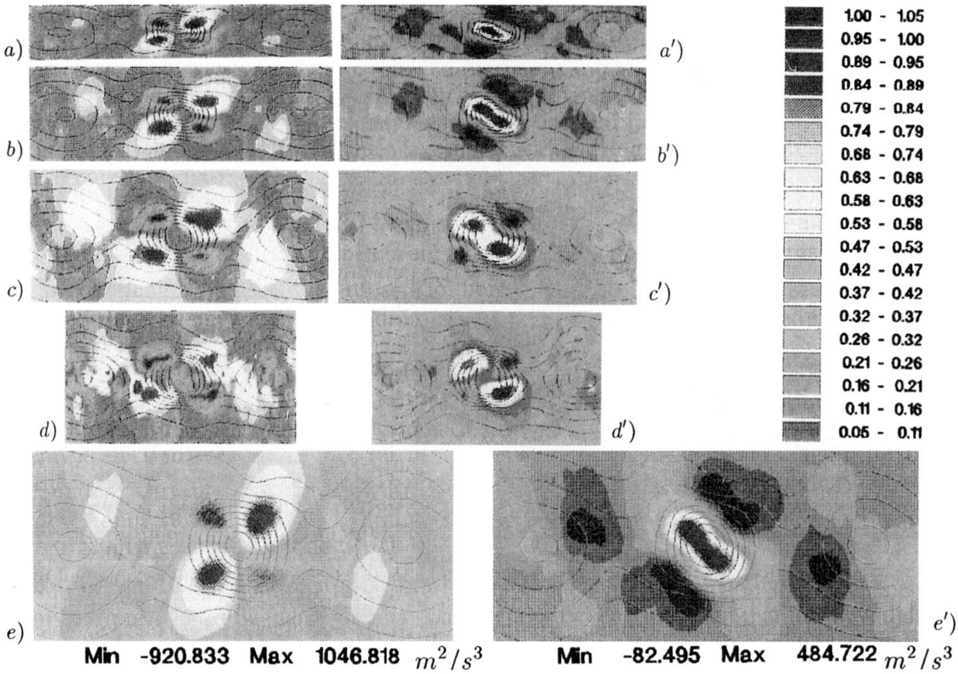

Figure 9: Triple Decomposition: Longitudinal and transversal productions for several thickness classes. *a*) $0.625\delta_\omega$, *b*) $0.875\delta_\omega$ *c*) $1.125\delta_\omega$ *d*) $1.375\delta_\omega$ *e*) all events.
Colour version in Colour Plates section, p. xix.

Triple decomposition

The efficiency of the DCS process and the possibility to select the criteria (size and/or shift, ...) allow to perform detailed double or triple decomposition as introduced in [10]).
The characteristics of the present mixing layer obtained from the triple decomposition will not be discussed here in details. Only some illustrations of the possibilities of the DCS are presented. Figure 8 shows the maps of two components of the coherent velocity, u_c and v_c from the triple decomposition. The iso-contours are obtained by cumulating all the detected events whatever the size or time shift, which here represents around 11 000 events. The structure of these two maps corresponds to the effect of a global rotation of a mass of fluid, as usually associated to CS.
Among others, the longitudinal and transversal productions can be determined from the triple decomposition: $- < u_r^2 > \partial u_c/\partial x$ and $- < v_r^2 > \partial v_c/\partial y$ [8] (the indices $_r$ stand for random and $_c$ for coherent parts). Figure 9 shows these results obtained by selecting different scales, ranging from 0.625 to 1.375 δ_ω. The selection of different scales allows to more precisely analyze the characteristics of different types of events. The "conventional" production terms obtained by cumulating all the events are given

in fig. 9-e for comparison. The cores of the larger CS are different when compared with the smaller ones: they exhibit very low transversal production, and large intensity of the longitudinal production.

Conclusion.

The Pseudo-Flow-Visualization is a powerful tool for a rapid analysis of the data issued from multiple simultaneous measurements in turbulent flows. Its ability to simply evidence the large scale structures in the flows is real, but some care should be taken in interpreting the results. Based on the results of the Pseudo-Flow-Visualization, a simple pre-determination of the period of probable presence of Coherent Structure is easily performed. A non conventional approach of Delocalized Conditional Sampling is proposed and applied to a plane subsonic mixing layer. It consits in an automatic selection of two optimum detector probes located anywhere inside the rake. Without any relevant threshold condition, the Coherent Structures of different sizes and locations can be identified. New statistics concerning the individual characteristics can be obtained, and a "morphological" conditional analysis can be performed. This technique has up to now been applied to relatively coarse HWR. New experiments are under development with finer rakes, namely 24 crossed wires.

Acknowledgements: this study is supported by DRET/DGA grant 90-171.

References

[1] Bellin S., Thèse Univ. Poitiers Sept. 1991
[2] Bernal L.P. *Phys. of Fluids*, vol. 31 No 9 pp. 2533-2543, 1988
[3] Browand F.K. & Troutt T.R. *J.Fluid Mech.* 917, 4 pp. 771-781, 1980
[4] Browand F.K. & Weidman P.D. *J.Fluid Mech.* 71, 1, 1976
[5] Delville J., Bellin S. & Bonnet J.P., in:*Advances in Turbulence II*, Fernholz & Fiedler ed., Springer, 1989
[6] Glauser M. & George R., Dubrovnick, 1991
[7] Hussain A.K.M.F. & Zaman K.M.B.Q.*J.Fluid Mech.* 159, pp. 85-104, 1985
[8] Hussain A.K.M.F. *J.Fluid Mech.* 173, pp. 303-356, 1986
[9] Tabatai M., Kawall J.G. & Keffer J.F. Dantec Information Nr 4, 1987
[10] Reynolds W.C. Hussain A.K.M.F. *J.Fluid Mech.* 54 part 2 pp. 263-288, 1972
[11] Roshko A.A. *AIAA Journal* 1', No 10 pp. 1349-1357, 1976
[12] Townsend A. A. *J.Fluid Mech.* 95 part 3 pp. 515-537, 1979
[13] Tso J. & Hussain F. *J.Fluid Mech.* 203, 1989.
[14] Zaman K.B.M.Q. & Hussain F. *JFM*, 112, 1984.

IDENTIFICATION OF LOW-SPEED SPOTS AND VORTICAL STRUCTURES IN THE WAKE OF A LONG CYLINDER IN NEAR-AXIAL FLOW.

M.K.Bull and W.A.Dekkers
Department of Mechanical Engineering
University of Adelaide
Adelaide, S.A. 5000
Australia.

1 Introduction

Our previous studies [1] of thick axisymmetric turbulent boundary layers on long cylinders have shown that frequently-occurring negative spikes are a characteristic feature in the time-histories of the streamwise (u) velocity fluctuations, at all radial positions throughout the boundary layer. The spikes represent "spots" of low-speed fluid, which appear to consist of inner-layer fluid that has been stripped away from the cylinder surface by large-scale turbulent cross-flows within the boundary layer itself. We have postulated [1,2] that this stripping process may well play a significant role in turbulence generation and maintenance in thick axisymmetric boundary layers.

Willmarth et al.[3], in experiments at cylinder Reynolds numbers $Re_d = U_1 d/\nu$ (where U_1 is the free-streamvelocity, d the cylinder diameter and ν the kinematic viscosity of the fluid) in the range 10^4 to 10^5, investigated the effect of inclining the axis of a very long cylinder at a small yaw angle β - of just a few degrees - to the main flow direction. Their results show that, while boundary layer flow persists (typically for $\beta \lesssim 1°$), the thick boundary layer developed on the cylinder becomes highly asymmetric: the layer thickness δ_W on the windward (upstream) side of the cylinder becomes extremely small and that on the leeward (downstream) side δ_L becomes very large. The same effect can also be seen in data obtained by Scott [4] at $Re_d = 910$ for the circumferential distribution of boundary layer thickness (defined here, for the purpose of illustration, as the distance from the cylinder surface at which the mean velocity is equal to 0.95 U_1, shown in figure 1. Also, in this latter flow, the u-velocity time-histories exhibit negative spikes. Thus the characteristic spots of low-speed fluid observed in axisymmetric boundary layers also occur in the asymmetric yawed boundary layer, at small yaw angles. They are detectable on the windward side of the yawed cylinder for β up to about 1° ($\delta_L/\delta_W \approx 20$), while on the leeward side

J. P. Bonnet and M. N. Glauser (eds.), Eddy Structure Identification in Free Turbulent Shear Flows, 103–114.

they can be clearly observed only for $\beta \lesssim 0.6°$ ($\delta_L/\delta_W \approx 4$) – apparently as a consequence of the disappearance of large-scale turbulence structures on the windward side of the cylinder and the cylinder becoming increasingly remote from the large-scale structures on its leeward side, as yaw angle is increased.

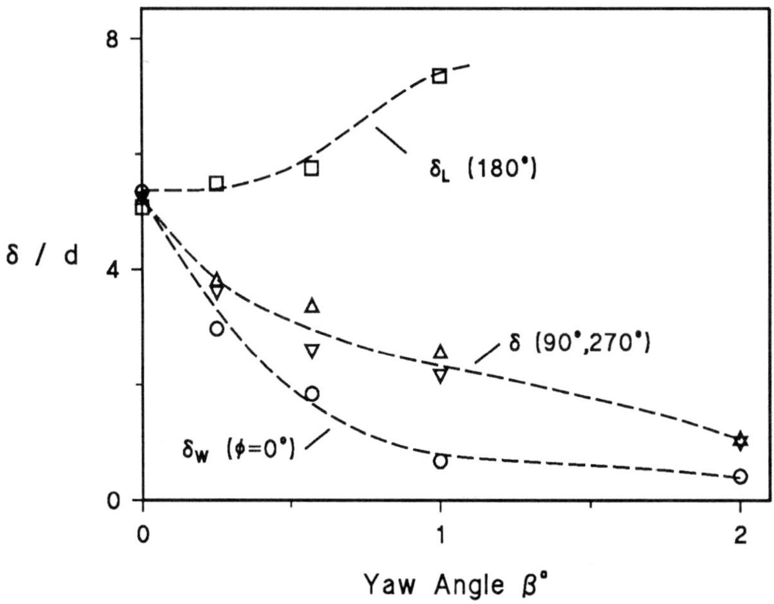

Figure 1: Effect of small angles of yaw on circumferential distribution of boundary layer thickness. $Re_d = 910$. ϕ = azimuthal angle measured from windward side of cylinder.

In the high-Reynolds-number experiments of Willmarth et al., the flow was found already to have developed a wake-like character at a yaw angle of 2.36°. Neither at this nor at higher yaw angles was periodic vortex shedding observed. On the other hand, our studies [2,5] at much lower cylinder Reynolds numbers have shown that, when the Reynolds number is below a well-defined limiting value, regular vortex-shedding from the cylinder can take place even at a yaw angle as low as 1°; and this discovery led us to the further speculation [2], that the displacement of low-speed fluid from the inner part of the turbulent boundary layer on a cylinder at zero or small non-zero angle of yaw, to form low-speed spots, may involve localised vortex shedding, as a result of turbulent cross-flows within the boundary layer passing over the cylinder.

Thus, apart from the possibility of laminar flow at very small Reynolds numbers, there are three possible regimes of flow over very long yawed cylinders, depending on the combination of yaw angle and Reynolds number: turbulent boundary layer flow, aperiodic turbulent wake flow, and turbulent vortex wake flow. The boundaries between the different regimes are shown in figure 2.

There are too few data to define the boundary between turbulent-boundary-layer flow and wake flow with any precision, but those available, as discussed in a later section, indicate that for both periodic and aperiodic wake flow it is in the region of $\beta = 1°$. For wake flows ($\beta \gtrsim 1°$), we have determined the Reynolds number at the boundary between the vortex-shedding and no-vortex-shedding regimes, over a range of yaw angles: it is about 300 at $\beta = 1°$, increasing to about 1500 at $\beta = 3°$, remaining essentially constant at this value over the range $3° < \beta < 6°$, and then increasing further at larger yaw angles.

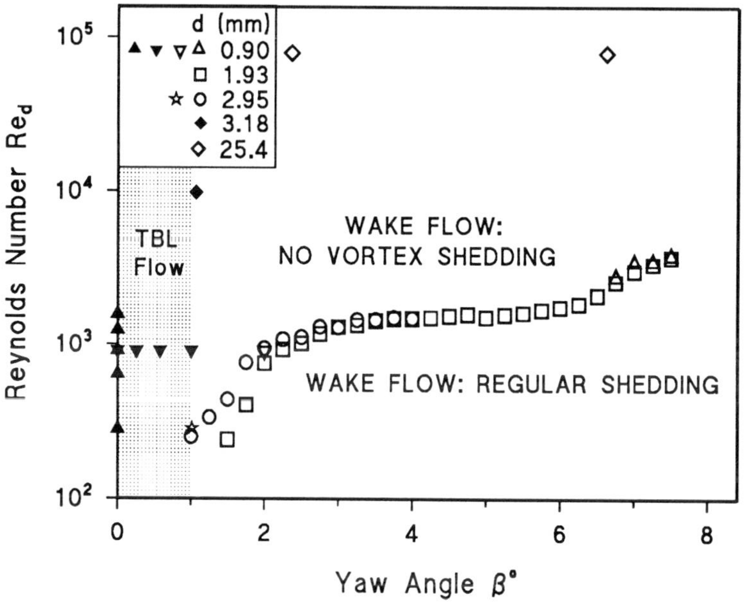

Figure 2: Dependence of flow regime on Reynolds number and yaw angle. Turbulent boundary layer flow: ▲, [1]; ▼, [4]; ◆, [3]. Transition between periodic and aperiodic flow: ▽, [4]; ✡, from flow visualisation (present investigation); △, □, ○, from hot-wire measurements (present investigation). Aperiodic wake flow: ◇, [3].

Thus the implication is that the imposition of a yaw angle of little more than 1° will transform a very thick axisymmetric turbulent boundary layer into an aperoidic turbulent wake at high Reynolds numbers or into a turbulent vortex wake at low Reynolds numbers. Here, in relation to the identification of eddy structures, the emphasis will be on axial and near-axial flows at low cylinder Reynolds numbers where the flow regime is either a very thick turbulent boundary layer or a vortex wake.

2 Identification Methods

Various methods have been used, in both axisymmetric and yawed flows, to obtain details of the flow processes referred to in the preceding section and to identify eddy structures associated with them. The methods include simultaneous and conditional sampling, making use of various combinations of:

(A) time-histories of streamwise velocity fluctuations, sensed by a single constant-temperature hot-wire;

(B) time-histories of temperature fluctuations, sensed by a cold-wire (resistance thermometer);

(C) marking of inner-layer fluid by local heating;

(D) air-flow visualisation by smoke-generation at the cylinder surface over a short axial segment of the cylinder;

(E) air-flow visualisation by plane sheets of smoke streaks generated by smoke-wires;

(F) water-flow visualisation by hydrogen bubbles; and

(G) simultaneous time-histories of streamwise and radial turbulence velocity fluctuations, obtained from radial arrays of hot-wire sensors (containing both u-wires and X-wires).

Some additional work is in progress using

(H) two-component laser-Doppler measurements of velocity fluctuations in both air and water flows.

3 Turbulent Boundary Layer Flow

3.1 Origin of Low-Speed Spots

To determine the location of the origin of low-speed spots, two procedures have been used, one involving conditional sampling of temperature and streamwise velocity fluctuations and the other sampling of streamwise velocity fluctuations synchronised with flow-visualisation recording.

3.1.1 Conditional sampling of temperature and streamwise velocity fluctuations

In this method (A+B+C), a small segment of a cylinder subjected to an air-flow was heated by directing an infra-red beam onto a short metal insert in a nylon cylinder [6]. Consequently, air passing over the metal insert, immediately adjacent to the cylinder surface, was heated. Temperature fluctuations and streamwise velocity fluctuations were detected at two closely-spaced points, at various locations in the boundary layer, by means of a cold-wire probe and a hot-wire probe respectively. The hot-wire was operated at a high overheat ratio to minimise its sensitivity to temperature fluctuations in the flow.

First, ensemble averages of time-signatures of negative velocity spikes were obtained from the hot-wire in isolation, for the condition $u_{min} < -nu'$ (where u_{min} is the maximum negative velocity of the spike, and u' the r.m.s. streamwise velocity fluctuation) for various values of n. These were found to have similar shapes, with only the magnitudes dependent on n. Next, time-histories of temperature fluctuations θ at the same point were obtained from the cold-wire in isolation. These exhibited positive temperature spikes, which on sampling and ensemble-averaging, subject to the condition $\theta_{max} > n\theta'$, yielded characteristic temperature spikes similar in shape to the velocity spikes.

Then simultaneous signals were obtained from the two probes in close proximity, and the streamwise velocity signal sampled on the condition of occurrence of a temperature spike exceeding a prescribed magnitude. The ensemble average of of the velocity samples reproduced the typical form of the unconditionally ensemble-averaged velocity spikes (figure 3), thus indicating that temperature-fluctuation spikes are associated with velocity spikes, and hence that velocity spikes sampled in this way originate at the same location as the temperature spikes i.e. in the wall-adjacent region of the boundary layer.

The reverse process, i.e. sampling the temperature-fluctuation signal on the condition of the occurrence of a prescribed high negative velocity excursion, leads to an ensemble-averaged temperature signature with a maximum which is not as large or as sharp as in the previous case. This is interpreted as indicating that the low-speed spikes making up the sequence observed at a given position in the boundary layer have their origins at various streamwise positions along the cylinder.

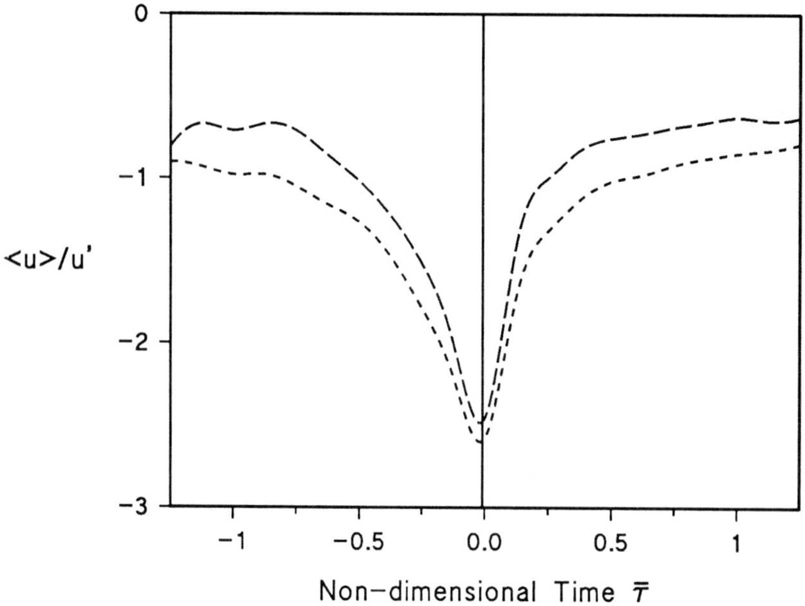

Figure 3: Ensemble averages of spikes, $y/\delta \approx 0.18$: - - - - - -, all u-velocity spikes; – – – –, temperature-associated u-velocity spikes. Discrimination level = $-2\ u'$. Non-dimensional time $\bar{\tau} = U_1\tau/\delta$.

3.1.2 Simultaneous sampling of streamwise velocity fluctuations and air-flow visualisation

In this case (A+D), a smoke-generator was incorporated as a segment of the cylinder itself: the segment was wound with resistance wire, smeared with oil, and smoke generated at the cylinder surface by electrical heating. The output of a single hot-wire sensing streamwise velocity fluctuations (and again operated at high overheat ratio to minimise sensitivity to temperature variations associated with the smoke producing visualisation) was displayed on an oscilloscope screen. By an optical arrangement, the smoke patterns and oscilloscope display were video-recorded simultaneously, each occupying half of the field of view of a Kodak Ekta Pro 1000 video camera. The recordings show frequent intervals in which fluid is swept radially outwards from the cylinder surface, and the arrival of such outward-moving fluid at the hot-wire location (in the outer layer) coincides with the occurrence of a low-speed spike in the oscilloscope trace. f The rate of spread of smoke can be roughly quantified, and appears to be consistent with the effective streamwise and radial velocities, U_L and V_L respectively, typical of low-speed spots, which are obtained from hot-wire-rake measurements described later. The results therefore support the conclusions drawn from the experiments described in the previous section.

In addition, the apparent circumferential rate of spread of smoke in these experiments appears to be rather greater than would be expected to be produced by the circumferential component of the turbulence fluctuations. This suggests that the effect may be due rather to large-scale undulation of the whole turbulent boundary layer with respect to the cylinder axis, and that gross motion of this type may be involved in the stripping process [5].

3.2 Eddy Structure associated with Low-Speed Spots

3.2.1 Simultaneous conditional sampling of streamwise and radial velocity fluctuations at various radial locations

A radial array of hot-wires (G), comprising four u-wires and one X-wire, was set up on one side of a cylinder in an air-flow; a single u-wire was used in conjunction with this array, and could be positioned independently to one side of the cylinder or, more usually, on the opposite side to the array. Signals from the array were simultaneously sampled on the condition of occurrence of a negative streamwise-velocity spike of magnitude greater than a prescribed value of nu' at any chosen u-wire sensor, and processed to obtain ensemble-averaged time histories.

The ensemble-averaged time histories show that spots of low-speed fluid moving in the streamwise direction cross a given radial plane at progressively earlier times as the radial distance from the cylinder increases - in the nature of streamwise movement of a "front" of low-speed fluid inclined forwards in the flow direction. They also allow the typical streamwise and radial components of velocity of the fluid within the spots themselves to be obtained as $U_L \approx 0.67\ U_1$ and $V_L \approx 0.06\ U_1$ respectively,implying that the spots of low-speed fluid follow trajectories which make a shallow angle with the cylinder axis, typically about 5°.

These results in combination with the hypothesis that the low-speed spots consist of inner-layer fluid stripped away from the cylinder by radial turbulence cross-flows lead to the model [6] illustrated in figure 4. It is envisaged that, during a period of high tranverse velocity, the resultant flow (which will be inclined at an angle of no more than a few degrees to the cylinder axis) progressively strips low-speed fluid from the cylinder, with the point on the cylinder at which stripping is occurring moving along the cylinder in the downstream direction at a velocity U_f. Such a process will result in low-speed fluid becoming concentrated along a front (as shown in the figure) which also advances with a velocity U_f parallel to the cylinder axis and which has an inclination to the axis determined by the relative values of U_f, U_L and V_L. From the measured data the propagation velocity of the front and its angle of inclination have been estimated to be about 0.6 U_1 and 45° respectively.

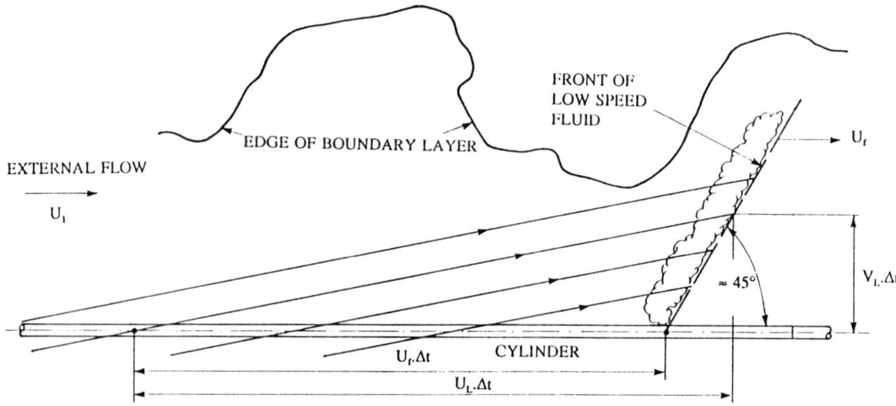

Figure 4: Formation and progression of fronts of low-speed fluid.

4 Transition from Boundary-Layer to Wake Flow

In the experiments of Scott [4] at $Re_d = 910$, boundary-layer flow was found to persist for β up to 1°. No measurements were made at yaw angles between 1° and 2°, but at $\beta = 2°$ the flow is a wake flow or is in the transitional state between a boundary layer and a vortex wake. There is a similar situation in the investigation of Willmarth et al. [3] at high Reynolds numbers, where a boundary-layer flow occurs at $\beta = 1.06°$, the flow is transitional between a boundary layer and an aperiodic wake at $\beta = 2.36°$, and there are no data at intermediate yaw angles. Our video recordings (F) of hydrogen-bubble visualisation of water flow over a cylinder at $\beta = 1°$ at $Re_d = 300$ show this flow to be in a transitional state between a boundary layer and a vortex wake. A typical video frame is shown in figure 5. It can be seen that the flow structure is not nearly as well-ordered on the leeward side of the cylinder as it is at higher yaw angles (compare figure 6 in section 5 below), and patches of turbulence appear intermittently on the windward side of the cylinder (whereas in wake flow no disturbance to the hydrogen-bubble lines can be detected upstream of the cylinder). This condition corresponds to the point shown as a star in figure 2. The evidence for determining the boundary between the turbulent-boundary-layer and wake regimes is thus, at best, very fragmentary, but it does seem to indicate that the boundary is roughly in the region of $\beta = 1°$.

It might also be noted that the patches of turbulent flow shown by figure 5 on the upstream side of the cylinder at $\beta = 1°$ and $Re_d = 300$ have an extent normal to the cylinder axis of several cylinder diameters (typically 3 to 4). In contrast, according

Figure 5: Hydrogen-bubble flow-visualisation, $Re_d = 300$, $\beta = 1°$. Flow is from right to left.

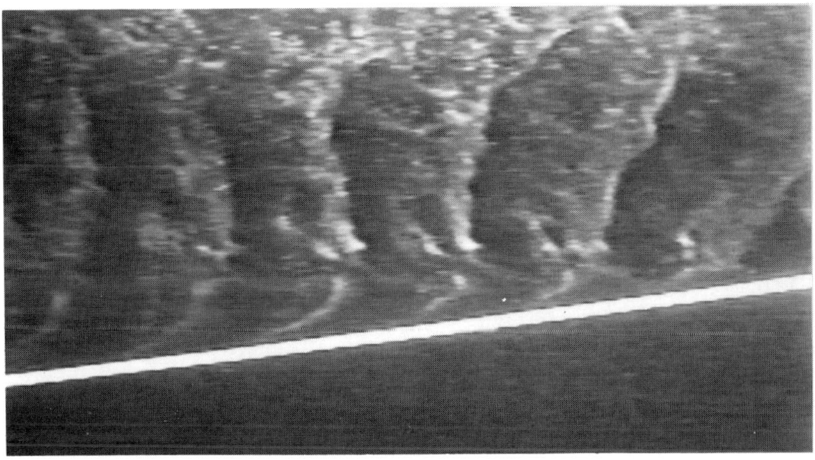

(a) $\beta = 6°$

(b) $\beta = 30°$

Figure 6: Hydrogen-bubble flow-visualisation, $Re_d = 300$. Flow is from right to left.

to figure 1, the value of δ/d for $\beta = 1°$ at $Re_d = 910$, obtained from mean- velocity measurements, is considerably smaller (≈ 0.7). Although the Reynolds numbers are not the same in the two cases, this comparison gives an indication that, even when mean-velocity measurements imply a very thin boundary layer, some very large excursions may be included in the averaging process.

5 Vortex Shedding from Yawed Cylinders

Vortex shedding from cylinders at very small angles of yaw at low Reynolds numbers was first detected [7] from periodic signals from a hot-wire in the cylinder wake (A), and confirmed by smoke-wire visualisation (E), in wind- tunnel experiments. Subsequently, the limiting Reynolds numbers at which shedding can occur at given values of yaw angle (figure 2), already briefly discussed in section 1, were determined from hot-wire signals.

More recently, video recordings of hydrogen-bubble visualisation (F) of vortex-shedding from yawed cylinders in a water flow have been made. For flow with $Re_d = 300$, these show that, for $\beta \gtrsim 20°$, vortices are shed from the cylinder at a single well-defined angle which is greater than β. This is consistent with the consideration that because the yawed cylinder is subject to a mean flow which has velocity components both normal and parallel to the cylinder axis, the boundary layer, before separation to form vortices, develops in both the axial and normal direction, and the resultant vorticity vector within it is consequently inclined to the main-flow direction at an angle greater than the yaw angle. Hence, as observed, the angle which shed vortices make with the free-stream direction is also expected to be greater than β. One frame from the video recording of a flow of this type, with $\beta = 30°$, is shown in figure 6(b). At smaller yaws angles, $\beta < 20°$, two vortex inclinations are observable; flow with $\beta = 6°$ is illustrated by figure 6(a). The angle β_{vi} at which vortices are initially inclined to the main-flow direction immediately after formation is, in accord with the preceding considerations, larger than the yaw angle itself. The angle β_{vf} which the vortices finally take up further downstream is larger again than β_{vi}, and considerably so at small β. The variation of vortex angle with yaw angle is shown in figure 7. The data indicate that, as $\beta \to 0°$, $\beta_{vi} \to 0°$ while $\beta_{vf} \to 180°$.

6 Concluding Remarks

A broad survey has been made of a number of phenomena which have been identified in the flow over long cylinders in axial and near-axial flow, and of various methods

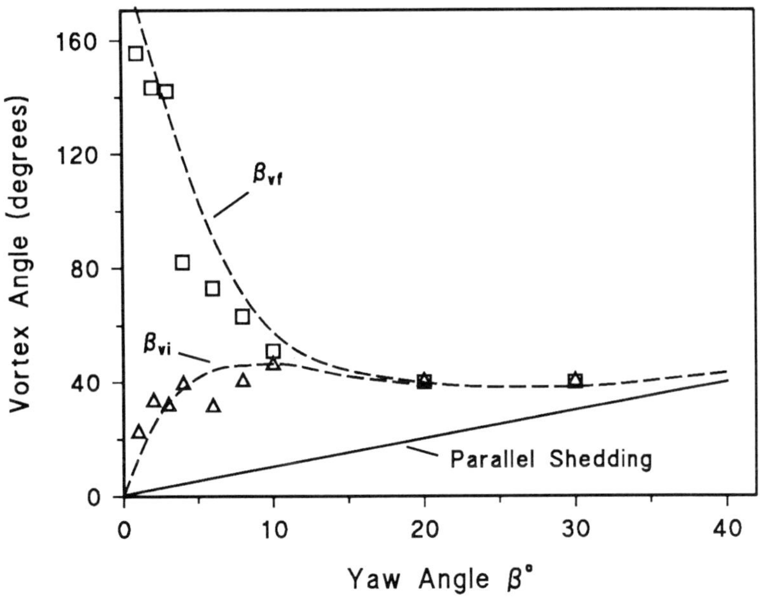

Figure 7: Dependence of shed-vortex inclination on yaw angle. $Re_d = 300$.

which have been applied to investigate them. Clearly many of these phenomena merit further investigation, and in particular: the speculation referred to in the introduction to this paper that the process of removal of low-speed fluid from the inner region of an axisymmetric boundary layer by large-scale turbulent cross-flows involves localised regions of vortex-shedding; the relationship between low-speed spots and wake vortices at low Reynolds numbers, and between low-speed spots and other wake structures at higher Reynolds numbers; and the details of the process of transition from a turbulent-boundary-layer to a wake flow under the influence of small angles of yaw, for both vortical and non-vortical wakes.

It is considered that a detailed understanding of these processes is fundamental to the understanding of turbulence generation and maintenance in thick axisymmetric turbulent boundary layers and of wake formation behind cylinders at small angles of yaw.

References

[1] R.E. Luxton, M.K. Bull and S. Rajagopalan. The thick turbulent boundary layer on a long fine cylinder in axial flow. *Aeron. Jl. of the R.Ae.S.* **88**, 186-199, 1984.

[2] M.K. Bull and R.E. Luxton. Vortex-shedding and the maintenance of thick turbulent boundary layers on long cylinders in axial flow. *Advances in Turbulence 3*, Springer Verlag, 1991.

[3] W.W. Willmarth, L.K. Sharma and S. Inglis. The effect of cross flow and isolated roughness elements on the boundary layer and wall pressure fluctuations on circular cylinders. O.N.R. Report on Contract No. N00014-76-C-0571, 1977.

[4] K.D. Scott. The turbulent boundary layer on a long slender cylinder in axial flow: effects of yaw. Final-year Honours Report, Dept. Mech. Eng., University of Adelaide, 1986.

[5] M.K. Bull and W.A. Dekkers. Vortex shedding from cylinders in near axial flow. Proc. 10th Australasian Fluid Mechanics Conference, Melbourne, pp.6.41 - 6.44, 1989.

[6] M.K. Bull, R.E. Luxton and S. Rajagopalan. Flow mechanisms in the thick axisymmetric turbulent boundary layer on a circular cylinder. Dept. Mech. Eng., University of Adelaide, Tech. Rept. 11/88, 1988.

[7] W.M. Azadegan. The turbulent boundary layer on a long thin cylinder in axial flow and vortex shedding at small yaw angles. Final-year Honours Report, Dept. Mech.Eng., University of Adelaide, 1987.

ON THE INFLUENCE OF INITIAL PARAMETERS ON SUBHARMONIC RESONANCE IN AN AXISYMMETRIC JET*

C.O. PASCHEREIT and H.E. FIEDLER
Hermann–Föttinger–Institut
Technische Universität Berlin, FRG

and

I. WYGNANSKI
Dept. of Aerospace and Mechanical Eng.
University of Arizona, Tucson, USA

Abstract. Hot wire measurements in an axisymmetric jet are presented. A resonant subharmonic interaction between the two axisymmetric travelling waves was induced in the shear layer of the jet by controlled sinusoidal perturbations with two frequencies separated by one octave. The phase averaging technique was applied to assess the evolution of the coherent waves. By Fourier decomposing the phase averaged data it was possible to obtain the magnitude as well as the phase of the measured signal.

In agreement with theory (Cohen & Wygnanski 1987, Michalke 1984) it was found that wherever the excited waves are non–dispersive they may interact in a manner that enhances the amplification rate of the subharmonic wave. The amplified subharmonic will exceed the fundamental's level to become the dominant instability component. It is shown that in addition to the requirement of equal phase speed of the two waves the initial phase difference between the subharmonic and the fundamental wave may lead to different amplification rates of the subharmonic wave.

The influence of the initial phase angle between the subharmonic and the fundamental wave has been investigated as well as the influence of the initial amplitude ratio, the overall forcing level and the conditions at the nozzle, i.e. the initial turbulence level and the initial momentum thickness. When the amplitude of the perturbation is low the initial phase difference has an effect on the amplification of the subharmonic. Depending on the phase angle the amplification of the subharmonic can be either suppressed or enhanced. A suppression of the subharmonic rate of amplification for certain initial phase differences could also be observed when flow in the nozzle was tripped and thus the spread of the shear layer was increased. However, an increase in the combined forcing level *reduces* the effect of the initial phase difference on the amplification of the subharmonic. Contourplots of the phase difference $(2\beta_{sub} - \beta_{fun})$ — which is a measure for the dispersion of the waves — clearly show the subharmonic being amplified in regions where the phase speeds are identical and suppressed in regions where the waves are dispersive. A stronger forcing level moves the phase locking of the two waves upstream and thus the influence of the initial phase difference vanishes.

By triple decomposition of the velocity signals (Hussain 1983) an estimate of the energy transfer to the subharmonic is obtained. The energy is mostly being supplied by the mean flow while the fundamental acts as a catalyst to initiate the enhanced energy transfer from the mean flow.

1. Introduction

The initial evolution of an axisymmetric jet has been carefully and continuously investigated ever since jet propulsion became technologically feasible. Nevertheless,

* Supported by the Deutsche Forschungsgemeinschaft under Grant No. Fi 178/34

J. P. Bonnet and M. N. Glauser (eds.), Eddy Structure Identification in Free Turbulent Shear Flows, 115–123.

the mechanisms affecting mixing and combustion are not yet completely understood, nor can they be fully predicted or controlled. This is particularly true for the resonant interaction between multiple axisymmetric modes.

Active control of the flow development can be achieved by external excitation. The spreading rate of the jet can be influenced by external forcing, which enhances the mixing on the molecular scale and therefore the rate of chemical reaction (Roberts 1984). The range of distances at which mixing is enhanced depends on the frequency and the amplitude of the excitation as well as on parameters of the unexcited motion. The first comprehensive study of the response of an axisymmetric jet to a controlled axisymmetric excitation ($m = 0$) was done by Crow and Champagne (1971). They observed that the instability wave increases the rate of spread as long as it grows in streamwise direction. Increasing the forcing amplitude beyond a level at which saturation is achieved further downstream, while maintaining all other flow parameters constant, does not enhance the growth of the fundamental instability wave beyond its saturation level and therefore has no additional effect on the mixing process (see also Oster & Wygnanski 1982), Fiedler & Mensing 1985, Weisbrot & Wygnanski 1988). The degree of control is therefore limited as long as only one frequency is excited.

Further enhancement of the mixing can be achieved by a concomitant excitation of a second instability wave, e.g., the subharmonic. When the fundamental becomes neutrally stable, it's subharmonic, which then amplifies at approximately its maximum rate, enhances the mixing (figure 1). Nonlinearities may lead to

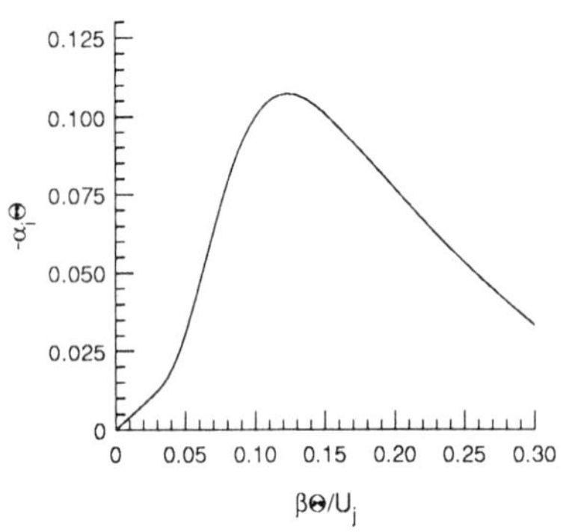

Fig. 1. Calculated growth rates of the axisymmetric component. Reprint from Cohen & Wygnanski (1987).

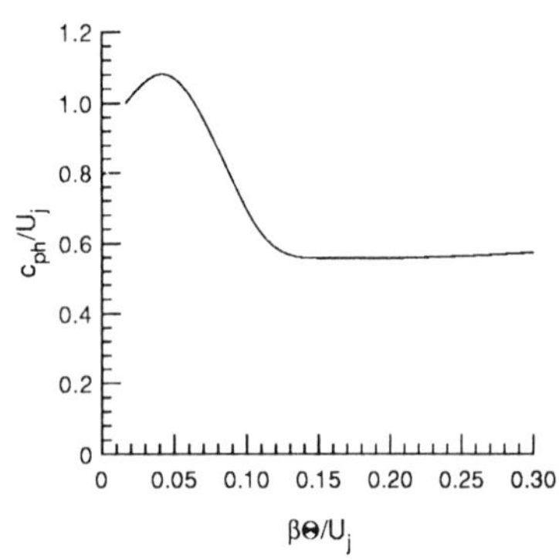

Fig. 2. Calculated phase speed of the axisymmetric component. Reprint from Cohen & Wygnanski (1987).

second-order interactions between the fundamental and the subharmonic wave which may resonate at times (Kelly 1967), resulting in an enhanced growth of the subharmonic. The augmentation of the subharmonic wave contributes to an additional enhancement of the mixing. Cohen & Wygnanski (1987) and Michalke (1984, only for $m = 0$) derived the conditions under which two disturbances can resonate in an axisymmetric shear layer. They found that for subharmonic resonance both waves must propagate with the same phase speeds (figure 2) and the growth rate of the fundamental must be close to zero. From figure 1 and figure 2 it becomes evident that subharmonic resonance is only possible when the fundamental is close to its neutral point. Monkewitz (1988) confirmed their theoretical results in his temporal stability analysis and showed that in addition the initial phase difference plays an important rôle on the growth of the subharmonic component.

Experimental investigations known so far seem to have examined only a few limited conditions (e.g. assessment of the evolution of the subharmonic from spectral measurements only in the centerline for a fixed amplitude ratio). There are more parameters which can influence the growth of the subharmonic. The present study focuses on the details of the interaction of fundamental and subharmonic wave by forcing both waves at the same time. The influence of the initial phase angle between the two waves has been investigated for different initial parameters, as the amplitude ratio of the two excited waves, the total forcing level and the momentum thickness. Spectral measurements at only one radial location are not sufficient to describe the evolution of an instability wave and the energy transfer to it. In the present work the conditional sampling technique has been used for a detailed investigation of the evolution of the coherent waves.

2. Experimental conditions

2.1. The jet facility

The experiments were carried out in an axisymmetric air jet facility. A cross section of the plenum chamber and the jet nozzle is shown in figure 3. The area ratio of the contraction is 36:1 with an exit diameter of $D = 50.8$ mm and a contour of two tangent arcs. The nozzle exit is extended by a parallel section of one diameter length after the end of the contraction. The exit velocity in the present experiment is $\overline{U}_{jet} = 8$ m/s, and the corresponding Reynoldsnumber, based on the nozzle diameter is 28,000. At 8 m/s, the freestream turbulence level is 0.097 %.

The facility was instrumented with a ring of eight normal hot wires spaced at equal circumferential angles (45° apart) and capable of measuring the streamwise component of the velocity. The traversing mechanism, controlled by the laboratory computer, was capable of moving all sensors simultaneously an equal distance in the radial direction.

Controlled axisymmetric excitation of the jet was accomplished by a loudspeaker located in the bottom of the plenum chamber. The excitation signal was generated on the laboratory computer providing the possibility for superimposing two signals separated by one octave and easily changing the phase between them.

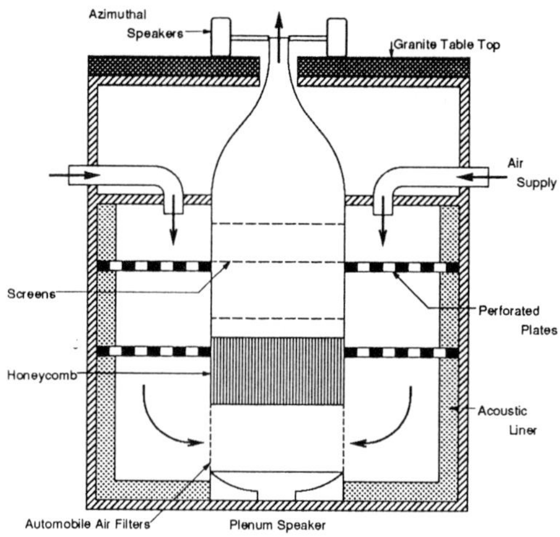

Fig. 3. The air jet facility.

2.2. Initial data processing

The low pass (10 KHz) filtered data signals from the eight hot wire probes, together with the reference signal, were passed through the A/D converters on the laboratory computer and converted to velocities with the calibration polynomial on the vector accelerator. Each buffer of data was then phase-averaged with respect to the reference signal and stored on disk.

In the presence of a coherent motion, an instantaneous variable (for example the streamwise velocity) is considered to consist of three components (Reynolds & Hussain 1970):

$$u_i(x_j,t) = \overline{U}_i(x_j) + \tilde{u}_i(x_j,\tau) + u_i'(x_j,t). \tag{1}$$

Here $\overline{U}$ is the mean or time averaged part, $\tilde{u}$ is the coherent part representing the periodic wave contribution, and u' represents incoherent turbulent fluctuations. The phase average is defined as

$$\langle u_i(x_j,t)\rangle = \lim_{N\to\infty} \sum_{n=1}^{N} u_i(x_j, t+nT), \tag{2}$$

where T is the period of the phase-coherent motion. The phase coherent component of the velocity is

$$\tilde{u}_i(x_j,\tau) = \langle u_i(x_j,t)\rangle - \overline{U}_i, \tag{3}$$

the phase–incoherent component of velocity is

$$u_i'(x_j,t) = u_i(x_j,t) - \langle u_i(x_j,t)\rangle. \tag{4}$$

For a more detailed description of the *triple–* and *double–decomposition* see Hussain (1983).

The data were then decomposed by calculating the Fourier coefficients, $F_{mn}(r)$, where

$$F_{mn} = \frac{1}{2\pi T}\int_0^{2\pi}\int_0^{T}\langle u(r_i,\phi,\tau)\rangle e^{i[m\phi-\omega\tau]}d\phi d\tau. \tag{5}$$

The F_{mn} are complex. It is therefore possible to deduct the magnitude as well as the phase of the measured signal.

3. Experimental results

3.1. The influence of the initial phase difference and the initial amplitude ratio

Prompted by the experiments by Arbey & Ffwocs–Williams (1984) the detailed analysis of the effect of the initial phase difference $\Delta\phi$ (defined with respect to the fundamental) was of particular interest. The jet was forced with two frequencies simultaneously and the phase relation $\Delta\phi$ between the two introduced waves was changed. For an initial phase difference of $\Delta\phi = 90°$, the amplification rate of the subharmonic was suppressed, as opposed to the cases with phase differences of either 0°, 180° or 270°. This effect was observed for three different amplitude ratios $\gamma = 0.1, 0.4$ and 1.6 and is displayed in figure 4 for an initial amplitude ratio of $\gamma = 1.6$.

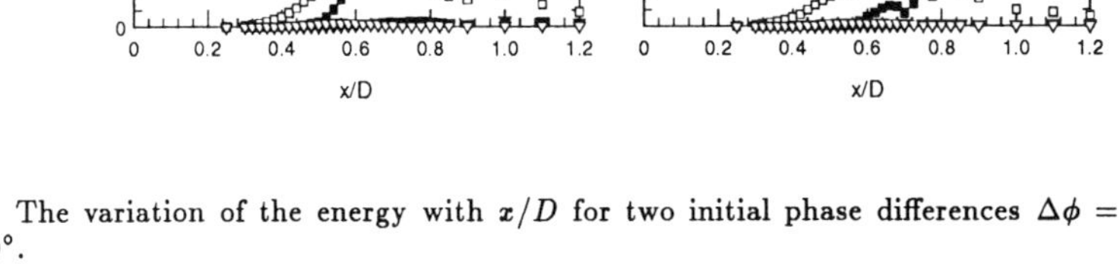

Fig. 4. The variation of the energy with x/D for two initial phase differences $\Delta\phi = 0°$ and $\Delta\phi = 90°$.

3.2. The influence of the overall forcing level and the initial momentum thickness of the mixing layer

All theoretical investigations are based on the assumption of parallel flow and small forcing amplitudes (e.g. Monkewitz 1988) and they suggest that a complete suppression of the subharmonic for certain phase angles is only possible for a truly parallel flow. Of special interest was therefore the influence of a higher forcing level — resulting in an increased spread of the mixing layer. Doubling the combined forcing level eradicated the effect of the initial phase difference between $\Delta\phi = 0°$ & $\Delta\phi = 90°$ (figure 5). In a highly forced mixing layer the assumption of parallel flow might become invalid and thus the influence of the initial phase angle may disappear. However, data taken in the flow exiting from the nozzle equipped with a trip–ring — which increased the spreading rate above that of the case with high amplitude forcing — at the low forcing level showed again an influence on the initial phase difference (figure 6).

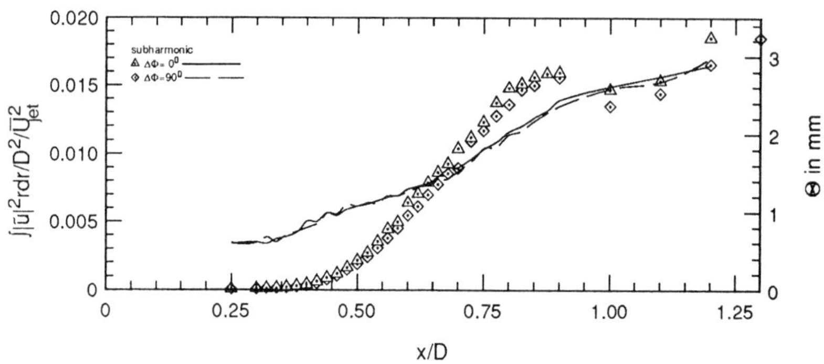

Fig. 5. The variation of the energy of the subharmonic with x/D for a higher forcing level and two different initial phase angles, $\Delta\phi = 0°$, $\Delta\phi = 90°$.

4. Discussion and Conclusion

With the conditional sampling technique it was possible to deduct the main features of the subharmonic resonance mechanism and to establish the rôle of initial parameters in influencing the resonant growth of the subharmonic. To start the resonance interaction the two waves must be non–dispersive and the fundamental must be close to its neutral point. In addition, for low forcing levels the initial phase difference between the two excited waves has an influence on the resonant growth of the subharmonic. An increase in the combined forcing level reduces this influence. Contour plots of the phase difference $(2\beta_{sub} - \beta_{fun})$ show that the in-

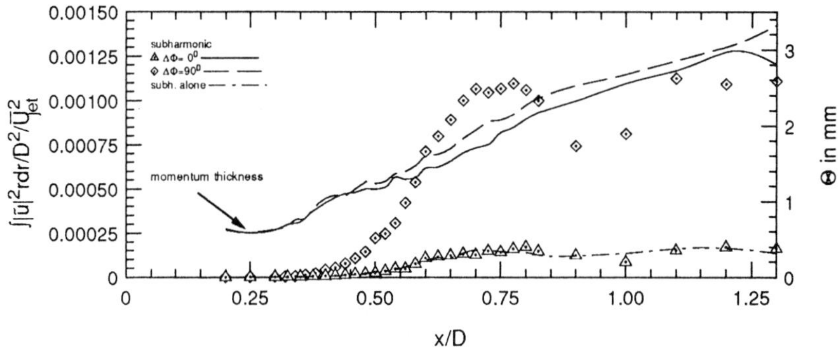

Fig. 6. The variation of the energy of the subharmonic with x/D for the tripped nozzle and two different initial phase angles, $\Delta\phi = 0°$, $\Delta\phi = 90°$.

creased forcing moves the location at which the two waves become non–dispersive upstream (figure 7).

Coherent and incoherent turbulent production terms were calculated using continuity condition. The energy balance shows that most of the energy gained by the subharmonic is supplied by the mean flow directly.

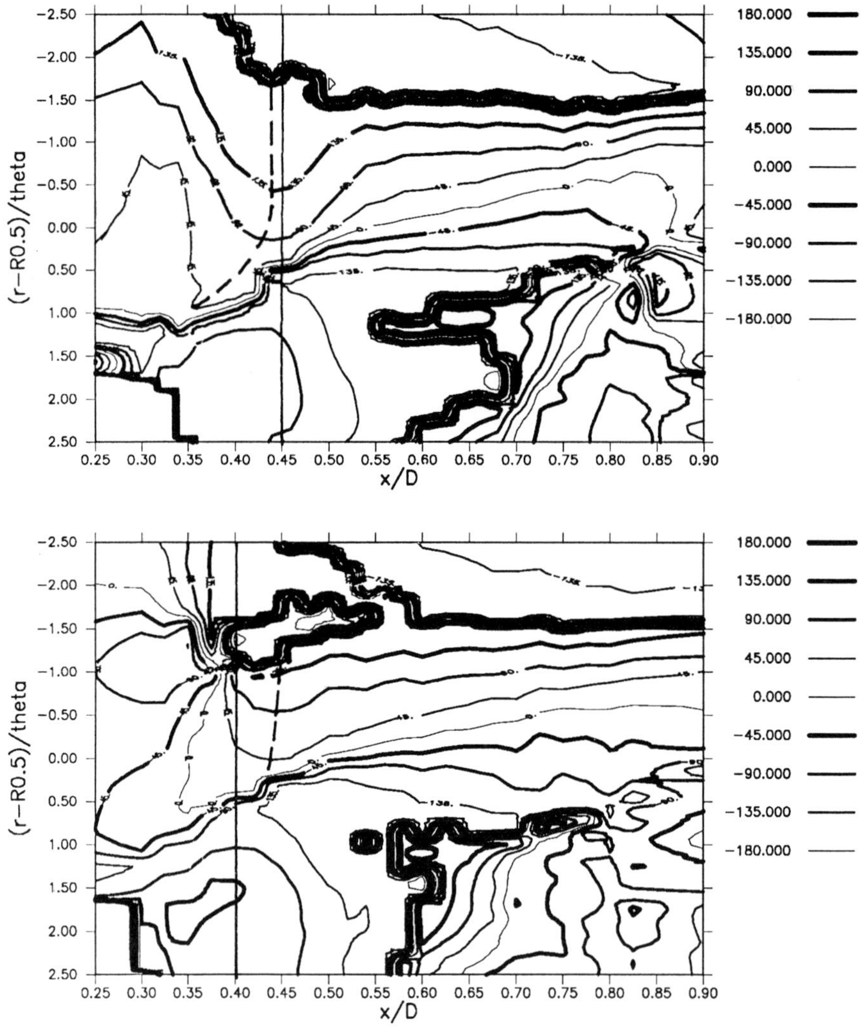

Fig. 7. Contourplots of the phase difference $(2\beta_{sub}-\beta_{fun})$ for the case with high amplitude forcing (*top*) and the tripped case (*bottom*). The solid line refers to the location where the fundamental is neutral, the dashed line to the location where the two waves become non-dispersive.

References

H. Arbey and J.E Ffwocs Williams. Active cancellation of pure tones in an excited jet. *Journal of Fluid Mechanics*, **149**:445–454, 1984.

S.C. Crow and F.H. Champagne. Orderly structure in jet turbulence. *Journal of Fluid Mechanics*, **48**:567, 1971.

J. Cohen and I. Wygnanski. The evolutions of instabilities in the axisymmetric jet. Part 2. The flow resulting from the interaction between two waves. *Journal of Fluid Mechanics*, **176**:221–235, 1987.

H.E. Fiedler and P. Mensing. The plane turbulent shear layer with periodic excitation. *Journal of Fluid Mechanics*, **150**:281–309, 1985.

A.K.M.F. Hussain and W.C. Reynolds. The mechanics of an organized wave in turbulent shear flow. *Journal of Fluid Mechanics*, **41**:241–258, 1970.

A.K.M.F. Hussain. Coherent structures — reality and myth. *Physics of Fluids*, **26**:2816–2850, 1983.

R.E Kelly. On the resonant interaction of neutral disturbances in inviscid shear flows. *Journal of Fluid Mechanics*, **31**:789, 1967.

A. Michalke. Survey on jet instability theory. *Prog. Aerospace Sci*, **21**:159–199, 1984.

P.A. Monkewitz. Subharmonic resonance, pairing and shredding in the mixing layer. *Journal of Fluid Mechanics*, **188**:223–252, 1988.

D. Oster and I. Wygnanski. The forced mixing layer between parallel streams. *Journal of Fluid Mechanics*, **123**:91–130, 1982.

C.O. Paschereit, D. Oster, T.A. Long, H.E. Fiedler, and I. Wygnanski. Flow visualization of interactions among large coherent structures in an axisymmetric jet. *Experiments in Fluids*, **12**:189–199, 1992.

C.O. Paschereit and I. Wygnanski. On instabilities in the axisymmmetric jet: Subharmonic resonance. In H. Branover, M. Mond, and Y. Unger, editors, *Current Trends in Turbulence Research*. 1990.

F. A. Roberts. *Effect of a periodic disturbance on structure and mixing in turbulent shear layers and wakes*. PhD thesis, California Institute of Technology, USA, 1984.

I. Weisbrot and I. Wygnanski. On coherent structures in a highly excited mixing layer. *Journal of Fluid Mechanics*, **195**:137–159, 1988.

Eduction of Vortical Structures by Pressure Measurements in Noncircular Jets

K. Toyoda, T. Okamoto and Y. Shirahama
Department of Mechanical Engineering
Hokkaido Institute of Technology
Sapporo 006, JAPAN

Summary

Vortical structures in the noncircular jets at the interaction mode were educed by measuring the phase-average fluctuating pressure, and the characteristics of the vortical structures were discussed in relation to the mixing mechanism of the noncircular jets. The contours of phase-average fluctuating pressure show clearly the vortical structures of the noncircular jets, which agree reasonably with those predicted by the flow visualization experiments. The vortical structures are characterized by partial merging, stretching and splitting to small vortices. The effects of the noncircular vortical structures on mixing enhancement were also discussed by the quantitative measurements of the velocity fields, the results suggesting that the interactions of vortical structures in the noncircular jets are very effective to enhance mixing.

1. Introduction

The dynamics of coherent structures has been noted as a key feature of turbulent shear flows, and many research works have been done on the structures. Since the coherent structures are characterized by vortical structures, it is crucial in the experimental works to educe the vortical structures by measuring vorticity usually from velocity data. However, the educed results have serious errors owing to the structure 'jitter' and to the differential process required in the vorticity measurements.

It is well known that the vortical structures are closely related to the pressure fluctuations [1]: the pressure fluctuations show direct footprints of the vortical structures better than the velocity fluctuations. Indeed the results of direct numerical simulation reveal that the low-pressure regions correspond to the vortical structures [2]. Thus the detailed pressure measurements in the turbulent shear flows are expected to give us useful information on the vortical structures. Moreover the pressure fluctuations are very important to consider the noise-generating mechanism related to the vortical structures [3]. Although a significance of the pressure fluctuations has been pointed out, there have been few works on the direct pressure measurements which are usually attended with serious

J. P. Bonnet and M. N. Glauser (eds.), Eddy Structure Identification in Free Turbulent Shear Flows, 125–136.

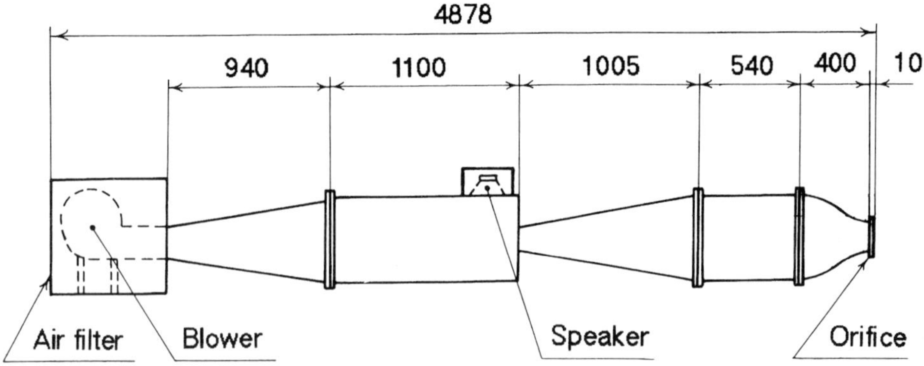

Fig.1 Wind tunnel.

errors owing to disturbance by a pressure probe inserted in the flow. The authors have developed a probe to measure the fluctuating static pressure in the turbulent flows [4], and confirmed that the direct pressure measurements with the probe were very effective to educe the large-scale vortical structures in a circular jet [5].

Noncircular jets have been noted as useful technique to control jet mixing. It is suggested that the mixing enhancement in noncircular jets is owing to the motions of the noncircular vortical structures evolved in the jets [6], [7]. The vortical structures are three-dimensionally complicated, and except flow visualization [8] there has been no way to detect the structures. However the visualized results have serious problems of misleading to incorrect structures, and we need other methods to detect the complicated vortical structures.

In the present study, the vortical structures in noncircular jets excited at the interaction mode were educed by measuring the phase-average fluctuating pressure, and compared with those predicted by the low-Reynolds number visualization experiments, and the characteristics of the vortical structures were discussed in relation to the mixing mechanism of noncircular jets.

2. Experimental apparatus and procedures

The wind tunnel used in the experiments is shown in Fig.1. The air issued from sharp-edged noncircular orifices attached at the end of the wind tunnel. In order to manipulate the vortical structures, the jets were excited by a loudspeaker at the side of settling chamber. The orifice shapes are triangular and rectangular of aspect ratios of 2 & 4. The equivalent diameter of each orifice De is 50 mm. The velocity at the jet exit Ue is 4 m/s and the Reynolds number Re(=UeDe/ν) is 1.3×10^4.

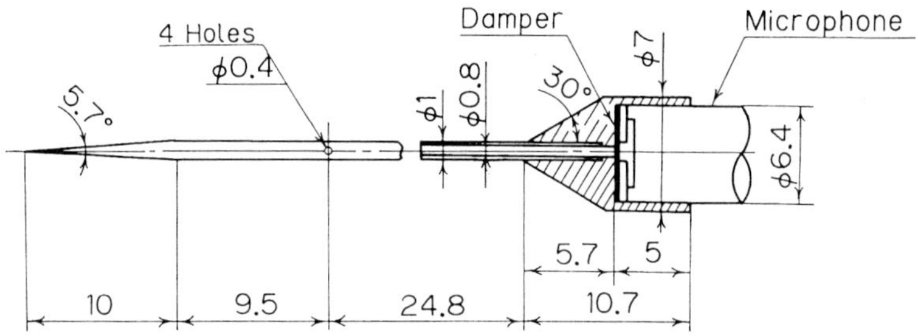

Fig.2 Pressure probe

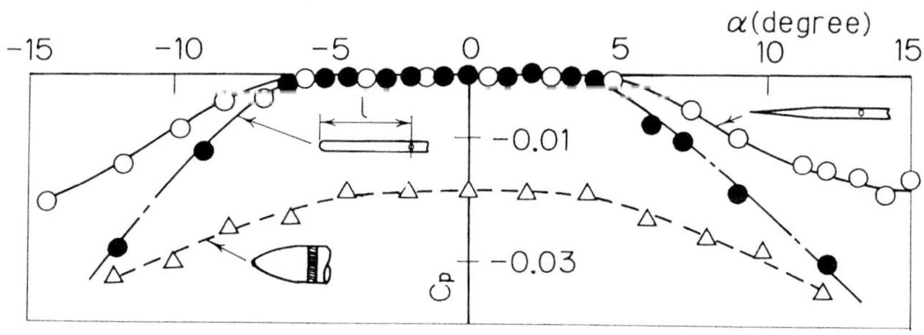

Fig.3 Errors of static pressure tubes in steady flow; $C_p = \Delta p / \frac{1}{2}\rho U_e^2$

The pressure probe used in the experiments is shown in Fig. 2. The static tube with four small holes is connected at the end to the condenser microphone. The structures and the dimensions of the probe were determined to minimize errors in the measurements of fluctuating static pressure. In order to damp the organ pipe resonance in the tube, a thin nylon gauze was inserted in front of the diaphragm of microphone, and the probe had a nearly flat response to 2.0 kHz. The cross-flow error of the static pressure tube was calibrated in the uniform flow by changing the flow attack angle to the tube. The result is shown in Fig.3 in comparison with those of a standard static pressure tube and a nosecone tube. The cross-flow error of the present tube is less than those of the others, and $|Cp|$ is less than 0.02 within the attack angles of $\pm$ 15 degrees. It means that the tube is applicable to the flow with an equivalent turbulence level of 27 % [9].

The dynamic calibration of the pressure probe was carried out in the wake of a circular cylinder by measuring pressure- and velocity-fluctuations simultaneously [10]. The true pressure was predicted with the measured velocity by the following equation [11],

$$p_f = -\rho(\bar{u} - u_c)u_f \tag{1}$$

where p_f and u_f are the fluctuating-pressure and -velocity, ρ is the density, $\bar{u}$ is the mean velocity and u_c is the convection velocity of disturbance. Since Equation (1) is deduced from the inviscid momentum equation, p_f and u_f should be measured in the potential flow

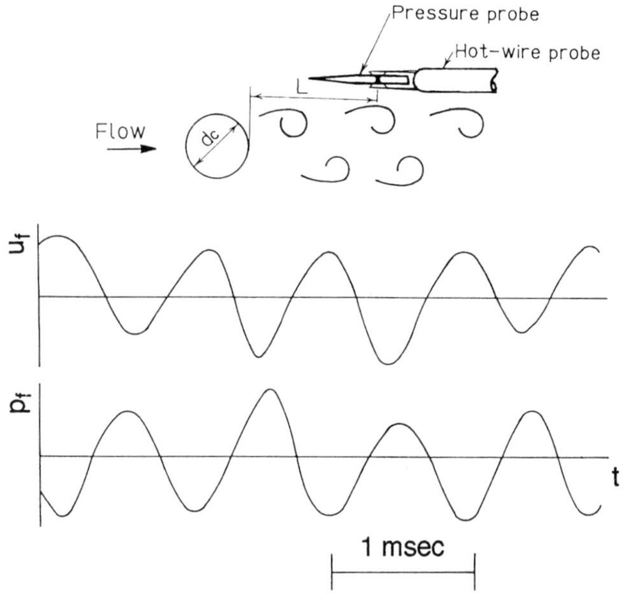

Fig.4 Calibration procedure of the pressure probe in fluctuating flow

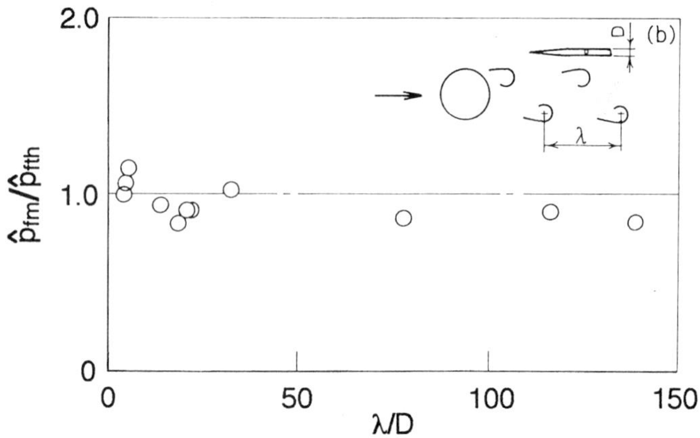

Fig.5 Calibration result of the pressure probe.

region outside Karman vortices behind the circular cylinder as shown in Fig.4. The cylinder diameters were 1-40 mm, the free-stream velocities were 8-25 m/s, and the frequencies of vortex shedding were 100-2000 Hz. The ratios of the measured rms value $\hat{p}_{fm}$ to the predicted rms value $\hat{p}_{fth}$ are shown in Fig. 5, where the ratios are plotted to the nondimensionalized wave length λ/D (λ: wave length of disturbance). The result shows that the errors of the measured values are below 18 % of the true values.

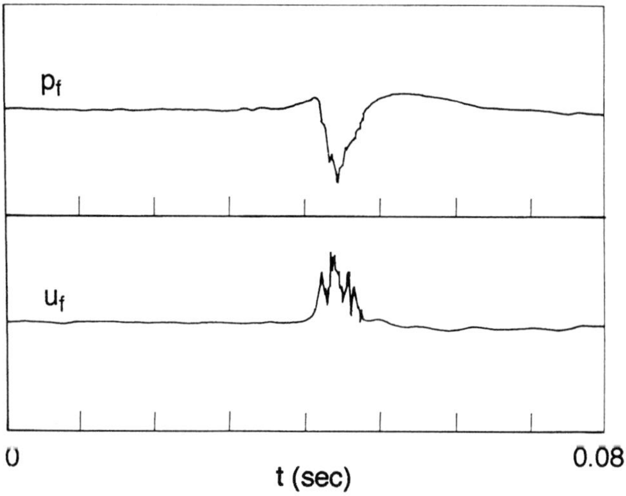

Fig.6 Response to vortical motion

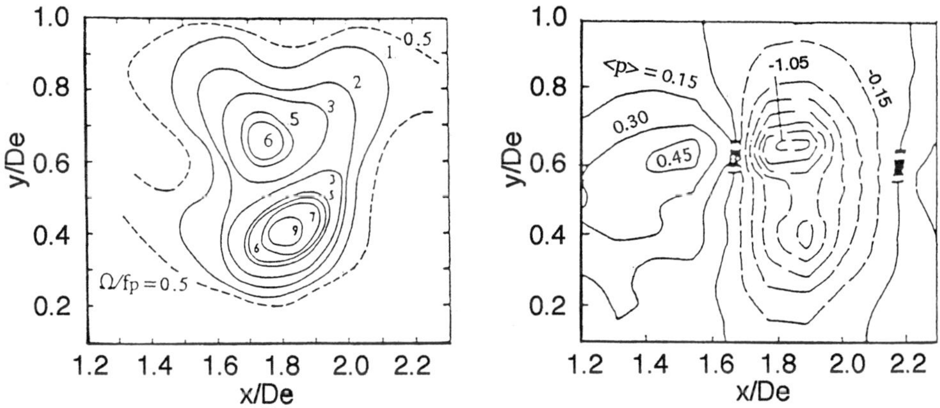

Fig. 7 Phase-average vorticity and pressure contours; Ω: vorticity; fp: excitation frequency; <p>: phase-average fluctuating pressure nondimensionalized by $\rho Ue^2/2$.

The response of the pressure probe to the vortical structure was also calibrated by measuring pressure- and velocity-fluctuations in a turbulent vortex ring generated by a single pulse jet. The fluctuation signals are shown in Fig.6, and the result shows that the pressure probe measures reasonably the pressure fluctuation in the vortical structure.

This type of pressure probe was also used to educe the vortical structures of a circular jet excited at the pairing mode [5], and the results were compared with those of the vorticity measurements [12]. An example of the comparison is shown in Fig.7, where the

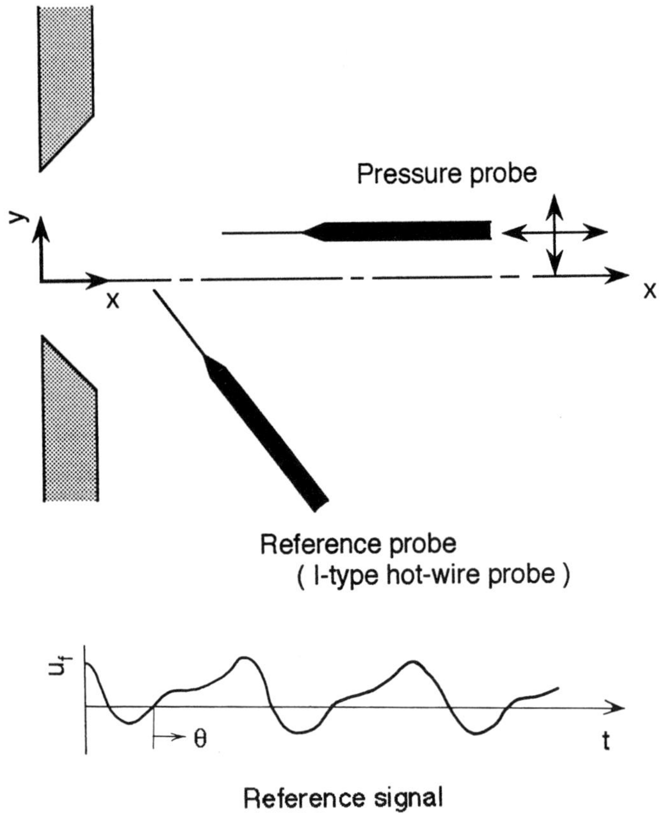

Fig.8 Arrangement of the probes for the phase-average measurement.

phase-average pressure contour is compared with the phase-average vorticity contour at nearly same phase. The negative pressure region corresponds to the vorticity concentrated region, and the pressure minima locate at the vorticity maxima of vorticity-concentrated region. The result suggests that the pressure measurements are very useful to detect the vortical structures in turbulent shear flows.

The arrangement of the probes for the phase-average measurements of fluctuating pressure is shown in Fig. 8. The reference velocity and the fluctuating pressure were detected respectively by a single hot-wire probe fixed near the jet center at x/De=0.8-1.0 and by the pressure probe moved at about 500 locations over $0.8<x/De<3.0$ and $1.2<y/De<1.2$. The phase-averaging was carried out at $\theta=0$ (zero-cross point) and $\theta=\pi/2$ of the reference signal as shown in Fig. 8. The traversing of the pressure probe, the data acquisition and the data processing were automatically controlled by the personal computer.

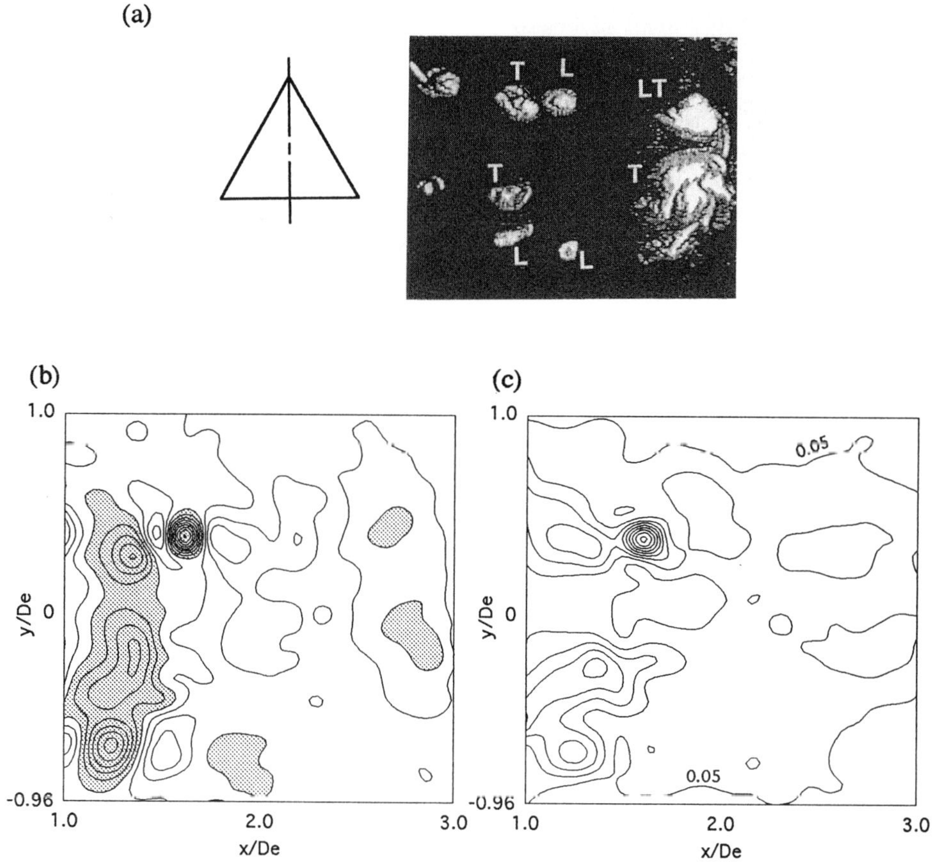

Fig.9 Triangular jet ($\theta = \pi/2$): (a) visualization; (b) <p> contour (dotted zone: <p> < -0.08; Δ<p> = 0.08); (c) <p'> contours (Δ<p'> = 0.05).

3. Results and discussions

Figure 9 shows the results of triangular jet: the visualized result, the contour of phase-average fluctuating pressure <p> and the contour of incoherent fluctuating-pressure intensity <p'>. The values of <p> and <p'> are nondimensionalized by the dynamic pressure $\rho Ue^2/2$ at the jet exit. The visualized result was obtained in the water jet experiments with Re=1.5×10^3 [8]. The negative regions (<p> <-0.08) in the <p> contour, shown by dotted zones, reveal characteristics of the vortical structures. The vortical structures shown in the <p> contour agree well with those in the visualized result: the merging of leading (L) and trailing (T) vortices on the upper side, and the passing-through of trailing vortex into leading vortex on the lower side. The <p'> contour

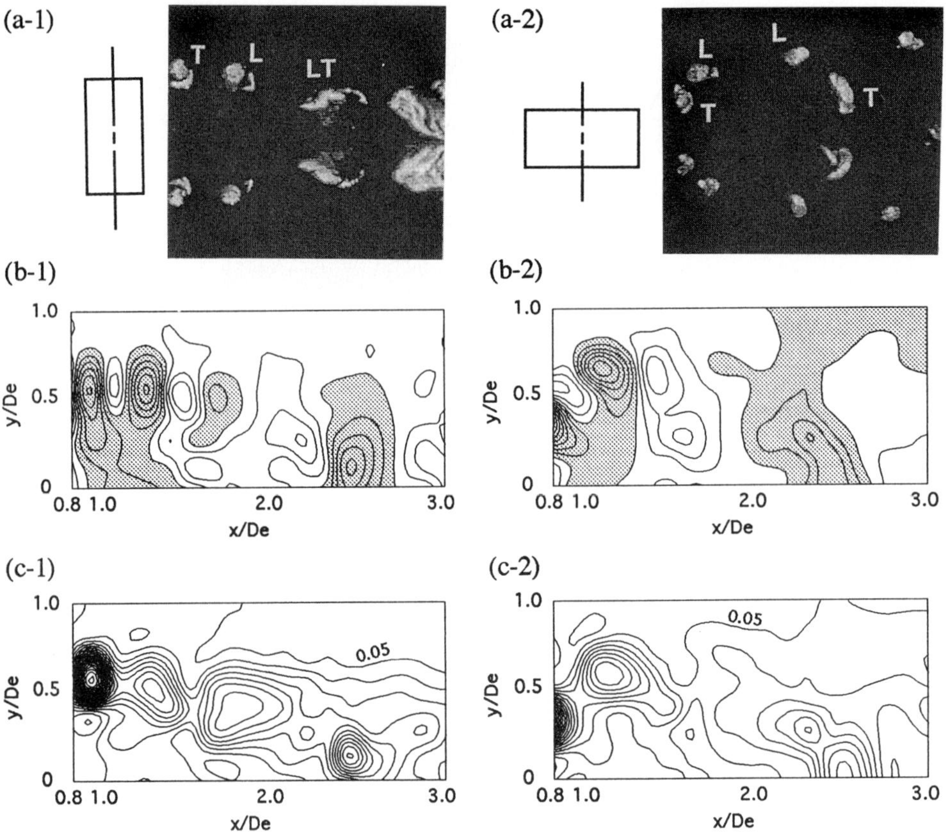

Fig.10 Rectangular jet (AR=2): (a-1) visualization in the major axis plane; (a-2) visualization in the minor axis plane; (b-1) <p> contour in the major axis plane (dotted zone: <p> < -0.08; Δ<p> = 0.08); (b-2) <p> contour in the minor axis plane; (c-1) <p'> contour in major axis plane (Δ<p'> = 0.025); (c-2) <p'> contour in the minor axis plane.

suggests that the maxima of <p'> locate at the centers of the vortical structures.

The results of rectangular jet (AR=2) are shown in Fig. 10. The vortical structures shown in the <p> contour on the major axis plane agree well with the visualized result which shows the merging of leading and trailing vortices. On the minor axis plane, the <p> contour shows the passing-through of trailing vortex into leading vortex, as suggested in the visualized result.

The vortical structures of the rectangular jet (AR=4) are very complicated as shown in Fig. 11, which was predicted by the visualization experiments [8]. Thus the phase averaging was carried out at two phases ($\theta=0$, $\pi/2$) to detect the sequence of the complicated vortical structures. Figure 12 shows <p> and <p'> contours at the two phases on the major axis plane together with the visualized results. The visualized result

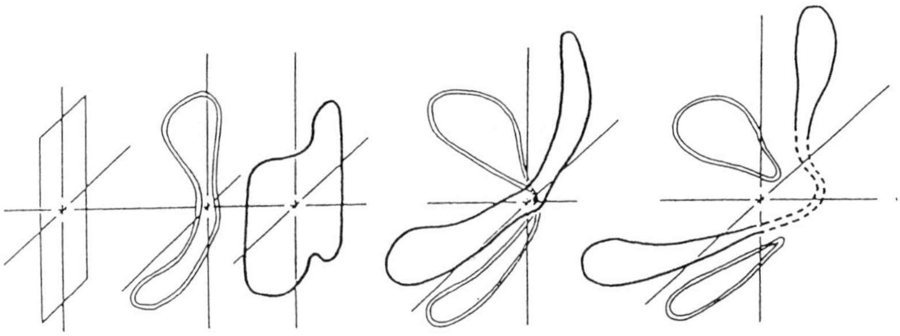

Fig.11 Interaction of vortical structures in rectangular jet (AR=4)

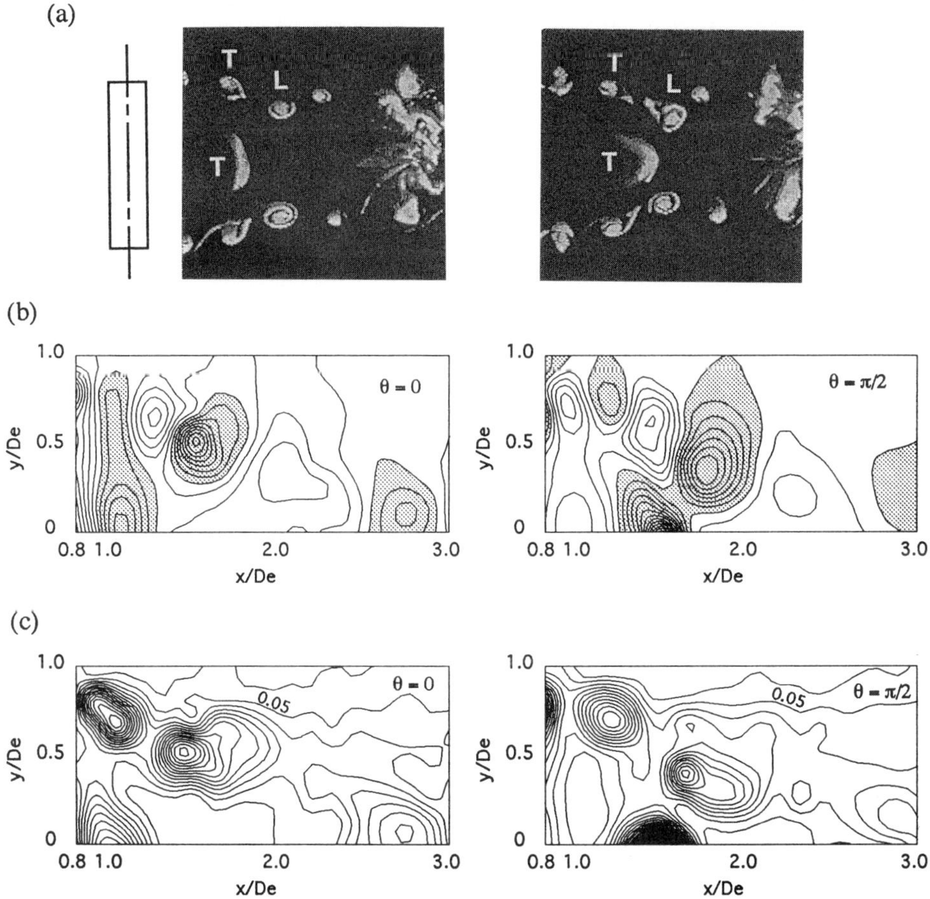

Fig.12 Major axis plane of rectangular jet (AR=4): (a) visualization; (b) <p> contour (dotted zone: <p> < -0.08; Δ<p> = 0.08); (c) <p'> contour (Δ<p'> = 0.025).

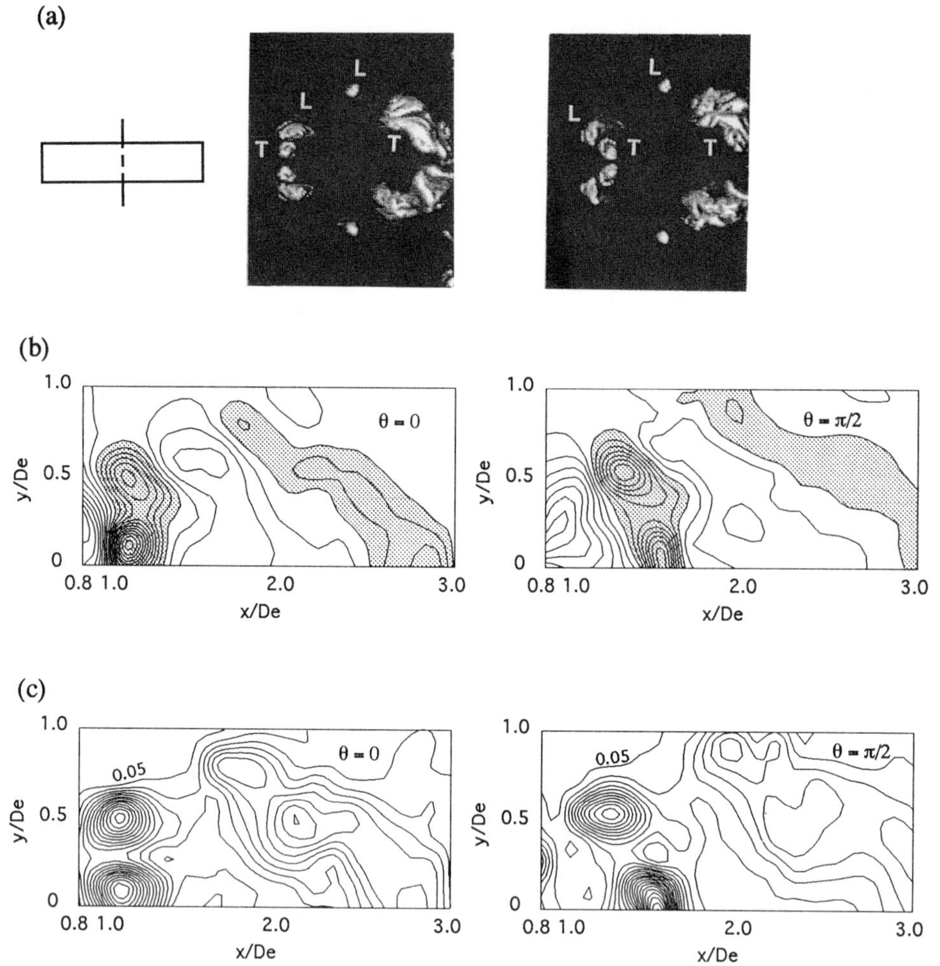

Fig.13 Minor axis plane of rectangular jet (AR=4): (a) visualization; (b) <p> contour (dotted zone: <p> < -0.08; Δ<p> = 0.08); (c) <p'> contour (Δ<p'> = 0.025).

illustrates how the vortical structures on the major axis plane change in the streamwise direction. The vortical structures in the <p> contour correspond to those in the visualization experiments. In Fig. 12(a), we can see clear dye lumps near the leading vortices. However there is no corresponding vortical structures in Fig. 12(b),(c). It suggests that the dye lumps are the remains of dye integrated by the upstream vortical structures and mislead to incorrect vortical structures. The results on the minor axis plane in Fig. 13 show that the vortical structures in the visualized result and in the <p> contour agree very well.

Table 1 Flow rates (Q/Qc at x/De=5).

Orifice shape	Unexcited jet	Excited jet
Circle	1.00	1.17
Triangle	1.07	1.21
Rectangle (AR=2)	1.10	1.40
Rectangle (AR=4)	0.99	1.35

To discuss entrainment and mixing in the noncircular jets, the flow rates of the jets were calculated by the equation

$$Q=\int_{A_{0.2}} U\, dA$$

where $A_{0.2}$ is defined as the jet cross-sectional area inside of the velocity ratio of 0.2. The flow rates at x/De=5 are shown in Table 1, where Q is nondimensionalized by the flow rate of an unexcited circular jet Qc. The flow rates are significantly increased in the excited rectangular jets, suggesting that the intense stretching and the splitting of noncircular vortical structures are very effective to enhance entrainment and mixing.

4. Conclusions

The present study is concluded as follows.

(1) The contours of phase-average fluctuating pressure reveal the complicated vortical structures in noncircular jets.

(2) The incoherent fluctuating pressure intensity is maximum at the center of vortical structure.

(3) The direct measurements of the pressure fluctuations are very effective to educe vortical structures in turbulent shear flows. Especially, since the pressure is scalar, the pressure measurements are expected to detect three-dimensional vortical structures in turbulent shear flows.

(4) The interactions of vortical structures in the noncircular jets are effective to enhance mixing.

References

[1] Bradshaw, P., Phys. Fluids 24 (1981), 777.

[2] Gyr, A., Structure of turbulence and drag reduction, IUTAM Symposium in Zurich (1989), Springer-Verlag, 50.

[3] Powell, A., J. Acoust. Soc. America 33 (1964), 177.
[4] Toyoda, K. and Shirahama, Y., Proc. 2nd Asian Congr. of Fluid Mech. (1983), 555.
[5] Toyoda, K. and Hussain, F., Proc. 5th Asian Congr. of Fluid Mech. Vol. 1 (1992), 587.
[6] Ho, C. M. and Gutmark, E., J. Fluid Mech., 179 (1987), 383.
[7] Hussain, F. and Husain, H.S., J. Fluid Mech., 208 (1989), 257.
[8] Toyoda, K. and Hussain, F., Proc. 4th Asian Congr. of Fluid Mech. (1989), 117.
[9] George, W. K. et al., J. Fluid Mech., 148 (1984), 155.
[10] Maekawa, T. et al., Proc. 15th Japan Nat. Congr. Appl. Mech. (1965), 159.
[11] Fuchs, H. V., J. Sound Vibr., 22-3 (1972), 361.
[12] Hussain, A.K.M.F. and Zaman, K.B.M.Q., J. Fluid Mech., 101-3 (1980), 493.

A STUDY OF FLOW PROPERTIES NEAR CRITICAL POINTS

Y. Zhou and R. A. Antonia

Department of Mechanical Engineering
University of Newcastle, N.S.W., 2308, Australia

Abstract

By applying the phase-plane technique to velocity data in the near-wake of a circular cylinder, three types of critical points are identified. Foci and saddle points occur most frequently, but a significant number of nodes is also found. Flow topology and properties associated with these points are examined in some detail. While foci and saddle points are associated with maxima of local vorticity and strain rate respectively, nodes are associated with a strong local divergence, indicating significant local three-dimensionality. The interrelationship between critical points is also discussed.

1 Introduction

When complicated three-dimensional flow patterns are viewed in an appropriate frame of reference (e.g. relative to an observer translating at a suitable convection velocity), they reveal a number of critical points. These are points where the velocity components are zero and the streamline slope is therefore indeterminate. In their review, Perry and Chong (1987) showed how the identification of these points can aid in describing and understanding flow patterns. They also provided examples of how critical point concepts can be applied to the study of flow patterns. Critical point theory has been extended (Chong et al., 1988) to provide a classification of three-dimensional flow patterns. Three-dimensional critical point methodology has been applied to direct numerical simulation data in various flows (Hunt et al., 1988; Chen et al., 1989,1990; Soria and Cantwell, 1992). The approach focuses on the invariants of the velocity gradient tensor.

With experimental data, a more restrictive approach is necessary since only incomplete information of the velocity gradient tensor is available. There is nevertheless interesting information that can be obtained in experiments, such as the one described here, where orthogonal arrays of X-probes provide instantaneous velocity data in two orthogonal planes ($u-v$ data in the (x,y) plane, $u-w$ data in the (x,z) plane). Such data allow critical points to be identified in both planes, thus permitting the interrelationship between these points to be examined with the possibility of gaining some insight into the three-dimensionality of the flow.

J. P. Bonnet and M. N. Glauser (eds.), Eddy Structure Identification in Free Turbulent Shear Flows, 137–149.

The present approach differs from previous experimental work in that the critical points are identified rigorously by application of the two-dimensional phase-plane technique on the computer. The main aim of this paper is to quantify flow properties in the vicinity of critical points and investigate the spatial relationship between different types of critical points which are identified in the two planes. The identification of critical points is essentially one way of detecting the flow organisation; a comparison with other detection methods has been presented in another paper (Zhou and Antonia, 1993a).

The flow considered here is the turbulent near-wake of a circular cylinder. It was chosen for several reasons. Firstly it has a high degree of organisation, and is therefore attractive for the purpose of clarifying the contribution of the organised motion to turbulent transfer processes. Secondly, a significant body of data is already available on the organisation of this flow. Additionally, the convection velocity of vortices for this particular flow has been estimated by Zhou and Antonia (1992); this knowledge is important in the present context since the flow field is viewed in a frame of reference translating with the vortices. Significant information has already been obtained on the organised motion of the flow (e.g. Cantwell and Coles, 1983; Kiya and Matsumura, 1985; Hussain and Hayakawa, 1987; Zhou and Antonia, 1993b). Nevertheless, our knowledge of this flow, such as the quantification of flow kinematics in regions surrounding critical points, is far from complete; this provided the main incentive for the present work.

2 Experimental Details

Experiments were carried out in an open return low turbulence wind tunnel with a 2.4 m long working section (0.35 × 0.35 m). The bottom wall was tilted to obtain a zero streamwise pressure gradient. A circular brass cylinder (d = 12.5 mm) was installed in the mid-plane and spanned the full width of the working section, 20 cm from the exit plane of the contraction. This resulted in a blockage of about 3.6% and an aspect ratio of 28. Measurements were made at a constant free stream velocity ($U_0 = 6.7$ m/s) at a distance of 20 diameters behind the cylinder. The corresponding Reynolds number Re ($\equiv U_0 d/\nu$) was 5600. In the free stream the longitudinal and lateral turbulence intensities were about 0.05% and 0.08% respectively.

Using two orthogonal arrays, each comprising eight X-wires (Figure 1), velocity fluctuations u, v in the (x, y) plane and u, w in the (x, z) plane were simultaneously obtained. The arrays were attached to separate traversing mechanisms and could be moved independently of each other. The eight X-wires in the (x, y) plane were fixed with the second X-wire (from the bottom) at the centreline so that they covered the transverse extent of vortices shed from the upper side of the cylinder, while the eight X-wires in the (x, z) plane could be displaced in the y direction and lay at $y/d \simeq 0.7$. The nominal spacing between X-wires in both planes was about 5 mm except for a relatively large gap of 9.1 mm between the fourth and fifth X-wires in the (x, z) plane. The physical blockage caused by the two arrays, cables and supports was estimated to be about 3%. Several types of measurements indicated that the interference to the flow due to the two arrays was negligible (Zhou and Antonia, 1993a).

The Wollaston (Pt-10% Rh) wires (dia. = 5 μm, length $\simeq$ 1 mm) were operated with constant temperature circuits. Signals from the circuits were offset, low

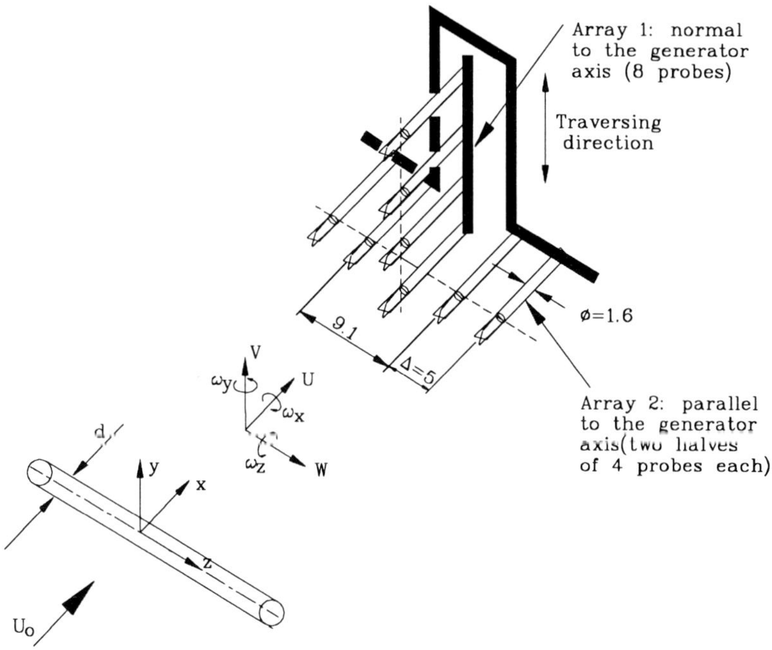

Figure 1: Experimental arrangement.

pass filtered (cut-off frequency = 1.7 kHz), amplified and then digitised using two 16 channel, 12 bit RC A/D boards respectively into two personal computers (NEC 386) at a sampling frequency of 3.5 kHz per channel. Data acquisition by the two computers was synchronised using a common external trigger pulse (the configuration is shown in Krogstad et al., 1992). The wires were calibrated for velocity and yaw and continuously checked for drift. Using calibration data, signals proportional to u, v and w, together with the local mean velocities $\overline{U}$, $\overline{V}$ ($\simeq 0$), $\overline{W}$ ($\simeq 0$), were formed and stored on digital tapes. The duration of the digital records was about 38 sec. Subsequent data processing was carried out on a VAX 8550 computer.

3 Determination of Critical Points

The classification of critical points with the aid of a $p-q$ chart is relatively well known (e.g. Kaplan, 1958; Perry and Chong, 1987). Although the following discussion is for the (x, y) plane, it is equally applicable to the (x, z) plane. With the origin of the co-ordinate system located at the critical point ($x_c = y_c = 0$), the velocity components U and V close to the origin can be approximated by first order Taylor series expansions

$$\begin{bmatrix} U \\ V \end{bmatrix} = \begin{bmatrix} a_{11} & a_{12} \\ a_{21} & a_{22} \end{bmatrix} \begin{bmatrix} x \\ y \end{bmatrix} = \underset{\sim}{A} \begin{bmatrix} x \\ y \end{bmatrix} \tag{1}$$

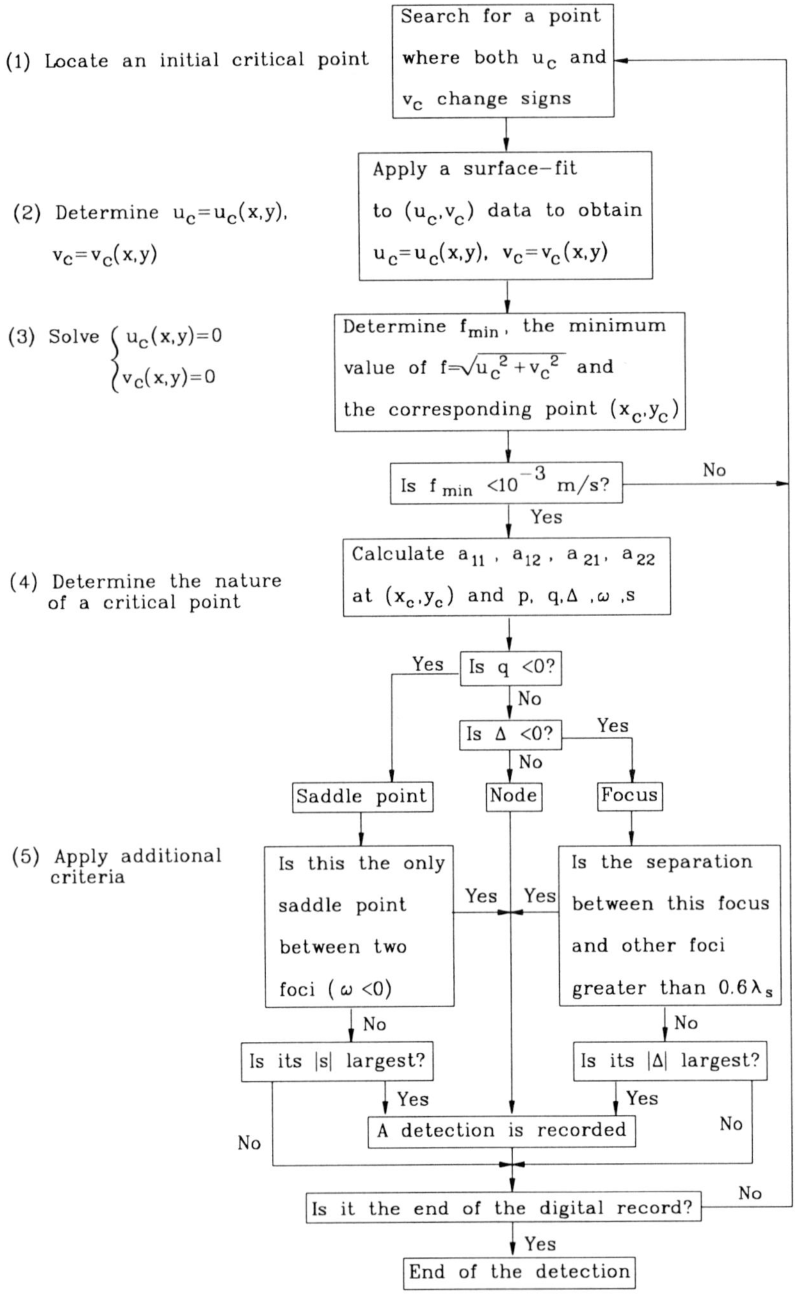

Figure 2: Flow chart for the detection scheme.

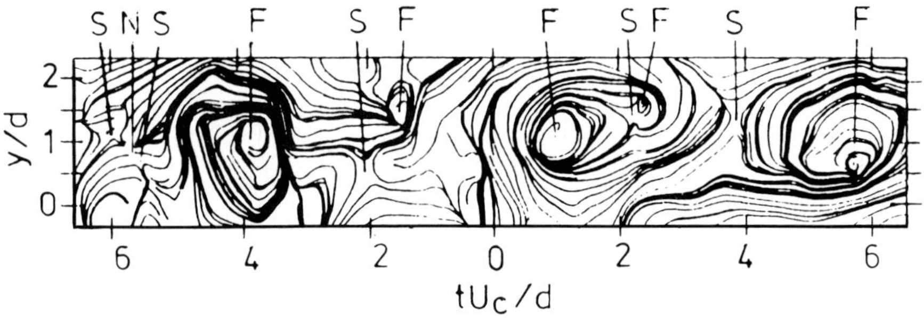

Figure 3: Instantaneous sectional streamlines in the (x, y) plane, ($U_c = 0.87U_0$, $t = 0$ is arbitrary). F denotes a focus, S a saddle point and N a node.

The eigenvalues which correspond to (1) are solutions of the characteristic equation

$$\lambda^2 - p\lambda + q = 0 \tag{2}$$

where $p = trace\,\underset{\sim}{A} = (a_{11} + a_{22})$ and $q = det\,\underset{\sim}{A} = (a_{11}a_{22} - a_{21}a_{12})$. The former may be referred to as the velocity divergence in the (x, y) plane. The three critical points of interest here are the saddle point ($q < 0$), node ($q > 0$, $\Delta < 0$) and focus ($\Delta < 0$), where Δ is the discriminant ($p^2 - 4q$) of Eq. (2). Note that the present convention is similar to that used in Kaplan (1958) and differs from Perry's (1984) in that the sign of p is inverted. In the present convention, streamlines spiral into or converge towards foci and nodes when p is negative; they spiral out of or emanate from these critical points when p is positive.

In a 2-D frame of reference which translates with a convection velocity U_c, the velocity components are now given by

$$\left.\begin{array}{lcl} u_c & = & U - U_c \\ v_c & = & V \end{array}\right\} . \tag{3}$$

The instantaneous signals U and V were formed by adding the local mean velocity values $\overline{U}$ and $\overline{V}$ ($\simeq 0$) to the digital time series of u and v. The convection velocity U_c ($= 0.87U_0$) along the vortex path was used.

The size of the experimental data grid is given by $\Delta x = 1.7$ mm and $\Delta y = 5$ mm (nominal spacing between X-wires). The value of Δx was identified, using Taylor's hypothesis (see Zaman and Hussain, 1981, for justification), with the product $(-\Delta t U_c)$, where Δt is the time interval between samples ($\equiv 3500^{-1}$ s). In view of the disparity in the magnitude of Δx and Δy, data were added at two equispaced y locations between each pair of existing rows of data. These extra data were obtained by interpolation, based on least-square cubic spline fits to the original data.

The detection procedure used to detect the three types of critical points is summarised in Figure 2. More details of this scheme are given in Zhou and Antonia (1993a). An inspection of instantaneous sectional streamlines (Figure 3) indicates that some critical points, e.g. foci at $(tU_c/d,\ y/d) = (-3.8,\ 1.0)$, $(1.0,\ 1.1)$ and saddle points at $(-2.1,\ 0.6)$, $(3.8,\ 1.4)$, appear to be quasi-periodic events. Others, e.g.

foci at $(tU_c/d,\ y/d) = (-1.5,\ 1.5),\ (2.4,\ 1.6)$ and a saddle point at $(2.2,\ 1.4)$, occur randomly and the inclusion of these points in detections would certainly smear conditional averages. Additional criteria were therefore introduced to exclude the latter events (see Figure 2). For foci (with the same sign of vorticity $\omega \equiv a_{21} - a_{12}$), the minimum detection wavelength λ_d was applied. Initially, λ_d was assumed to be equal to the average shedding wavelength λ_s but, after inspecting instantaneous sectional streamlines, λ_d was reduced to $0.6\lambda_s$. When several foci occurred within this distance, only the one associated with the largest magnitude of Δ (negative) was chosen. This choice was based on the observation that quasi-periodic foci usually have a larger $|\Delta|$ than randomly occurring ones. In fact, a focus with a large $|\Delta|$ is generally associated with a large $|\omega|$ (Zhou and Antonia, 1993a). For saddle point, only one detection with the largest magnitude of the strain rate $s \equiv a_{21} + a_{12}$ between two foci was retained. Approximately two-thirds of the number of original focus or saddle point detections survived these additional criteria. No additional condition was used for nodes since their occurrence is somewhat random.

Once critical points are detected, the conditional average of an instantaneous quantity F is given by

$$<F>_\tau = \frac{1}{N} \sum_{m=1}^{N} F_{j_m+\tau} \tag{4}$$

where τ represents time (in samples, positive or negative) relative to the detection points j_m and N is the total number of detections.

4 General Properties of Critical Points

Since the location of a critical point ($u_c = v_c = 0$) is clearly affected by the choice of the convection velocity [see Eq. (3)], it was important to assess the sensitivity of this location to the variation of U_c. At $x/d = 20$, U_c increases from about $0.85U_0$ at the wake centreline to $0.92U_0$ near the edge of the wake (Zhou and Antonia, 1992). The effect of this variation on location of the critical points was found to be quite small (Zhou and Antonia, 1993a). The effect was further minimised by using the value of U_c ($= 0.87U_0$) along the vortex path. Unless otherwise stated, results in both (x, y) and (x, z) planes are only presented for critical points associated with negative vorticity since critical points associated with positive vorticity exhibit essentially the same features.

The total numbers of critical points which remained following the application of the additional criteria are shown in Table 1. Foci and saddle points make up most of the population, although the number of nodes is not small. There are almost twice as many nodes in the (x, z) plane as in the (x, y) plane. This is probably because the eight X-wires in the (x, z) plane covered a larger active area than those in the (x, y) plane. [Note that a few X-wires in the (x, y) plane were close to the free stream and thus detected fewer critical points than those close to the centreline. This was confirmed by examining the lateral distribution of detections (not shown here)]. Not surprisingly, the normalised frequency $(N/T)d/U_0$ for detections of foci is close to the expected Strouhal number ($\simeq 0.2$, for the present Reynolds number). In the (x, y) plane, the average location of the saddle points ($y_c \simeq 1.06d$) is significantly larger than that of foci ($y_c \simeq 0.64d$); both values are in good agreement with those obtained by Hussain and Hayakawa (1987) at $x/d = 20$. Conditional averages,

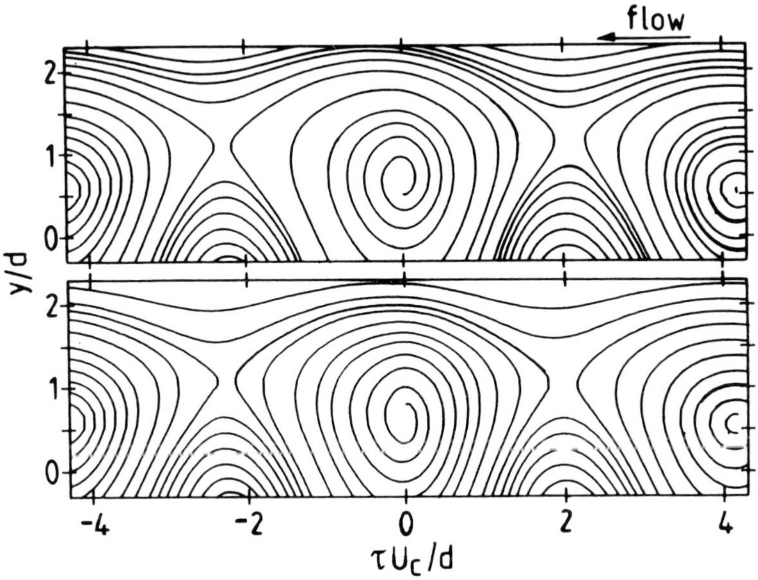

Figure 4: Conditional sectional streamlines based on foci. ($U_c = 0.87U_0$). (a) $p > 0$; (b) $p < 0$.

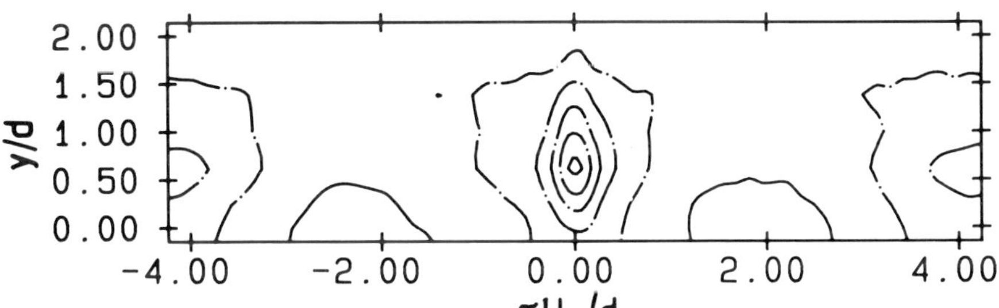

Figure 5: Spanwise vorticity contours based on detections of foci. Solid and dashed lines represent positive and negative values of $<\omega>\ d/U_0$ respectively; the values range from –0.9 to 0.1 with a step of 0.2.

defined by Eq. (4), allow various flow properties to be examined in the vicinity of critical points. A few such properties, evaluated at $\tau = 0$, are given in Table 1. As expected, the conditional vorticity is largest at foci while the conditional strain rate is greatest at saddle points. The values are larger — but only slightly — in the (x, y) plane than in the (x, z) plane. The magnitude of the divergence is largest at the nodes. The separatrix angles $\theta_{1,2} = \tan^{-1}[(\lambda_{1,2} - a_{11})/a_{12}]$ (Perry, 1984) have approximately the same magnitude ($\simeq 50°$) in both planes. For irrotational flow, the

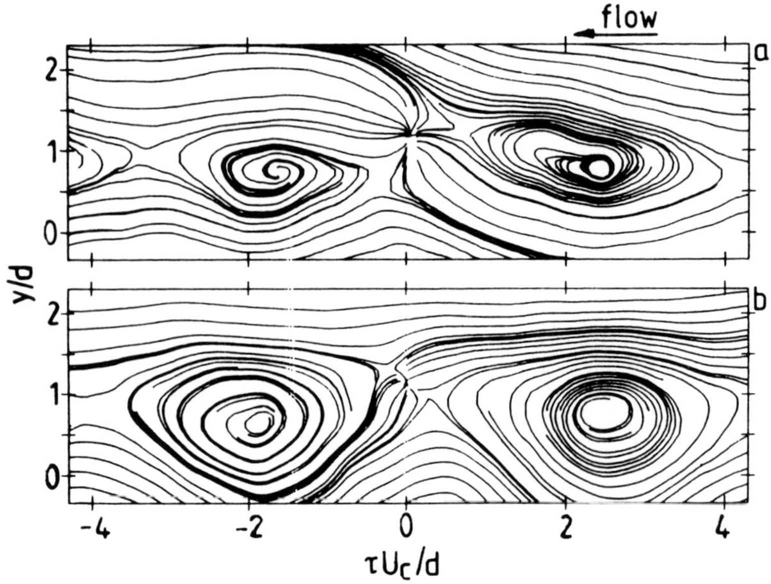

Figure 6: Conditional sectional streamlines based on node detections. (a) $p > 0$; (b) $p < 0$.

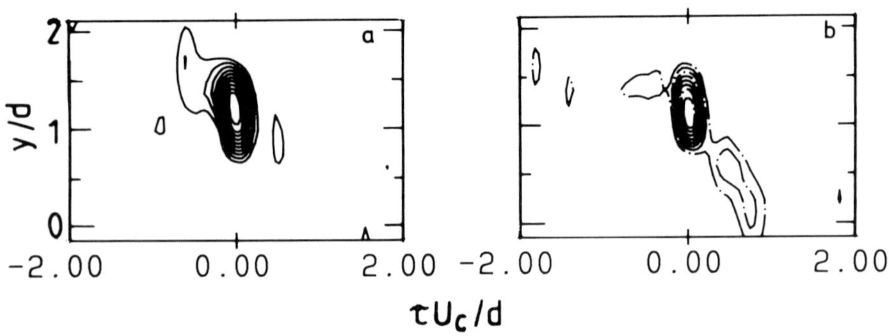

Figure 7: Contours of the velocity divergence $<p> d/U_0$ based on node detections. (a) 0.05 to 0.5; (b) –0.5 to –0.05 (step = 0.05).

converging and diverging separatrices through the saddle point should be orthogonal (Perry, 1984). The fact that the value of $(<\theta_1> - <\theta_2>)$ is slightly greater than 90° is probably associated with the existence of weak vorticity at the saddle points.

	(x,y) Plane			(x,z) Plane		
	Focus	Saddle	Node	Focus	Saddle	Node
N	3571	2774	767	3605	2862	1675
$(N/T)d/U_0$	0.18	0.13	0.04	0.18	0.13	0.08
$<y_c>/d$	0.64	1.06	0.95			
$<\|\omega\|> d/U_0$	1.78	0.25	0.31	1.44	0.25	0.29
$<\|s\|> d/U_0$	0.42	0.92	0.36	0.41	0.81	0.31
$<\|p\|> d/U_0$	0.35	0.25	0.88	0.31	0.23	0.83
$<\theta_1>$		48°			50°	
$<\theta_2>$		−51°			−50°	

Table 1 Averaged Values of Properties at Critical Points

5 Topology Based on Critical Points

Conditional sectional streamlines based on detected foci are shown in Figure 4 (the same scale is used in the x and y directions in order to avoid any distortion of the physical space). These streamlines spiral in or out depending on the sign of the velocity divergence at the focus and exhibit the familiar Kármán vortex street topology, in close similarity to those based on vorticity detections (Zhou and Antonia, 1993b). Conditional sectional streamlines spiral out of foci for $p > 0$ (Figure 4a) and into foci for $p < 0$ (Figure 4b). For $p > 0$, the continuity equation $\partial u/\partial x + \partial v/\partial y + \partial w/\partial z = 0$ requires that $\partial w/\partial z$ is negative. Conversely, for $p < 0$, $\partial w/\partial z > 0$. The former situation may be consistent with vortices undergoing axial compression while, for the latter, the vortices are probably being stretched axially (Perry and Steiner, 1987). The flow is therefore three-dimensional when p is non-zero and locally two-dimensional when p is zero. The conditional vorticity contours corresponding to Figure 4a (those corresponding to Figure 4b are identical and therefore not shown) are shown in Figure 5. Evidently foci coincide with the vorticity peak. The elongated contours near $\tau U_c/d = 0$ are due to the lateral dispersion of foci.

The conditional sectional streamlines (not shown) based on saddle points exhibit a topology similar to that of Figure 4, except for a streamwise displacement. The corresponding contours of $<s> d/U_0$ (not shown) show a maximum at the saddle point, as observed by Hussain and Hayakawa (1987).

Conditional sectional streamlines based on nodes detected in the region $0.9 \lesssim y/d \lesssim 1.3$ are shown in Figures 6a and 6b. The two saddle points and the node near $\tau = 0$ seem to be aligned with the converging separatrix when $p > 0$ (Figure 6a) and with the diverging separatrix when $p < 0$ (Figure 6b). The patterns in Figure 6 are also observed instantaneously, e.g. $tU_c/d \simeq -6.0 \sim -5.6$ and $y/d = 0.6 \sim 1.1$ in Figure 3. Conditional contours (Figure 7) of p corresponding to Figure 6 show a maximum at $\tau = 0$. As previously noted, non-zero p corresponds to a local three-dimensionality. The concentration of $<p>$ near the origin therefore indicates a strong three-dimensionality near the node. Independently of the sign of p, the contours in Figure 7 tend to be aligned with the diverging separatrix (cf. Figure 6). It has been suggested that ribs, or structures which lie in the (x,y) plane and

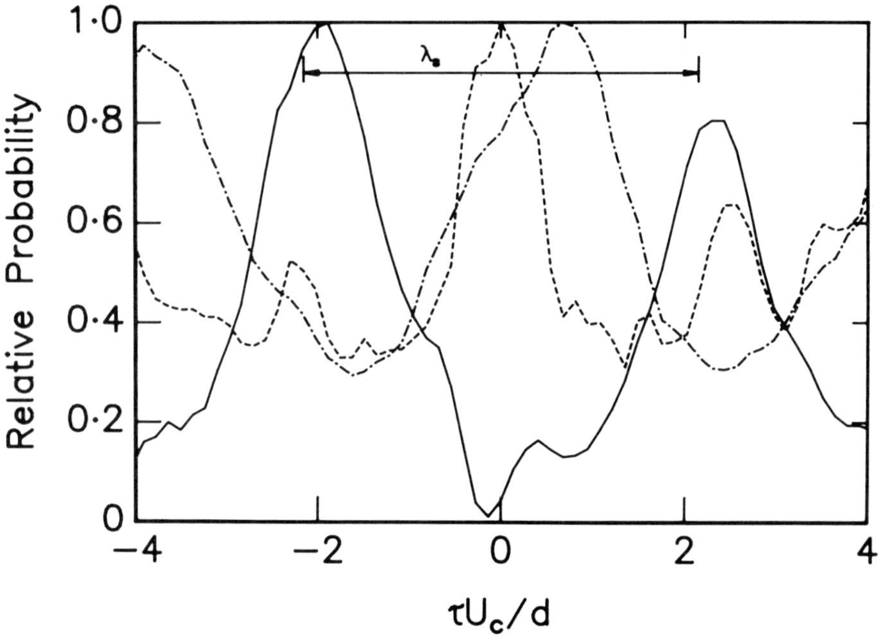

Figure 8: Relative probabilities (smoothed) of duration between critical points in the (x,y) plane, i.e. saddle points and foci (——), saddle points and nodes (----), and between saddle points in the (x,y) plane and foci in the (x,z) plane (— - —). ($\tau = 0$ is the instant at which saddle points are detected).

have vorticity components in the x and y directions, are generally oriented along the diverging separatrix (e.g. Hayakawa and Hussain, 1989). This suggestion has been strengthened by Zhou and Antonia's (1993a) observation that ribs tend to have an inclination to the x axis of about 55°, which is only marginally larger than the slope ($\simeq 50°$) of the diverging separatrix through the saddle point. This near-coincidence in orientation suggests a possible connection between nodes and ribs, which is corroborated in the following section.

6 Interrelationship Between Critical Points

Figure 8 presents the relative probabilities of the duration between critical points in the (x,y) plane and between saddle points in the (x,y) plane and foci in the (x,z) plane, where $\tau U_c/d = 0$ corresponds to the instant at which saddle points in the (x,y) plane are detected. These probabilities have been normalised to have a maximum value equal to 1. As expected, the peak probability for foci in the (x,y) plane occurs at $\tau U_c/d \simeq \pm 0.5\lambda_s$. For nodes the main peak appears near $\tau U_c/d = 0$, although two minor peaks are identifiable at $\tau U_c/d \simeq \pm 0.5\lambda_s$. Nodes are therefore more likely to occur at the same location as saddle points than at the

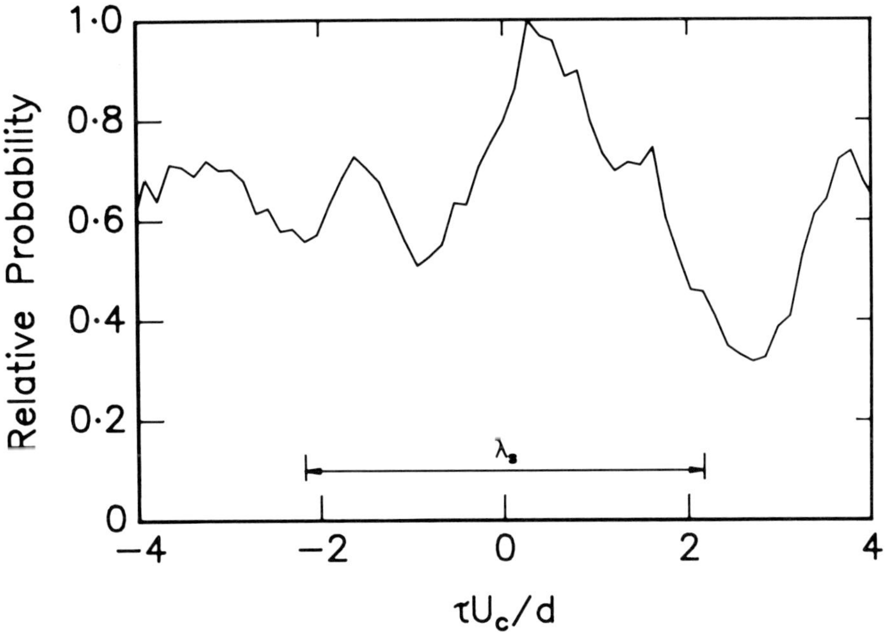

Figure 9: Relative probability (smoothed) of duration between nodes in the (x, y) plane and those foci in the (x, z) plane corresponding to the saddle region (window width $\simeq \pm 0.25\lambda_s$) in the (x, y) plane. ($\tau U_c/d = 0$ is the instant at which nodes are detected).

focus location. Foci in the (x, z) plane occur with a relatively high probability near saddle points in the (x, y) plane. The foci corresponding to the peak probability can be interpreted in terms of rib-like structures (details are given in Zhou and Antonia, 1993a where the three-dimensional aspects of the motion are discussed in greater detail) and their distribution (Figure 9) relative to nodes in the (x, y) plane exhibits a relatively strong peak near $\tau U_c/d = 0$ (i.e. the instant at which the nodes are detected), indicating a tendency for nodes to occur at approximately the same instant as these foci. If it is assumed that ribs are associated with relatively intense vortex stretching, it is not unreasonable to expect them to be accompanied by strong three-dimensional activity, thus accounting for the presence of nodes. It would appear that although the importance of foci and saddle points is relatively well established in the context of studying organised motions, the significance of nodes should not be underestimated.

7 Conclusions

The phase-plane technique was used to determine critical points from simultaneously sampled measurements in the (x, y) and (x, z) planes in the near-wake of a circular cylinder. Foci and saddle points occurred most frequently, but a significant number of nodes was detected. While foci and saddle points are associated with local maxima of vorticity and strain rate respectively, nodes correspond to the maximum of the velocity divergence, implying a local three-dimensional activity. The interrelationship between critical points indicates that nodes in the (x, y) plane are most likely to occur in the saddle regin and be associated with rib-like structures.

Acknowledgement

The support of the Australian Research Council is gratefully acknowledged.

References

CANTWELL, B. and COLES, D. : 1983. An Experimental Study of Entrainment and Transport in the Turbulent Near Wake of a Circular Cylinder, *J. Fluid Mech.*, **136**, 321-374.

CHEN, J., CANTWELL, B. and MANSOUR, N. N. : 1989. The Topology and Vorticity Dynamics of a Three-Dimensional Plane Compressible Wake, *Proc. Tenth Australasian Fluid Mechanics Conference*, Melbourne, p.5.1.

CHEN, J. H., CHONG, M. S., SORIA, J., SONDERGAARD, R., PERRY, A. E., ROGERS, M., MOSER, R. and CANTWELL, B. J. : 1990. A Study of the Topology of Dissipating Motions in Direct Numerical Simulations of time-Developing Compressible and Incompressible Mixing Layers, *CTR Proc. Summer Program 1990*, 139-141.

CHONG, M. S., PERRY, A. E. and CANTWELL, B. : 1988. A General Classification of Three-Dimensional Flow Patterns, Stanford, SUDAAR 572.

HAYAKAWA, M. and HUSSAIN, F. : 1989. Three-Dimensionality of organised Structures in a Plane Turbulent Wake, *J. Fluid Mech.*, **206**, 375-404.

HUNT, J. C. R., WRAY, A. A. and MOIN, P. : 1988. Eddies, Streams, and Convergence Zones in Turbulent Flows, *CTR Proc. Summer Program 1988*, Report CTR-S88, 193-208.

HUSSAIN, A. K. M. F. and HAYAKAWA, M. : 1987. Eduction of Large-Scale Organized Structures in a Turbulent Plane Wake, *J. Fluid Mech.*, **180**, 193-229.

KAPLAN, W. : 1958. *Ordinary Differential Equations*, Reading, Mass., Addison-Wesley Pub. Co., pp.414-423.

KIYA, M. and MATSUMURA, M. : 1985. Turbulence Structure in Intermediate Wake of a Circular Cylinder, *Bulletin of JSME*, **28**, 2617-2624.

KROGSTAD, P-Å., ANTONIA, R. A. and BROWNE, L. W. B. : 1992. Structure Investigation in a Turbulent Boundary Layer Using Orthogonal X-Wire Arrays, *Proc. Eleventh Australasian Fluid Mechanics Conf.*, Hobart, 251-254.

PERRY, A. E. : 1984. A Study of Degenerate and Nondegenerate Critical Points in Three-Dimensional Flow Fields, *Forschungsbericht*, Institut für Experimentelle Strömungsmechanik, Göttingen, p.63.

PERRY, A. E. and CHONG, M. S. : 1987. A Description of Eddying Motions and Flow Patterns Using Critical-Point Concepts, *Ann. Rev. Fluid Mech.*, **19**, 125-155.

PERRY, A. E. and STEINER, T. R. : 1987. Large-Scale Vortex Structures in Turbulent Wakes Behind Bluff Bodies. Part 1. Vortex Formation Process, *J. Fluid Mech.*, **174**, 233-270.

SORIA, J. and CANTWELL, B. J. : 1992. The Identification and Classification of Topological Structures in Free Shear Flows, *IUTAM Symposium "Eddy Structure Identification in Turbulent Shear Flows"*, Poitiers, VII.12.1-VII.12.5.

ZAMAN, K. B. M. Q. and HUSSAIN, A. K. M. F. : 1981. Taylor Hypothesis and Large-Scale Coherent Structures, *J. Fluid Mech.*, **112**, 379-396.

ZHOU, Y. and ANTONIA, R. A. : 1992. Convection Velocity Measurements in a Cylinder Wake, *Expts. in Fluids*, **13**, 63-70.

ZHOU, Y. and ANTONIA, R. A. : 1993a. Critical points in a Turbulent Near-Wake, *J. Fluid Mech.* (submitted).

ZHOU, Y. and ANTONIA, R. A. : 1993b. A Study of Turbulent Vortices in the Near-Wake of a Cylinder, *J. Fluid Mech.* (to appear).

IV. Wavelets

WAVELET ANALYSIS OF FRACTAL SIGNALS APPLICATION TO FULLY DEVELOPED TURBULENCE DATA

A. ARNEODO, E. BACRY and J.F. MUZY

Centre de Recherche Paul Pascal, Avenue Schweitzer
33600 Pessac, France

The recently developed multifractal formalism[1] has proven particularly fruitful in the characterization of singular measures arising in a variety of physical situations[2]. Notable examples include the invariant probability distribution on a strange attractor, the distribution of voltage drops across a random resistor network, the distribution of growth probabilities on the boundary of a diffusion limited aggregate and the spatial distribution of dissipative regions in a turbulent flow. This formalism[1] accounts for the statistical scaling properties of these measures through the determination of their singularity spectrum f(α) or their generalized fractal dimensions D_q. The f(α) singularity spectrum provides a rather intuitive description of a multifractal measure in terms of interwoven sets of Hausdorff dimension f(α) corresponding to singularity strength α. This spectrum was shown to be intimately related, by means of a Legendre transform, to $\tau(q) = (q-1)D_q$. Actually, the concept of multifractal originated from a general class of multiplicative cascade models introduced by Mandelbrot[3a] to account for the intermittent nature of the rate of turbulent energy dissipation. Along this line, measurements of the f(α) spectrum based on the local dissipation have been recently reported[4] ; these experimental results bring conspicuous evidence for the multifractal nature of the dissipation field. An alternative description of the fine structures in fully developed turbulence consists in working exclusively with inertial range quantitities as proposed by Parisi and Frisch[5]. Basically their multifractal approach relies on the determination of the spectrum $D(h)$ of Hölder exponents h of the velocity field from the inertial scaling properties of structure functions of variable order: $S_p(\ell) =< (\delta v_\ell)^p >\sim \ell^{\zeta_p}$, ($p$ integer > 0), where δv_ℓ is a longitudinal velocity increment over a distance ℓ. $D(h)$ is essentially the Legendre transform of the set of scaling exponents ζ_p. There are, however, some fundamental drawbacks to this method. Indeed, it generally fails to fully characterize the singularity spectrum $D(h)$; actually only the part of the spectrum corresponding to $0 < h < 1$ is amenable to the structure function method. Moreover, regular behavior may introduce drastic bias in the estimate of the ζ_p's.

Our purpose here is to elaborate on a novel strategy which will allow us to extract the whole singularity spectrum $D(h)$ directly from any experimental signal. This strategy is based on the use of a new tool introduced in signal analysis, *the wavelet transform*[6] (WT), which has proved very powerful to characterize the scaling properties of multifractal measures[7]. Beyond the reported applications of our method to various recursive,

J. P. Bonnet and M. N. Glauser (eds.), Eddy Structure Identification in Free Turbulent Shear Flows, 153–158.

Brownian and turbulent signals, our ambition is actually to establish the fundations for a thermodynamical formalism for singular signals[8–10].

The WT is a space-scale analysis[6] which consists in expanding signals in terms of *wavelets* which are constructed from one single function, the analyzing wavelet g, by means of dilations and translations. The WT of a signal $s(x)$ is generally a complex valued function which is defined as:

$$T_g(a, x_0) = \frac{1}{a}\int_{-\infty}^{+\infty} s(x) g^\star\left(\frac{x - x_0}{a}\right) dx \ , \qquad a > 0 \tag{1}$$

where $g^\star$ is the complex conjugate of g. Provided g be well localized around $x = 0$ and have a vanishing integral, no information about s is lost since this transformation is invertible for a large class of functions s. The WT can be used as a mathematical microscope[7] to analyze the local regularity of functions; g charaterizes the optics, while x_0 is the position analyzed and $1/a$ the magnification. In fact, if the signal $s(x)$ has, at the point x_0, a local scaling (Hölder) exponent $h(x_0)$ in the sense that, in the limit $x \to x_0$, $|s(x) - P_n(x_0)| \sim |x - x_0|^{h(x_0)}$ ($P_n(x_0)$ is a polynomial of degree n and $n < h(x_0) < n + 1$), then $T_g(a, x_0) \sim a^{h(x_0)}$ for $a \to 0$, provided the first $(n + 1)$ moments of g are zero[11]. Thus one can extract, from a log-log plot of the WT amplitude versus the scale a, the local scaling exponent $h(x_0)$ for any fixed position x_0. The situation is somewhat more intricate when investigating fractal signals[2]. The characteristic feature of these singular signals is the existence of a hierarchical distribution of singularities. Locally the Hölder exponent $h(x_0)$ is then governed by the singularities which accumulate at x_0. This results in unavoidable oscillations around the expected power-law behavior of the WT amplitude. The exact determination of h from log-log plots on a finite range of scales is therefore somewhat uncertain. Moreover, there exist fundamental limitations (which are not intrinsic to the WT technique) to the measure of Hölder exponents from local scaling behavior in a finite range of scales. Therefore the determination of statistical quantities like the singularity spectrum $D(h)$ requires a method which is more feasible and more appropriate than a systematic investigation of WT local scaling behavior as previously experienced in Refs. 12b and 13.

The classical multifractal description[1] in terms of thermodynamic quantities supposes, more or less explicitly, the existence of an underlying multiplicative process. Previous applications[7] of the wavelet microscope to multifractal measures lying on Cantor sets have demonstrated its fascinating ability to reveal the construction process (renormalization operation) of recursive singular measures. Since the analyzing wavelet g can be chosen to be orthogonal to polynomial behavior of arbitrarily high order, the WT capability to capture the intricate singularity arrangement is likely to extend to fractal signals. As proved by Mallat and Hwang[14], the WT local maxima detect all the singularities of a large class of signals. Thus the representation in the (a, x) half-plane of the local maxima of $| T_g |$ is likely to contain all the information on the hierachical distribution of singularities in the signal. As illustrated in Fig. 2b, the WT modulus maxima are disposed on connected curves called maxima lines. Le us define $\mathcal{L}(a_0)$ the set of all the maxima lines l that exist for $a \leq a_0$. Our method[8–10] to computing the singularity spectrum of a fractal signal consists in taking advantage of the space-scale partitioning given by the maxima line skeleton to define a partition function which scales, in the limit $a \to 0$, in

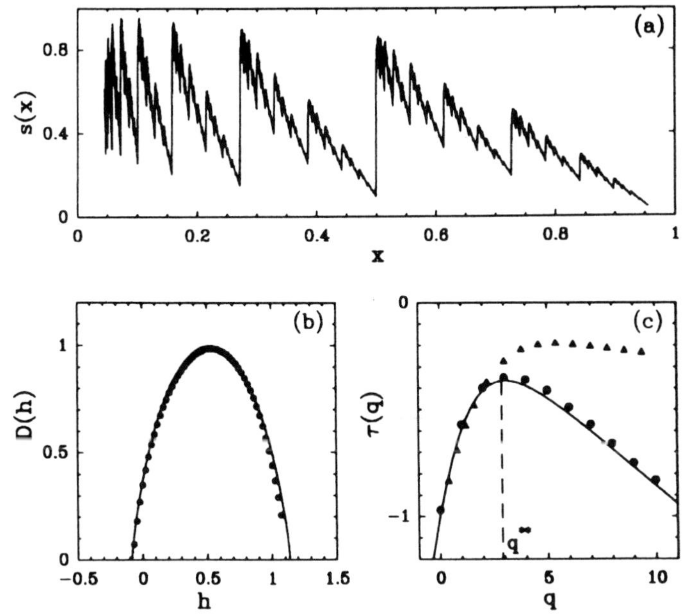

Figure 1: WTMM and SF analysis of a signal that posseses singularities of negative Hölder exponent. (a) The multifractal recursive signal. (b) $D(h)$ vs h. (c) $\tau(q)$ vs q. In (b) and (c) : (•) WTMM method using the Mexican hat as analyzing wavelet; (▲) SF method; (—) theoretical spectra.

the following way:

$$\mathcal{Z}(a,q) = \sum_{l \in \mathcal{L}(a)} \left(\sup_{(x,a') \in l} |T_g(a',x)| \right)^q \sim a^{\tau(q)}, \tag{2}$$

where $q \in I\!R$. By using the behavior of the wavelet coefficient along the maxima lines ($|T_g(a,x_0)| \sim a^{h(x_0)}$), one can show, in the limit $a \to 0$ (steepest descent method) that the $D(h)$ singularity spectrum can be computed from the exponents $\tau(q)$ via a Legendre transform:

$$D(h) = \min_q (qh - \tau(q)). \tag{3}$$

It is tempting to relate the exponents $\tau(q)$, defined from the wavelet transform modulus maxima (WTMM) method, to the exponents ζ_p of the structure functions (SF). A simple comparison[10] gives immediately $\tau(q) = \zeta_q - 1$; but as discussed in Ref. 10, this relationship does not hold for every value of q. In fact, it can be shown that $\tau(q)$ is a more general spectrum than the ζ_p's, in the sense that $\tau(q)$ is the exact Legendre transform of the $D(h)$ singularity spectrum in most situations[9,10].

In Fig. 1, we show the results of both the WTMM and the SF analysis of a deterministic multifractal recursive signal that posseses some singularities of negative Hölder exponents[10] (the jumps in Fig. 1a). The analytical $D(h)$ spectrum, represented by a solid line in Fig. 1b, actually extends below the value $h = 0$, over the range [-0.1, 1.36]. The numerical data issued from the WTMM analysis are found in remarkable agreement with

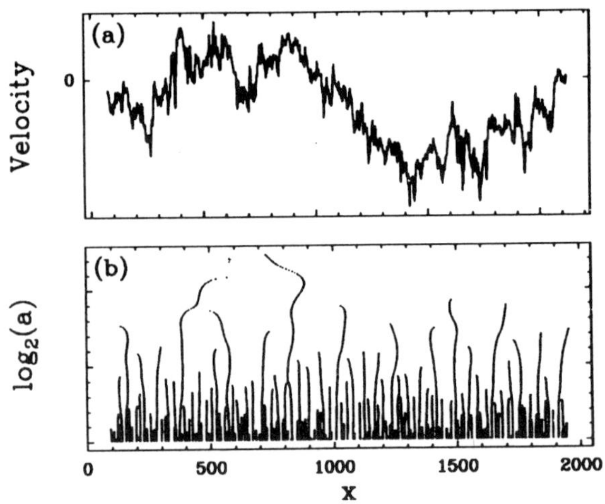

Figure 2: Continuous wavelet transform analysis of fully developed turbulence from wind-tunnel data. (a) The turbulent velocity signal over about one integral length scale. (b) WT local maxima ; the analyzing wavelet is the Mexican hat.

the theoretical spectrum. The comparison of the WTMM and the SF methods is reported in Fig 1c where the data for $\tau(q)$ obtained from both methods are compared to the analytical spectrum. The SF numerical data systematically depart from the theoretical curve for $q > q^{**}$, where q^{**} corresponds to the critical value where $\tau(q)$ starts to decrease, i.e., where negative Hölder exponents are dominating the behavior of the structure functions (from the properties of the Legendre transform, h corresponds to the derivative of $\tau(q)$ with respect to q). This example illustrates one of the deficiencies of the SF method, i.e., its unability to resolve negative Hölder exponents. Fig. 1 shows that the WTMM method, which involves smooth analyzing wavelets, does not posseses this drawback and provides a reliable quantitative characterization of the multifractality of the considered recursive signal[8–10].

In recent years, there have been several attempts to apply the wavelet analysis to fully developed turbulence data[12]. A preliminary investigation of the velocity field at inertial range scales has revealed a multifractal branching process in the wavelet representation that is clearly different from the fractal branching hierarchy observed with Gaussian processes[12a]. Moreover, dynamically significant events having strong localized gradients have been identified in the turbulent signal using local wavelet analysis[12b]; such events are not in the fractional Brownian signal. However, because of the operational limitations of the local scaling analysis, there has been thus far no reliable determination of the singularity spectrum $D(h)$ directly from the velocity signal without recourse to dissipation-type quantities. In Figs. 2 and 3, we present the results of our wavelet analysis[8] of a turbulent velocity signal recorded in the wind tunnel S1 of ONERA at Modane[15]. The Taylor scale based Reynolds number is $R_\lambda = 2720$ and the extent of the inertial range following approximately the Kolmogorov $k^{-5/3}$ law is almost three decades. The results reported here concern the analysis in the inertial range of about 100 integral lengthscales of the recorded turbulent signal (Fig. 2). The analysis of a fractional Brow-

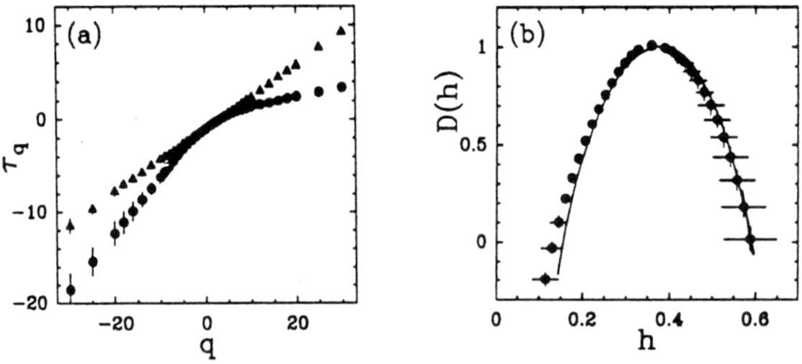

Figure 3: WTMM measurement of the singularity spectrum of fractionnal Brownian motion (▲) and tubulent velocity signal (•). (a) $\tau(q)$ vs q. (b) $D(h)$ vs h. In (b) the solid line corresponds to the average singularity spectrum obtained from dissipation field data (Ref. 4) via the Kolmogorov scaling relation.

nian signal[3b] having a $k^{-5/3}$ spectrum is shown for comparison (Fig. 3). When plotted versus q, the scaling exponents $\tau(q)$ obtained for the fractional Brownian motion remarkably fall on a line $\tau(q) = q/3 - 1$ of slope $h = 1/3$. From the Legendre transform (Eq.(3)) one gets $D(h) = 1$. Thus, as expected theoretically, we find that the Brownian signal is almost everywhere singular with a unique Hölder exponent $h \simeq 1/3$. In Fig. 3a, the $\tau(q)$ curve extracted from the WTMM method of the turbulent signal unambiguously deviates from a straigth line. The values of h obtained when varying q from - 30 to 30 range in the interval [0.10, 0.62]. The corresponding singularity spectrum $D(h)$ is shown in Fig. 3b. It displays the characteristic single-humped shape of multifractal signals. Its maximum $D(h(q = 0)) = -\tau(0) = 1.003 \pm 0.005$ does not deviate substantially from $D_F = 1$. This strongly suggests that the turbulent signal is almost everywhere singular[8]. The manifest part[3c] of $D(h)$ (> 0) is compared to a solid curve which actually corresponds to short-terms statistics data of dissipation field[4] (at lower Reynolds numbers) assuming the validity of the Kolmogorov scaling relation[16] $\epsilon_\ell \sim (\delta v_\ell)^3/\ell$ between the instantaneous velocity increments and local dissipation. Despite some systematic deviations, the fact that one cannot discriminate between these two singularity spectra, within the experimental uncertainty, can be interpreted a posteriori as an experimental verification of the Kolmogorov hypothesis. This observation can also be understood as an experimental confirmation of the universality of the multifractal singularity spectrum of fully developed turbulence[4]. However, it is clear that considerable further work is needed for reliable quantitative conclusions. In particular, long-term statistical analysis have to be carried out in order to capture more accurately the latent part[3c] ($D(h) < 0$) of the singularity spectrum including possible violent rare events corresponding to negative Hölder exponents[12b]. This analysis is currently in progress. It is likely to provide fundamental informations about the true role played by the vortex filaments (of the sort observed in numerical 3D simulations and recent experiments) in the intermittency phenomenon of the fine structures of fully developed turbulent flows[17].

This work was supported by the D.R.E.T. under contract (DRET N°89/196).

References

[1] T.C. Halsey, M.H. Jensen, L.P. Kadanoff, I. Procaccia and B.I. Shraiman, Phys. Rev. A **33**, 1141 (1986).

[2] Essays in honour of B.B. Mandelbrot, *Fractals in Physics*, edited by A. Aharony and J. Feder, Physica **38** D (1989).

[3] (a) B.B. Mandelbrot, J. Fluid Mech. **62**, 331 (1974); (b) B.B. Mandelbrot, C.R. Acad. Sci. Paris **260**, 3274 (1965); (c) B.B. Mandelbrot, Pure Appl. Geophys. **131**, 5 (1989).

[4] C. Meneveau and K.R. Sreenivasan, J. Fluid Mech. **224**, 429 (1991).

[5] G. Parisi and U. Frisch, in *Turbulence and Predictability in Geophysical Fluid Dynamics and Climate Dynamics*, edited by M. Ghil, R. Benzi and G. Parisi (North-Holland, Amsterdam, 1984) p. 84.

[6] (a) *Wavelets*, edited by J.M. Combes, A. Grossman and P. Tchamitchian (Springer-Verlag, Berlin, 1989); (b) Y. Meyer, *Ondelettes* (Herman, Paris, 1990); (c) *Les Ondelettes en 1989*, edited by P.G. Lemarié (Springer-Verlag, Berlin, 1990).

[7] (a) A. Arneodo, G. Grasseau and M. Holschneider, Phys. Rev. Lett. **61**, 2281 (1988); (b) A. Arneodo, F. Argoul, J. Elezgaray and G. Grasseau, in *Nonlinear Dynamics*, edited by G. Turchetti (World Scientific, Singapore, 1988) p. 130.

[8] J.F. Muzy, E. Bacry and A. Arneodo, Phys. Rev. Lett. **67**, 3515 (1991); (b) A. Arneodo, E. Bacry and J.F. Muzy, *Wavelets and Turbulence* (Springer, Berlin, 1992) to appear.

[9] E. Bacry, J.F. Muzy and A. Arneodo, J. Stat. Phys. **70** (1992) to appear.

[10] J.F. Muzy, E. Bacry and A. Arneodo, "Multifractal formalism for fractal signals. The structure function method versus the wavelet transform modulus maxima method", preprint (1992).

[11] S. Jaffard, C.R. Acad. Sci. Paris **308**, 79 (1989); M. Holschneider and P. Tchamitchian, in Ref. 6c, p. 102.

[12] (a) F. Argoul, A. Arneodo, G. Grasseau, Y. Gagne, E.J. Hopfinger and U. Frisch, Nature **338**, 51 (1989); (b) E. Bacry, A. Arneodo, U. Frisch, Y. Gagne and E.J. Hopfinger, in *Turbulence and Coherent Structures*, edited by O. Metais and M. Lesieur (Kluwer, Dordrecht, 1991) p. 203.

[13] M. Vergassola and U. Frisch, Physica **54** D, 58 (1991).

[14] S. Mallat and W.L. Hwang, I.E.E.E. Trans. on Inf. Th. **38**, 617 (1992).

[15] Y. Gagne, Thesis, University of Grenoble (1987); B. Castaing, Y. Gagne and E.J. Hopfinger, Physica **46** D, 177 (1990).

[16] A.N. Kolmogorov, J. Fluid Mech. **13**, 82 (1962).

[17] U. Frisch and S. Orszag, "Turbulence: Challenges for Theory and Experiments", Physics Today (1990), p. 24.

EXPERIMENTAL ANALYSIS OF COHERENT MOTIONS IN A TURBULENT BOUNDARY LAYER

A. Benaïssa, F. Anselmet, F. Moret-Bailly, J. Liandrat
I.M.S.T. 12, Avenue Général Leclerc, 13003 Marseille, France.

Abstract

This paper reports results obtained from the combination and the comparison of three methods of coherent motion investigation using temperature as a passive scalar. These procedures are used for the analysis of experimental data obtained in a turbulent boundary layer from multi-point measurements. Two methods, commonly used for the eduction of organized structures in various shear flows, Vita and Wag, were first tested and thoroughly compared. As a remedy to the need for these detection schemes of a priori knowledge of some of the characteristic features of the events to be analyzed, the wavelet transform is then applied to these temperature signals. This provides local multi-scale decompositions which allow the determination of new characteristics of these structures, such as the estimation of the spectral distributions conditioned on their occurence.

Introduction

The major role played by organized coherent structures in the dynamics of various turbulent shear flows has now been widely documented (e.g. Hussain, 1986; Robinson, 1991). These motions are particularly important for the development of turbulent boundary layers but, in spite of numerous experimental and numerical studies devoted to the subject, the classification of various types of structures inferred from different data analyses, and the precise understanding of their mutual interactions and of their spanwise characteristics are still requiring efforts. In particular, the inclusion of this motion contribution into turbulent engineering modelling could not so far be achieved despite the recent development of kinematic models (Zhou, 1993). However, it is now well established that these structures significantly contribute to the transport of momentum and scalars. Indeed, for instance, Antonia and Fulachier (1989) have quantitatively maped their properties in a slightly heated turbulent boundary layer using temperature as a passive marker of this flow through the eduction of spatially coherent events classified in terms of heatings and coolings. The former ones are related with fluid coming from regions close to the wall and may be associated with hairpin vortices, whereas the latter ones are related to the engulfment of cold fluid originating from the outer regions of the boundary layer.

J. P. Bonnet and M. N. Glauser (eds.), Eddy Structure Identification in Free Turbulent Shear Flows, 159–168.

This paper is reporting results extending that work. It is also based on organized structure detection from temperature measurements. Since it is well known that, in order to investigate spatially organized events, an efficient detection procedure can not be relative to one point measurements, the extensions of the general purpose Vita (Blackwelder and Kaplan, 1976) and Wag (Antonia et al., 1987) algorithms to multi-point procedures were applied to data obtained from rakes of wires or 3D arrangements of fewer probes. After a comparison of these eduction techniques - which will not be described in detail -, local spectral characteristics provided by the wavelet transform (Farge, 1993; Moret-Bailly et al., 1991) are used to obtain new information about the scale distributions associated with the typical events previously investigated, the so-called heatings and coolings.

Experimental arrangement and signal analysis procedures

The turbulent boundary layer develops on the working section floor of the low-speed wind tunnel (Fig. 1). The wall is heated from the beginning to a constant uniform temperature relative to the ambient one of 10K and wall suction can be applied through a porous plate from the downstream position X=3.05m. The mean free-stream velocity U_e is 12.6 m/s. Measurements are performed at the downstream station X'=3.7m where the boundary layer thickness is 57mm and the momentum thickness Reynolds number is 4100 when the suction rate $A=-V_w/U_e$ (where V_w is the wall suction velocity) is equal to 0.003. The properties of this flow have already been studied quite extensively. In particular, the investigation of the topology of the organized motion performed by Antonia and Fulachier (1989) in terms of coolings and heatings has shown that the stabilizing influence of suction resulted in an increase of the flow organization: in particular, the contribution of the coherent structures to the heat transport was significantly increased by suction.

For results reported here, these high gradient regions are identified in simultaneous temperature signals obtained from an array of seven 0.63µm wires, lying in the same plane normal to the mean flow direction and parallel to each other.

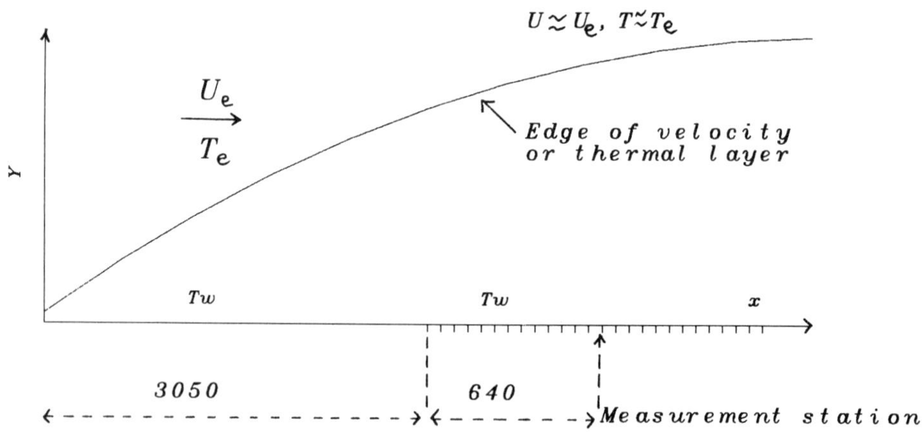

Fig. 1 Sketch of experimental arrangement.

The nominal separation between wires is 0.9mm and their active length about 0.8mm. Data presented in this paper are relative to only one position of the rake, where the wire closest to the wall is located at $y^+=6$ and that furthest from it is at $y^+=207$ ($y^+=yu_*/\nu$, with $u_*=0.67$m/s, friction velocity and ν, kinematic viscosity). Signals are digitized at a frequency of 37.5KHz/channel onto a PC computer and analyzed either on the PC or on the IMST Sun network.

The Vita technique is based on the evaluation of a normalized short-term r.m.s. value of one of the signals for a specified length of time. If this quantity oversteps a specified threshold, an event is detected. An additional slope condition allows one to discriminate between heatings - associated with a relatively sudden increase in temperature - and coolings. These latter fronts are very coherent in space and can be interpreted as the boundary between ejections and sweeps. Since their contributions to Reynolds stresses, temperature variances and heat fluxes are more important than those of heatings (Antonia and Fulachier, 1989), they are more specifically studied here. However, a few results are also reported for heatings. Spatial coherence is determined by thresholding the windowed correlations of the selected portions of signals with the corresponding adjacent temperature traces.

The Wag method is quite similar, but the algorithm is based on the computation of a local average gradient evaluated on a sliding window moving point by point through the entire signal. A detection is recorded for each continuous patch for which the average gradient exceeds a prescribed threshold when the window is centered on the first point of the patch and retains a positive value when the window is centered on every point in the patch. A sign condition is also available to discriminate between positive and negative temperature fronts, and correlations similar to those applied with the Vita scheme are used. Following a systematic study of these two procedures (not reported here), Wag was finally preferred to Vita since it was found that its sensibility to the prescribed parameters (thresholds and window sizes) is smaller as far as temperature ensemble averages are concerned. It was also found that Wag responds equally well to a large-scale event and a burst of high-frequency signal, whereas Vita filters out frequencies of period significantly shorter than the averaging window. The number of selected coolings was typically about 100 (at $y^+\approx150$) over the record duration used for the present study corresponding to 20 blocks of 4096 data (i.e. 2.2 sec.). However, two threshold levels associated with two window sizes were selected and the main results are briefly discussed in the next section. Attention is mainly paid to a typical set of data containing only 5 blocks.

As a remedy to the need for these eduction schemes of a priori knowledge of some of the characteristic features of the events to be studied, and in order to allow more refined analyses of their properties, the multiscale approach provided by the wavelet decomposition is applied to the same temperature signals. These results will be most thoroughly examined hereafter. Using a generating function $\psi(t)$, the wavelet coefficients $T_\psi(a,b)$ are computed as the scalar product of the signal data sequence with the discretized versions of dilated (factors a_m) and translated (factors b_n) functions $\psi_{a_m b_n}(t)=(1/\sqrt{a_m}).\psi((t-b_n)/a_m)$ deduced from the generating one. In the framework of this study, $\psi(t)$ is a zero mean function and figure 2a presents the two analysing functions that have been tested, the so-called Haar, H(t), and Morlet, M(t), wavelets. The former one was first used since it is providing a multi-scale extension of the Wag function $W_l(j)$ defined over a window containing l points since, for $a=l$, $T_H(l,j)=W_l(j)$. But, in order to define more precisely local properties, the use of a regular wavelet such as the latter one was required since it is known that regularity and oscillations are needed to correctly characterize signals (Meyer, 1990). The various scales studied are defined from the smallest one a_0 using a geometrical progression $a_m=A^m a_0$; here, A is equal to 1.185 and a_0 to 23 points corresponding to about 0.06 times the boundary layer thickness (or about 35 times the Kolmogorov length scale) at $y^+\approx150$. When local distributions of the wavelet coefficients are examined, these coefficients are evaluated for each of the scales and at each position of the signal (i.e. b_n corresponds to n points). On the other hand, when only the

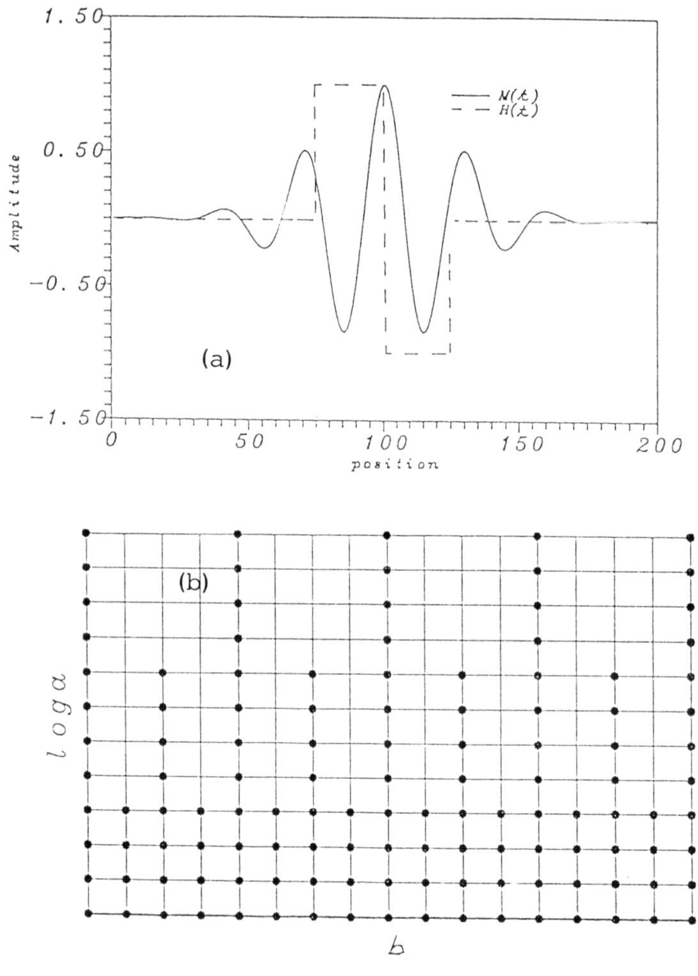

Fig. 2 Characteristic features of the wavelet functions.
a). Sketch of the Haar (H(t)) and Morlet (M(t)) generating wavelets.
b). Phase space lattice corresponding to the wavelet frame
for the evaluation of averaged properties.

distribution of averaged quantities such as local energies - which are the mean values computed, at one position, over all scales - or mean energies at one scale - which are the mean values computed, at one scale, over all positions of the signal - is investigated, the coefficients are evaluated on a coarser mesh in order to increase the calculation efficiency. This space lattice (Fig. 2b) is defined, following a scheme proposed by Daubechies (1990), such that b_n varies with the scale a_m: b_n is doubled once every fourth scale (starting with b_0 corresponding to one point for a_0). The missing coefficients ψ_{ambp} are then evaluated from the linear interpolation of the two computed adjacent ones ψ_{ambn} and ψ_{ambn-1} at the same scale a_m.

Results and discussion

Figure 3 presents an example of local scale decomposition obtained by the Haar (3b) and Morlet (3c) wavelets for a 1200 point portion of the temperature signal (3a) obtained at the position closest to the wall ($y^+=6$). The range of scales analysed extends from 23 to 7400 points (this range of scales includes 40 wavelets) so that, when computing the largest scale coefficients, signal portions not visible on the graph are taken into account. It is worth recalling that gradient regions are not detected on the same number of points for the two wavelet decompositions; thus, in order to provide a more meaningful comparison, figure 3b is truncated for the largest scales whereas figure 3c is truncated towards the smallest ones. However, both sketches reveal an almost regular succession of zones of large positive and negative coefficients extending over quite large ranges of scales, the former ones being generally coincident with the cooling events and the latter ones extending over smaller scales. However, the regions of "most intense activity" have characteristic scales of about the same amplitude. In addition (Fig. 3d), the positions of detected cooling events using Wag and two different thresholds (0.65 and 0.8 times the temperature standard deviation at that distance from the wall) together with two window sizes (45 and 75 points) are pointed with arrows and comparison is provided with the modulus of the wavelet coefficients for these two scales, determined with the Morlet algorithm. It is worth noticing that, for both scales presented here, a quite good correspondence is found between the positions of the Wag detected events and those associated with large values of the energy. However, Wag detects additional events in regions where the modulus of T is quite small. In fact, for a more quantitative comparison, and for the definition of a detection criterion based on these energy levels, one should take into account the contribution from the neighbouring points and the smoothing effect of the Morlet function on gradients due to the presence of the two side lobes. One should also try to work out a criterion using information obtained at more than one point in the flow.

From this decomposition, when a regular function such as the Morlet wavelet is used, one can compute the local scale distributions of the turbulent energy contained in the analyzed signal. These energies E(a,b) are computed from the modulus of $T_\psi(a,b)^2 / a^2$ for each point of the signal: in order to account for the fact that the neighbouring points are involved in the computation of the wavelet coefficients at one position b, it is necessary to use a dumping function such that $E(a,b) = \int [T_\psi(a,b)^2 / a^2]\,[1/a\ \chi((b'-b)/a)]\,db'$ where χ is chosen as the gaussian distribution (Moret-Bailly et al., (1991)). Then, various local spectral characteristics such as the local mean scale or the local energy C(b) can be inferred from integration among all scales. Figures 4b (Haar) and 4c (Morlet) show that, for the same portion of signal as on figure 3, the quantity C(b) presents significant increases when strong

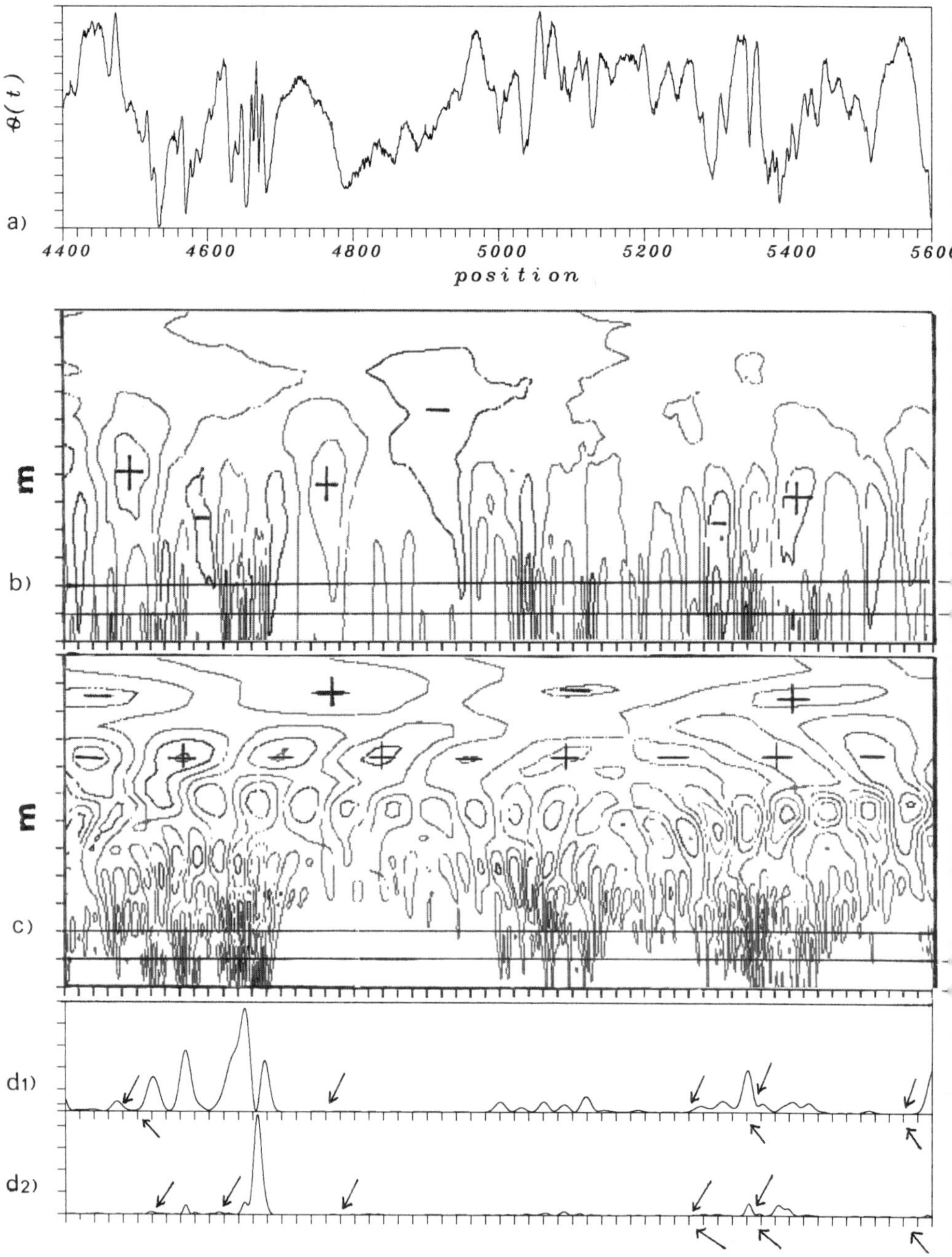

Fig. 3 Distributions of wavelet coefficients for a typical temperature signal obtained at $y^{+}=6$.
a). Temperature trace.; b). Haar wavelet.; c). Morlet wavelet.; d). Coefficients for one particular scale together with positions of Wag detected coolings using that window size: 1- Scale corresponding to 45 points; 2- 75 points.

Colour version in Colour Plates section, p. xx.

temperature gradients of either sign are occuring. It is interesting to note that, since these local energies are integrated over all scales a, the differences - in terms of the positions and the levels of the largest fluctuations - between Haar and Morlet results are much less apparent than on figure 3. More systematic studies are in progress to determine whether a quantitative detection criterion based on the value of C, or the ratio of C to the local mean scale, or the local spectral distribution E(a,b), or any combination of these quantities can be sorted out, and to investigate the variations with the distance to the wall. Indeed, it is clear that the procedure - based on the local extent of the energy distribution - which was proved to be effective for analyzing the transition process in the flow over a rotating disk (Moret-Bailly et al., 1991) is not adequate for the present turbulent boundary layer signals corresponding to a much more intricate situation.

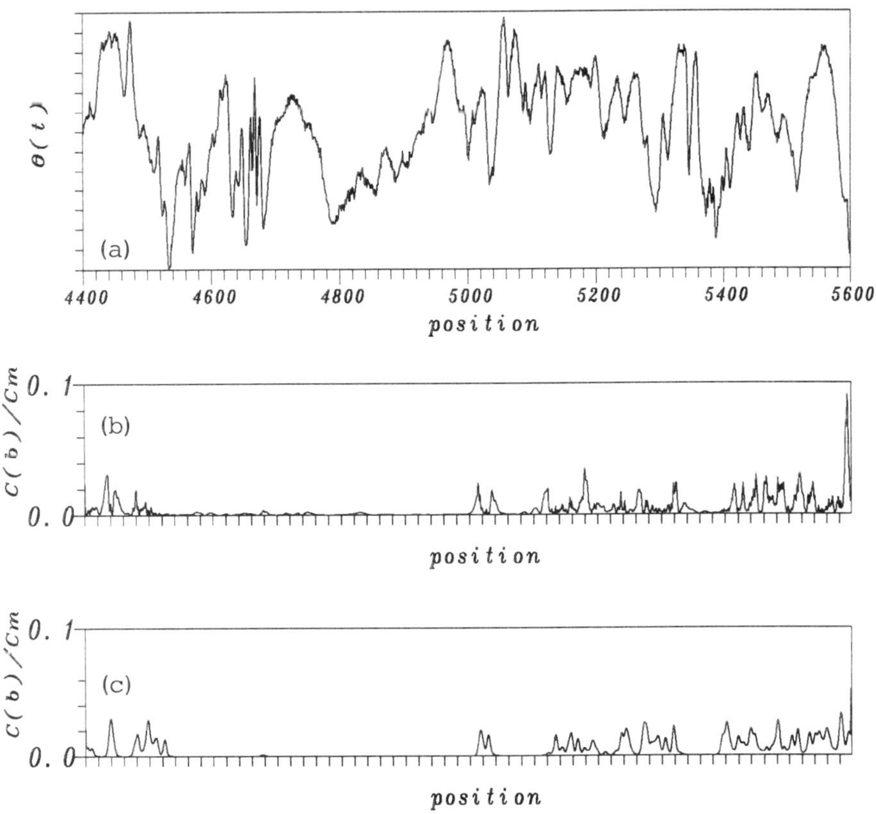

Fig. 4 Local energies associated with temperature fluctuations obtained at $y^+=6$.
a). Sketch of the analyzed portion of signal.
b). Local energy inferred from the Haar wavelet.
c). Local energy inferred from the Morlet wavelet.

Figure 5 reports results for mean local energy distributions at several distances from the wall, over the range covered by the rake, using only 5 blocks of data for each of them. Usual energy distributions obtained using the FFT algorithm (with 1024 points) are presented on figure 5a with the usual semi-logarithmic scales such that the surfaces under the curves are proportional to the temperature variances. These spectral densities are in fact normalized to unity for easier comparison and the well-known hierarchy of distributions is observed, with lower energetic frequencies associated with lower mean velocities in the region very close to the wall. For the same set of data, the distributions obtained with the Morlet wavelet analysis are reported on figure 5b, with slightly different multiplicative factors applied to the energies in order to allow more distinction between the four positions. The agreement with those of figure 5a is very good, but it is worth noticing that the repartition of the computed frequencies is then regular when frequencies are plotted on a logarithmic scale since they are defined from a geometrical progression in the wavelet analysis. However, even with the use of the coarser mesh to imrove the calculation efficiency, such an analysis requires about 20 times much more CPU time than the FFT. But, of course, its main interest lies in its ability to provide mean scale decompositions conditioned on the occurence of coherent events detected by Wag, such as coolings or heatings. Figure 5c shows that, generally speaking, these energy contributions are significantly shifted towards larger frequencies than the global ones (Fig. 5a and 5b). In addition, for three of the positions presented here, two energy peaks can be observed, showing that coolings are associated with preferred frequencies, the most energetic and largest one being obviously characteristic of the gradient region. However, for the position closest to the wall, the trend is quite different since the energy seems to be more regularly distributed, which may be due to the fact that, very close to the wall, the influence of the organized motion is quite smaller. Ensemble spectral averages associated with heatings, reported here for only two positions, present characteristic features quite similar to those for coolings. It is also interesting to note that, when the relative contributions to the temperature variances are computed from the distributions on figure 5c, good agreement is found with the results reported by Antonia and Fulachier (1989). In addition, it is worth noticing that these spectral distributions are giving a posteriori support to the typical window sizes usually used for the estimation of ensemble averages since those sizes correspond to frequencies of about 200Hz: indeed, the contributions of lower frequencies (or larger window sizes) appear to be negligible. A more quantitative study of these spectral distributions is still in progress, in particular for comparison with the usual bursting characteristic frequencies.

Conclusion

It was shown that the wavelet analysis is a powerful tool to provide quantitative new information about the turbulence structure in shear flows and their spatially coherent organization. However, additional work is required for better comparison with usual results obtained by more classical methods, and especially since the boundary layer organized motion is quite intricate and not yet completely resolved. Future development of this work includes extension to Reynolds stress and heat flux properties, as well as to the linkage between small and large scales of the flow through the analysis of simultaneous signals of temperature and its dissipation. In particular, the development of a procedure allowing the determination of cross-wavelet charateristics of two signals, in a way similar to that leading to co-spectra, phases, and so on, seems quite promising.

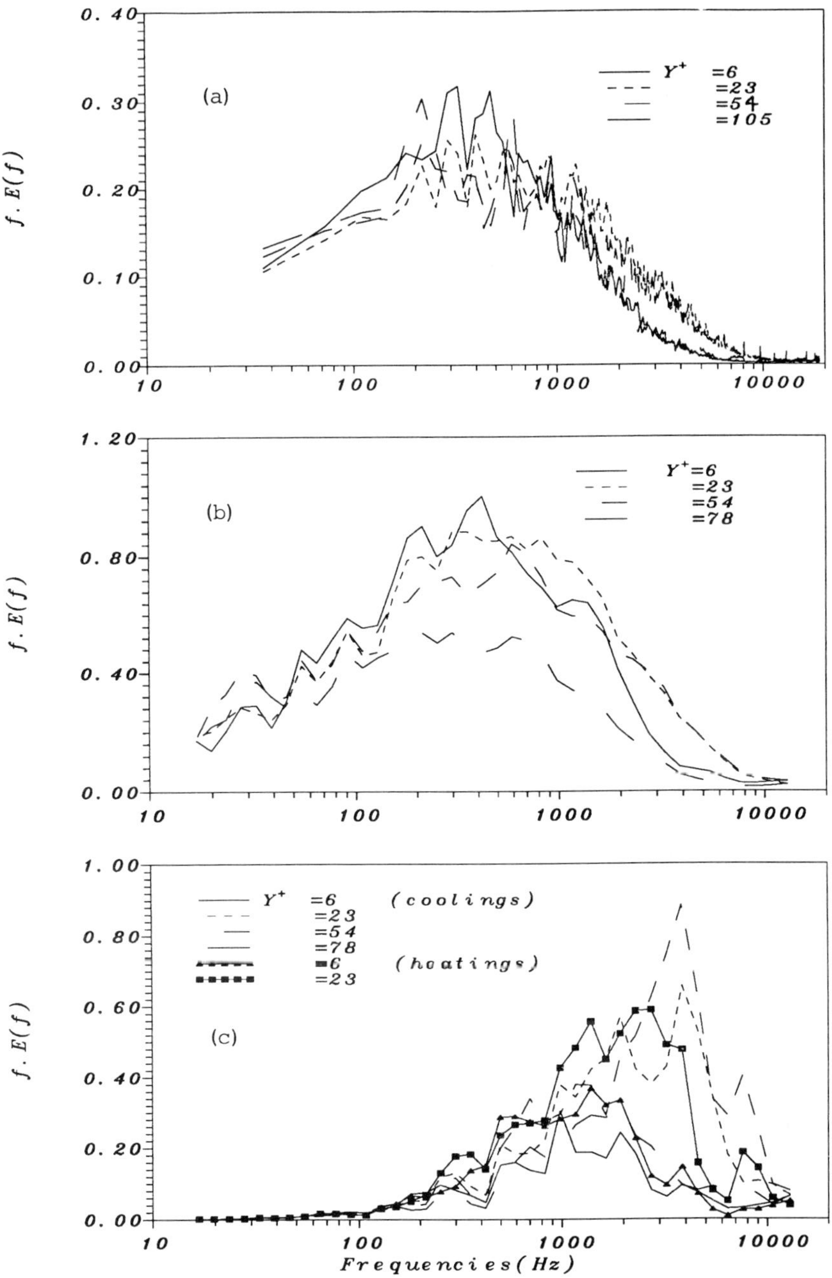

Fig. 5 Spectral distributions for various distances from the wall.
a). Energy spectra obtained with the Fourier analysis.
b). Energy spectra obtained with the Morlet wavelet.
c). Conditional spectra obtained with Morlet for coolings and heatings.

Fruitful discussions with L. Fulachier are gratefully acknowledged.

References

ANTONIA, R.A., BROWNE, L.W.B. and BISSET, D.K. 1987
Symmetric and antisymmetric modes in the turbulent far-wake.
In proc. 6th T.S.F., Toulouse, p 10-3-1.

ANTONIA, R.A. and FULACHIER, L. 1989
Topology of a turbulent boundary layer with and without wall suction.
Journal Fluid Mech., vol. 198, p 429.

BLACKWELDER, R.F. and KAPLAN, R.E. 1976
On the wall structure of the turbulent boundary layer.
Journal Fluid Mech., vol. 76, p 89.

DAUBECHIES, I. 1990
The wavelet transform, time-frequency localization and signal analysis.
IEEE Trans. Inf. Theory, vol. 36, p 961.

FARGE, M. 1992
Wavelet transforms and their applications to turbulence.
Annual Rev. Fluid Mech., vol. 24, p 395.

HUSSAIN, A.K.M.F. 1986
Coherent structures and turbulence.
Journal Fluid Mech., vol. 173, p 303.

MEYER, Y. 1990
Ondelettes et opérateurs.
Editions Herman.

MORET-BAILLY, F., CHAUVE, M., LIANDRAT, J. and TCHAMITCHAN, P. 1991
Déterminations du nombre de Reynolds de transition dans une étude de couche limite sur un disque tournant.
C.R.A.S., vol. 313, Série II, p 591.

ROBINSON, S.K. 1991
Coherent motions in the turbulent boundary layer.
Annual Rev. Fluid Mech., vol. 23, p 601.

ZHOU, Y. 1993
Organised motion in a turbulent wake.
Ph. D. Thesis, University of Newcastle, Australia.

On the Use of 2D Wavelets for the Analysis of Coherent Structures

Y. Lebret [1], R. Murenzi[2], D. Vandromme [1]

1 Introduction

A common difficulty of experimental and numerical works is the complete exploitation of hude amount of data. With the improvement of experimental techniques, a lot of time dependent data are now available in fluid dynamics studies, and efficient tools are required to examine the physics. In this paper we consider two flow configurations. The first is the interaction between a shock wave and a fluid interface and the second is the naturally unsteady compressible turbulent shear layer. Since the precursor work of TOWNSEND [1], the existence of two kinds of turbulence is now well accepted. Indeed, in turbulent shear flows, coherent large scale structures are observed whereas smaller scales are more random. Experiments of PAPAMOSCHOU and ROSHKO[2] have shown that these structures exist also in compressible flows. Thus, the turbulence can be splitted into coherent vortices which are dynamically active, and advected by the mean flow and the "random" turbulence. Various tools have been developped to study these particular flows such as, the proper orthogonal decomposition, the pattern recognition, and the wavelet transform. The latter has been used in the present work. This technique is well adapted to capture a lot of flow features. For example, in image processing the mexican hat wavelet is well adapted to detect positions and characteristic scales, but directions are not well resolved. The Morlet wavelet is more suitable to detect directional information. In the first part of this article, the wavelet transform is summarized. Then the second part will describe the flows and analyze the results.

[1]LMFN-URA 230, INSA de Rouen, 76131 Mont Saint Aignan, B.P. 8, France
[2]CERFACS, 42 avenue G. CORIOLIS, 31400 Toulouse, France, presently at CTSPS, Clark-Atlanta University, James P. Brawely Drive at Fair Street, S.W. Atlanta Georgia 30314 USA

J. P. Bonnet and M. N. Glauser (eds.), Eddy Structure Identification in Free Turbulent Shear Flows, 169–178.

2 2D Wavelets Transform

The wavelet transform (WT) is developped here on 2D fields, but extension to three dimensions is straightforward [3].

2.1 Definition

- **2D Scalar fields**

In general we consider scalar fields as complex-valued functions on the real space $\Re^2$, $f : \Re^2 \rightarrow C$ and square integrable, i.e $f \in L^2(\Re^2, d\vec{x})$:

$$\|f\|^2 = \int_{\Re^2} d\vec{x} | f(\vec{x}) |^2 < \infty \qquad (1)$$

which corresponds to the finite energy condition The Fourier transform of the field f is defined, as usual, by

$$\hat{f}(\vec{k}) = (2\pi)^{-1} \int_{\Re^2} d\vec{x}\ e^{-i\vec{k}.\vec{x}} f(\vec{x}) \qquad (2)$$

where $\vec{k} \in \Re^2$ is the spatial frequency and $\vec{k}.\vec{x} = k_1.x_1 + k_2.x_2$ is the euclidian scalar product.

- **2D Wavelets**

A 2D wavelet [4][5] is a function ψ verifying the following conditions:

- $\psi \in L^2(\Re^2, d\vec{x})$
- there exist a constant c_ψ such that:

$$c_\psi = (2\pi)^2 \int_{\Re^2} d\vec{k} \frac{| \hat{\psi}(\vec{k}) |^2}{| \vec{k} |^2} < \infty \qquad (3)$$

where $\hat{\psi}$ is the Fourier transform of ψ.

2.2 Examples of wavelets

- 2D Mexican hat

$$\psi(\vec{x}) = (2 - \vec{x}.A\vec{x})\ e^{-\frac{1}{2}\vec{x}.A\vec{x}} \qquad (4)$$

$$\hat{\psi}(\vec{k}) = \det(B).(\vec{k}.B\vec{k})e^{\frac{-1}{2}\vec{k}.B\vec{k}} \qquad (5)$$

where $\vec{x}.A\vec{x} > 0$ and $B = A^{-1}$. ψ is isotropic if the matrix A is proportional to the identity matrix, otherwise it is an anisotropic wavelet.

- 2D Morlet wavelet

$$\psi(\vec{x}) = e^{i\vec{k}_0.\vec{x}} e^{\frac{-1}{2}\vec{x}.A\vec{x}} - e^{\frac{-1}{2}\vec{k}_0.A\vec{k}_0} e^{\frac{-1}{2}\vec{x}.A\vec{x}} \tag{6}$$

$$\hat{\psi}(\vec{k}) = \det(B)(e^{\frac{-1}{2}(\vec{k}-\vec{k}_0).B(\vec{k}-\vec{k}_0)} - e^{\frac{-1}{2}\vec{k}_0.B\vec{k}_0} e^{\frac{-1}{2}\vec{k}.B\vec{k}_0}) \tag{7}$$

where $\vec{x}.A\vec{x} > 0$ and $B = A^{-1}$

2.3 Wavelet transform

Let us consider a wavelet ψ and its dilated, rotated and translated versions:

$$\psi_{(a,\theta,\vec{b})}(\vec{x}) = a^{-1}\psi(a^{-1}r_{-\theta}(\vec{x} - \vec{b})) \tag{8}$$

where:

- $a \in \Re^*_+$ is the dilation parameter,
- $r \in SO(2)$ is a rotation matrix and θ is the angle,
- $\vec{b}$ the translation vector.

In Fourier space we obtain,

$$\hat{\psi}_{(a,\theta,\vec{b})}(\vec{k}) = ae^{i\vec{k}.\vec{b}}\hat{\psi}(ar_{-\theta}\vec{k}) \tag{9}$$

One can show that the span of $\{\Omega(a,\theta,\vec{b})\psi)\}$ is dense in $L^2(\Re^2, d\vec{x})$. This enables to decompose every $f \in L^2(\Re^2, d\vec{x})$ on a such family [4][5].

The wavelet transform of f related to the wavelet ψ is then defined as

$$F(a,\theta,\vec{b}) = \frac{1}{\sqrt{C_\psi}} < \Omega(a,\theta,\vec{b})\psi \mid f > \tag{10}$$

$$= \frac{1}{\sqrt{C_\psi}} \int_{\Re^2} d\vec{x} a^{-1}\bar{\psi}(a^{-1}r_{-\theta}(\vec{x} - \vec{b}))f(\vec{x}) \tag{11}$$

$$= \frac{1}{\sqrt{C_\psi}} \int_{\Re^2} d\vec{x} a e^{i\vec{k}.\vec{b}}\bar{\hat{\psi}}(ar_{-\theta}\vec{k})\hat{f}(\vec{k}) \tag{12}$$

It is a convolution at each scale and in each direction. It can also be interpreted as a mathematical microscope with *magnification* $\frac{1}{a}$, *position* $\vec{b}$, *directional selectivity* given by the rotation parameter θ. If the wavelet is *isotropic*, i.e $SO(2)$ invariant, then the wavelet transform F depends on three variables, $F(a,\vec{b})$(one scalar and one vector).

3 Applications

3.1 Interaction between shock wave and fluid interface

3.1.1 Flow Description

A propagating shock wave meets a density discontinuity between two gases at rest in a shock tube, yielding the development of a Richtmyer-Meshkov instability [6]. This instability is visualized by a Schlieren technique and the instability is characterized by a mushroom shape (figure 1).

3.1.2 Results

The aim is to detect turbulent time scale attached to particular flow structures. Thus mexican hat is used to detect discontinuities without any directional selectivity, and the Morlet wavelet is used to capture discontinuities according to a particular orientation. Results obtained with the mexican hat is given on figure 2, and with the Morlet wavelet on figure 3-4. Discontinuities are selected when the wavelet goes from a minimum to a maximum. Choices of the dilation coefficient allow to select different scales in the analyzed signal. For a given value, the significant wavelet support gives the minimal structure width which can be detected. With this value, and the local Mach number M, a turbulent time scale associated to this structure can be estimated by :

$$t = \frac{width}{M * speed\ of\ sound}$$

For that flow, and a Mach number of 1.5 (which is a global Mach number), the characteristic time was $t = 10^{-6}s$ [7].

3.2 Study of turbulent shear layer

3.2.1 Flow Description

The compressible mixing layer is made of two turbulent supersonic parallel flows. The Mach numbers of the upper and of bottom flow are 1.5 and 0.4 respectively. The static pressure in both streams is equal to 10^5Pa. The total temperature is $T_t = 300K$. The computational domain dimensions are $1.5x1.0$ m high. The theoritical convective Mach number [2] is 0.46.

3.2.2 Results

The wavelet analysis is used to detect coherent structures. To do so, the vorticity field is prefered to the other fields. The Morlet wavelet is used here, because it has the property that the maximum value of the modulus of the wavelet transform corresponds to the maximum peak of enstrophy, whereas the imaginary part can select preferential orientations of the signal. Our aim is to locate the core of the structures. These cores evolve as solid rotating bodies, so their detection is equivalent to the selection of the fluid portion which possesses all orientations. On the figures, the imaginary part of Morlet wavelet for two different times flow is presented. The directional selectivity property of the wavelet is obvious on the pictures. For each field, six different angles are used, and for each angle, the maxima of the imaginary part are selected. After doing intersection of all the maxima fields, it is straightforward to locate common core for each time. Then, knowing the displacement of these maxima and the time step between each field, the convective velocity can be obtained.

4 Conclusion

2D wavelet algorithms have been implemented and have been used to select different physical aspects of flows. Two different properties of the wavelet were used. The first one is the ability to detect characteristic scaled structures, allowing to find time scale attached to these structures. The second one is the directional selectivity of the Morlet wavelet. This property was used to detect coherent structures cores. In the future, we intend to use these tools to study turbulent coherent structures (scale, orientation,...), and to build an objective recognition pattern tool.

We are grateful to F.X. Keller for hisfruitful work.

5 References

[1]Townsend, *The structure of turbulent shear flows*, 1956, Cambridge university press, 1st Ed, 2nd Ed 1976.

[2]Papamoschou and Roshko, *Observation of supersonic free shear layers*, AIAA paper n° 86-0162, January 86

[3] D. Astruc, L. Plantie, R. Murenzi, Y. Lebret, D. Vandromme, *on the use of 3D wavelet transform for the analysis of computational fluid dynamics results*, International conference wavelets and applications, Toulouse, 8-13 June 92.

[4] R. Murenzi, *Ondelettes Multidimensionnelles et Applications à l'analyse d'images*, PhD thesis January 1990, Université Catholique de Louvain.

[5] J-P Antoine, R. Murenzi, *Isotropic and Anisotropic Multidimensional Wavelet: Applications to the Analysis of 2D Fields..* To appear in the proceedings of the colloquium, *Ondelettes et Turbulences*, Princeton University June 2-June7, 1991.

[6] C. Cavailler, H. Croso, P. Gandeboeuf, J.F. Haas, G. Rodriguez, *Results from the Vaujours vertical shock tube*, C.E.A., 3rd International Workshop on the physics of compressible turbulent mixing, Royaumont (France), 17-19 juin 1991

[7] Keller F.X., *Application des ondelettes à la détection de structures cohérentes en turbulence*, Rapport de D.E.A., CORIA, Septembre 1992

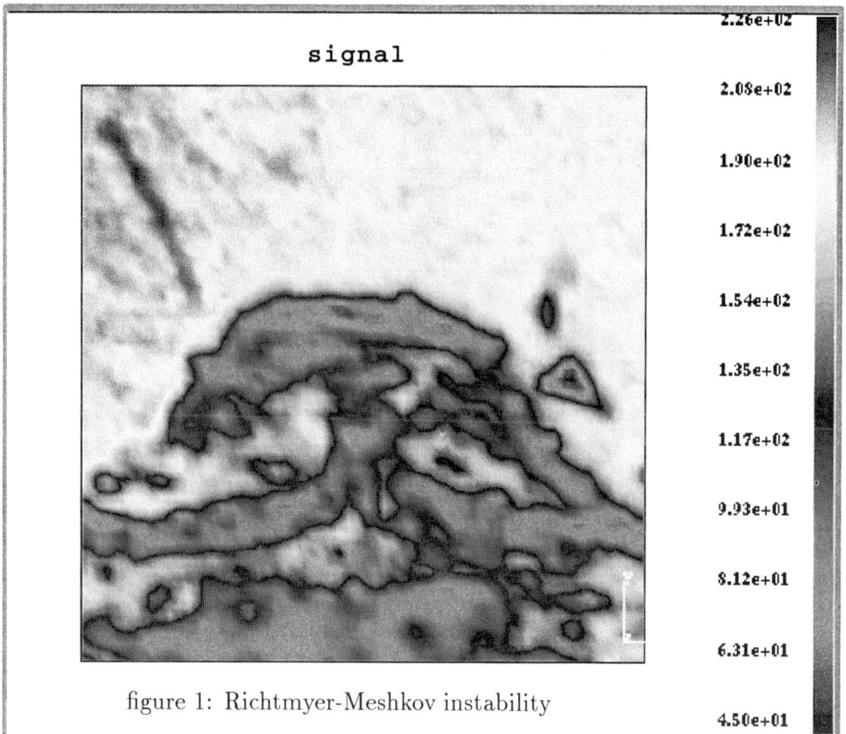

figure 1: Richtmyer-Meshkov instability

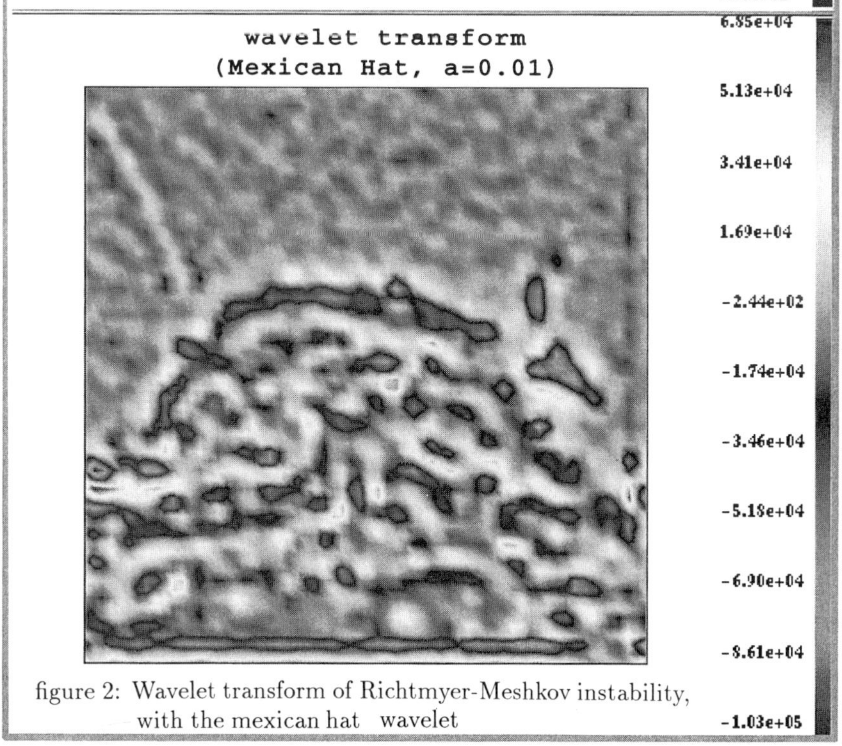

figure 2: Wavelet transform of Richtmyer-Meshkov instability, with the mexican hat wavelet

Colour version in Colour Plates section, p. xxi.

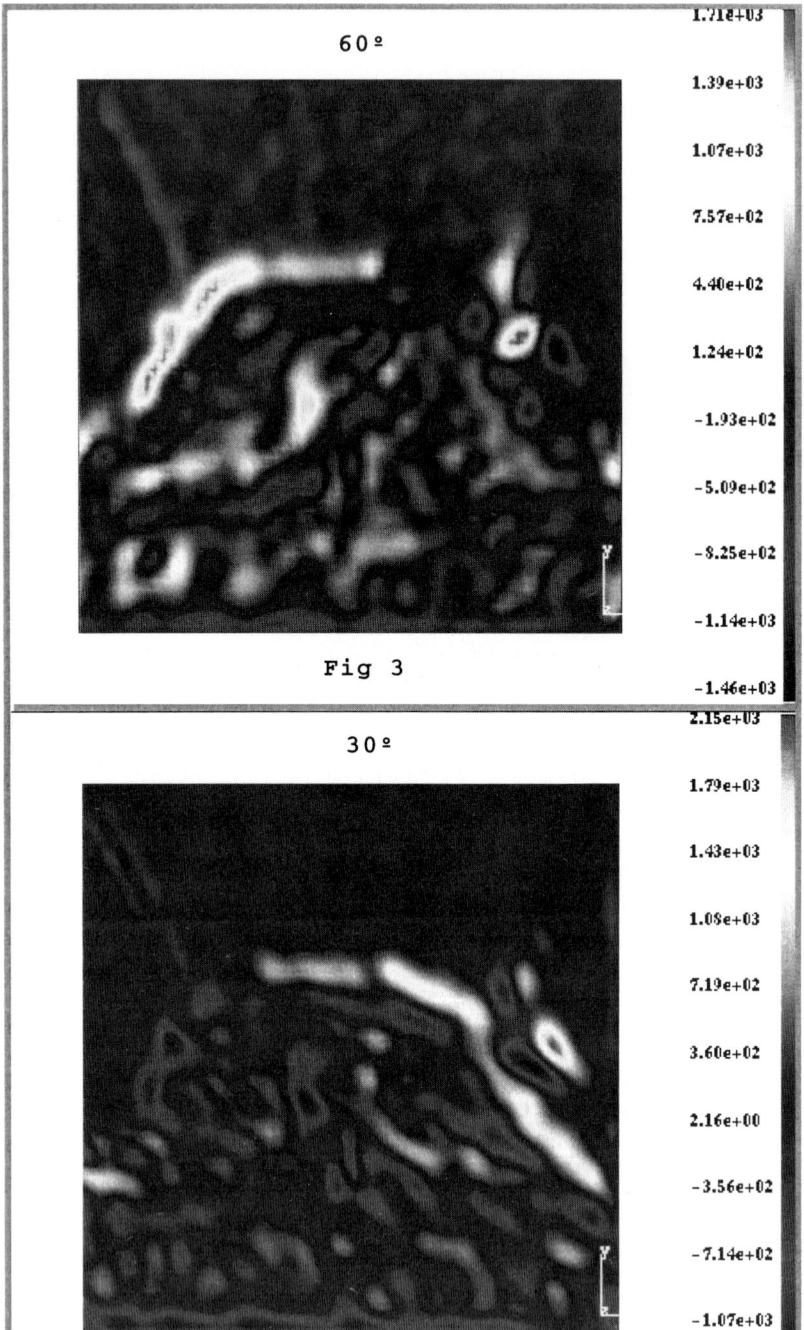

figure 3-4: Imaginary part of the Morlet wavelet transform of Richtmyer-Meshkov instability

Colour version in Colour Plates section, p. xxii.

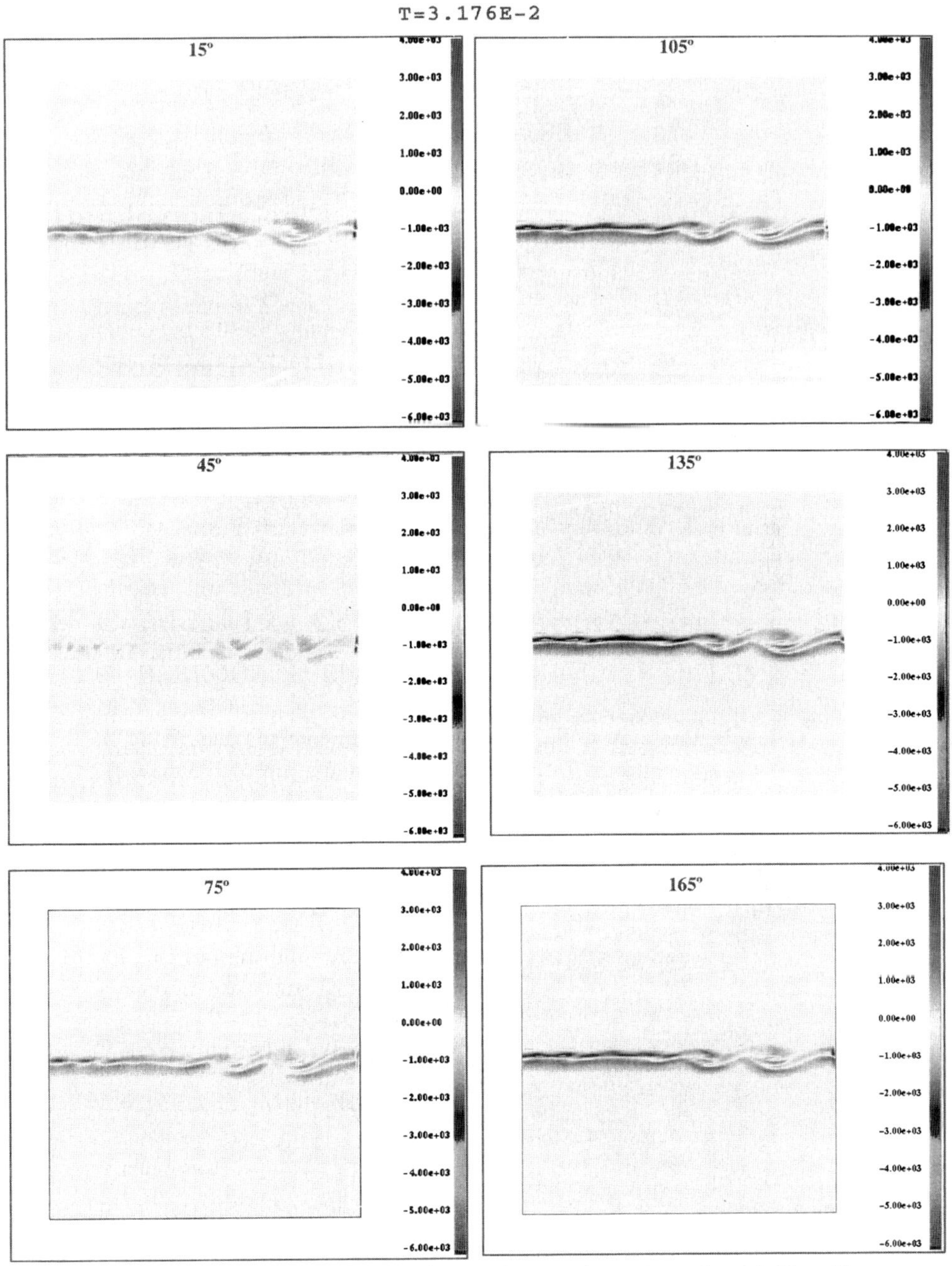

figure 5: Imaginary part of the Morlet wavelet transform of rotational fields, with different orientations, at $t = 3.176\ 10^{-2}\ s$

Colour version in Colour Plates section, p. xxiii.

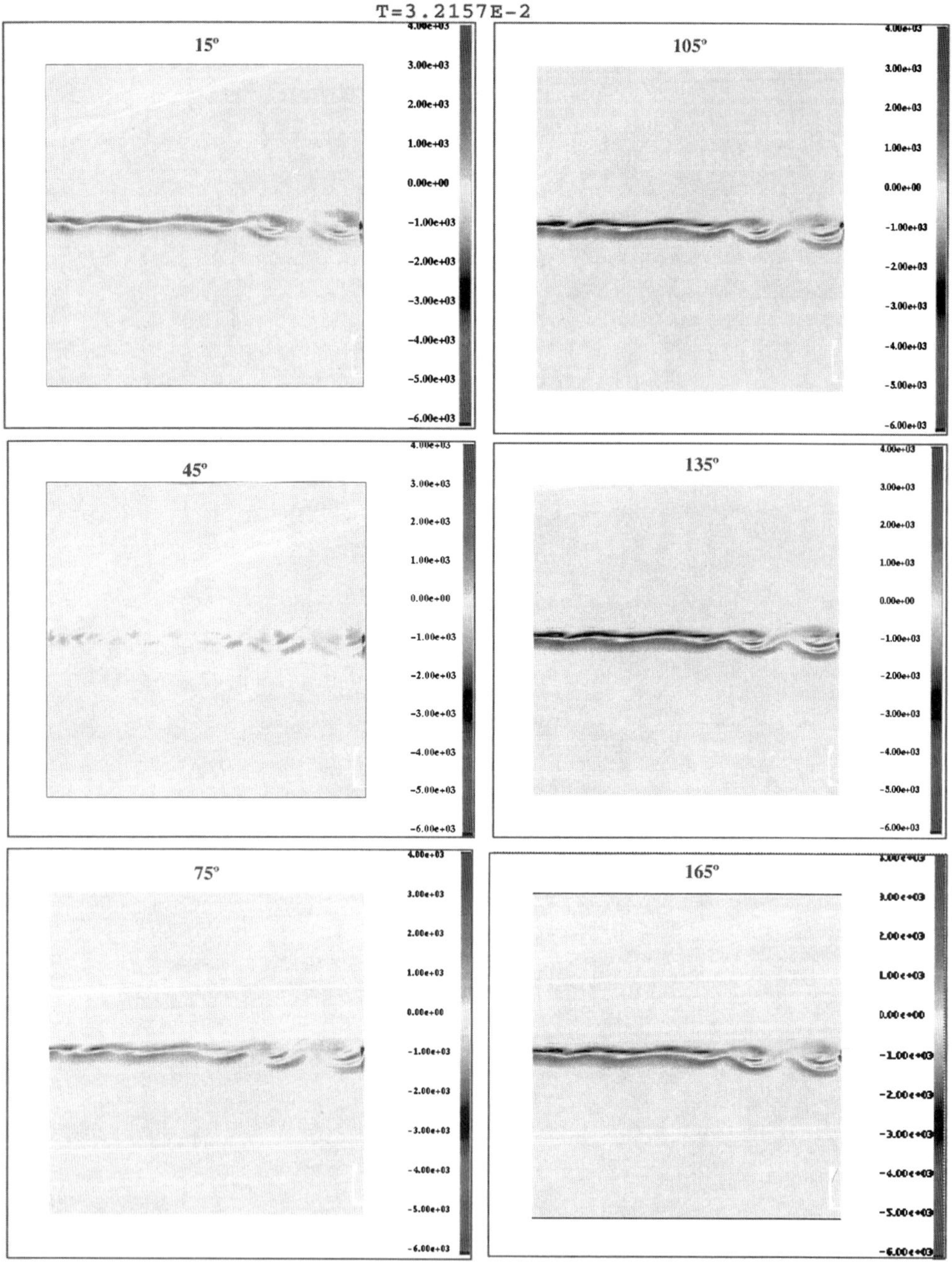

figure 6: Imaginary part of the Morlet wavelet transform of rotational fields, with different orientations, at $t = 3.2157\ 10^{-2}\ s$

Colour version in Colour Plates section, p. xxiv.

V. Pattern Recognition Analysis

ANALYSIS OF TURBULENT SIGNALS

J.A.Ferré and Francesc Giralt

Escola Tècnica Superior d'Enginyeria
Universitat Rovira i Virgili
Carretera de Salou s/n
43007 Tarragona, Catalunya, Spain

SUMMARY

The fundamentals of principal component analysis and pattern recognition techniques are presented and their ability to analyze features of turbulent signals is compared. Both techniques are developed in a general framework, independent of the turbulent flow being investigated. Their performance is evaluated from the analysis of a single synthetic signal as well as of the velocity field measured by multiple probes in a fully developed turbulent wake behind a cylinder.

INTRODUCTION

The computation of auto and cross-correlation coefficients, spectra and coherence functions were a common approach to analyze turbulence in the past. This statistical description of turbulent flows was satisfactory while turbulence was considered mainly as "statistics". However, the visualization of ordered motions in boundary layers and in shear layers showed that correlation measurements had been hiding very important features of turbulence, as pointed out by Laufer (1975). Since then, there have been several attempts to design experimental and analytical techniques capable of yielding quantitative information of turbulent flows similar to that obtained by visualization.

The first step beyond a correlation coefficient was made by Corrsin and Townsend in the fifties when they introduced the concept of intermittency. Some interesting results concerning flow structure in wakes have been obtained by LaRue and Libby (1974) and Fabris (1979) using intermittency analysis. Quadrant analysis, a detection technique which has been extensively used in boundary layers, has also been successfully applied by Keffer et al.(1982) to analyze turbulent transport in free flows.

One of the best spatial and analytical description of organized motions in near wakes has been obtained by conditional sampling or phase averaging (Perry and Watmuff, 1981; Cantwell and Coles, 1983). LaRue and Libby (1974) also obtained useful information in wakes using spectral analysis. However, Budny et al. (1979) showed that even spectral analysis using coherence functions could not detect the large-scale structure of turbulence beyond x/D=60. This is the reason why the results obtained by Townsend (1956) and Grant (1958) from the analysis of correlation measurements were the only descriptions of large eddies in far wakes published in the literature for a long period of time. None of the

J. P. Bonnet and M. N. Glauser (eds.), Eddy Structure Identification in Free Turbulent Shear Flows, 181–194.

techniques used in other flows (visualization, VITA-Variable Integral Time Average, conditional sampling, etc) were able to reveal the structure of self-preserving wakes.

Three different and basic approaches have been used to analyze the structural features of turbulent flows and, in particular, of plane turbulent wakes:

i) Orthogonal decomposition, which was introduced by Lumley (1965).
ii) Conditional sampling, which was first successfully used by Browne et al. (1986) to identify key features of the temperature and velocity fields in turbulent wakes. This technique was reviewed by Antonia (1981).
iii) Pattern recognition (PR), a technique that was first formulated by Mumford (1982) and later modified and used by Ferré and Giralt (1989a,b) to study several wake flow configurations.

There are other techniques, such as that used by Hayakawa and Hussain (1989) to detect the peaks of large-scale vorticity, that combine conditional sampling and PR. It is important to note that conditional sampling results obtained by Antonia et al. (1987) are consistent with those reported by Ferré et al. (1990) using PR. A discussion on the role of organized motions and their dynamics, as well as about the capabilities of different analytical identification techniques is given by Lumley (1990).

Irrespective of how data is obtained, turbulent signals can be treated and analyzed like time series. Therefore, the problem of discovering the behavior of turbulent flows is equivalent to the problem of characterizing hidden features of time series. Conditional sampling has a very sound physical foundation but it is difficult to imagine how it could be applied to characterize time series in a general context or in fields other than turbulence. On the other hand, orthogonal decomposition is also a standard tool in many areas of statistics and image processing, where it is better known as Principal Component Analysis (PCA), covariance diagonalization or Karhunen-Loève expansion (see chapter 7 of Watanabe, 1985). Because data in a computer is always discrete, the simplest formulation of the orthogonal decomposition approach is PCA, which is a technique well documented in any multivariate analysis textbook (see Chatfield and Collins, 1980, for example). Other techniques or modifications of those previously mentioned have been recently proposed to analyze the large multivariate data files generated by the direct simulation of turbulence (Adrian and Moin, 1988).

This paper studies the fundamentals of PCA and PR. The capabilities of these two standard techniques when analyzing hidden features of time series, i.e. turbulent velocity and scalar fields, are then examined using a single synthetic signal and the velocity filed measured with a set of eight x-wire anemometers in the far wake of a circular cylinder.

IDENTIFICATION TECHNIQUES: Principal Component Analysis

Let us suppose that a sensor plugged into an A/D converter has sampled a turbulent quantity, such as the instantaneous velocity or temperature, and that the t_x samples obtained are stored into the time series $D(t)$, $t=1,\ldots,t_x$. The passage over the sensor of any large-scale structure contained in this turbulent flow will leave repetitive patterns $p_i(\tau)$, $\tau=1,\ldots,\tau_x$ into the time series. To illustrate this experiment and for the sake of simplicity, let us assume that the embedded patterns have a length of $\tau_x=3$ samples and occur at random positions, in accordance to the scheme

	$p_1(1)$	$p_1(2)$	$p_1(3)$			$p_2(1)$	$p_2(2)$	$p_2(3)$	...
D(1)	D(2)	D(3)	D(4)	D(5)	D(6)	D(7)	D(8)	D(9)	...

In this example, the first pattern left by the passage of one structure over the sensor appears at samples {D(2), D(3), D(4)} and the second at samples {D(7), D(8), D(9)}. Since the patterns are embedded into a turbulent signal, the individual realizations are not identical. However, if two patterns $p_i(\tau)$ and $p_j(\tau)$ belong to the same family they should be similar and, thus, strongly cross-correlated

$$\rho = \frac{\sum_{\tau=1}^{\tau_x} p_i(\tau)\, p_j(\tau)}{\sqrt{\sum_{\tau=1}^{\tau_x} p_i^2(\tau)}\ \sqrt{\sum_{\tau=1}^{\tau_x} p_j^2(\tau)}} \approx 1 \qquad [1]$$

If the time series D(t) contains a total of i_m patterns, then the ensemble averaged pattern $P(\tau)$, $\tau=1,\ldots,\tau_x$, could be computed as

$$\begin{aligned} P(\tau) &= \frac{1}{i_m}\sum_{i=1}^{i_m} p_i(\tau) = \frac{1}{i_m}\left[p_1(\tau)+p_2(\tau) + \ldots + p_{i_m}(\tau)\right] = \\ &= \frac{1}{i_m}\left[D(1+\tau) + D(6+\tau) + \ldots \text{ up to } i_m \text{ terms}\right] \end{aligned} \qquad [2]$$

In a real situation, however, this ensemble average cannot be computed because the locations of the patterns in D(t) are not known. It is necessary to examine all the data windows of length τ_x=3 samples contained into the time series to locate the patterns, or at least to obtain a good estimate of their ensemble average. The procedure could be depicted by the scheme

	$p_1(1)$	$p_1(2)$	$p_1(3)$			$p_2(1)$	$p_2(2)$	$p_2(3)$	...
D(1)	D(2)	D(3)	D(4)	D(5)	D(6)	D(7)	D(8)	D(9)	...
$W_1(1)$	$W_1(2)$	$W_1(3)$							
	$W_2(1)$	$W_2(2)$	$W_2(3)$						
		$W_3(1)$	$W_3(2)$	...					

where data windows are written as $W_k(\tau)$, $\tau=1,2,3$, with k ranging from k=1 to $k=t_x-2$ because $\{D(t_x-2), D(t_x-1), D(t_x)\}$ is the last data window that can be examined. The data windows of this example are vectors in a three-dimensional space, while in a more general formulation they would be vectors in *n*-dimensional vector space. It is obvious that the data windows W_2, W_7, etc, matching the depicted patterns p_1, p_2, etc, respectively, will group along the same direction, which in turn should barely coincide with that of the ensemble averaged pattern $P(\tau)$. All other data windows will randomly fill the vector space. The grouping along a single direction of the data windows that match patterns is a direct consequence of the assumption stated before that the patterns belonging to the same family have a high cross-correlation coefficient. The reason is that when the patterns are viewed as vectors, their cross-correlation coefficient can also be computed by the expression $\rho=\cos\alpha$, where α is the angle formed by the two vectors. Whenever $\alpha\approx 0$, then $\rho\approx 1$, i.e. strongly cross-correlated patterns are almost parallel.

Turbulent flows may contain different types and sizes of structures and, therefore, different patterns will be embedded into the time series sampled by any sensor. As a consequence, it is reasonable to accept that a *n*-dimensional vector space could have up to *n* directions along which different patterns could concentrate. If these directions are arbitrarily forced to be orthogonal, so that they could be interpreted as the footprints of independent flow structures, then the question is to determine the set of orthogonal directions that best represents the actual distribution of a set of data vectors in a *n*-dimensional space.

In the example stated above, the candidates to contain replicas of the embedded pattern are the windows of three data points, that are the rows of V

$$V = \begin{pmatrix} W_1(1) & W_1(2) & W_1(3) \\ W_2(1) & W_2(2) & W_2(3) \\ W_3(1) & W_3(2) & W_3(3) \\ \cdots\cdots & \cdots\cdots & \cdots\cdots \\ \cdots\cdots & \cdots\cdots & \cdots\cdots \end{pmatrix} = \begin{pmatrix} D(1) & D(2) & D(3) \\ D(2) & D(3) & D(4) \\ D(3) & D(4) & D(5) \\ \cdots\cdots & \cdots\cdots & \cdots\cdots \\ D(t_x-2) & D(t_x-1) & D(t_x) \end{pmatrix} \quad [3]$$

According to the previous discussion, the rows of V that contain patterns belonging to the same family will point to a single direction, i.e. will be almost parallel with $\rho \approx 1$, while all the other vectors will be randomly distributed. Thus the identification of the patterns embedded into D(t) turns out to be the location of the feature axis along which most of the vectors of V concentrate, and the *identification problem*, i.e. the identification of the repetitive patterns embedded into the time series at random positions can be formulated as an *analysis problem*, i.e. the location of the feature axis along which most of the vector of V concentrate.

The solution to the analysis problem is given by PCA through the diagonalization of a covariance matrix S with elements s_{pq}, where s_{pq} is the covariance between the columns p and q of the matrix V (see Chatfield and Collins, 1980). The relationship between the columns of V and the data in the time series D(t) indicates that S can be written as

$$S = \begin{pmatrix} s_{11} & s_{12} & s_{13} \\ s_{21} & s_{22} & s_{23} \\ s_{31} & s_{32} & s_{33} \end{pmatrix} = \begin{pmatrix} \sigma(0) & \sigma(1) & \sigma(2) \\ \sigma(1) & \sigma(0) & \sigma(1) \\ \sigma(2) & \sigma(1) & \sigma(0) \end{pmatrix} = \sigma(0) \begin{pmatrix} \rho(0) & \rho(1) & \rho(2) \\ \rho(1) & \rho(0) & \rho(1) \\ \rho(2) & \rho(1) & \rho(0) \end{pmatrix} \quad [4]$$

where $\sigma(k)$ is the autocovariance of the time series at lag k and $\rho(k)=\sigma(k)/\sigma(0)$ is the autocorrelation. S is a symmetric and positive definite matrix with τ_x real eigenvalues and mutually orthogonal eigenvectors. The feature axis along which most of the vectors of V concentrate is the eigenvector of S with the largest eigenvalue. The following eigenvectors could be interpreted as the set of independent patterns that several structures have embedded into the time series. It should be noted that the trace $(S) = \tau_x\, \sigma(0)$ is also equal to the sum of the eigenvalues (Chatfield and Collins, 1980). Then, the eigenvector with eigenvalue λ_k can be interpreted to account for a fraction of the variance of the data equal to

$$\lambda_k \Big/ \sum_{\tau=1}^{\tau=\tau_x} \lambda_\tau \quad [5]$$

and the eigenvectors, and the patterns they represent, can be ordered from the largest to the smallest contribution to the rms of the data.

In order to evaluate which features PCA can actually extract of a turbulent signal, a synthetic time series was generated by placing the deterministic pattern $\varphi/5\, \exp(-(\varphi/5)^2)$ at non-overlapping random positions over a zero-noise background. The pattern was evaluated for $\varphi=-13,\ldots,0,\ldots,13$, i.e. over a length of $\tau_x=27$. The pattern and the autocorrelation function of the time series are shown in Fig. 1. A matrix V with data windows of 27 points was formed, and a 27×27 covariance matrix S was computed from the autocorrelation function of the time series. The six eigenvectors of the covariance matrix S with the largest eigenvalues are also shown in Fig. 1, together with the fraction of the variance and the cumulative fraction of the variance explained by each eigenvector. It is interesting to note that the first eigenvector accounts for 28% of the variance, although it bears no resemblance to the synthetic pattern. The second eigenvector (multiplied by -1) is almost identical to the embedded pattern but, surprisingly, it accounts for only 27% of the variance. Four extra eigenvectors are needed to reach 97% of the variance.

The analysis of the synthetic data has shown that PCA is not able to extract unambiguously the pattern embedded into the time series. The reason for this failure is that the set V contains few "well-aligned" patterns and many "misaligned" patterns. For example, in Eq. [3] the second row of V contains a well aligned pattern. Then the first, third, fourth, etc, rows should contain misaligned patterns. In fact, the eigenvectors obtained by PCA are the best set of vectors that could be used to recover any row of V, but not just those that contain well-aligned patterns. Following with the example of the synthetic time series, Fig. 2a presents a well-aligned pattern and its reconstruction using only the first six eigenvectors. However, the same eigenvectors are capable of reconstructing any misaligned pattern, like that presented in Fig. 2b. This observation leads us to the conclusion that the analytical procedures that, like PCA, are based on the formulation of an eigenvalue problem for the correlation function cannot be used to represent the actual patterns embedded into a time series because they are not able to discern between well-aligned and misaligned patterns. The reason is that the *analytical problem* that has been solved does not correspond, in fact, to the *identification problem* that has been posed.

One way to improve the performance of PCA would be to select, among the eigenvectors with the largest eigenvalues, only those that have good replicas of themselves into the time series D(t). If this procedure had been applied to the example presented in Fig. 1, eigenvectors 4, 5 and 6 would probably have been discarded. Moreover, the eigenvectors that could attain a high cross-correlation when time-displaced one respect to the other, should be classified into the same family because they represent "misaligned" versions of the same embedded pattern. In the previous example of Fig. 1, this procedure would classify eigenvectors 1 and 2, and probably 3, into the same family.

There is a second reason for the failure of PCA, as demonstrated in the examples of Fig. 3a and 3b, that present the uv-quadrant analysis of synthetic data. Both sets of data have been generated by the superposition of two families of patterns, the only difference is that in Fig. 3a the two families have the same number of points, while in Fig. 3b the ratio is 5:1. In the first case, the correlation coefficient is about zero, ρ=-0.077, while in the second case the most populated family dominates and ρ=0.614. When PCA is used to extract the feature axis along which most of the points concentrate, the results shown in Fig. 3a and 3b are obtained. In Fig. 3a, the eigenvectors almost coincide with the coordinate axis indicating no structure in the data, and only in the second case, one eigenvector follows the larger family, but still the smaller one is not represented by the second eigenvector. Many intermediate positions of the eigenvectors could be reached changing the ratio of individuals in each family form 5:1 to 1:1, being the topological description of the data always the same.

The conclusion is that any technique based solely on correlation measurements will never be adequate to extract the patterns embedded into a turbulent field because the actual values of the correlation tensor depend not only of the large eddies present in the flow, but also of their number. To say it with an example, suppose that a turbulent flow is the superposition of a noisy background and two different types of eddies, say δ and ε eddies, and that the ratio of δ to ε eddies depends on some experimental conditions, for example resonance. Then, the actual values of the correlation tensor could change form experiment to experiment depending on the actual ratio of δ to ε eddies. Therefore, if the technique that is used to identify the organized motions is based only on correlation measurements, it will yield a different description of the typical eddies of the flow, depending wether resonance is on or off, when the only thing that has actually changed in the flow is the ratio of δ to ε eddies, but not eddies themselves.

IDENTIFICATION TECHNIQUES: Pattern Recognition

Analyzing again the example of Fig. 3, the question that arises is whether it is possible to conceive a method that can succeed to detect the true feature axis, capturing the shape of the distribution but being insensitive to their actual density. In order to reach this objective, the raw data will be used through an iterative PR procedure that can be outlined in six steps.

(i) An arbitrary axis (unit vector) is selected to be used as initial template.
(ii) All the point vectors are projected over the template, and those showing an angle smaller (a correlation larger) than some threshold value are ensemble averaged.
(iii) The ensemble average replaces the initial template selected at step (i), and step (ii) is repeated until convergence is achieved, i.e., until the template and the ensemble average are almost parallel. The last ensemble average obtained after iteration is the first feature axis.
(iv) An axis perpendicular to the first feature axis is selected again as initial template.
(v) All the point vectors are projected over the template, and those showing an angle smaller (a correlation larger) than some threshold value are ensemble averaged.
(vi) The ensemble average replaces the initial template selected at step (iv), and step (v) is repeated until convergence is achieved, i.e., until the template and the ensemble average are almost parallel. The last ensemble average obtained after iteration is the second feature axis.

The results that are obtained when this procedure is applied are also included in Fig. 3a and 3b, respectively. The comparison of the feature axis that PCA and PR are able to extract from the data highlights the superior performance of PR over PCA.

The PR approach can also be used to detect the patterns $p_i(\tau)$ embedded into the time series D(t). The main difference between this problem, which was first addressed with the example in Fig. 1, and the identification problem stated in the uv-quadrant example is that now the order in which the patterns appear is important, while in the example of Fig. 3 it was irrelevant. Suppose that E(τ) is a template that is expected to represent a family of patterns $p_i(\tau)$, then the location where the individual patterns of this family may appear within D(t) can be determined using the cross-correlation coefficient between E(τ) and D(t)

$$\rho(t) = \frac{\sum_{\tau=1}^{\tau_x} E(\tau)\, D(t+\tau-1)}{\left(\sum_{\tau=1}^{\tau_x} [E(\tau)]^2\right)^{1/2} \left(\sum_{\tau=1}^{\tau_x} [D(t+\tau-1)]^2\right)^{1/2}} \quad ; \; t=1,\ldots,(t_x-\tau_x+1) \qquad [6]$$

as an indicator function. A maximum in the correlation coefficient ρ(t) will indicate that E(τ) is aligned with an embedded pattern. Therefore, the maxima of ρ(t) identify unambiguously the locations of the hidden patterns that are similar to $E(\tau)$. Once a set of patterns has been detected their ensemble average $P(\tau)$ can be easily computed with Eq. (2). In this procedure the selection of a well aligned replica of $E(\tau)$ implies the rejection of all the "misaligned" versions of this pattern, as opposed to PCA. If the ensemble average is very similar to the template the search is finished. However, if the ensemble is different from the initial template, the procedure can continue iteratively, using as a new template the last ensemble average until both the template and the ensemble average are reasonably similar. Since the procedure is iterative, the initial template is not critical and the final pattern obtained depends only very weakly on it. However it is reasonable to check the consistency of the educed pattern using different initial templates.

Fig. 4 shows the performance of this technique when it is applied to identify the pattern of Fig. 1 embedded in a noisy time series. The actual length of the time series is 1000 points and the signal to noise ratio is only 2.2. A rough estimate of the embedded pattern is used as initial template. Had we not known the true pattern, its shape could be suggested by the change of sign in the autocorrelation function of the time series (not presented here). The best averaged pattern that is displayed in Fig. 4 has been obtained averaging the data windows that are known to contain the embedded pattern. When the iterative procedure is applied, two iterations are needed to achieve convergence and to educe the pattern. The qualitative behavior of the correlation coefficient given by Equation [6] is also displayed in Fig. 4. There is a good agreement between the best possible average and the pattern educed after two iterations.

The PR procedure is easily extended to situations where velocities are sensed by an aligned set of sensors. Then, the data are written as a two dimensional array $D(i,t)$, $i=1,\ldots,i_x$, $t=1,\ldots,t_x$, where each row (index i) corresponds to a different spatial location (different sensor) and each column (index t) contains the data sampled simultaneously by the set of sensors at each time step. If the template $E(k,\tau)$, $k=1,\ldots,k_x$, $\tau=1,\ldots,\tau_x$ has not the same width than the data, i.e $i_x \neq k_x$, then the correlation coefficient has also two index, $\rho(i,t)$, corresponding the i index to the transverse alignment and the t index to the time alignment between the template and the data (for more details see Ferré and Giralt, 1989a). PCA can also be applied to two-dimensional data without modification if the two-dimensional windows of data are renumbered into one-dimensional ones. The covariance between columns p and q of the data matrix V that results form this reordering should be computed now from the covariance between rows r and s of the two-dimensional data $D(i,t)$ at lag k, i.e., $\sigma_{rs}(k)$. The details are straightforward but tedious, and are left to the reader.

ANALYSIS OF EXPERIMENTAL DATA

The performance of the PR procedure and the PCA technique is evaluated comparing their capacity to educe the large-eddy activity present in two-component velocity signals measured in the wake of a circular cylinder at $x/D=420$ and $Re=1600$ and previously analyzed by Giralt and Ferré (1993) using PR. Here only the *u* and *w* velocity data obtained from a slice through the plane parallel to the cylinder axis at the lateral position of maximum shear stress are again analyzed. A sketch of the flow and the sampling plane is included in Figure 5. Fig. 6a presents the template that was used to search for large scale activity using PR. In this plot $z^*=z/l_0$ and $x^*=U_0t/l_0$, where t is time, U_0 is a convection velocity and l_0 is the half-width of the wake. The x* coordinate increases from right to left, so the flow is observed flowing from left to right. Fig. 6b is the ensemble average obtained by the PR procedure after three iterations. It can be observed that the template is almost identical to the last iteration. However Fig. 6b presents some additional circulatory motions that indicate further organization or packing of the double roller eddies in the streamwise coordinate. The significance level of the double-roller footprints shown in Fig. 6b is very high because the *u* peak signal is 1.1 rms units and the *w* peak signal is 0.6 rms units after averaging several hundreds of individual patterns. More than one third of the data recorded during the experiment contributed to this ensemble average, therefore this sort of structure fills at least 33% of the flow in the far wake.

It has been shown that the double-roller eddies not only exist as an average, but also in an instantaneous basis (Giralt and Ferré, 1993). However the correlation coefficient between these instantaneous patterns and their ensemble average is seldom higher that 0.75. This means that the angle between the average and the individual patterns is about 45°, a value that at first glance seems too high because 45° is half a way between any pair of orthogonal directions. However, the volume filled by n-double cones of 45° in a n-dimensional space decreases an n increases, as shown in Fig. 7. In two dimensions, two double cones fill the whole plane. However, in three dimensions, three double cones of 45° are mutually tangent but do not fill all the space because some "holes" appear between them. As the dimensionality of the supporting space increases the fraction of volume occupied by n double-cones of 45° decreases in such a way that it falls below 1% for $n>20$. It has to be noted that the pattern presented in Fig. 6 is a vector in a n-dimensional space with $n>100$. If patterns were randomly distributed in this highly dimensional space, the probability that some of them concentrate around a single axis should be very low. Therefore, even a correlation coefficient as low as $\rho\approx0.7$ reveals a noticeable concentration of patterns around an axis.

The *u* and *w* velocity data sampled in the horizontal plane were also analyzed using PCA. The data were first written in a matrix with 16 rows, alternating the *u* and *w* velocities. Then, two-dimensional windows of 16×30 data points were renumbered into one dimensional ones and used to construct a vector set V, similar to that in Equation [3]. Finally, the eigenvectors and eigenvalues of the covariance matrix S associated with V were computed. Fig. 8 presents the cumulative fraction of the variance of the first 100 eigenvectors. It is worth noting that there is not a clear cut in the eigenvalues, and the results of PCA would suggest that there is no dominant structure in the flow. However, as it has been shown previously, it is possible to find double-roller eddies at least in 33% of the recorded data. Another aspect that was also investigated is whether some of the eigenvectors of S coincide with the double-roller eddy identified. Fig. 9 shows the absolute value of $\cos(\alpha)$, where α is the angle between the double roller and the first 100 eigenvectors It is interesting to note that the highest value attained is $\cos(\alpha)\approx0.5$, i.e. $\alpha\approx60°$. Even when the double roller is reconstructed from the eigenvectors closer to it, i.e. with the largest $|\cos(\alpha)|$ values, more that 10 eigenvectors are needed to reach an angle of 30°, as shown in Fig. 10. This lack of fit, that was first observed with the synthetic data of Fig. 1, support again the assertion that the results obtained from the diagonalization of the covariance matrix are not the best choice to represent the actual patterns in the flow.

FINAL REMARK

The most important conclusion that derives from the examples presented in the paper is that the procedures that are based on the solution of an eigenvalue problem for the correlation tensor can not be trust to represent the actual patterns embedded into turbulent signals. The reason is twofold. First they are not able to discern between aligned and misaligned patterns (Fig. 1) and, then, the results obtained by the diagonalization of the covariance matrix cannot be interpreted directly as the elemental eddies of the flow. Secondly, the description of the typical eddies of a flow that is obtained through the correlation tensor depends not only of the type of eddies, but also of their number (Figure 3). Again, this feature invalidates the use of the correlation tensor to educe the typical eddies of a turbulent flow. It should be realized that this two features will be shared by any procedure based on correlation measurements. PR is a technique that can overcome these drawbacks with a theoretical basis that is flow independent. Its actual performance has been demonstrated analyzing data sampled in a turbulent wake. However, it has yet to be tested in other free-shear or wall-bounded flows.

ACKNOWELEDGEMENTS

This work was financially supported by DGICYT, projects PB88-0218 and PB91-0551. Computer support was granted by Servei de Tecnologia Química, Universitat Rovira i Virgili, Tarragona.

NOMENCLATURE

D	cylinder diameter
D(t)	time series
$E(\tau)$	template
l0	half-width of the wake
$p_i(\tau)$	embedded pattern
$P(\tau)$	ensemble averaged pattern, defined in Eq. [2]
Re	Reynolds number = DU_0/ν
s_{pq}	covariance between columns p and q of the vector set V
S	covariance matrix, defined in Eq. [4]
t	time or time index
U_0	convection velocity
u	streamwise fluctuating velocity
V	set of vectors defined in Eq. [3]
w	spanwise fluctuating velocity
$W_k(\tau)$	data window extracted from a time series
x	streamwise coordinate
x*	U_0t/l_0
z	spanwise coordinate
z*	z/l_0
α	angle between two vectors
λ_i	eigenvalue
ρ	correlation coefficient
$\rho(t)$	correlation coefficient at lag t
$\sigma(k)$	autocovariance of D(t) at lag k
σ_{rs}	autocovariance of D(i,t) at lag k
τ	time or time index

REFERENCES

- Adrian R.J., Moin, P. (1988) "Stochastic estimation of organized turbulent structure: homogeneous shear flow" J. Fluid Mech. **190**, 531.
- Antonia, R.A. (1981) "Conditional sampling in turbulence measurement", Ann. Rev. Fluid Mech., **13**, 131.

- Antonia, R.A.,Browne, L.W.B., Bisset, D.K., Fulachier, L. (1987) "A description of the organized motion in the turbulent far wake of a cylinder at low Reynolds number", J. Fluid Mech. **184**, 423.
- Browne, L.W.B., Antonia, R.A., Bisset, D.K. (1986) "Coherent structures in the far field of a turbulent wake", Phys. Fluids **29**, 3612.
- Budny, R.S., Kawall, J.G., Keffer, J.F. (1979) "Vortex street evolution in the wake of a circular cylinder", Proc. Second Symp. on Turb. Shear Flows, London.
- Cantwell, B., Coles, D. (1983) "An experimental study of entrainment and transport in the turbulent near wake of a circular cylinder", J. Fluid Mech. **136**, 321.
- Chatfield, C., Collins, A.J. (1980) *Introduction to Multivariate Analysis*, Chapman and Hall, London.
- Fabris, G. (1979) "Conditional sampling study of the turbulent wake of a cylinder. Part 1", J. Fluid Mech. **94**, 673.
- Ferré, J.A., Giralt, F. (1989a) "Pattern-recognition analysis of the velocity field in plane turbulent wakes", J. Fluid Mech. **198**, 27.
- Ferré, J.A., Giralt, F. (1989b) "Some topological features of the entrainment process in a heated turbulent wake", J. Fluid Mech. **198**, 65.
- Ferré, J.A., Mumford, J.C., Savill, A.M., Giralt, F. (1990) "Three-dimensional large-eddy motions and fine-scale activity in a plane turbulent wake", J. Fluid Mech. **210**, 371.
- Giralt, F. Ferré, J.A. (1993). "Structure and flow patterns in turbulent wakes". Submitted to Phys. of Fluids.
- Grant, H.L. (1958) "The large eddies of turbulent motion", J. Fluid Mech. **4**, 149.
- Hayakawa, M., Hussain, F. (1989) "Three-dimensionality of organized structures in a plane turbulent wake", J. Fluid Mech. **206**, 375.
- Keffer, J.F., Kawall, J.G., Giralt, F., Béguier, C. (1982) "Characteristics of coherent structures in complex turbulent shear flows", in *Turbulence in Heat and Mass Transfer*, Hemisphere Publ. Corp., Washington, 177.
- LaRue, J.C., Libby, P.A. (1974) "Temperature and intermittency in the turbulent wake of a heated cylinder", Phys. Fluids, **17**, 873.
- Laufer, J. (1975) "New trends in experimental turbulence research", Ann. Rev. Fluid Mech., **7**, 307.
- Lumley, J.L. (1965) "The structure of inhomogeneous turbulent flows", Proc. Intern. Colloq. on Atmospheric Turb. and Radio Wave Propagation, Moscow. pp 166.
- Lumley, J.L. (1990) Comment 1 to Session One, in *Whither Turbulence? Turbulence at the Crossroads*, ed. J.L. Lumley, Springer-Verlag, Berlin.(Lecture Notes in Physics, **357**, pp 49-57)
- Mumford, J.C. (1982) "The structure of the large eddies in fully developed turbulent shear flow. Part 1. The plane jet", J. Fluid Mech. **118**, 241.
- Perry, A.E., Lim, T.T., Chong, M.S. (1980) "The instantaneous velocity fields of coherent structures in coflowing jets and wakes", J. Fluid Mech.,**101**, 243.
- Perry, A.E., Watmuff, J.H. (1981) "The phase-averaged large-scale structures in three-dimensional turbulent wakes", J.Fluid Mech. **103**, 33.
- Townsend, A.A. (1956) "The Structure of Turbulent Shear Flow (1st ed.)", Cambridge University Press, Cambridge.
- Watanabe, S. (1985). *Pattern Recognition: Human and Mechanical.* John Wiley & Sons, New York

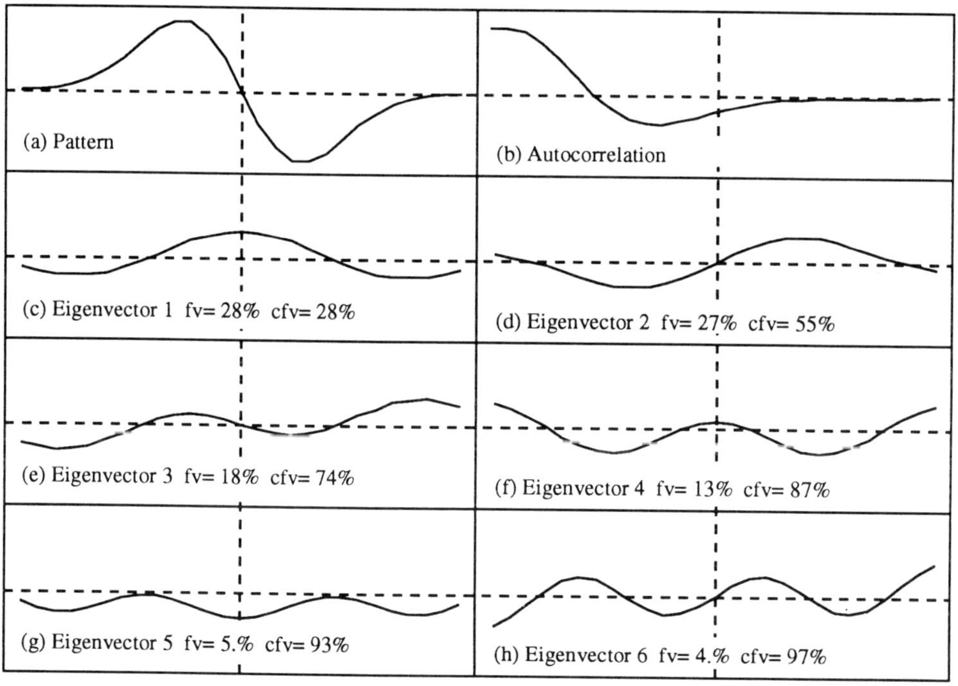

FIGURE 1. Principal component analysis of a synthetic time series . (a) basic pattern. (b) autocorrelation. (c)-(h) eigenvectors with the largest eigenvalues

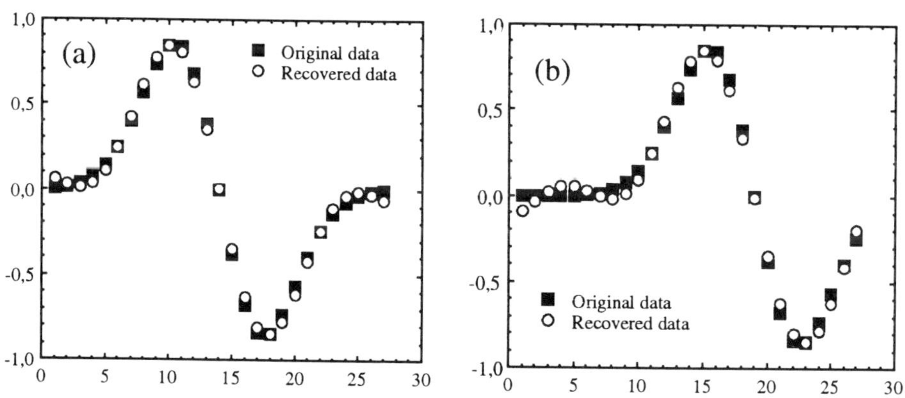

FIGURE 2. Reconstruction of two different data windows with the eigenvectors displayed in Figure 1. (a) aligned pattern. (b) misaligned pattern.

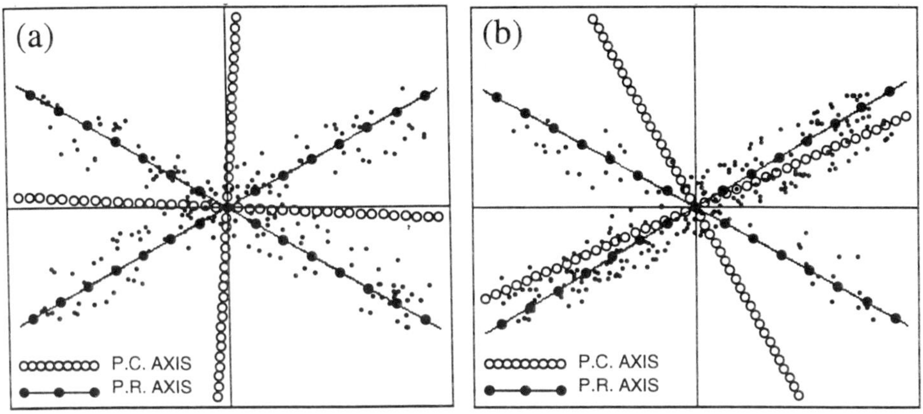

FIGURE 3. PCA and PR analysis of synthetic uv data. (a) two populations with an equal number of points. (b) two populations with an unequal number of points and with a ratio 5:1.

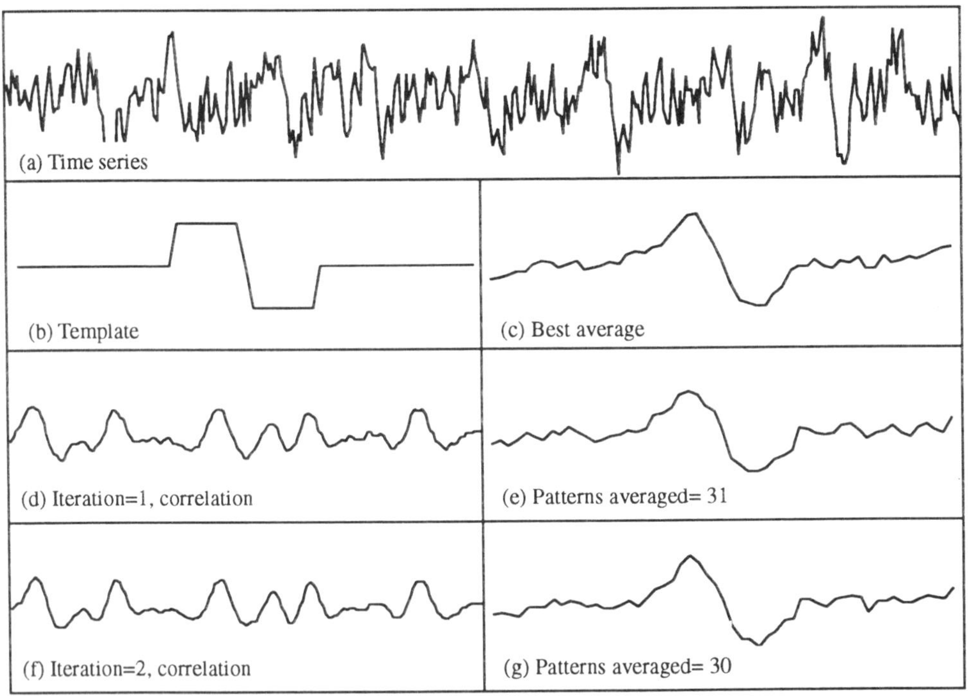

FIGURE 4. PR analysis of a synthetic time series. (a) time series. (b) initial template. (c) best averaged pattern. (d) correlation coefficient at iteration 1. (e) pattern educed after iteration 1. (f) correlation coefficient at iteration 2. (e) pattern educed after iteration 2.

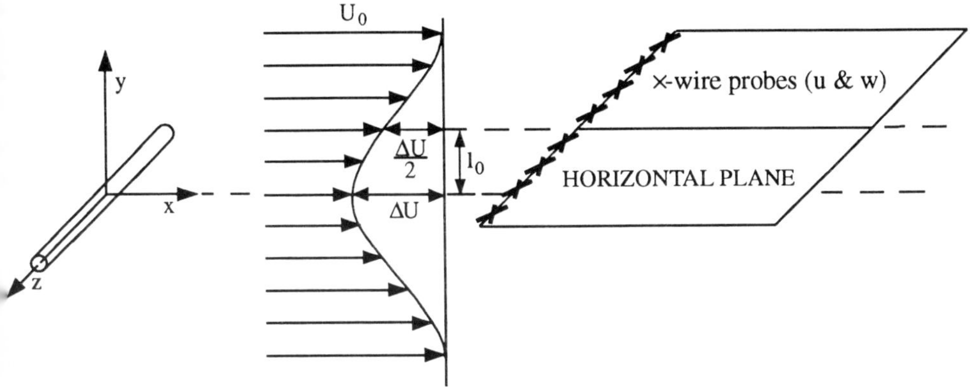

FIGURE 5. Sketch of the flow and sampling plane.

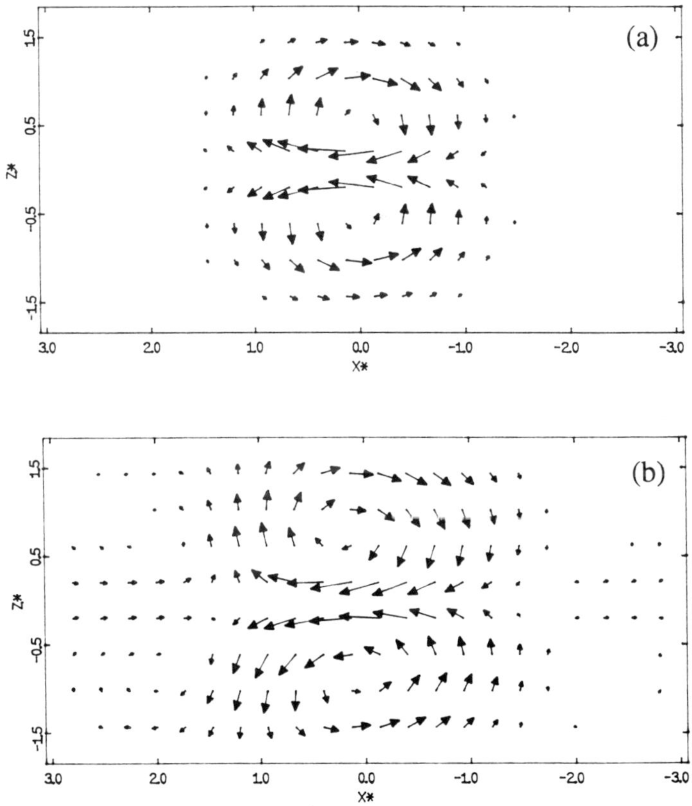

FIGURE 6. PR analysis of the u and w data in the horizontal plane. (a) initial template. (b) ensemble average after three iterations.

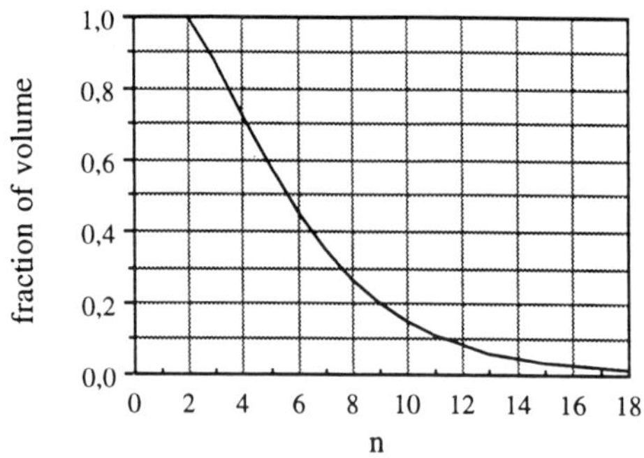

FIGURE 7. Fraction of the volume of a n-dimensional sphere filled by n double cones of 45° as a function of n.

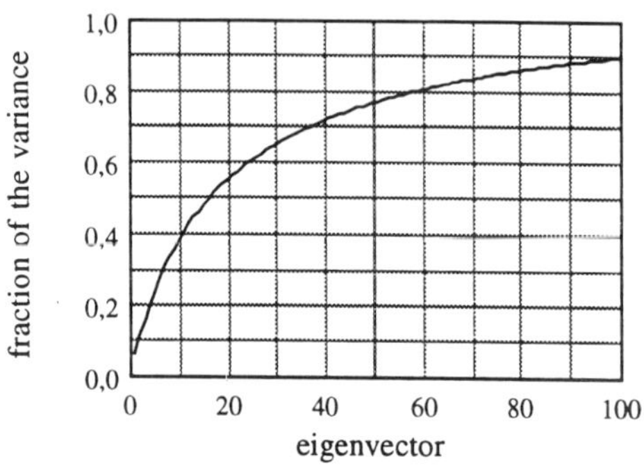

FIGURE 8. Cumulative fraction of the variance explained by the first 100 eigenvectors obtained by PCA.

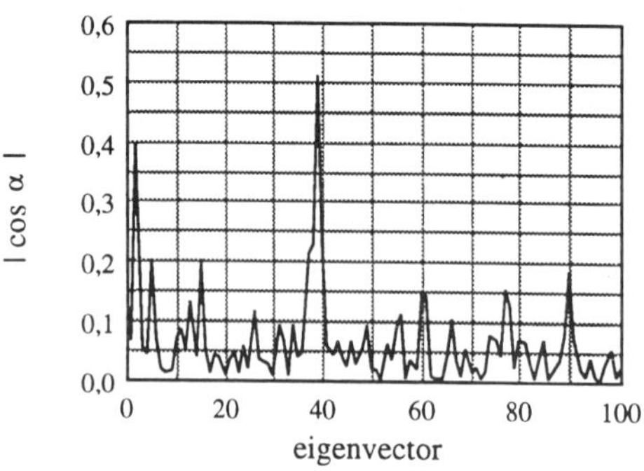

FIGURE 9. Angle between the double roller eddy displayed in Fugure 6b and the first 100 eigenvectors obtained by PCA.

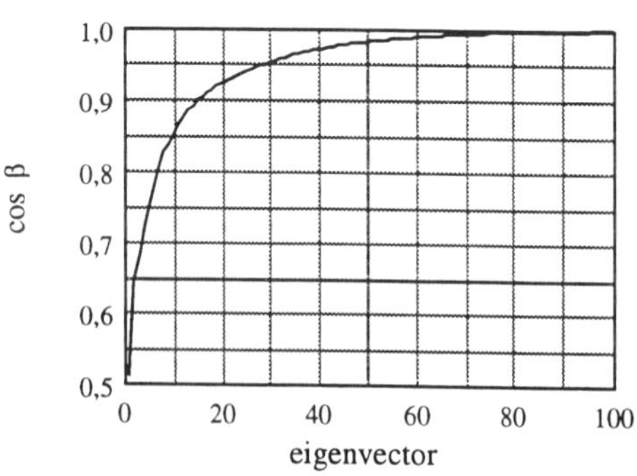

FIGURE 10. Angle between the double roller eddy displayed in Figure 6b and its reconstruction using the eigenvectors closer to the double roller.

PATTERN RECOGNITION ANALYSIS OF STRUCTURE IN MANIPULATED CHANNEL FLOW SIMULATIONS.

A.M.Savill[1], H.Klein[2] & R.Friedrich[2]

1:Engineering Department
University of Cambridge, England.
2:Institute of Fluid Mechanics
Technical University of Munich, Germany.

Abstract. The possibility of applying Pattern Recognition Analysis (PRA) techniques, originally developed for analysing experimental multipoint hot-wire probe data, to Large Eddy Simulation output for fully-developed plane and manipulated channel flows is discussed. It is apparent that patterns can be detected in the Simulated streamwise velocity fluctuation fields which are similar to the ensemble-averaged patterns derived from PRA of experimental data for plane and manipulated boundary layers. This means that the detailed velocity vector and pressure field information provided by the Simulations can be used to examine the typical 3D structure of the large eddies producing the patterns. In practice it appears that these eddies have a similar vortex-loop structure to that inferred from the more limited experimental data for both plane wakes and boundary layers. However their scales are such that they appear more like squatter, wake 'horse-shoe' vortices than the thinner 'hairpin' eddies found in boundary layers. Furthermore, just as the existence of horse-shoe vortices in the wake seems to be closely associated with stronger entrainment, the eddies in each half of the channel flow appear to be reinforced by cross-flow from the opposite wall region. When manipulators are introduced into the flow these not only influence the flow near the channel walls, but also serve to damp such cross-flow interactions. There would thus appear to be rather more similarity between internal and external flow manipulation than has previously been suspected; as is made clear by further comparison with plane and manipulated boundary layer Simulation results. This finding has important implications for the scaling of internal manipulator parameters, as well as attempts to predict their optimum values using turbulence models developed for external flow application. These points are discussed together with the possibility of using the Simulations to resolve outstanding questions regarding the manner in which internal flows recover from manipulation, and how more beneficial control of the flow structure might be achieved.

Key words: Simulations, Pattern Recognition, Manipulated Flow Structure

1. Introduction

The concept of applying passive turbulence manipulators to external turbulent boundary layers is now well established and optimum configurational parameters (non-dimensionalised by the boundary layer thickness, δ) have been determined [eg. see 1]. Although it now appears unlikely that manipulators can provide practical nett drag reductions, a great deal is known about their effect on the flow structure and kinematics [eg. see 2]; several drag reduction mechanisms have been identified; and there is considerable current interest in their alternative use as very low drag penalty devices for controlling heat transfer, separation and vortex-dominated flows [see 1].

J. P. Bonnet and M. N. Glauser (eds.), Eddy Structure Identification in Free Turbulent Shear Flows, 195–211.

Rather less attention has been paid so far to possible internal flow applications. Four experimental studies have been reported in the litterature, three in pipe flows and one in a channel [see 3,4].However only a limited number of manipulator configurations have been tested to-date and so the optimum device parameters and consequent achievable performance levels are still uncertain; particularly under fully developed flow conditions when the pipe or channel width D replaces δ as the scaling parameter.

In developing pipe flow, where a δ can still be defined, experiments have indicated that the optimum tandem device parameters are almost identical to those for external application (viz. chord length l=1.2D, spacing s=10δ, height h=0.7δ, and thickness t=.01l, for thin profiled devices), but again no nett drag reduction seems feasible.More optimistic results have been obtained in fully developed pipe flows where it appears possible to at least recover the device drag of both single and tandem ring manipulators. Turbulence modelling techniques, first successfully applied to manipulated boundary layers and then shown to reproduce initial pipe flow manipulation results, are now being used to perform parametric optimisation studies - see [3]. So far the suggested prefered parameters, again for thin profiled plates, appear to be: l=1.2(D/2), s=10l & t<1.5%l, and these are consistent with the findings of the only fully developed channel flow manipulation experiment, although different optimum values for h/(D/2) have been suggested.

Neither the experiments or the modeling have shed much light on the mechanisms involved in internal flow manipulation. It has generally been assumed that these would be essentially the same as those found to be acting in manipulated external boundary layers, since the near wall structure is known to be similar. However it is known that manipulators reduce boundary layer growth by suppressing entrainment [5], thereby partially interrupting the eddy structure regeneration cycle, and this process is absent in fully developed internal flows. It has therefore been inferred [5,6] that manipulators would be less efficient under such conditions. Recently a number of Large Eddy Simulations have been performed for both plane and manipulated channel [7] and boundary layer flows [8], with increasing resolution and streamwise extent [9,10]. These have provided very much more detailed information than is available from the most extensive external flow experimental studies, but this Simulated 'data' has yet to be subjected to any in-depth analysis. By comparison, the far more limited instantaneous multi-point u' hot-wire data recorded in various manipulated boundary layers has already been analysed in considerable detail [eg.see 11]. Most recently PRA procedures (originally developed by Townsend, Savill & Mumford [12],but later greatly refined by Ferre & Giralt [13], and used by these two groups collaboratively to deduce the full 3-D eddy structure of the plane wake [14]) have been applied to similar data for both plane and manipulated boundary layers [15]. The results indicated that very simple trial patterns are sufficient to extract the typical (ensemble-averaged) large eddy structure of the boundary layer and to show that this is asymmetric; being composed of approximately equal numbers of opposite-sign rotation isolated inclined roller eddies, rather than symmetric hairpin loops, as also suggested by studies of individual structures detected in Direct Numerical Simulations [16]. Subsequent analysis of the manipulated flow data indicated that the same PRA techniques could be used to distinquish the superimposed structure of the manipulator wake; thus complimenting earlier single and two-point measurements, by revealing how the wake interacts with the boundary layer structure and suggesting that in recovery from manipulation the flow has a higher degree of near-wall organisation. The purpose of this paper is to present a first assessment of the Large Eddy Simulation data using similar advanced graphical analysis techniques.

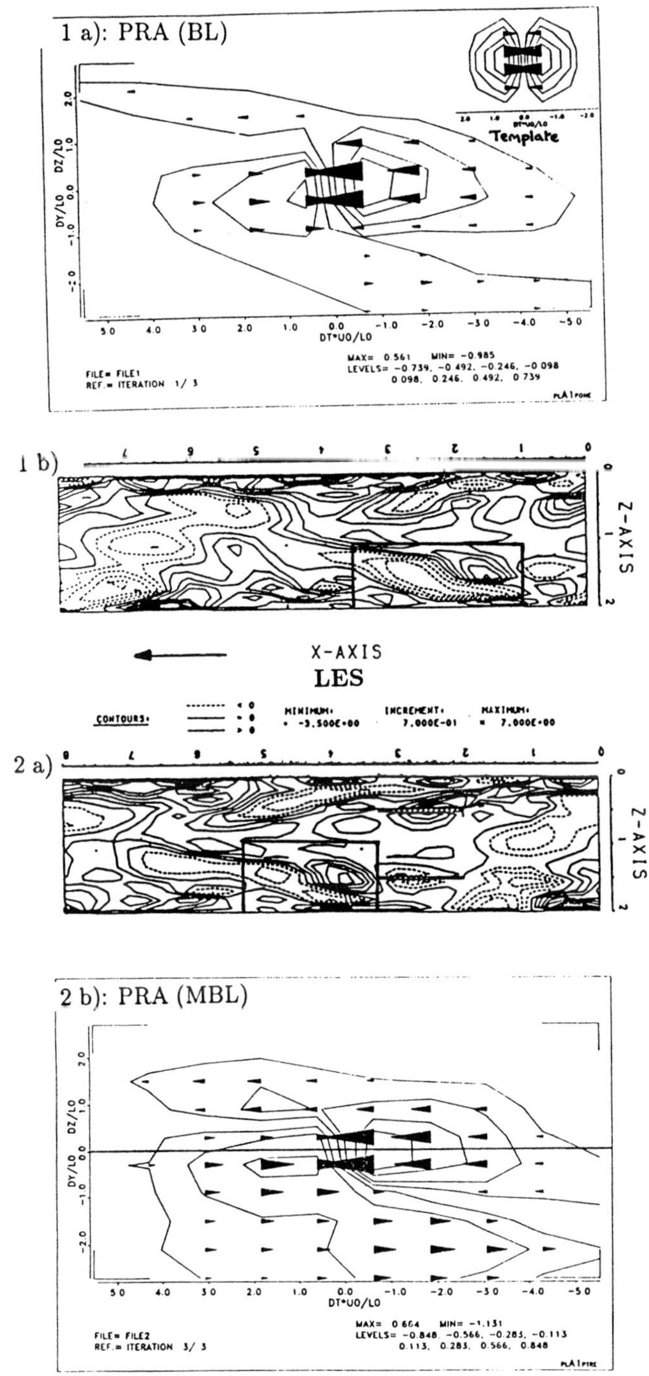

Fig.1 & Fig.2 : Contour lines of longitudinal velocity fluctuations

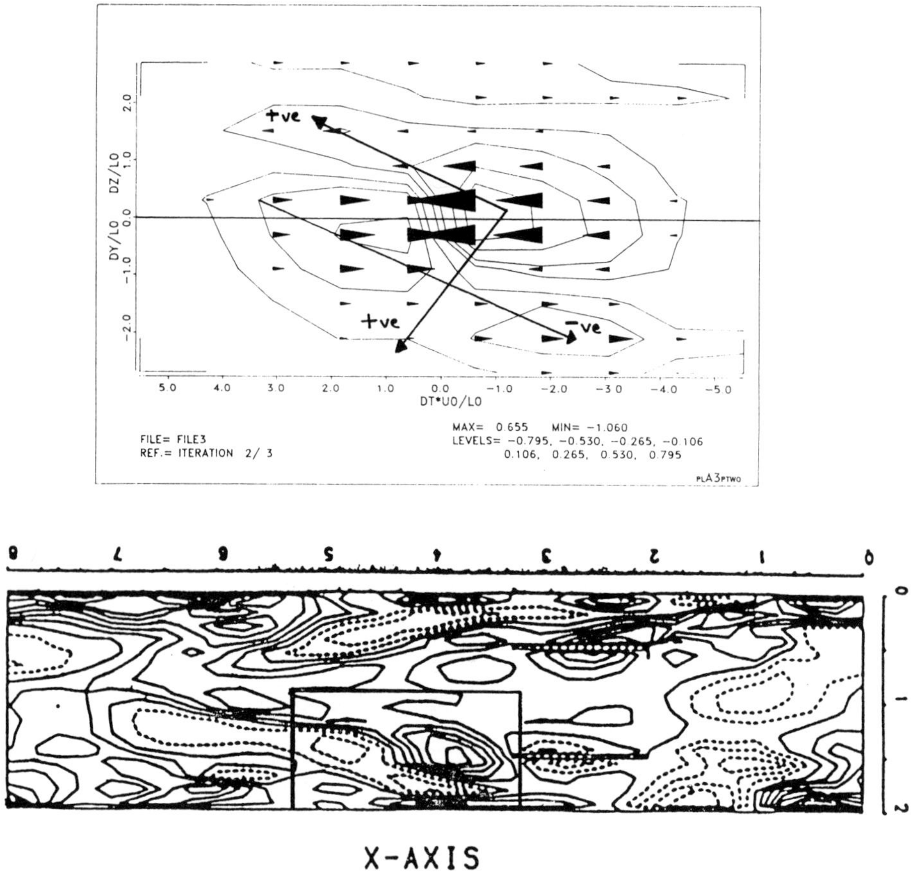

Fig.3 : Comparison of superimposed u' velocity correlation patterns, derived from PRA using an alternative 'velocity front' template, with LES.

2. The Simulations

The Large Eddy Simulations have all been performed by HK and RF at the Lehrstuhl fur Stromungsmechanik of the Technische Universitat Munchen using a CRAY X-MP or CRAY Y-MP. Initial low-resolution (32x16x16 grid) Simulations, for plane and manipulated (l=1.25(D/2), h=0.5(D/2)) channel flows at ReD=150,000, have been followed by higher resolution (64x32x32 grid) Simulations for these flows (with l=1.5(D/2), h=0.31(D/2)) and for corresponding plane and manipulated(l=1.5δ, h=0.31δ) boundary layers (Reθ=8700). In each case the flow domain considered was 8x4x2 d(=D/2) or δ. However more recently comparative results have also been obtained for the latter cases using a similar resolution 192x32x32 grid in order to examine a larger 24x4x2 d flow region. It is important to note that in all cases the manipulators were introduced at a distance D downstream from the inlet plane and were assumed to be infinitely thin, although the normal wall boundary conditions were applied along both their surfaces so that the boundary layers developing there were effectively assumed to be fully turbulent. This would not be true in reality for such short manipulators at the Reynolds numbers under consideration, but a comparison of experimental results for different Rel suggests that this is relatively unimportant for very thin devices, and that the damping of the vertical velocity fluctuations by the plates was correctly accounted for.

3. The Analysis

The PRA/graphical analysis was performed by AMS in the Engineering Department of Cambridge University and at EPF Lausanne. In order analyse the complete velocity vector and pressure field data from selected realisations of the Large Eddy Simulations the corresponding datasets were transfered over existing JANET/EARN e-mail links, from Munich to Cambridge and the EPF Lausanne ERCOFTAC Pilot Centre, using standard MAILFILE routines. The data files were then downloaded and translated into a format acceptable for analysis, using the UNIRAS suite of general 2D graphics routines and a specialist 3D CFD Graphics software package: CACTUS (developed at EPFL), on either HP 340 or HP 835 Turbo, SRX Graphics Workstations, respectively.

4. Results & Discussion

It is clear from a comparison of Figures 1(a)&(b) and 2(a)&(b) that the ensemble averaged u'field shear-plane contour patterns detected using PRA [1] in both a plane boundary layer, and close behind (2.25δ from the trailing edge of) a single manipulator plate with l=1.8δ, h=0.5δ, are remarkably similar to instantaneous u' contours from individual realisations of the initial low resolution (32x16x16) Large Eddy Simulations for the equivalent plane and manipulated channel flows, taken from [7]. This is despite the rather distinct form of the manipulated flow pattern, which has been attributed [15] to the superposition of a typical wake eddy 'pattern' - see [14] - on top of the normal unmanipulated flow pattern, as illustrated by Figure 3. The LES pattern is perhaps better illustrated by Figure 4 in which the local velocity excess (yellow) and defect (blue) regions have been colour coded differently to the mean zero convected velocity background (green). This close similarity of ensemble averaged and instantaneous patterns suggests that one can take the latter to be

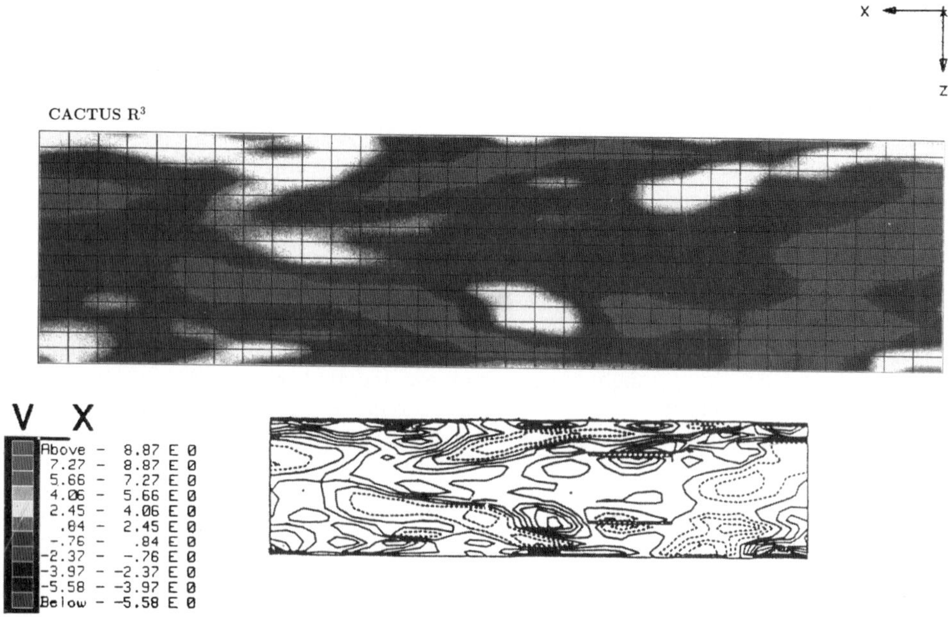

Fig.4 : Colour-coded u' intensity map, with original LES contour plot.

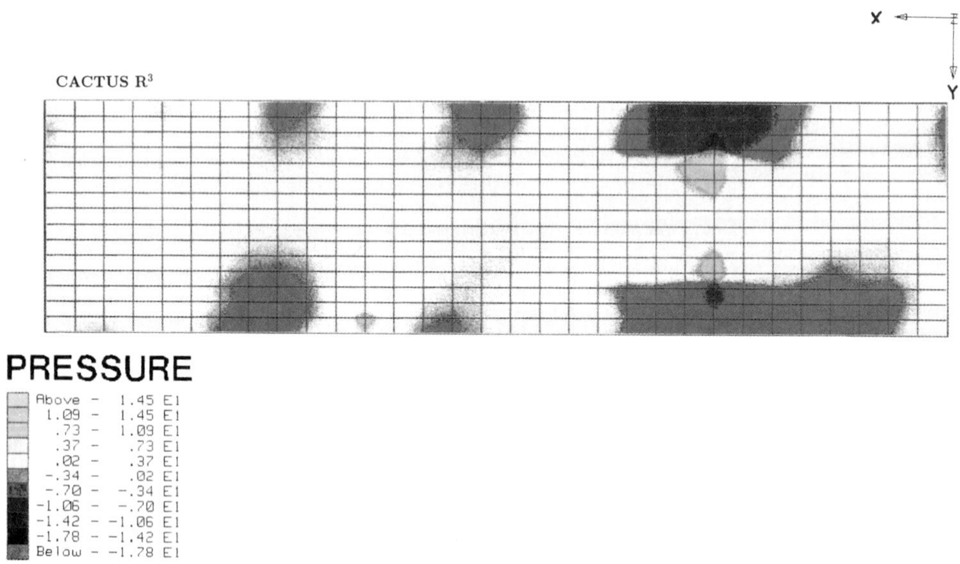

Fig.5 : Vertical section through the pressure field corresponding to Fig.4

Colour version in Colour Plates section, p. xxv.

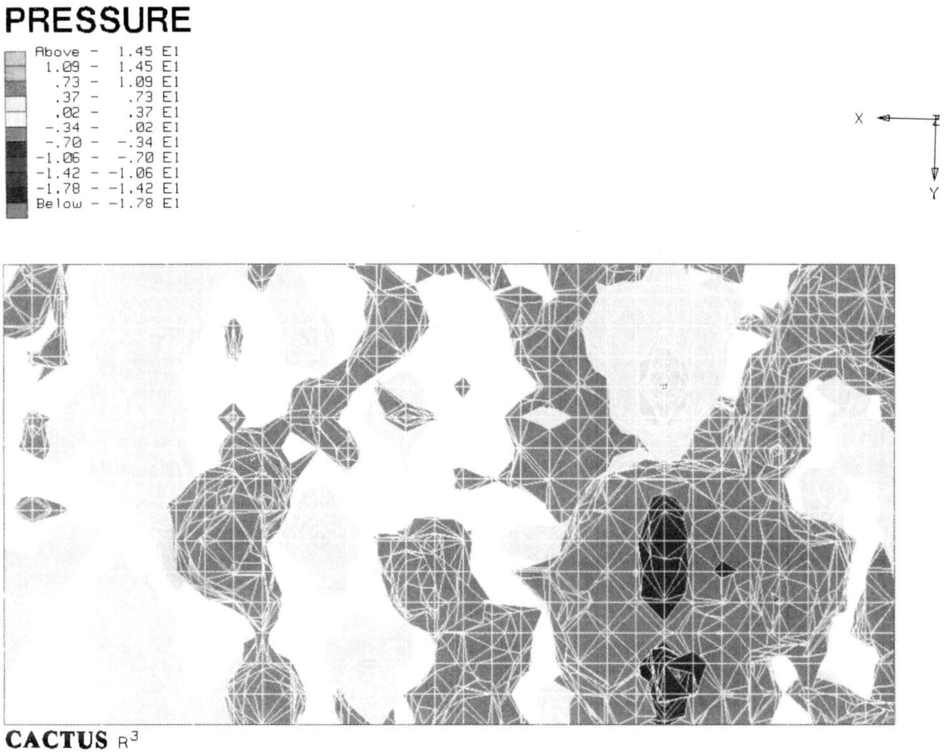

Fig.6 : Spanwise section through the pressure field between the lower manipulator and the bottom wall of the channel.

Colour version in Colour Plates section, p. xxvi.

representative of the shear-plane u' 'footprint' of the typical large eddy structure close behind a manipulator, and use the full velocity field information available from the

Simulation to examine the complete 3D form of this structure even prior to applying PRA to the Simulations themselves. In fact Robinson et al. [16] have shown that the pressure field is a better discriminator of eddy structure than either the velocity or vorticity fields; low p being associated with the cores of vortices. Since pressure field information was also available from the simulations this possibility was therefore examined first. One problem in the manipulated flow case, as illustrated by Figure 5, was that the pressure field was dominated by very high positive peaks at the leading edge of the manipulators and low pressure regions immediately beneath these (as also found in turbulence model predictions for external flow manipulators - see [1]), so that that it was difficult to distinquish the relatively smaller pressure variations across the majority of the computational flow domain. Nevertheless by studying spanwise (xz) sections above the plane of the manipulators or between these and the walls, and by carefully selecting the PRA acceptance thresholds, patterns such as those presented in Figure 6 could be detected. These showed a close similarity to the low p', partial loop section structures, and isolated structure-less high p regions, detected by Robinson et al. [16] in much lower Reynolds number Direct Numerical Simulations of both plane channel and boundary layer flows. Loop-like structures were detected both above and below planes of manipulators, but were squatter (with similar transverse and spanwise dimensions, scaled in terms of d rather than wall units) than boundary layer hairpin eddies (see [17] for further discussion of the latter). They therefore bore closer resemblance to the horse-shoe vortices found in a wake [14].

Initial studies of 2D sections of the corresponding instantaneous uv, vw and uw vector fields also revealed that the eddy structure associated with the u' patterns of Fig.2 was partially driven (or at least reinforced) by a flow into its core originating from the other side of the channel; much as the wake horse-shoes are driven by entrainment. This was a rather unexpected finding, although Antonia & Tietel [18] have recently shown that such cross-flow plays a significant role in the dynamics of fully developed channel flow. Indeed a subsequent examination of typical spanwise velocity field contour plots from the higher resolution LES [8] revealed correlated flow regions extending across the centre-line of the unmanipulated channel flow, and indicated that when manipulators were introduced these tended to align the instantaneous flow velocity vectors with the mean flow, and suppress vortices in both the core and wall regions whilst separating these from one another.

A more detailed examination of the uv field was therefore conducted in order to confirm this picture. Figure 7 clearly shows flow right across the channel upstream of the manipulator being curtailed by the influence of the manipulators, while the flow downstream in the core is remarkably rectilinear for several chords downstream and does not interact significantly with the near wall flow over this region. Progressively stepping through cross-stream (yz) sections of the vw vector field provided an even more instructive view of the influence of the manipulators and their wakes on the flow structure. Just upstream of the manipulator leading edge interactions between the two halves of the flow were clearly evident - Figure 8(a), whereas immediately downstream of the leading edge this cross-flow was directed back towards the channel centre; any cross-stream vorticity being redirected into the stream direction - see Figure 8(b). By the centre plane of the devices it appeared these two processes had been completed (Figure 8(c)) and thereafter the continued presence of the plates appeared to suppress the intensity and scale of any longitudinal motions; this process continuing even immediately behind the manipultots trailing edge where the core and near-wall flow were as distinctly separate as when divided by the imposed solid surface - see Figure 8(d). Even at several chord lengths further downstream - see Figures 8(e)&(f) - although there was a progressive growth of a weak

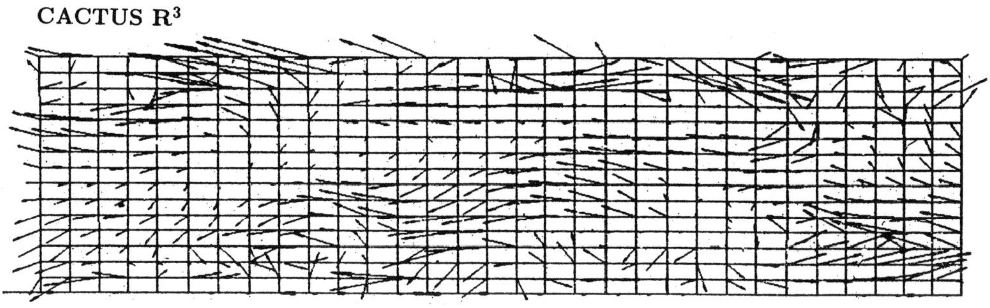

Fig.7 : Shear-plane section of the uv velocity vector field along the centre-line of the channel.

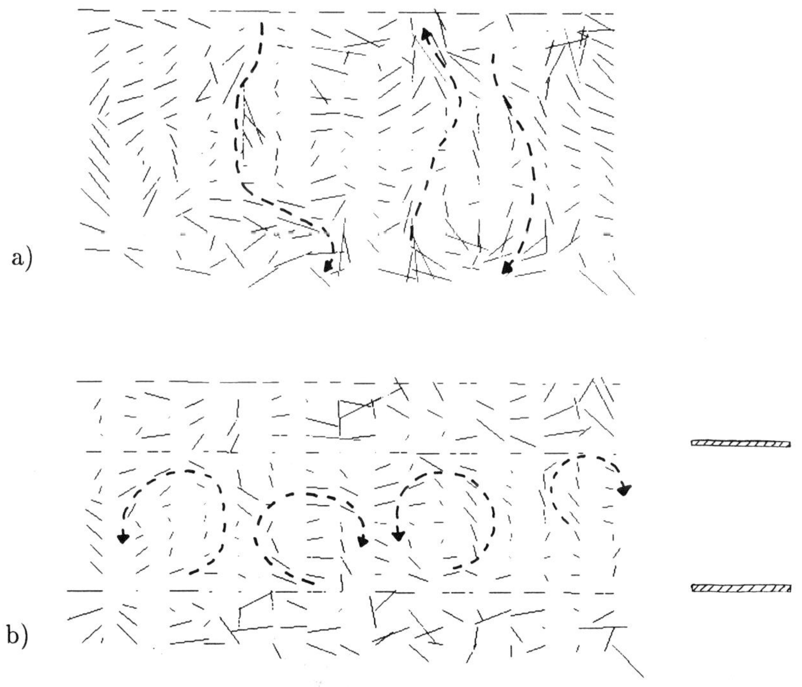

Fig.8(a) & (b) : Cross-stream sections through the vw vector field (a) immediately upstream and (b) immediately downstream of the manipulators leading edge.

CACTUS R³

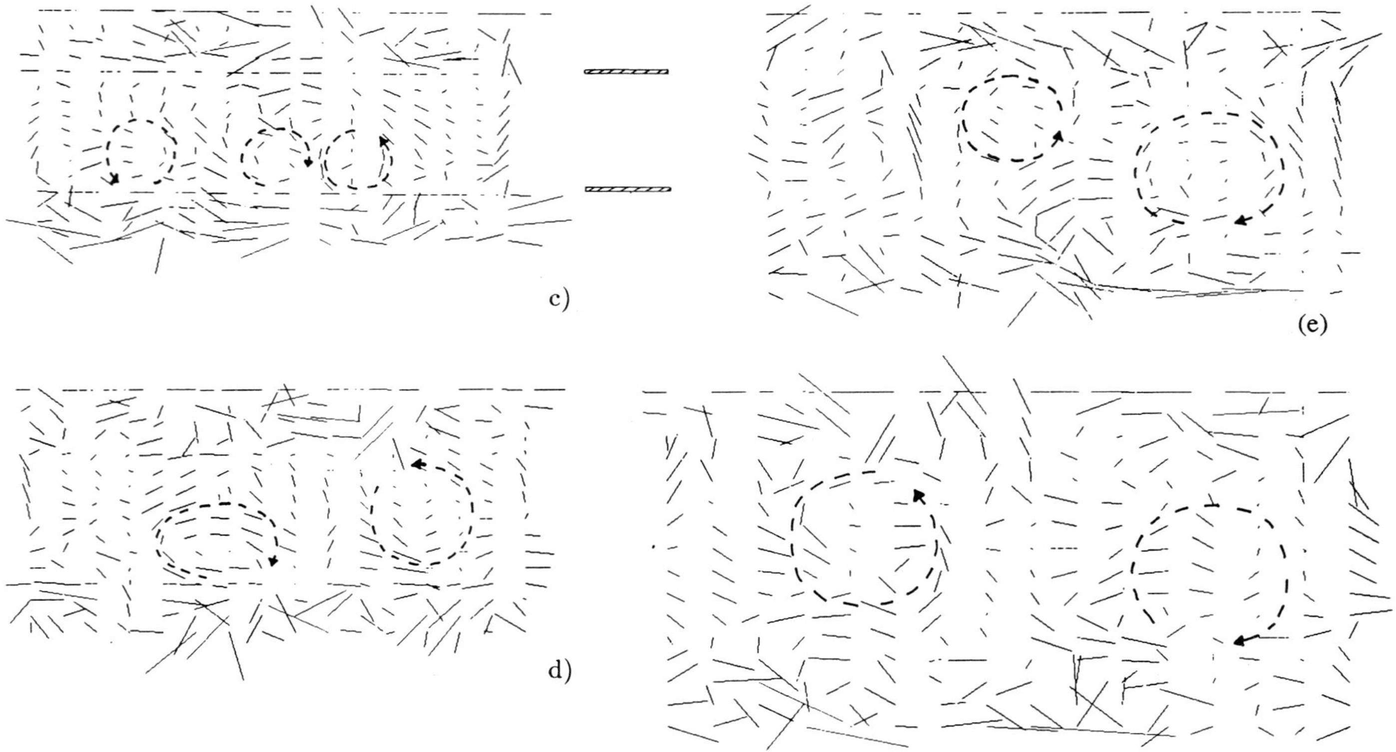

Fig.8 (c) & (d) : Cross-stream sections through the vw vector field (c) through the centre plane of the manipulators and (d) 0.5(D/2) downstream of the manipulators trailing edge.

Fig.8 (e) & (f) : Cross-stream sections through the vw vector field at stations (e): 4(D/2), and (f): 6(D/2), downstream of the manipulators.

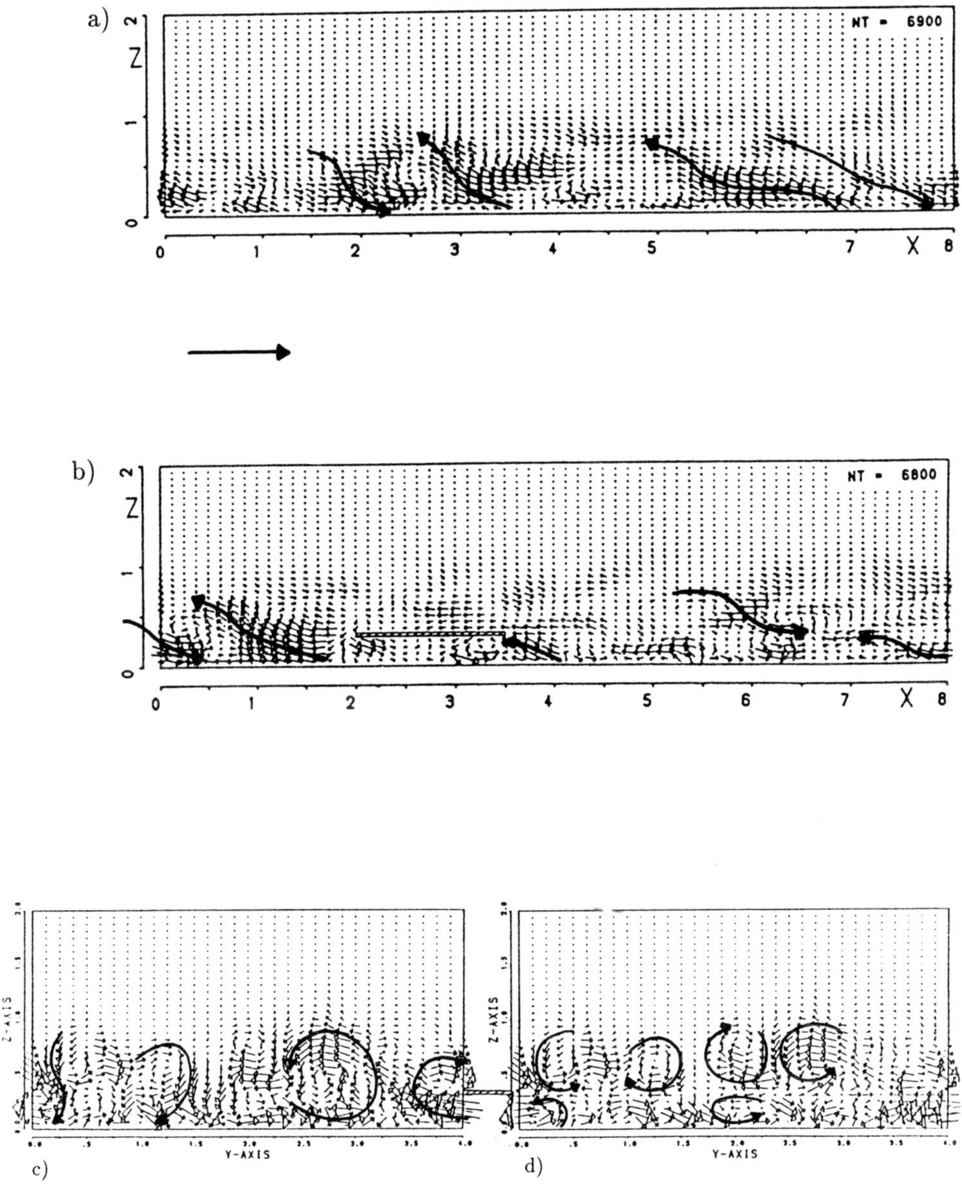

Fig.9 : Fluctuating velocity-vectors in an (x,y) plane: (a) Boundary Layer [BL] (b) Manipulated Boundary Layer [MBL], and in an (x,z) plane: (c) BL (d) MBL.

interaction between the core and wall flows, the scale of the core eddies remained small and these had little effect on the wall region. There was also no evidence of any significant cross flow in this region.

The Large Eddy Simulations for the manipulated boundary layer have yet to be studied in detail, but the relevance of the above observations to these can be gauged from the fact that typical uv velocity vector and v' fluctuation maps (as presented in [9]) clearly show a similar suppression of entraining sweep-type motions to that illustrated in Figure 9, as well as longitudinal vortices in the outer layer above the device location.

The uw field did not reveal any significant differences between the flow upstream and downstream of the devices. This is to be expected if, as indicated by the p' field, the manipulators tend to cut through (but not destroy) the original horseshoe structures; particularly if these are later replaced by similar 'footprint' upper wake structures, as indicated by the experimental studies of Savill & Mumford [19]. However there was no clear evidence for the type of re-connection of isolated horse-shoe structures reported to occur behind manipulators by Makita et al. [20]. Instead the two halves of the structure appeared to remain de-correlated as they decayed; a view supported by the more recent flow visualisation studies conducted by Shengxia et al. [21].

Close to the wall there was insufficient resolution to detect any subtle changes in either the typical eddy structure or its longer range order. Clearly the next step in such analysis must be to study the available higher resolution channel flow LES datasets, particularly those covering a larger streamwise extent. It has already been established that these do also contain 'typical' u' footprints (see example of Figure 10) and exhibit similar Cf reduction distributions to those measured experimentally behind both thick and thin equivalent devices ($l=1.2D$, $h=0.45\delta$, $t=1$ or 0.1mm, $t=0.02$ or 0.004δ) in external boundary layers ($Re\theta=4000$) - see Figure 11. In addition it would appear that the development of the mean flow velocity defect associated with the wake of the manipulators is almost the same for similar devices in both the internal and external flows (see Figure 12). This is despite the fact that, because there is no entrainment in the fully developed internal flow; the streamlines remaining aligned parallel to the walls so that the manipulators are not effectively placed at a small positive angle of attack (as in the boundary layer) and the wake is therefore not subjected to 'downwashing'. Indeed, just as in external flow, the minimum Cf location is essentially co-incident with the point where the wake defect reaches the wall.

Overall then it would appear there are rather more similarities between manipulated external and internal flows than has previously been suspected. It will be interesting to see if further PRA of the extended higher resolution Simulations can shed more light on the flow structure in the near-wall region at the maximum reduction location, where external flow studies indicate a strong influence of imposed 2D effects from the wake; and in the recovery region, where the same studies suggest a more stable near-wall structure develops due to a persistent shielding influence of the wake in blocking external disturbances. A better understanding of these structural modifications in such non-equilibrium flows is required in order to improve current turbulence models, which all predict too rapid a recovery for both external and internal flows (largely due to the use of equilibrium approximations [22]) by turning to new '2D limit' closure approaches [23] .

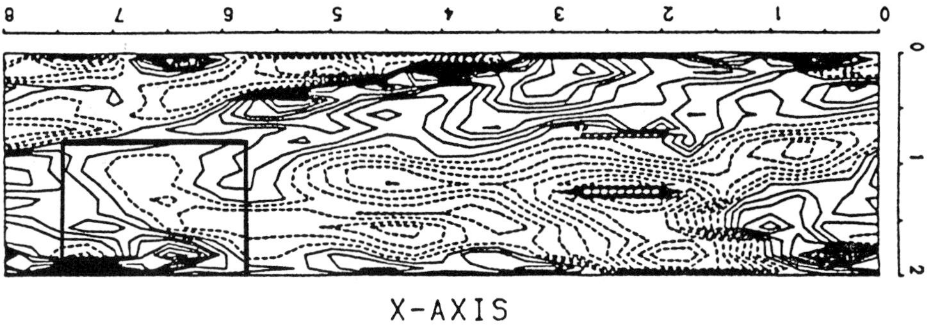

Fig.10 : Example of opposite sign typical eddy correlation contour 'footprint' in u' field from a different LES dataset.

Contours: – – – – – < 0, ———— ≥ 0

Minimum: - 5.694, maximum: 5.753, increment: 0.6025

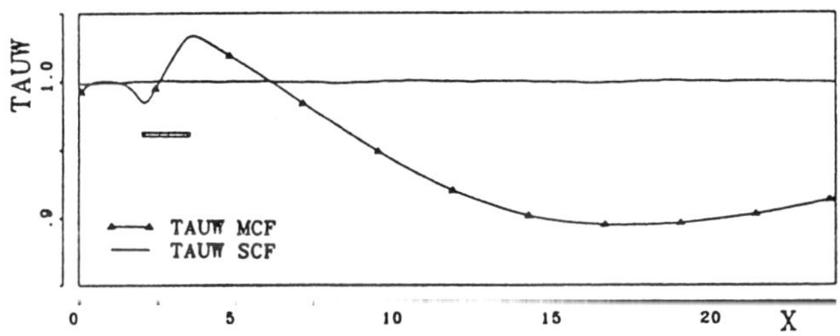

Comparison of wall shear-stress in the CF and MCF (LES)

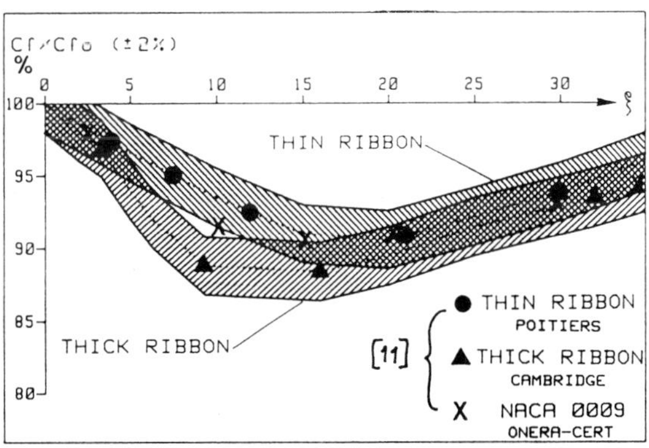

Fig.11 : Comparison of (a) wall shear-stress in the Channel Flow [CF] & Manipulated Channel Flow [MCF] from LES, and (b) Percentage of skin friction reduction versus normalized streamwise location in BL from Experiment.

5. Concluding Remarks

The present analysis has shown that even artificially generated, relatively coarse resolution, Large Eddy Simulations can provide valuable new information regarding the manipulation of turbulent eddy structure, and demonstrated that experimentally-derived Pattern Recognition Analysis techniques can indeed usefully be applied to such Simulations. Of course the results have to be interpreted with care, since, although the Reynolds numbers

considered are similar to experiments, no information is provided concerning small scale interactions due to the relatively low-resolution employed and the use of sub-grid scale models. Fortunately in the case of control by turbulence manipulators one is primarily concerned with their effect on the Large Eddy structure.

Other work, performed at CTR by Stretch et al. [24] using a 3D template PRA approach, developed as a direct extension to the 2D/1-component procedure of Ferre & Giralt [13], has already indicated the value of applying PRA to Direct Numerical Simulations. However it seems clear from the present study that it is not in fact necessary to employ a 3D PRA techniques in order to extract the full 3D eddy structure from the 3D velocity vector and pressure (and or scalar) fields provided by Turbulence Simulations. Indeed this may well not be the best strategy. Instead much simpler 2D/1-component [13,14], or the more recently developed 2D/2-component [25] template approaches are sufficient to detect typical large eddy structures in such 'data'. The 3D approach produces unambiguous results, but due to natural variations in the shape, intensitity, and orientation of individual structures, the degree of correlation between the detected patterns and the template or the frequency of detection of the patterns will be low (depending on whether a detection threshold is specified or not). Using a simpler 2D template to search for the 'footprint' patterns typical of the planes of circulation of the eddies, and then storing the full 3D fields associated with these, allows the detection of larger numbers of structures with a higher degree of correlation between individual patterns and the template. It is also far easier to account for a range of orientations and scale variations when working with 2D rather than 3D patterns. Both these benefits are particularly important when applying PRA to Simulated datasets where the record lengths and hence numbers of structures to be detected will be considerably smaller than for experimental databases. With care, experience and by re-processing with different templates, the sign and degree of organisation of single and even multiple structures can be evaluated just as unambiguously as with the 3D approach; and, as we have seen, even distorted or superimposed structures can be distinguished.

At the same time the apparent power of such simple 2D detection techniques raises the possibility of performing PRA not just to post-process Simulations, but actually within the Simulations themselves - in real time. This would help overcome one of the basic limitations of current Simulations: that most of the generated data has to be overwritten due to storage limitations. By detecting structures as they evolved it would be possible to ensure that only the most relevant data from successive time-steps was recorded and thus to study their dynamics in detail. It would also be possible to investigate the effects of artificially manipulating individual structures in order to test new flow management concepts. Indeed some progress in this direction has already been made at NASA Ames by applying simple constraints to the wall boundary conditions [26] or flow properties [27] specified for Simulations.

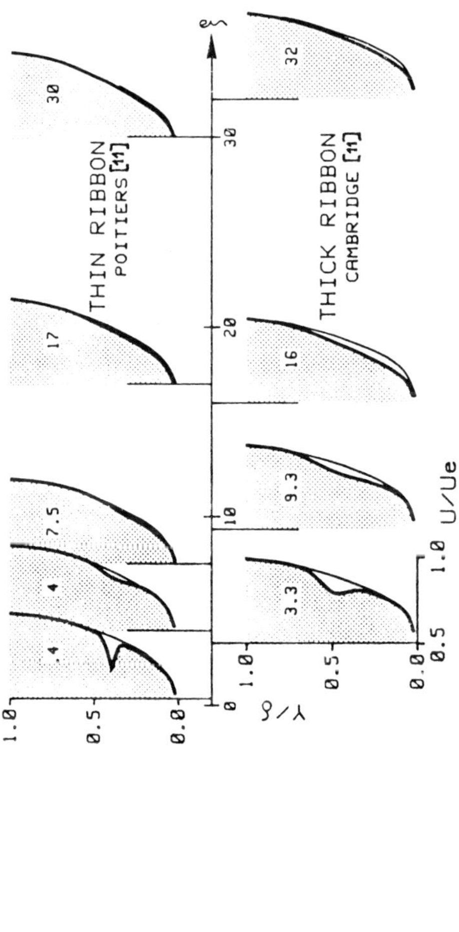

Fig.12 : Streamwise component of mean velocity U in BL (expt) & LES Momentum deficit Δ<u> = <u>MBL - <u>BL downstream of manipulator.

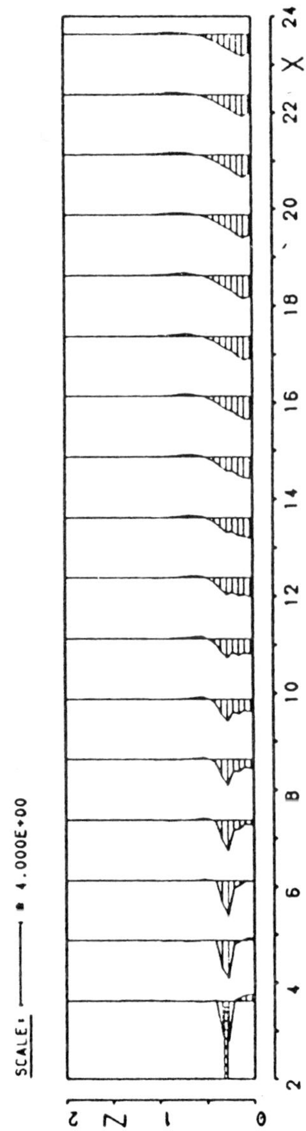

LES Momentum deficit $\Delta\langle u\rangle = \langle u\rangle_{MBL} - \langle u\rangle_{BL}$ downstream of the manipulator.

6. Acknowledgements

This work was performed as part of a Research Project on Pattern Recognition Analysis and Turbulence Simulations conducted at the EPF Lausanne ERCOFTAC Pilot Centre; closely linked to the activities of the ERCOFTAC Large Eddy Simulation Special Interest Group (LESIG) and the Drag Reduction SIG. Much of the analysis was performed while Dr.Savill was an ERCOFTAC Visiting Researcher at EPFL in 1990 & 1991. He is grateful to EPFL for funding these visits and providing access to the CACTUS/HP 835 Turbo SRX Graphics Analysis system, and to Rolls-Royce plc for supporting and encouraging this research under Research Brochure PVA3-100D. Additional thanks are due to Dr.Spyros Gavrilakis and Francois Thibaud of EPFL for assistance with writing LES data conversion/selection routines for UNIMAP and CACTUS respectively. The LES Simulations were generated by Dr.Klein & Professor Friedrich as part of work supported at T.U.Munich by Stiftung Volkswagenwerk.

7. References

[1]Savill, A.M. : 1989, Drag Reduction by Passive Devices - a Review of some Recent Developments. *Structure of Turbulence and Drag Reduction*, IUTAM Symp., Zurich (Springer-Verlag).

[2] Tardu, S. & Binder, G. : 1990, Review: Effect of the OLDs on near wall coherent structures; Discussion and Need for Future Work. In *Recent Developments in Turbulence Management* (Ed: K.-S.Choi, Kluwer Academic Publishers).

[3] Pollard, A., Thomann, H. & Savill, A.M. : 1990, Manipulation and Modelling of Turbulent Pipe Flow: Some Parametric Studies of Single and Tandem Ring Devices. In *Turbulence Control by Passive Means* (Ed: E.Coustols, Kluwer Academic Publishers)

[4] Mah, Y.A., Chew, Y.T. & Khoo, B.C. : 1991, Blade manipulators in fully developed turbulent pipe flow. Proc. *1st ASME/JSME Joint Fluids Engineering Conference*, Portland Paper No.GTOP-91-6.

[5] Savill, A.M., Truong, T.V. & Ryhming, I.L. : 1988, Turbulent drag reduction by passive means: a review and report on the first European drag reduction meeting. *J. de Mecanique Theorique et Appliquee* 7 (4), 353-378.

[6] Chang, S. & Blackwelder, R.F. : 1990, Modification of large eddies in turbulent boundary layers. *J.Fluid Mech.* 213, 419-442.

[7] Friedrich, R. & Klein, H. : 1989a, Large-Scale Turbulence Structure in a Manipulated Channel Flow. In *Proc. IUTAM Symposium on Structure of Turbulence and Drag Reduction* (Ed:A.Gyr, Springer-Verlag) p.483-493

[8] Klein, H. & Friedrich, R. : 1989 Manipulating large-scale turbulence in a channel and a boundary layer. *Proc. 7th Turbulent Shear Flows Symposium* 25.4.1-6.

[9] Klein , H. & Friedrich, R. : 1990, Large-Eddy Simulation of Manipulated Boundary layer and Channel Flows. In *Turbulence Control by Passive Means* (Ed: E.Coustols, Kluwer Academic Publishers).

[10] Klein, H. : 1990, *Numerische Simulation manipulierter turbulenter Scherstromungen.* Ph.D. thesis. Technicsche Universitat Munchen.

[11] Lemay, J., Savill, A.M., Bonnet, J-P. & Delville, J. : 1987, Some Similarities Between Turbulent Boundary layers Manipulated by Thin and Thick Flat Plate Manipulators. *Proc. 6th Turbulent Shear Flows Symposium* TSF6 (Ed.J.Cousteix & R.Houdeville, Springer Verlag).

[12] Townsend, A.A., Savill, A.M. & Mumford, J.C. : 1980, Identification of flow patterns in turbulent flows. *EUROMECH 132 on Hot-wire, hot-film and Conditional Measurements,* Lyon (reviewed in *J.Fluid Mech* 110).

[13] Ferre, J.A. & Giralt , F. : 1989, Pattern-recognition analysis of the velocity field in a plane turbulent wake. *J. Fluid Mech.* 198, 27-64.
[14] Ferre, J.A., Mumford, J.C., Savill, A.M. & Giralt , F. : 1990, Three-dimensional large-eddy motions and fine-scale activity in a plane turbulent wake. *J.Fluid Mech.* 210, 371-414.
[15] Savill, A.M. & Ferre, J.A. : 1989, Extension of Pattern Recognition Analysis Techniques to Distorted and Manipulated Flows. *Proc. 10th Australasian Fluid Mechanics Conference*, Melbourne I, A6.1-5.
[16] Robinson, S.K., Kline, S.J. & Spalart , P.R. : 1988, Quasi-CoherentStructures in the Turbulent Boundary Layer. Parts I & II. *Proc.Zoran Zaric Memorial International Seminar on Near-Wall Turbulence* (Ed. E.Ganic; Hemisphere Publishing Corporation).
[17] Savill, A.M. : 1990, A structural model for turbulent boundary layers with implications for their prediction and control. In *Model Concepts: Turbulent Boundary Layer Structure Workshop,* NASA Langley Research Centre.
[18] Antonia, R.A. & Teitel , M. : 1989, Organised motion in a fully developed turbulent duct flow. *Proc. 10th Australasian Fluid Mechanics Conference*, Melbourne, I, D.35-38.
[19] Savill , A.M.& Mumford, J.C. : 1988, Manipulation of turbulent boundary layers by outer-layer devices: skin friction & flow-visualization results.*J. Fluid Mech.* 191, 389-418.
[20] Makita, H.,Sassa, K., Abe, M. & Itabashi, A.: 1989, Decay Process of a Manipulated Large-scale Horse-shoe Vortex in a Boundary Layer. *AIAA J.* 27 (2) 155-160.
[21] Shengxi, S., Changhui, D. & Zhou M.D. : 1989, A Visualisation Around Plate Manipulators and Discussion of the Mechanism of Drag Reduction. *Proc. 4th Asian Congress of Fluid Mechanics*, Hong Kong Vol. II, p.C146-149.
[22] Savill, A.M. : 1987, Algebraic and Reynolds Stress Modelling of Manipulated Boundary Layers Including Effects of Free-Stream Turbulence. *Proc. RAeS International Conference on Turbulent Drag Reduction By Passive Means*, I, 89-143.
[23] Savill , A.M.: 1990, Tutorial on Turbulence Modelling. In *Simulation of turbulent and multi-phase flows.* Proc. 2nd ERCOFTAC Summer School, Oxford (CUP to appear).
[24] Stretch, D., Britter, R.E. & Kim, J. : 1990, Patterns in Simulated Turbulent Channel Flow. *Proc. 3rd European Turbulence Conference* (see also*CTR Annual Research Briefs*).
[25] Ferre, J.A., Giralt, F. & Antonia, R.A. : 1989, Evidence forDouble-Roller Eddies in a Turbulent Wake from Two-Component Velocity Measurements. *Proc. 7th Turbulent Shear Flows Symposium*, Stanford Vol.2,p.24.2.
[26] Moin, P., Ki, J. & Choi, H. : 1989, On the Active Control of Wall-Bounded Turbulent Flows. *AIAA-89-0960.*
[27] Choi, H. Moin, P. & Kim, J. : 1990, Turbulence Control for Drag Reduction in Wall-bounded Flows. *APS Bulletin* 35 (10) 2232.

VI. Proper Orthogonal Decomposition

DETECTION OF SELF-SIMILAR MODES IN TURBULENCE: APPLICATION TO A WAKE FLOW

Nian-Zheng Cao and Nadine Aubry
The Levich Institute and Department of Mechanical Engineering
The City College of the City University of New York
New York, NY 10031

Abstract. By using the space-time scaling invariance of the Navier-Stokes equations, the spatio-temporal structure of fully developed turbulence, in the homogeneous or inhomogeneous cases, can be understood in terms of a spatio-temporal self-similarity. The latter consists in a hierarchy of spatial and temporal modes which are all similar, namely related to each other through a dilation symmetry group. Our analysis uses bi-orthogonal decompositions which decompose the flow into spatial and temporal orthogonal modes for which there is a one-to-one correspondence. It also permits the derivation of (exponential) spectrum laws which, for the kinetic energy of the flow, coincide with Kolmogorov's $k^{-5/3}$ power law when the flow is both homogeneous and incompressible. Such self-similarity, or fractal structure in a spatio-temporal sense, originates in the *space-time* dynamics of the flow and is unlikely to be observed from the analysis of a one-dimensional signal (obtained by a single point measurement or an instantaneous picture of the flow). It is, however, recovered by the spatial and temporal two-point correlations and therefore should appear in the structure of the eigenvectors and eigenvalues of the Proper Orthogonal Decomposition in which the ensemble average is a time average. The analysis of spatial two-point correlations, experimentally measured in a turbulent far-wake flow, supports the previous theory.

1. Space-time scaling symmetry of the Navier-Stokes equations

Utilizing the scaling invariance of the Navier-Stokes equations, decay laws for the 'inertial' (or scaling) range of the turbulence spectrum have been derived [1] for both incompressible and compressible, homogeneous and inhomogeneous turbulence. In the incompressible case, the Navier-Stokes equations, in nondimensional variables,

$$\begin{aligned} &\frac{\partial u}{\partial t} + (u.\nabla)u = -\nabla p + \frac{1}{Re}\nabla^2 u \\ &\nabla.u = 0 \end{aligned} \qquad (1)$$

where u denotes the velocity field, p the pressure and Re the Reynolds number are invariant under the scaling group of transformations:

J. P. Bonnet and M. N. Glauser (eds.), Eddy Structure Identification in Free Turbulent Shear Flows, 215–224.

$$\begin{aligned} x &\rightarrow \lambda x \\ t &\rightarrow \lambda^{1-h} t \\ u &\rightarrow \lambda^{h} u. \end{aligned} \tag{2}$$

An interesting remark is that the Reynolds number is invariant through the transformation so that the kinematic viscosity needs to be rescaled. As well-known, it is reasonable to assume that the fully developed turbulent solutions are themselves invariant under one of these groups. As noticed by Frisch [2], this is precisely one of Kolmogorov's hypotheses [3], at least in a statistical sense (we will return to this point below). First, let us assume that the flow is indeed invariant under the scaling transformation Eq. 2 for a particular value of h (λ, chosen smaller than 1, being the group element), namely

$$u(x, \lambda^{1-h} t) = \lambda^{h} u(\lambda^{-1} x, t). \tag{3}$$

The appropriate value of h, h = 1/3, is determined by the additional, physical assumption that the energy transfer is constant through the scale transformation Eq. 2 [3]. Under the additional assumption that the flow is homogeneous and isotropic, the Fourier spectrum decays as $k^{-5/3}$ [3]. This law has been strongly supported by experimental and numerical results. Scaling is, however, limited to the so-called inertial subrange of the spectrum, being broken in the energy containing range by the influence of the boundary conditions and in the dissipative range.

2. Biorthogonal decomposition and space-time symmetries

Using the same arguments as Kolmogorov [3], it is possible to derive spectrum laws for inhomogeneous turbulence. For this purpose, it is convenient to use the spectrum of the operator U (as defined below), introduced in the derivation of 'bi-orthogonal decompositions' [4]. Bi-orthogonal decompositions consist in expansions of a space-time function - or signal - u(x,t) (here the velocity) into spatial orthogonal modes in a Hilbert space H(X) ($x \in X$) and temporal orthogonal modes in a Hilbert space H(T) ($t \in T$) for which there is a one-to-one correspondence between both sets of modes. The operator U is defined as

$$U: H(X) \rightarrow H(T)$$

such that

$$\forall \varphi \in H(X), \quad (U\varphi)(t) = \int_X u(x,t)\varphi(x)d\mu(x) \tag{4}$$

where $d\mu(x)$ (resp. $d\tilde{\mu}(t)$) denotes the measure defining the scalar product in H(X) (resp. H(T)). The spectral analysis of the operator U gives a biorthogonal decomposition of u(x, t), which, if U is a compact operator, can be written as

$$u(x,t) = \sum_{n=1}^{N} A_n \overline{\varphi_n(x)} \psi_n(t) \tag{5}$$

$$\text{with} \quad A_1 \geq A_2 \geq \ldots \geq A_N > 0,$$

$$\text{and} \quad (\varphi_n, \varphi_m)_{H(X)} = (\psi_n, \psi_m)_{H(T)} = \delta_{n,m},$$

where the parentheses denote the scalar products in the respective Hilbert spaces. The treatment of cases where U is non-compact can be found in [1]. Two particular examples of such a decomposition are the two-dimensional Fourier decomposition for plane waves in an infinite domain or a spatially and temporally periodic domain and the probability theory tool called the proper orthogonal decomposition (POD), Karhunen-Loève expansion or principal component analysis viewed from a deterministic approach (which uses $L_2(X)$) when the statistical average is chosen to be the time average only (in all other cases, the connection with the dynamics, essential in all our derivations, cannot be made). The application of the latter to turbulence was first proposed and developed by Lumley [5], discussed more recently in a certain number of works and applied to a rather large number of flows (see e.g. [6] as well as contributions to this symposium). The POD, however, is formally statistical and consists in decomposing the two-point correlation only, namely U^*U or UU^* where U^* is the adjoint operator of U. Returning to the (deterministic) expansion of U itself, it requires the introduction of two Hilbert spaces H(X) and H(T). Although $L_2(X)$ and $L_2(T)$ are most often used, other Hilbert spaces may be useful in certain cases, in particular when the hypothesis of Mercer theorem concerning the square integrability of the two-point correlation is not satisfied. Furthermore, our viewpoint through biorthogonal decompositions is deterministic, concentrated on the space-time dynamics and the presence of space-time symmetries. The spatio-temporal dynamics can be studied in the temporal configuration space $\chi(T) = \mathrm{Ker}(U^*)^{\perp}$ spanned by the ψ_n's and the spatial one $\chi(X) = \mathrm{Ker}(U)^{\perp}$ spanned by the φ_n's, these two spaces being related to each other via the isomorphism U. This spatio-temporal dynamics tool permits the study of bifurcations (or instabilities) in both space and time [7, 8]. The definition of the turbulence spectrum as that of the spectrum of the operator U for an inhomogeneous flow is natural since each eigenvalue represents the square root of the norm of the corresponding term in Eq. 5 (in the sense of the two Hilbert spaces defined above). This is well known in the POD context.

In the biorthogonal decomposition framework, a space-time symmetry naturally appears as a pair of operators $(\tilde{S}, S)$ (or a group of such operators), $\tilde{S}$ acting on H(T) and S on H(X), which intertwines the operator U defined in Eq. 4, i.e. $\tilde{S}U = US$ (the same is true with the adjoint pair $(\tilde{S}^*, S^*)$: $\tilde{S}^*U = US^*$) [7]. Then each eigenspace in $\chi(X)$ is S-invariant and, similarly, each eigenspace in $\chi(T)$ is $\tilde{S}$-invariant. Well-known examples of S and $\tilde{S}$ are the translation symmetries characterizing traveling waves [7, 8]. The same argument can be generalized to any space-time symmetry satisfying

$$US = \gamma \tilde{S} U \text{ and } US^* = \gamma^{-1} \tilde{S}^* U, \tag{6}$$

(where γ is a real number, strictly smaller than 1), in which case the corresponding part of the spectrum decreases exponentially fast, namely

$$A_{n+1} = \gamma A_n \tag{7}$$

and the associated spatial and temporal modes satisfy the properties $\varphi_{n+1} = S\varphi_n$ and $\psi_{n+1} = \tilde{S}\psi_n$.

3. Self-similar structure of inhomogeneous flows and spectrum law

Applying the previous remarks to a fully developed turbulent flow and using the scale invariance (Eq. 2), the commutation relation (Eq. 6) becomes

$$\tilde{S}_\lambda U = \lambda^{-h} U S_\lambda, \tag{8}$$

where the two symmetry operators are the spatial and temporal dilations:

$$(S_\lambda \varphi)(x) = \varphi(\lambda^{-1} x),$$
$$(\tilde{S}\psi)(t) = \psi(\lambda^{-(1-h)} t). \tag{9}$$

The exponentially decaying spectrum law is then

$$A^2_{n+1} = \lambda^{-2h} A^2_n. \tag{11}$$

The modes φ_n and the modes ψ_n can then all be deduced from one by the recurrence relations $\varphi_{n+1} = S_\lambda \varphi_n$ and $\psi_{n+1} = \tilde{S}_\lambda \psi_n$ and are similar to orthonormal wavelets in space and time. However, the fractal geometry is not satisfied by the flow u(x, t) itself but only by its two-point correlations. This can be seen by considering the spatial correlation operator, U^*U, and the temporal one, UU^*, which satisfy the relations:

$$S_\lambda^{-1} U^* U S_\lambda = \lambda^{2h} U^* U \tag{12}$$

and

$$\tilde{S}_\lambda^{-1} U U^* \tilde{S}_\lambda = \lambda^{2h} U U^*. \tag{13}$$

If, however, the scale transformation is assumed in a statistical sense, namely it is satisfied only by the spatial two-point correlation

$$\int_X u(\lambda x, t) u(\lambda x', t) d\tilde{\mu}(t) = \lambda^{2h} \int_X u(x, t) u(x', t) d\mu(t), \tag{14}$$

i.e. Eq. 12 is valid but not Eq. 13, the spectrum law is unchanged and there still exists a temporal symmetry $\tilde{S}$ such that the commutation relation Eq. 8 (where $\tilde{S}_\lambda$ has to be replaced by $\tilde{S}$) holds. Finally, it may be worth mentioning that if the norms previously introduced are not the L_2-norms, the above derivations can still be performed leading in any case to an exponential decaying spectrum. The decay rate, however, may be different from that given in Eq. 11.

4. Application to a turbulent wake flow

We now apply these ideas to a turbulent wake flow behind a circular cylinder at a Reynolds number Re = 5000. Some preliminary remarks can be found in [10]. The nine components of the two-point spatial correlations were measured by Guezennec and Gieseke [9] over a 17 × 17 grid perpendicular to the flow direction, spanning the entire width of the wake by using two probes and making the assumptions of (spanwise and streamwise) homogeneity and stationarity. All measurements being performed at a downstream distance x / D = 100 where D is the diameter of the cylinder, a Taylor hypothesis was applied to recover the streamwise variation of the correlations.

A classical POD technique, decomposing the correlations thus measured into Fourier modes in the streamwise (x) and spanwise (z) directions, and POD modes in the inhomogeneous direction (y)

$$R_{ij}(x-x',y,y',z-z')=<u_i(x,y,z,t)u_j(x',y',z',t)>$$
$$=\sum_n \sum_{l,k}[A^{(l,n,k)}]^2 \varphi_i^{(l,n,k)}(y)\varphi_j^{(l,n,k)}(y') \qquad (15)$$

(where < > denotes the temporal average) was applied to determine statistically relevant spatial modes $\varphi_i^{(l,n,k)}(y)$ and energy spectra $[A^{(l,n,k)}]^2$. The first term in Eq. 15 (n=1) was found to include approximately 20% of the kinetic energy and 90% of the latter was contained in about the first twelve terms, as shown in Figure 1. Although too many assumptions were made to enable complete identification with the deterministic biorthogonal decomposition which requires *simultaneous* measurements at all spatial points, a rough comparison, based on the stationarity assumption, seems reasonable. Then, spectra should display an exponential decay, which they approximately do for each pair of wavenumbers, at a rate which seems independent of the wavenumber pair (see Figure 2).

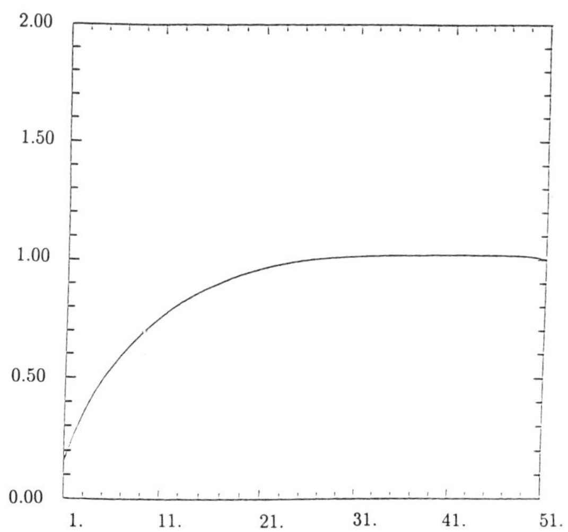

Figure 1. Cumulative normalized kinetic energy $(\sum_{n=1}^{M}\sum_{l,k}[A^{(l,n,k)}]^2)/E$ (E being the total kinetic energy of the flow) as a function of the number M of terms retained in the expansion.

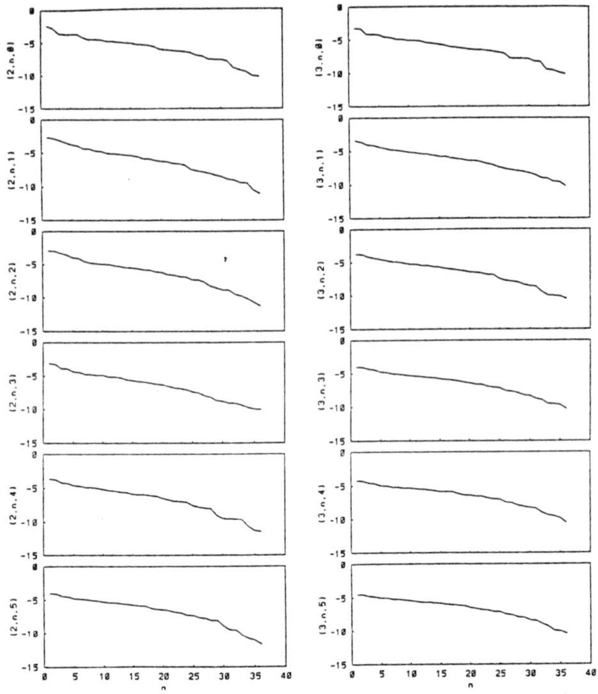

Figure 2. Spectra $A^2_{l,n,k}$ on a logarithmic scale versus the mode index n, for various wavenumber pairs. l, n, k on the left of each plot refer to the index of the streamwise Fourier mode, the POD mode and the spanwise Fourier mode, respectively.

At this point, it is, however, difficult to draw any conclusions regarding the three-dimensional structure of the symmetry which needs further investigations relaxing, in particular, the streamwise and spanwise homogeneity assumptions. In order to avoid this difficulty, we now concentrate on the structure of the flow in the inhomogeneous direction. For this purpose, we consider the two-point correlations at x = x' and z = z', namely $R_{ij}(0,y,y',0)$, so that Eq. 15 becomes

$$R_{ij}(0,y,y',0) = \sum_n \lambda^{(n)} \varphi_i^{(n)}(y) \varphi_j^{(n)}(y'). \tag{16}$$

The spectrum, displayed in Figure 3 shows a tendency to decay exponentially fast. This tendency, however, is partially hidden by the presence of other symmetries, such as the statistical reflection symmetry which splits the space of spatial functions into two orthogonal subspaces, one spanned by eigenvectors of the varicose type, the other one spanned by eigenvectors of the sinuous type, both being invariant under the dilation symmetry. Such an additional symmetry has the effect of accumulating the eigenvalues by packets in the spectrum. A simple observation of the form of the modes allows the extraction of the various subgroups of self-similar modes. Modes inside the

same subgroup, denoted as $\varphi^{(n_l)}$, l = 1, 2, ... show self-similar features and their corresponding spectra decay exponentially fast, as illustrated in Figures 4 and 5.

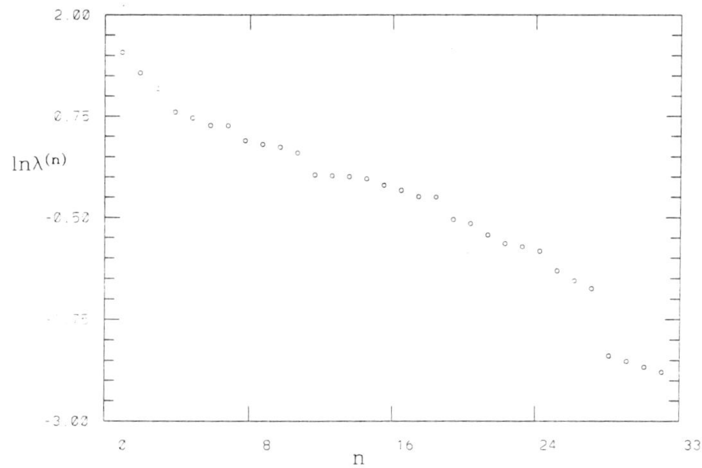

Figure 3. One-dimensional spectrum $\lambda^{(n)}$ on a logarithmic scale versus the mode index n.

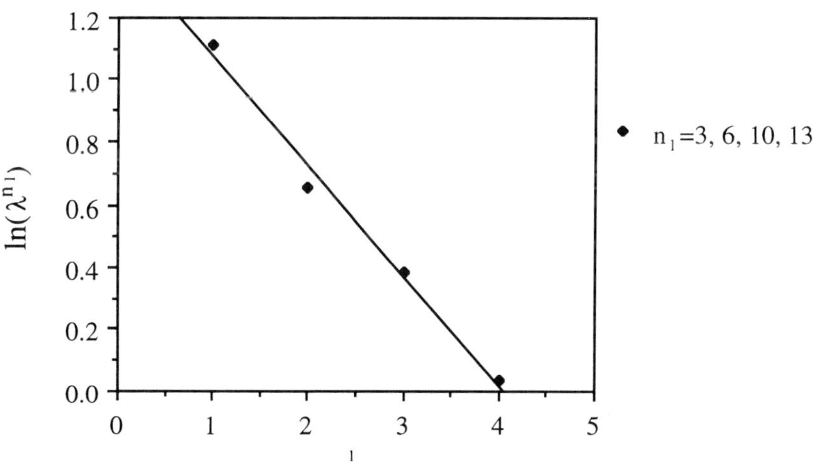

Figure 4a. Spectrum $\lambda^{(n_l)}$ of a series of self-similar modes (n_l = 3, 6, 10, 13).

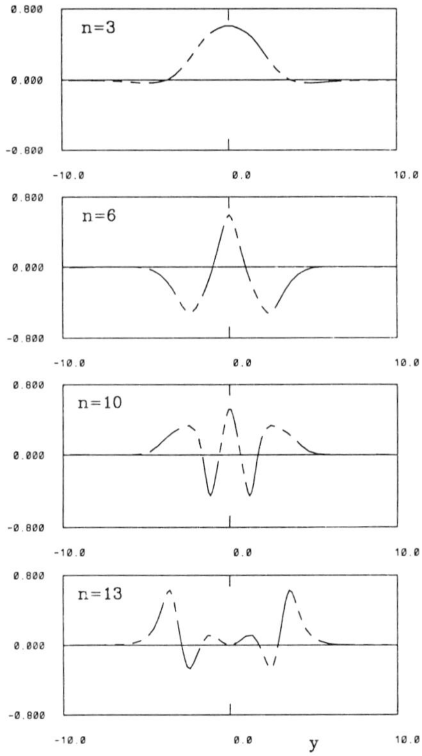

Figure 4b. Series of self-similar modes $\varphi_i^{(n_1)}$, ——— i = 1, — · — i = 2, — ·· — i=3 (n_1 = 3, 6, 10, 13) whose spectrum is displayed in Fig. 4a.

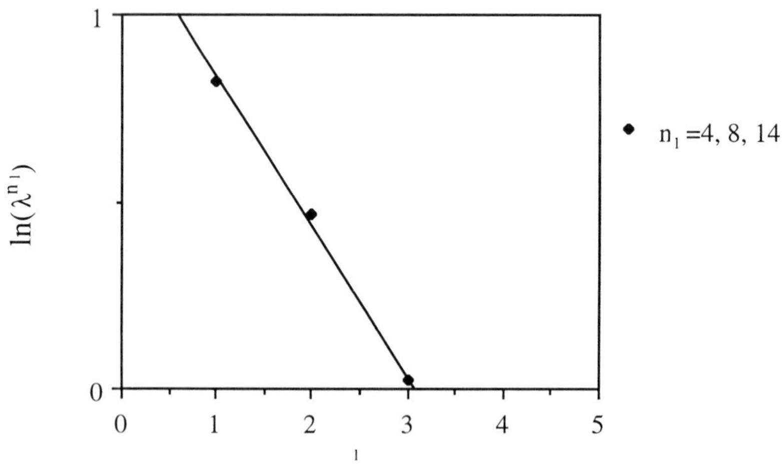

Figure 5a. Spectrum $\lambda^{(n_1)}$ of a series of self-similar modes (n_1 = 4, 8, 14).

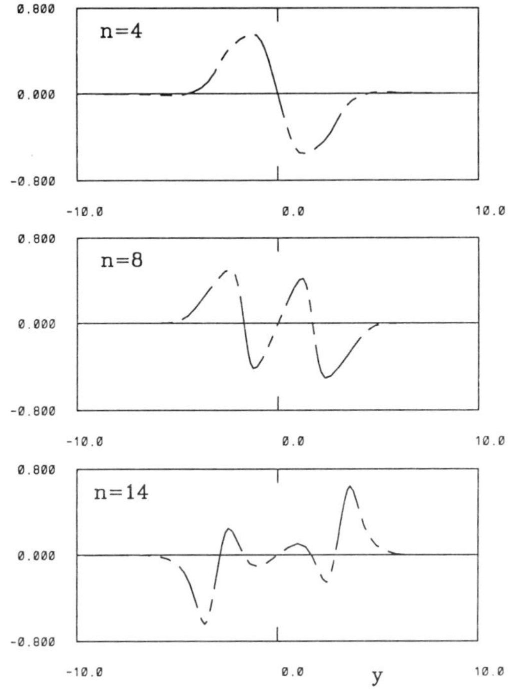

Figure 5b. Series of self-similar modes $\varphi_i^{(n_1)}$, ——— i = 1, —·— i = 2, ——--—— i=3 (n_1 = 4, 8, 14) whose spectrum is displayed in Fig. 5a.

Therefore, all (spatial) modes within a subgroup can be deduced from one by dilation symmetry. (The restriction to a few modes within each group is obviously due to the limited spatial resolution imposed by experimental constraints.) We then know that there exists a temporal symmetry of the same type for which, however, we cannot get any insight from the previous experimental data.

Conclusion. An analysis of the Navier-Stokes equations and their scaling symmetries shows that the structure of turbulent flows should be self-similar in space and time, leading to orthogonal spatial modes (and similarly orthogonal temporal modes) which are related one to another by a dilation symmetry group. We have shown that two-point correlation data, experimentally measured in a turbulent far wake behind a cylinder, support the theoretical deductions. However, instantaneous data at multiple positions, which should be provided by novel experimental techniques such as Particle Image Velocimetry, should be more adequate (for reasons explained above) and provide a better resolution. Finally, we would like to emphasize that the possibility of decomposing turbulent flows into sets of orthogonal modes which are all similar should open myriads of possibilities for a better undertanding, modeling and computation of turbulence. Particularly for wake flows, the combination of this approach with that of reference [11] where the body is omitted should be very powerful.

Acknowledgements

We gratefully acknowledge the support of ONR grant No. N00014-90-J-1554 (Fluid Dynamics, Code 1132F), NSF/PYI award MSS89-57462 and NATO grant No. 900265.

References

[1] N. Aubry, R. Guyonnet and R. Lima, *J. Stat. Phys.* **67**, 203 (1992). Lian, W. and Aubry, N. Self-similarity of compressible turbulence, in Transitional and Turbulent Compressible flows, eds. L. D. Kral and T. A. Zhang, New York: ASME (1993).

[2] U. Frisch, Special 1991 issue on *Turbulence and Stochastic Processes: Kolmogorov's ideas 50 years on* of *Proc. Roy. Soc. A.*, 89 (1991).

[3] A. N. Kolmogorov, *Dokl. Akad. Nauk SSSR* **30**, 301 (1941).

[4] N. Aubry, R. Guyonnet and R. Lima, *J. Stat. Phys.* **63**, 683 (1991).

[5] J. L. Lumley, in *Atmospheric Turbulence and Radio Wave Propagation*, A.M. Yaglom and V.I. Tatarski: eds., Moscow: Nauka, 66 (1967). J. L. Lumley, *Stochastic Tools in Turbulence*. Academic (1970).

[6] N. Aubry, P. Holmes, J. L. Lumley and E. Stone, *J. Fluid Mech.* 192, pp.115-173 (1988). A. Deane, G. Karniadakis, I. G. Kevrekidis and S. A. Orszag, *Phys. Fluids A* 3, p. 2337 (1991). M. Glauser, X. Zheng and C. R. Doering, A low dimensional dynamical systems description of the axisymmetric jet mixing layer, Clarkson University preprint (1991). S. Sanghi and N. Aubry, *J. Fluid Mech.* 247, pp. 455-488 (1993). L. Sirovich, *Q. Appl. Math.* 5, p. 561 (1987).

[7] N. Aubry, R. Guyonnet and R. Lima, *J. Nonlin. Sci.* **2**, 183 (1992).

[8] N. Aubry, *Theor. and Comp. Fluid Dyn.* **2**, 339 (1991). N. Aubry and W. Y. LianExploiting and detecting space-time symmetries, in *Exploiting symmetries in Applied and Numerical Analysis*, Lectures in Applied Mathematics, E. Allgower, K. Georg and R. Miranda, eds.(1992) (to appear). S. Slimani, N. Aubry, P. Kolodner and R. Lima, Binary fluid convection, in *Bifurcation Phenomena and Chaos*, eds. H. H. Bau, L. Bertram and S. A. Korpela. New York: ASME, pp. 39-46 (1992).

[9] Y. Guezennec and T. J. Gieseke, *Bull. Amer. Soc.* vol. **36** (10), p. 2683 (1991).

[10] N. Aubry and N. Cao, Role of the dilation symmetry in turbulence, in *Wavelets and its Applications*, Proc. of the International Conference Wavelets and Applications, Toulouse (France), June 8-13, 1992 (to appear).

[11] N. Cao and N. Aubry, Reduced numerical simulation of wake flows behind a circular cylinder, in *Separated flows*, eds. J. Craig-Dutton and P. Purtell, New York: ASME (1993).

CHARACTERIZATION OF THE ORGANIZATION IN SHEAR LAYERS VIA PROPER ORTHOGONAL DECOMPOSITION.

Delville J.
C.E.A.T Laboratoire d'Etudes Aérodynamiques
43 route de l' Aérodrome
F-86036 Poitiers cedex

1. Introduction.

A lot of attention has been given recently to the identification of the so-called Large Scale, Coherent, Dominant, Characteristic or Eddy structures which play a dominant role in the dynamical properties of the flow. The aim of the present paper is to precise, from an experimental investigation, what kind of information can be inferred from long time averaged stochastic quantities such as the space or space-time correlations. Two ways have become more and more used: the Proper Orthogonal Decomposition (POD) introduced by Lumley (1967) [9] and the Linear Stochastic Estimation (LSE), introduced by Adrian (1977) [1]. The POD approach will be here only developed.
In this paper, the POD is discussed in terms of "characterization of the organization of the flow". The POD is applied for various scalar or vectorial approaches. An analysis is then performed for the convergence efficiency, the turbulent energy representation and the modes/scales relationship. The consideration of the instantaneous contribution of the first modes of the POD allows to precise, on selected samples, the contribution of these modes to the organization of the flow. By a statistical approach, the temporal coupling of the first modes of the POD allows to introduce a dynamical behavior for the organization of the flow.

2. Proper Orthogonal Decomposition.

The POD was proposed by Lumley as an unbiased way for extracting structures from turbulent flows. He proposed that the coherent structure be the structure that has the largest mean square projection on the velocity field. This maximization leads to a Fredholm integral value problem, where the kernel is the correlation tensor.

The POD is applied here for scalar configurations, involving only one component of the flow as u (longitudinal), v (transversal) or w (spanwise). Vectorial application are also considered where the instantaneous velocity vector (u, v) or (u, w) are taken into account. The scalar applications of POD will be noted POD_u, POD_v and POD_w

J. P. Bonnet and M. N. Glauser (eds.), Eddy Structure Identification in Free Turbulent Shear Flows, 225–237.

respectively, while the vectorial POD's will be noted POD_{uv} or POD_{uw}.

In the present study, only the y (mean gradient or inhomogeneous) direction is considered, the time being removed by Fourier Transform (invoking stationarity of the flow). Of course this approach is limited by the fact that the tridimensionnal aspect of the flow cannot be accessed: only one slice of the flow is viewed here. Hence the full tri-dimensional behavior of the flow is not analyzed. However, useful informations on the global organization of the flow can be outlined.

Following the approach of Lumley [9], the dominant structure of the flow can be determined from the following equation:

$$\sum_{j=1}^{nc} \int \Psi_{ij}(y, y'; f) \tilde{\Phi}_j^{*(n)}(y'; f) dy' = \lambda^{(n)}(f) \tilde{\Phi}_i^{(n)}(y; f),$$

where nc is the number of velocity components on which POD is performed and where the cross-spectrum $\Psi_{ij}(f)$ is the temporal Fourier transform of the two-point space-time correlations $R_{ij}(y, y'; \tau) =< u'_i(y, t) u'_j(y', t + \tau) >$. Here, the number of components nc is 1 or 2 depending on the scalar or vectorial approach used. The components of velocity effectively used can be either (u), (v), (w), (u, v) or (u, w) following the experiment under consideration.

This approach is not unique and is only one among the different ways POD may be performed. For example, POD can be derived also directly from the spatial correlations, then no frequency dependence is involved, [8] or from conditional measurements [7].

From the POD, statistical quantities of the flow field can be rebuilt: energy, spectra, space time correlations. Moreover, the use of rakes of probes, allows for reconstruction of the instantaneous contribution of each mode to the velocities [3]. Then the well known problem of the phase loss of the POD due to the use of two-point correlations is avoided because in this case, the velocity field is known by the experimental process simultaneously over the whole inhomogeneous direction.

Four main levels of POD can be introduced:

— For the uni-dimensional scalar POD, only one quantity (e.g. u) is considered,

— The vectorial POD simultaneously takes into account two components. In these two cases only the inhomogeneous direction of the flow is considered.

— A third level could be to take into account the cross homogeneous direction of the flow (e.g. spanwise direction in the case of the mixing layer). In this case, a spanwise wave number dependency is added to the previous PODs [4].

— Lastly, the most general POD is to consider all the components of the velocity.

A striking feature of the POD is that the first mode becomes more and more representative when the "level" of the POD is increased. However, when POD is applied, several aspects remain to be precised: what is the meaning of the modes we get?, can the first mode be directly linked to the "structure(s)"?,...

In a sense, the POD can be looked as an extension of the Fourier transform where the direction for which it is performed is inhomogeneous. The modes of POD in case of homogeneity or periodicity are found to be nothing else than the Fourier modes. It exists then an analogy between modes of POD and modes of harmonic decompositions.

3. Experimental apparatus and flow configuration.

This study involves multi-probes (single or X-hot wires) measurements in a subsonic plane turbulent mixing layer. Rakes of sub-miniature hot wire probes are used. A set of up to 48 single or 24 X-wires can be simultaneously sampled at 10 kHz and gives a good resolution of the temporal-spatial behavior of the flow. A complete description of the experimental arrangement can be found in [2]. The subsonic turbulent plane mixing layer under investigation has the following characteristics: external velocities U_a=42.8 and U_b=25.2 m.s^{-1}, fully turbulent co-flowing boundary layers; the test cross section is 30x30 cm^2. The convection velocity U_c is defined as $0.5(U_a + U_b)$. Four experiments are performed at two downstream distances X from the trailing edge of the splitting plate, in the asymptotic part of the mixing layer (see the following table). The hot wire rakes are lying in the mean gradient direction Y. The rake extent L_y is for each experiment greater than twice the vorticity thickness δ_ω. The separation between probes Δ_Y goes from half to twice the Taylor micro scale λ_f.

Exp. #	probes configuration	X mm	Δ_Y mm	L_y/δ_ω	δ_ω mm	λ_f mm	Re_θ $(U_a - U_b)\theta/\nu$
I	12 X wires (u, v)	600	6.0	2.2	27.6	3.6	4.9 10^5
II	48 single wires (u)	870	1.8	2.4	35.8	4.0	7.6 10^5
III	22 X wires (u, v)	870	3.6	2.1	35.8	4.0	7.6 10^5
IV	12 X wires (u, w)	600	6.0	2.2	27.6	3.6	4.9 10^5

4. POD approach.

Scalar and vectorial POD. Figure 1 compares three applications of the POD performed on the same data set. The eigen-value spectra $\lambda^{(n)}(f)$ take into account the distribution of energy over the whole spatial extent of the experiment. They are representative of the occurrence of the given mode (n) in the flow realizations. Figures 1-a), b) and c) correspond to the scalar POD$_u$, POD$_v$ and vectorial POD$_{uv}$ respectively. The eigen value found from these PODs show quite different characteristics. In the case of the POD$_u$ (fig. 1-a) the first mode is maximum for $f\delta_\omega/U_c \sim 0.15$

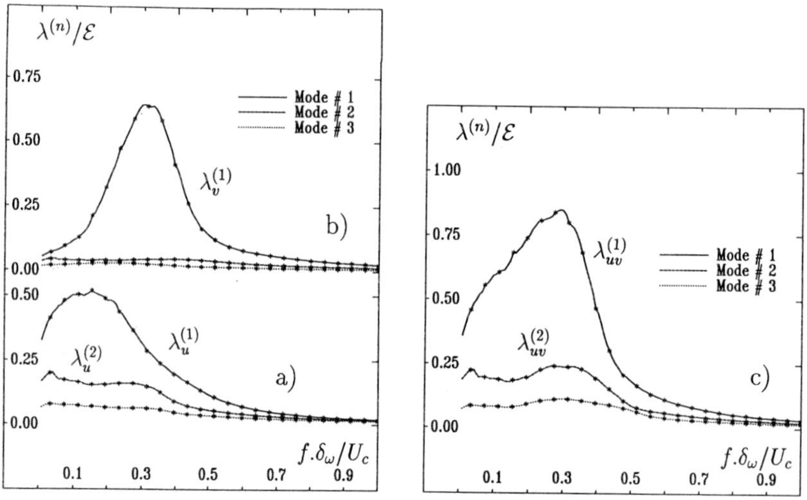

Figure 1: Eigen Values distribution for *a*) u scalar POD, *b*) v scalar approach and *c*) vectorial u, v approach. Experiment I

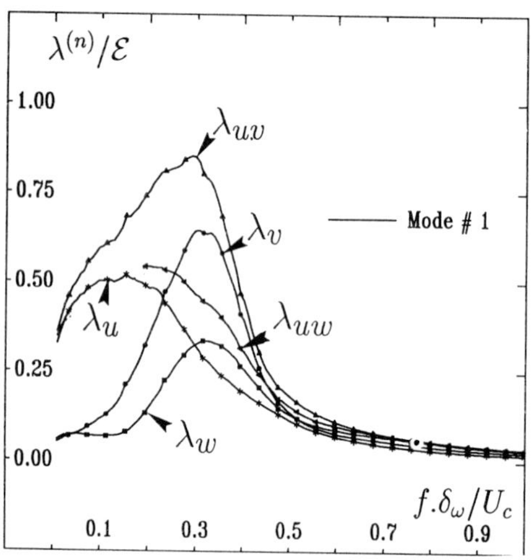

Figure 2: Eigen Values: First modes for *a*) POD_u, *b*) POD_v, *c*) POD_w, *d*) POD_{uv} and *e*) POD_{uw}. Experiment I et IV.

that is only half the value of the corresponding frequency found for the POD$_v$ (fig. 1-b). In this flow, the typical Strouhal number found for the structure passage (from spectral analysis at the edge of the mixing layer) is $S_\omega \simeq 0.3$. In this context of the scalar POD$_u$ the first mode seems then not to be related to the "structure concept". The first mode of POD$_u$ is in fact proportionally less representative than the one of POD$_v$, as proved by the significant level of energy remaining in the second mode, $\lambda_u^{(2)}$ when compared to $\lambda_v^{(2)}$. This behavior has to be related to the main organization of the flow: the structures passage is translated on the u component by phase oppositions of alternate signs in the external parts of the mixing layer, while the v component remains dominantly in phase over the whole y extent and presents then a simpler organization.

For the vectorial POD$_{uv}$, the first eigen-values $\lambda_{uv}^{(1)}(f)$ correspond mainly to the envelope of the two first modes of the scalar PODs: for each frequency $\lambda_{uv}^{(1)}(f) \geq \max(\lambda_u^{(1)}(f), \lambda_v^{(1)}(f))$. In fact the $\lambda_{uv}^{(1)}(f)$ evolution can be split into two frequency ranges: for $f\delta_\omega/U_c < 0.2$, the u component dominates while for higher frequencies, the v component becomes dominating.

The same features can be found when the POD$_{uw}$ is considered (figure 2). In the POD$_{uw}$ the general shape and level of the eigen-value remain close to the one of the POD$_u$. This is due to the relative levels of $\lambda_u^{(1)}$ and $\lambda_w^{(1)}$: for each frequency, $\lambda_w^{(1)}(f)$ remains always less than $\lambda_u^{(1)}(f)$.

This overall behavior is in agreement with the definition of POD which maximizes energy representation: then for each range of frequency, the POD will "select" the corresponding more energetic component. Moreover, because the vectorial POD$_{uv}$ contains informations on both u and v, it seems to be more suited than the scalar ones for rendering the "structures" aspect of organization of the flow.

Convergence. The dominant POD mode is by definition the best representation of the turbulent energy in the flow. The energy contained in the first mode is generally found to be of the order of 40-50% [3, 5] for uni-dimensional POD and up to more than 70% for higher levels of POD applications [10]. By analogy with the Fourier modes, one can expect that the level of representation of one mode of the POD is directly linked to the number of spatial points used. The convergence of the POD can be defined as the ratio between the energy contained in n modes and the global energy $\mathcal{E}$ contained within the flow:

$$c_n = \sum_{i=1}^{n} \mathcal{E}^{(i)} / \mathcal{E}.$$

Here $\mathcal{E}^{(i)} = \int \lambda^{(i)}(f)\, df$ is the energy contained in the whole flow field for mode (i), integrated over all frequencies. This criterion is drawn on figure 3-a) for the scalar POD POD$_u$. On this figure, the extent of the rake is kept constant while Ny the number of probes used in the rake is decreasing from 48 to 11 (here the number of modes one can obtain is directly Ny). For mesh sizes varying in a ratio of 4, the

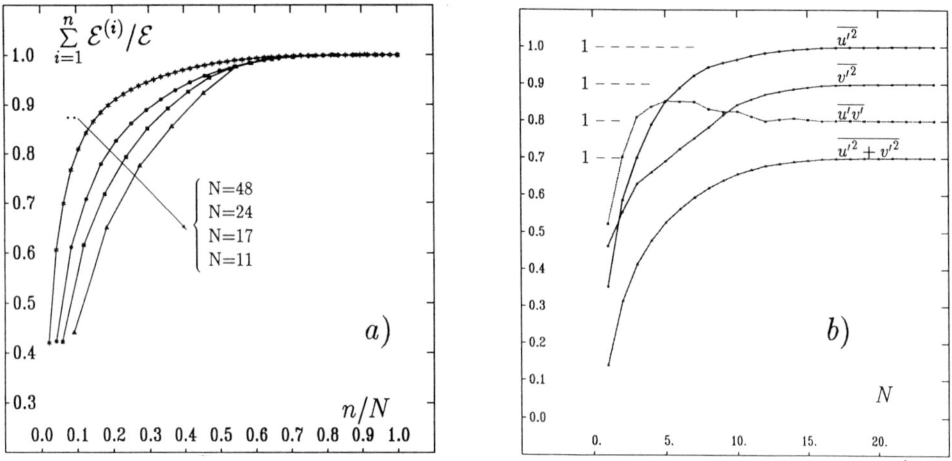

Figure 3: Convergence of POD: $a)$ Influence of number N of spatial nodes on POD convergence. Scalar u POD. Experiment II . $b)$ Convergence of $< u_i'u_j' >$ Vectorial POD$_{uv}$ experiment I

energy contained in the first mode remains quasi constant (about 42%). The level of representation of this mode is then a characteristic of the flow itself and is not dependent on the mesh size. However, one given mode becomes more and more biased when N decreases, as denoted by the increase of the convergence rate. For example, the energy contained in the first three modes is 0.7 for N=48 and reaches 0.8 for N=11.

The same kind of study can be performed in the vectorial case POD$_{uv}$. Similar results are obtained for the convergence of the global energy $\overline{u'^2+v'^2}$. However in the vectorial case, one can precise the convergence of each term of the Reynolds stresses individually (figure 3-b). In this case, one can define the convergence as:

$$c_l = \int \overline{u_i'(y)u_j'(y)}^{(l)}\, dy / \int \overline{u_i'(y)u_j'(y)}\, dy$$

where

$$\overline{u_i'(y)u_j'(y)}^{(l)} = \int \lambda^{(l)}(f)\tilde{\Phi}_i^{(l)}(y,f)\tilde{\Phi}_j^{*(l)}(y,f)\, df$$

The components $\overline{u'^2}$ and $\overline{v'^2}$ converge uniformly toward their asymptotical value. However the $\overline{u'v'}$ Reynolds stress start to converge very quickly and then exhibits an overshoot from the third mode. This overshoot is associated with an underestimation of the $\overline{v'^2}$ component as indicated by the linear slope of the curve appearing between modes 3 and 10.

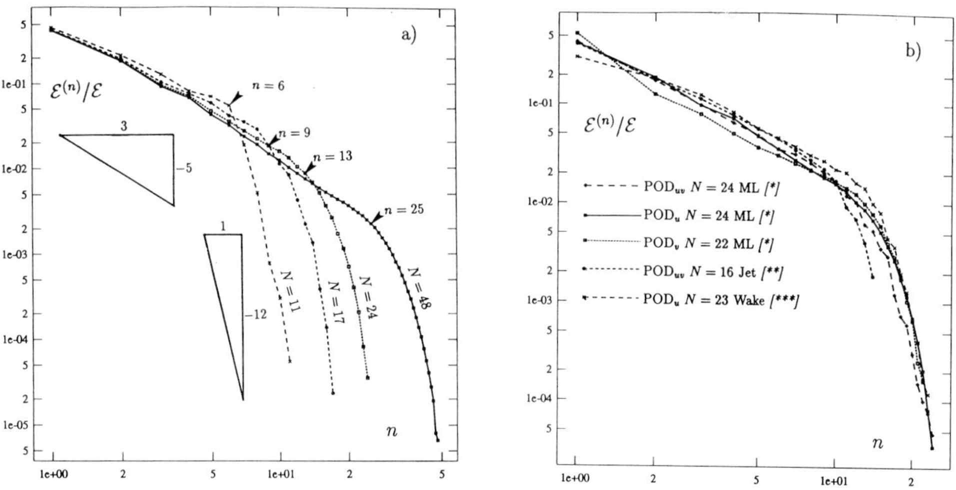

Figure 4: Distribution of relative energy (integrated over frequencies and space) for each mode of the POD. *a*) Experiment *II*. *b*) Same distribution for various application of POD (scalar or vectorial) and for various flows: *[*]* present study, *[**]* from axisymmetric jet data [5], far wake of a flat plate *[***]*.

The number of modes obtained from an application of POD is $N = nc \times Ny$ where Ny is the number of spatial nodes used. This number is always greater than the Nyquist number of the spatial mesh ($Ny/2$) and one can ask about the significance of the extra number of modes obtained. In order to precise this point, the distribution of the relative energy $\mathcal{E}^{(n)}/\mathcal{E}$ contained in one mode is plotted on fig. 4 in a log-log representation. The convergence of POD seems to follow a quasi universal behavior. For the first modes, up to $n \sim N/2$, a first slope close to -5/3 is found. The higher modes follow an other much higher slope ~ -12. Moreover, for the modes before $N/2$, an overshoot can be found (fig. 4-a). This behavior may be analyzed by analogy with the harmonic decomposition: the modes of POD can be compared with wave numbers or frequencies and then this kind of plot can be compared with a classical spectrum. The first slope can then be related to a scale power-law and the overshoot to an aliasing effect which is unavoidable for spatial measurements. The second slope is then more related to numerical effects. The quasi-universality of this behavior is evidenced if fig. 4-b. For different flow configurations (mixing layer, wake, axisymmetric jet), the corresponding distributions are plotted for various applications of POD (scalar or vectorial) where N is of the same order of magnitude. It can be noticed that the mode $N/2$ plays the same role of cut-off mode for the scalar PODs as well as for the vectorial ones. It should be recalled that in this last cases the number of modes obtained is twice the number of spatial nodes. Contradictorily, the cut-off mode seems then to be related to N rather than to Ny.

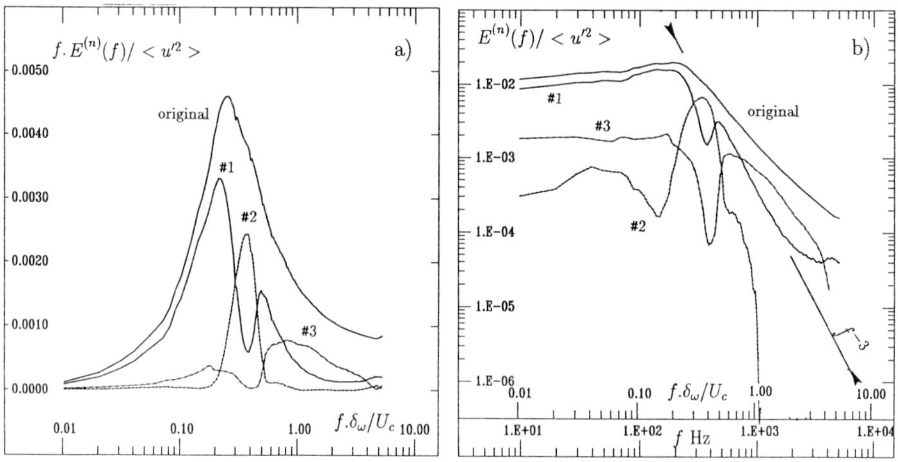

Figure 5: Reconstruction on the mixing layer axis of the spectrum of the u component velocity. Vectorial POD experiment *III*. *a*) $f.E(f)$ vs $\log(f)$ representation. *b*) log-log representation.

Local spectra reconstruction. The link between mode of the POD and spatial scale which appears from the analysis of the convergence of the POD can be precisely analyzed. For this purpose, the contribution of the modes to the conventional frequency spectrum is studied on the mixing layer axis:

$$E_u^{(l)}(y,f) = \lambda^{(l)}(f)\tilde{\Phi}_1^{(l)}(y,f)\tilde{\Phi}_1^{*(l)}(y,f).$$

The contribution of the three first modes of the vectorial POD to the spectrum of the longitudinal component u is plotted on figure 5 for a probe lying near the mixing layer axis. By using a $f.E(f)$ vs $\log(f)$ representation, the frequencies where the modes are dominant are pointed out (fig. 5-a). It is then obvious that the higher modes contributes to the higher frequencies and thus to smaller scales. In fact these modes toggle to preserve a continuous slope in the total spectrum (fig. 5-b). When looking at the specific behavior of the first two modes, it appears that they are interlaced: as the frequency increases, the first mode is more representative then the second one, then the first one again. The resulting spectrum obtained by keeping only these two modes corresponds in fact to the one that can be obtained in the case of the POD_u. A typical slope (f^{-3}) is obtained [3]) in this case.

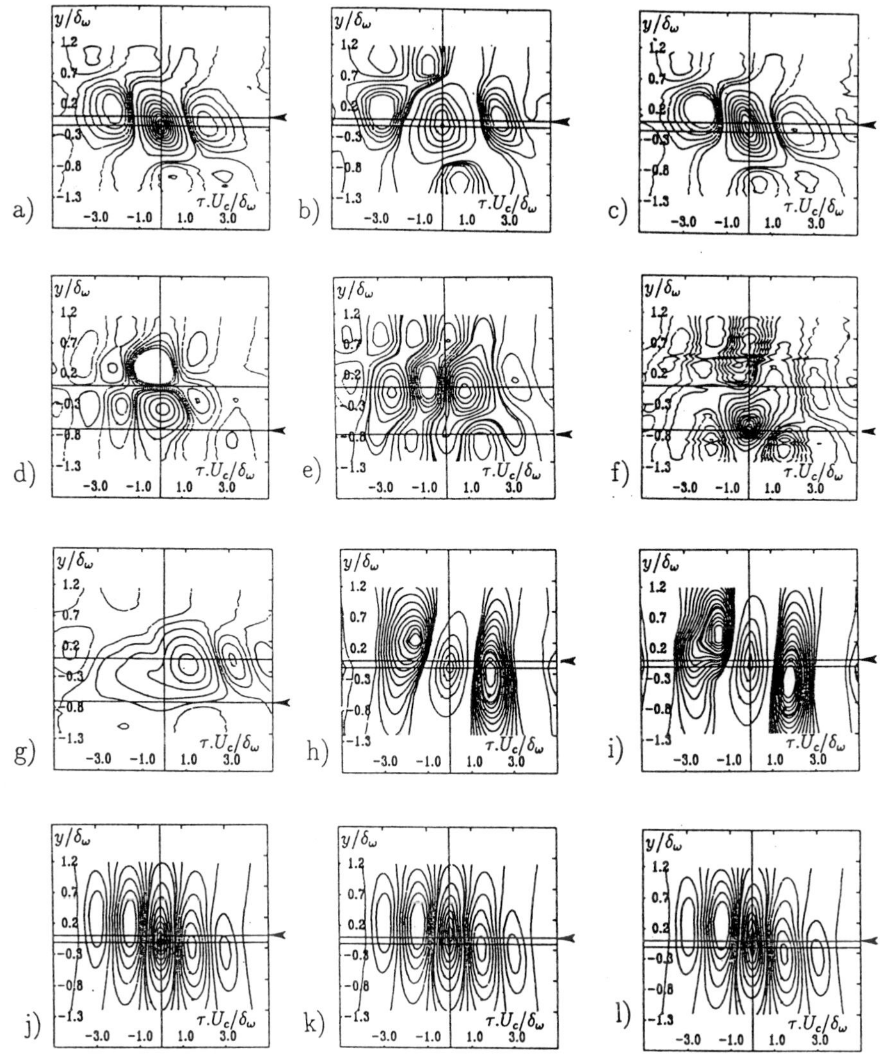

Figure 6: Comparison of various POD reconstruction of the space time correlations in the mixing layer. (← : reference probe).

fig.	a	b	c	d	e	f	g	h	i	j	k	l
R_{ij}	uu	uu	uu	uu	uu	uu	uu	uv	uv	vv	vv	vv
Y/δ_ω	0	0	0	-0.8	-0.8	-0.8	-0.8	0	0	0	0	0
POD	u	uv	orig	u	uv	orig	u	uv	orig	v	uv	orig
mode	1	1	1	2	1	1	1	1	1	1	1	1

Space time correlations. Space time correlations (STC) rebuilt from different POD application are plotted on figure 6. These STC are obtained from the eigen modes by the inverse time Fourier Transform of $\Psi_{ij}^{(l)}(y, y', f)$ where:

$$\Psi_{ij}^{(l)}(y, y', f) = \lambda^{(l)}(f)\tilde{\Phi}_i^{(l)}(y, f)\tilde{\Phi}_j^{*(l)}(y, f).$$

In each case of POD, the first mode always predicts the correlations quite well, when the reference probe is located near the mixing layer axis (figures a,b,c and h to l). The prediction of R_{uu} obtained from POD_{uv} is less satisfactory than from POD_u, but the overall organization of the STC is preserved. This is due to the contribution of v (fig.b). At contrary, R_{vv} and even R_{uv}, predicted by the first mode of the various PODs are quite comparable (figures. h to l). In fact, this last characteristic can be observed, whatever the location of the reference probe is within the mixing layer.

When the reference probe is located near the external part of the mixing layer, the R_{uu} correlation is not correctly predicted by the first mode of the scalar POD_u (fig. f and g) and additional modes are required for a correct reconstruction (fig. d). Particularly, the antisymmetrical organization which is generally considered as being the trace of the structures passage can not be retrieved from only the first mode. One can then ask the question of the significance of the first mode of POD_u in terms of "conventional" structures. However the POD_{uv} allows to get a better representation (fig. e) at this location. Here, the spatial behavior of the scalar POD_u is weighted by the contribution of the v component and the first mode is now able to catch some new informations as the phase opposition between the external parts of the flow.

Instantaneous reconstruction. When hot wire rakes are used, the velocity distribution is known simultaneously at each time step for the direction where the POD is applied (y). The instantaneous velocity contribution of each mode can then be examined. The Fourier transform of the velocity can be retrieved from the eigen functions $\tilde{\Phi}_i$:

$$\widetilde{u_i}(y, f) = \sum_{l=1}^{N} a_n(f)\tilde{\Phi}_i^{(l)}(y, f)$$

where $a_n(f)$ is found from:

$$a_n(f) = \int \widetilde{u_i}(y, f)\tilde{\Phi}_i^{(l)}(y, f)\, dy$$

In terms of structure identification, the participation of any mode to the instantaneous flow field can be considered. Figure 7 shows, for a selected sample an example of the instantaneous velocity field plotted in a frame moving with the convective velocity U_c and using a Taylor hypothesis based on U_c for the longitudinal direction. The sample size is $\tau U_c/\delta_\omega \simeq 18$ which corresponds to about 5.5 structures separation based on the typical Strouhal number $St_\omega = 0.3$. The organization which can be visually found from the original velocities (fig. a) is relatively well reconstructed by the first mode of the vectorial POD_{uv} (fig. b). However the spatial extent of

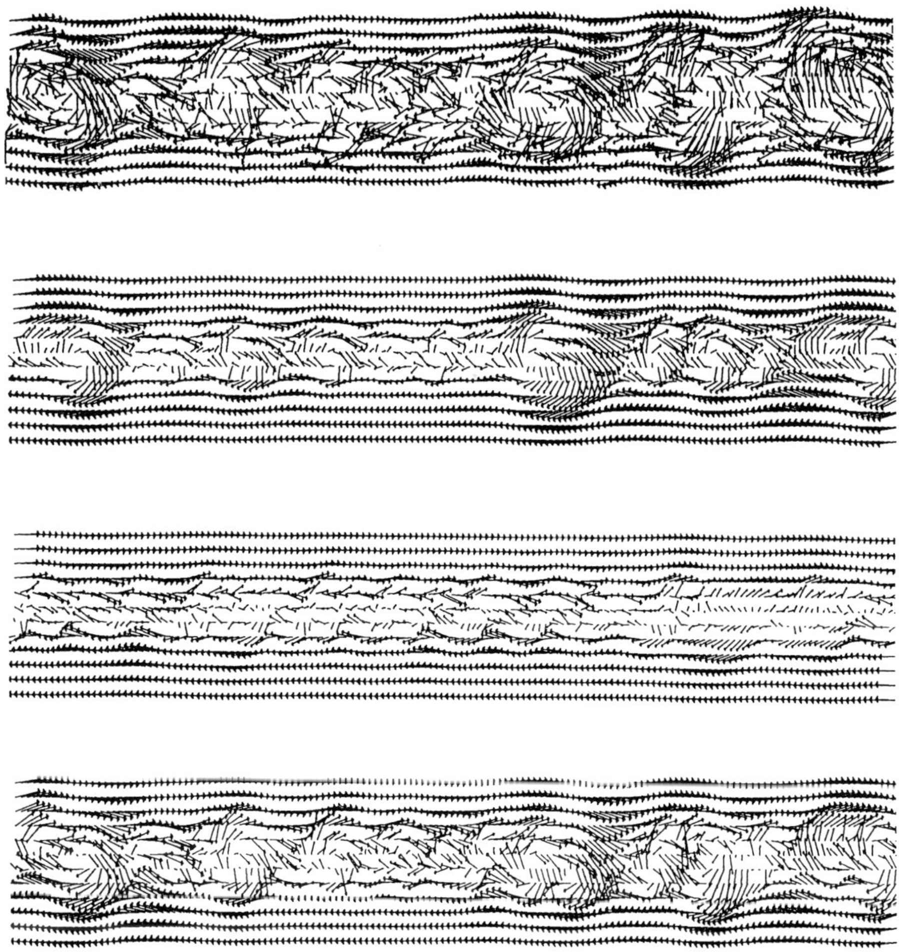

a)

b)

c)

d)

Figure 7: Instantaneous flow patterns in the mixing layer, experiment I: *a*) Original directly measured velocity field. *b*) Contribution of first mode of the POD_{uv}. *c*) Contribution of the second mode. *d*) Contribution of the first three modes.

the events is generally underestimated and other modes are needed to improve this spatial estimation.

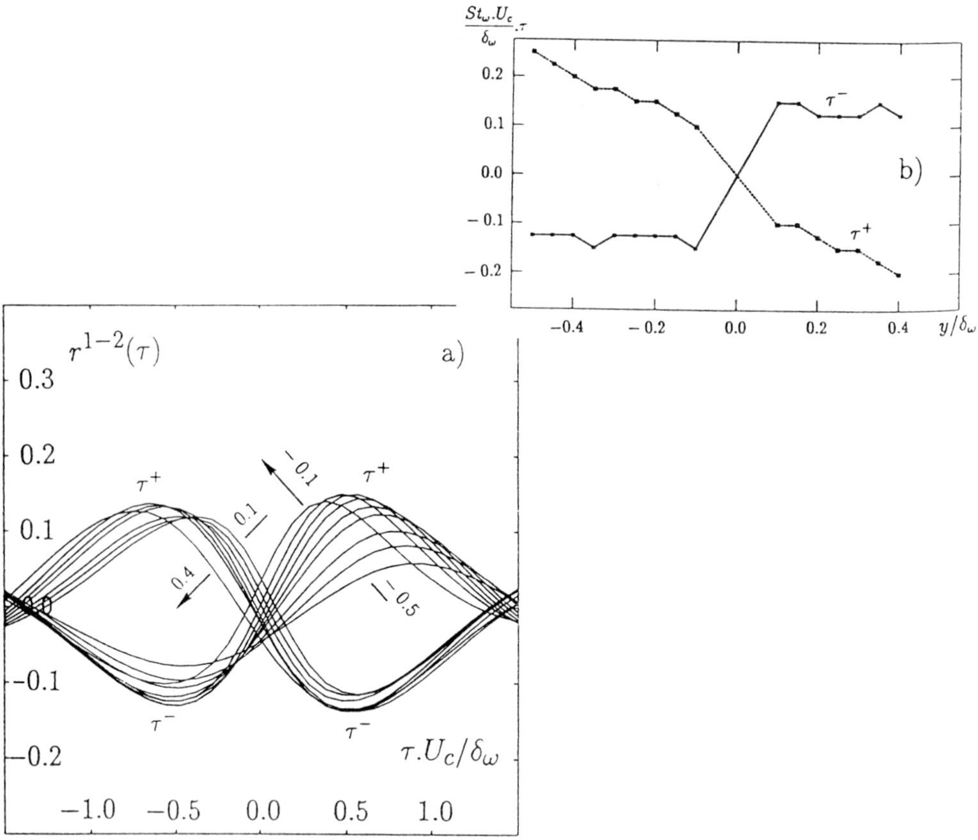

Figure 8: Coupling between modes 1 and 2. Mixing layer, scalar u POD, experiment II. a) Evolution of the temporal correlation coefficient for $y/\delta_\omega = -0.5$ to -0.1 and 0.1 to 0.4. b) Corresponding evolution of the optimum correlation time delay.

Coupling between modes. For each time step (or for each realization of the flow), according to the definition of the POD, each mode is statistically non-correlated with the other modes. However a dynamical relationship can be introduced by considering "time correlation" between the contribution of the modes to the components of the velocity. Here an example of the coupling between the modes (p) and (q) of the longitudinal component u_1 of the velocity can be written:

$$r_{11}^{p-q}(\tau) = < u_1^{(p)}(y,t).u_1^{(q)}(y,t+\tau) > / \sqrt{< u_1^{(p)}(y,t)^2 > . < u_1^{(q)}(y,t)^2 >}.$$

In other terms, a temporal coupling between modes can then be introduced and is plotted on fig. 8. By this way, a typical behavior is exhibited for r_{11}^{1-2} (fig. 8-a). Optimal time delays can be introduced: τ^+ and τ^- correspond respectively to the maxima and minima of correlations respectively. A different behavior is found for the positive and negative correlations. The two modes are in phase opposition for a time delay of the order of 1/8 of the structure passage period T_p (τ^- in fig. 8-b) conversely they appear to be in phase for a time delay $\tau^+ \sim 0.5Y/\delta_\omega T_p$.

5. Conclusion.

The convergence of POD under different scalar or vectorial applications and for various flow configurations seems to follow a quasi-universal behavior. A power law can be introduced for the modes up to a Nyquist cut-off mode number. This feature, combined with the frequency contribution of the modes to the spectra on the mixing layer axis shows that the higher modes of POD can be related to smaller spatial scale characteristics. From the analysis of the reconstruction of space time correlations and of instantaneous velocity profiles, the vectorial POD is found to gain a better representation of the organization of the flow in terms of "structures". Lastly, the temporal coupling between the first two modes of POD exhibits an asymmetric behavior when in phase or opposite phase couplings are considered.

Acknowledgements: This study has been performed under the Grant DRET/DGA 90-171.

References

[1] Adrian, R.J. 1977. On the role of conditional averages in turbulence theory. *Turbulence in liquids.* Science Press, Princeton

[2] Bellin S., Delville J., Vincendeau E., Garem H. & Bonnet J.P IUTAM Poitiers 1992.

[3] Delville, J. Bellin, S. & Bonnet, J.-P. 1990. In " Turbulence and coherent structures. Grenoble." M. Lesieur and O. Métais ed. Kluwer 1990.

[4]J. Delville & L. Ukeiley. To be presented in TSF 9 Kyoto 1993

[5]Glauser M. PhD Thesis Buffalo 1987.

[6]Glauser M. George W. Application of multipoint measurements for flow characterization. Exp. Therm. and Fluid Science. 5,pp 617-632. 1992.

[7]Glezer A., Kadioglu Z. & Pearlstein J. 1989 Phys. of Fluids A, vol. No 8

[8] Herzog S. PhD. Thesis, Cornell University 1986

[9] Lumley, J.L. 1967. In *Proceedings of the international colloquium on the fine structure of the atmosphere and its influence on radio wave propagation, Moscow, June 15-1965*, edts. A.K. Yaglom & U.I. Tatarsky p. 166-176.

[10]Moin P. & Moser R.D. 1989. J. Fluid Mech., vol. 200 pp. 451-509

NEAR-WALL RECONSTRUCTION OF HIGHER-ORDER MOMENTS AND LENGTH SCALES USING THE POD

M. N. Glauser
Clarkson University, Potsdam, NY 13699, USA
and
T. B. Gatski
NASA Langley Research Center, Hampton, VA 23681-0001, USA

ABSTRACT

An analysis of the near-wall behavior of the proper orthogonal decomposition (POD) eigenfunctions (see Lumley 1967) derived from direct numerical simulation (DNS) of channel flow of Dinavahi and Zang (1993) is performed. This work is driven in part by a desire to include *structure* in turbulence models. Although this particular application is in the near wall region (because of availability of the data base), the ideas discussed here are applicable in free shear flows as well. Consistent with previous studies, a low order multi-mode reconstruction of the kinetic energy and Reynolds shear stress suffices. A similar reconstruction of the isotropic dissipation rate is shown to be insufficient, however. An analysis is performed of the multi-mode composition of the dissipation rate in the near-wall region, and it is shown that a significant number of higher-order modes are required to achieve the correct asymptotic consistency in the near-wall region. In an attempt to avoid this problem, a length scale definition is proposed in terms of an integration of the two-point correlation tensor which factors in the presence of the wall. The wall is accounted for by only integrating out to $2y^+$ and not over the entire domain. Viscous and inviscid estimates for the dissipation rate are used in the near-wall and core regions, respectively, in conjunction with this length scale representation, to obtain an estimate of the dissipation rate throughout the domain. The resulting dissipation rate exhibits the proper behavior near the wall and in the inertial layer. Higher-order POD mode estimates of the length scale are computed and compared to that computed using the full correlation tensor from the DNS results of Dinavahi and Zang (1993). The three mode estimate is found to agree quite well with the length scale obtained when the entire correlation tensor is used, whereas a one mode estimate differs somewhat.

INTRODUCTION

It is desirable in the development of turbulent closure models to incorporate as much physical information about the flow as possible. The ability to incorporate turbulent structure information into the development of closure models, including single

J. P. Bonnet and M. N. Glauser (eds.), Eddy Structure Identification in Free Turbulent Shear Flows, 239–249.

point closures, would be beneficial since there is an abundance of such information available on a wide variety of flows. Unfortunately, such a methodology has not been developed up to this time, although DNS simulations and POD eigenfunctions have been used to reconstruct some of the Reynolds stress moments in channel flows.

The POD technique has been applied to direct simulation results for a channel flow by Moin and Moser (1989). Their intent was to study the coherent structures within this wall-bounded flow. Distributions of the Reynolds stress components were reconstructed across the channel using differing numbers of modes. This was done to determine the modal makeup of the various turbulent stress components. It was found that in the near-wall region ($y^+ \leq 40$) five modes were necessary in order to get an accurate representation of the DNS results. They also found that as the dimensionality of the characteristic eddy increases, the energy content of the first mode increases over its one-dimensional counterpart.

Gilbert and Kleiser (1991) have used direct simulation data to examine the gradient diffusion hypothesis and the eddy viscosity relation in the near-wall region. They used the DNS data to compute the turbulent diffusion term in both the kinetic energy and dissipation rate equations and used their usual gradient diffusion models to determine the damping behavior in the near-wall region.

In order to effectively develop models based on the POD eigenfunctions, it is necessary to evaluate whether the various moments can be accurately reconstructed. This specifically applies in the near-wall region of turbulent flows, where the correct asymptotic behavior is crucial. In this paper, mode reconstructions for the turbulent kinetic energy, shear stress and dissipation are evaluated for their asymptotic consistency in the near-wall region using a one-dimensional, multi-mode reconstruction. It is shown that a 7-mode reconstruction closely mimics the turbulent kinetic energy and Reynolds shear stress in the near-wall region when compared to the recent DNS simulations of channel flow by Zang (see Dinavahi and Zang, 1993). The POD reconstruction of the dissipation rate is shown to require a significant number of additional modes in close proximity to the wall ($y^+ \leq 5$). In an attempt to avoid this problem, a length scale definition is proposed in terms of an integration of the two-point correlation tensor which factors in the presence of the wall. The wall is accounted for by only integrating out to $2y^+$ and not over the entire domain. Viscous and inviscid estimates for the dissipation rate are used in the near-wall and core regions, respectively, in conjunction with this length scale representation, to obtain an estimate of the dissipation rate throughout the domain. The resulting dissipation rate exhibits the proper behavior near the wall and in the inertial layer. Higher-order POD mode estimates of the length scale are computed and compared to that computed using the full correlation tensor from the DNS results. The three mode estimate is found to agree quite well with the length scale obtained when the entire correlation tensor is used, whereas a one mode estimate differs somewhat.

Flow Field Decomposition using the POD

The proper orthogonal decomposition technique has been available for over twenty-five years (e.g. Lumley, 1967). It suffices here to just outline the pertinent aspects of the theory as it pertains to the development of the turbulent moments of interest. Lumley suggested that the coherent structure should be that structure which has the largest mean square projection on the velocity field. This process involves maximizing the mean square energy via the calculus of variations and leads to the following integral eigenvalue problem:

$$\lambda\phi_i(\vec{x}) = \int R_{ij}(\vec{x},\vec{x}')\phi_j(\vec{x}')d\vec{x}'. \tag{1}$$

The symmetric kernel of this Fredholm integral equation is the two–point correlation tenser R_{ij} defined by

$$R_{ij}(\vec{x},\vec{x}') = \overline{u_i(\vec{x})u_j(\vec{x}')}, \tag{2}$$

where $\vec{\phi}$ is the candidate structure, and $\vec{x}$ and $\vec{x}'$ represent different spatial points in the inhomogeneous directions and different times if the flow is non–stationary.

From the Hilbert–Schmidt theory, it can be shown that the solution of a Fredholm integral equation of the first kind for a symmetric kernel and a finite energy domain (i.e., statistically inhomogeneous) is a discrete set, hence Eq.(1) can be written as

$$\lambda^n\phi_i^n(\vec{x}) = \int R_{ij}(\vec{x},\vec{x}')\phi_j^n(\vec{x}')d\vec{x}'. \tag{3}$$

The eigenfunctions of the Fredholm equation are orthogonal over the interval, so that

$$\int \phi_i^n(\vec{x})\phi_i^m(\vec{x})d\vec{x} = \delta_{nm}, \tag{4}$$

for normalized eigenfunctions. The eigenvalues of the Fredholm equation with a real symmetric kernel are all real and uncorrelated,

$$\overline{a^n a^m} = \lambda^n\delta_{nm}, \tag{5}$$

and the fluctuating random field u_i can be reconstructed from the eigenfunctions in the following way:

$$u_i(\vec{x}) = \sum_{n=0}^{\infty} a^n\phi_i^n(\vec{x}). \tag{6}$$

The random coefficients are calculated from

$$a^n = \int u_i(\vec{x})\phi_i^{n*}(\vec{x})d(\vec{x}), \tag{7}$$

where the ϕ_i^{n*} are the complex conjugates of the eigenfunctions obtained from Eq. (3). The turbulent kinetic energy is the sum over n of the eigenvalues λ^n, and each structure makes an independent contribution to the kinetic energy, Reynolds stress and dissipation.

In the work reported here, the DNS results of Zang (Dinavahi and Zang, 1993) are used to supply the correlation tensor R_{ij}. A 1-D POD similar to that applied by Moin and Moser (1989) will be applied to the DNS simulations. In this case, the spanwise and streamwise information is suppressed and only the wall normal direction y^+ is studied. With these assumptions, Eq. (3) then becomes

$$\lambda^n \phi_i^n(y^+) = \int R_{ij}(y^+, y^{+'}) \phi_j^n(y^{+'}) dy^{+'}. \tag{8}$$

The kinetic energy and Reynolds stress can be reconstructed in terms of the eigenvalues and eigenfunctions as

$$k^+(y^+) = \frac{1}{2} \sum_{n=0}^{\infty} \lambda^n (\phi_i^n(y^+))^2 \tag{9}$$

and

$$\overline{u_1 u_2}^+(y^+) \equiv \overline{uv}^+(y^+) = \sum_{n=0}^{\infty} \lambda^n \phi_1^n(y^+) \phi_2^n(y^+), \tag{10}$$

respectively.

The isotropic dissipation can be written as

$$\overline{\left[\frac{\partial u_i}{\partial x_j}\right]^2} = \overline{\frac{\partial u_i}{\partial x_j} \frac{\partial u_i}{\partial x_j}}. \tag{11}$$

Upon substituting Eq. (6) into the above expression, the equation for the dissipation in terms of the eigenfunctions and coefficients becomes

$$\overline{\left[\frac{\partial u_i}{\partial x_j}\right]^2} = \sum_n \sum_m \overline{a^n a^m} \delta_{mn} \frac{\partial \phi_i^n}{\partial x_j} \frac{\partial \phi_j^m}{\partial x_j}. \tag{12}$$

This equation can be simplified further by inserting Eq. (5). It then becomes

$$\overline{\left[\frac{\partial u_i}{\partial x_j}\right]^2} = \sum_n \lambda^n \left[\frac{\partial \phi_i^n}{\partial x_j}\right]^2. \tag{13}$$

Since a 1-D POD was performed, it is only possible to obtain three terms of the dissipation tensor – those associated with the derivatives in the y^+ direction directly The contribution to the dissipation from the remaining terms is found by utilizing the assumptions of local axisymmetry (as described by George and Hussein 1991) and local isotropy. This approach is described in some detail in Gatski and Glauser (1992).

From the analysis of Gatski and Glauser (1992) it is shown that a significant number of higher-order POD modes are required to achieve the correct asymptotic consistency of the dissipation in the near-wall region. A length scale formulation, which contains knowledge of the structure in the turbulent flow, may provide a way around this problem. In the work reported here, a length scale is formulated from the two-point correlation tensor in two different ways: a scale obtained by accounting for the wall and one obtained by integrating out over the entire domain. The former is represented by

$$L_{ij}(y^+) = \frac{1}{R_{ii}(y^+, y^+)} \int_0^{2y^+} R_{ij}(y^+, y^{+'})dy^{+'}, \tag{14}$$

and the latter by

$$L_{ij}(y^+) = \frac{1}{R_{ii}(y^+, y^+)} \int_0^{Y} R_{ij}(y^+, y^{+'})dy^{+'}, \tag{15}$$

where Y is the channel half-width. The rational used in accounting for the presence of the wall by integrating out to $2y^+$ is based on the fact that turbulent eddy size is constrained by the wall itself. The factor of two is incorporated into the definition for comparison with the integration over the entire channel half-width. In calculating the dissipation rate, viscosity must play a role in the near wall region. The relationship can be shown to be,

$$\epsilon^+(y^+) = C_1 k^+ / L_{ii}(y^+)^2. \tag{16}$$

In the core region, the more familiar inviscid estimate is used which is given by,

$$\epsilon^+(y^+) = C_2 k^{+3/2} / L_{ii}(y^+). \tag{17}$$

These two forms are used to obtain $\epsilon^+(y^+)$ from the length scales which are then compared to DNS results. A possible approach for turbulence modeling, using these ideas, may be to develop evolution equations for the POD eigenfunctions and compute the length scale from them. The dissipation can then be computed from this length scale and used to close the evolution equation for the kinetic energy, hence eliminating the need for the dissipation equation.

RESULTS

The DNS results (Dinavahi and Zang, 1993) for fully developed turbulent channel flow at an Re_τ of 317 were analyzed using the POD technique outlined in the previous section. These DNS calculations were initiated in the laminar regime and carried through transition into the fully turbulent regime. As indicated earlier, the intent here is to analyze the near-wall behavior of the reconstructed turbulent kinetic energy, shear stress, dissipation rate and length scales for their asymptotic consistency.

The near-wall asymptotic consistency of the POD reconstruction of the kinetic energy and shear stress are shown in Figure 1. As can be seen, even a first mode reconstruction has the correct functional behavior in both cases. Higher-order modes contribute to the correct amplitude level so that with the seven mode reconstruction the near-wall asymptotic behavior is closely mimicked.

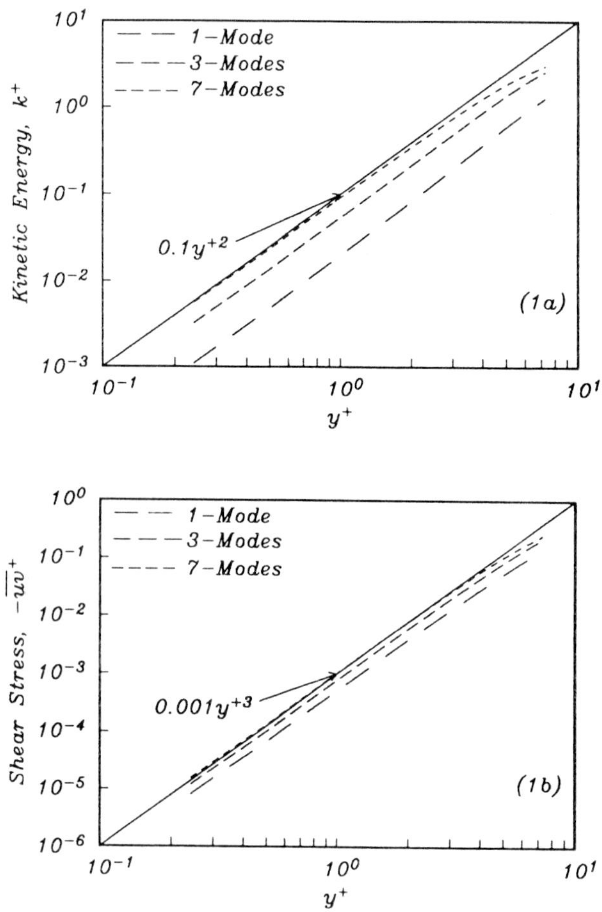

Figure 1: Asymptotic consistency in near-wall region: (a) kinetic energy, (b) shear stress.

Figure 2 shows the reconstruction of the turbulent dissipation rate ($\epsilon^+ = \overline{\left[\frac{\partial u_i}{\partial x_j}\right]^2}$). The figure shows only the contribution from the terms which are derivatives of distance from the wall. The 1D POD reconstruction can only handle these terms directly while the remaining terms must be obtained using symmetry assumptions. It suffices here to present only the contribution that can be obtained directly. As the figure shows several modes are needed to accurately reproduce the near-wall behavior from the DNS results.

Figure 3 shows a plot of the length scale obtained from the trace of $L_{ij}(y^+)$. Note that the latter definition for $L_{ii}(y^+)$ diverges at the origin suggesting the need to factor in the presence of the wall. Figure 4 shows the calculation of the dissipation rate from the length scale (which factors in the wall) and the kinetic energy, compared to the dissipation rate computed directly from the DNS results. Note the proper behavior for $y^+ < 5$ and $y^+ > 30$ and the discrepancy for $5 < y^+ < 30$. This is not surprising since the viscous definition is necessarily constrained to the region in close proximity to the wall, and the inertial definition in regions removed from the wall. Clearly an overlap region exists where a distinct definition for the dissipation rate needs to formed. A formal matching procedure may be a useful approach for this region.

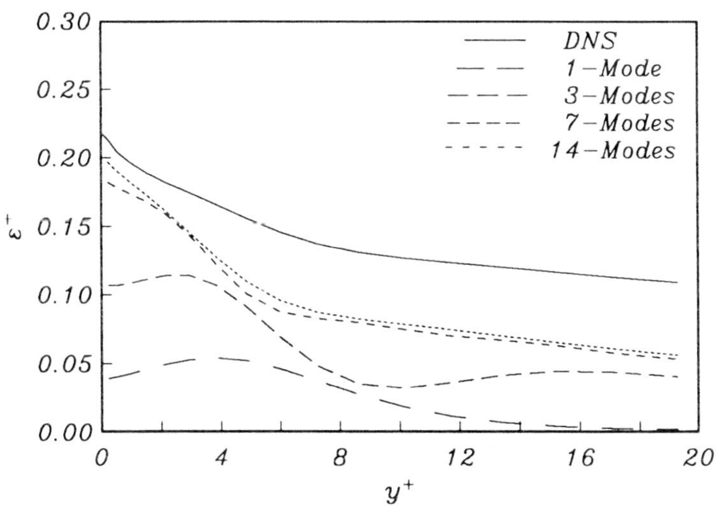

Figure 2: Effect of mode addition on ϵ^+ in close proximity to wall.

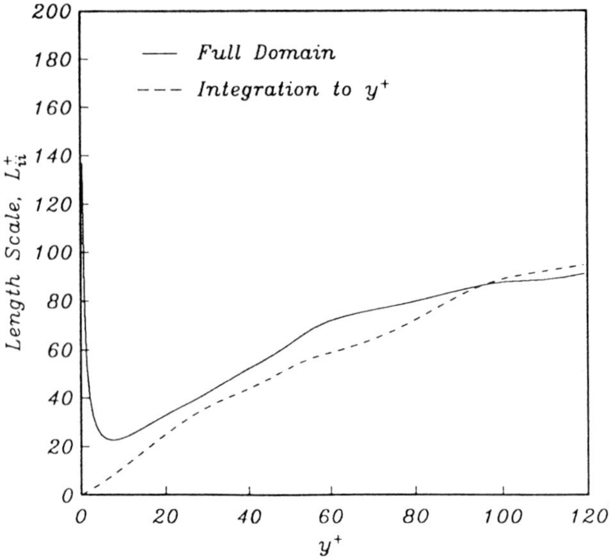

Figure 3: Length scales computed with and without accounting for the presence of the wall

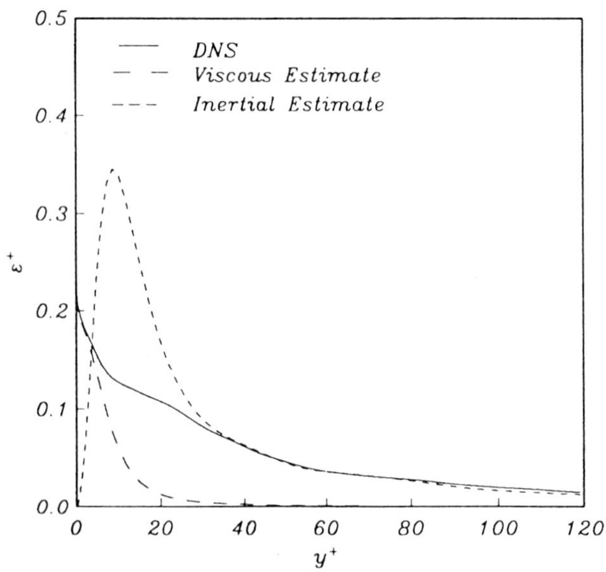

Figure 4: Dissipation rates computed from length scale definitions and DNS results.

Figures 5 and 6 show the one and three POD mode estimates of the length scale compared to that computed earlier using the DNS results. In both of these plots the POD estimates of the length scale are computed in two ways: the first involves normalizing the integration of the POD estimates of the correlations by the POD estimates of $R_{ii}(y^+, y^+)$ and the second estimate involves normalizing by the full $R_{ii}(y^+, y^+)$ – essentially estimating the integral only. Notice how the one mode estimate, normalized by the one mode estimate of R_{ii}, compares quite well with the length scale from the full DNS out to $y^+ = 20$, but then deviates substantially; whereas, the one mode estimate normalized by the total R_{ii} compares quite well beyond $y^+ = 60$. This is a manifestation of the convergence characteristics of R_{ii}, i.e., more modes are needed further out in the boundary layer to represent R_{ii}. Both of the three mode estimates of the length scale, presented in Figure 6, compare quite well out to about $y^+ = 90$. This is not surprising since the various second moments are represented sufficiently with three modes as shown in Figure 1. The comparison is remarkable for both the one and three mode estimates when compared to the number of modes required to obtain the dissipation rate as shown in Figure 4.

CONCLUSIONS

The results presented confirm the ability of the POD to properly reconstruct the second moments in the near-wall region of a turbulent flow. It appears that the present one-dimensional reconstruction is insufficient to properly account for the near-wall treatment of the turbulent dissipation rate. In an attempt to avoid this problem, a length scale definition has been proposed in terms of an integration of the correlation tensor which factors in the presence of the wall. The dissipation rate computed from this length scale and the kinetic energy exhibit the proper behavior near the wall and in the inertial layer. One and three POD mode estimates of the length scale are computed and compared to the length scale obtained when the entire correlation tensor is used. The 3 mode estimate compares quite well whereas the 1 mode estimate deviates somewhat from the full length scale.

The question that naturally arises at this point is whether the extra processing of simulation data is necessary since the various turbulent stresses and budgets can be directly obtained from simulation data. The answer lies in the relatively low Reynolds number range of the simulations to date. It has recently been shown by Sirovich and Rodriguez (1987) for the Ginzburg-Landau equation that the lower-order eigenfunctions are somewhat Reynolds number independent. If this were to hold for the Navier-Stokes equations, then the model development, based on the eigenfunction reconstruction would have a wider range of applicability. A reasonable approach may be to develop evolution equations for the eigenfunctions and compute the length scale from them. The dissipation can then be computed from this length scale and used to close the evolution equation for the kinetic energy.

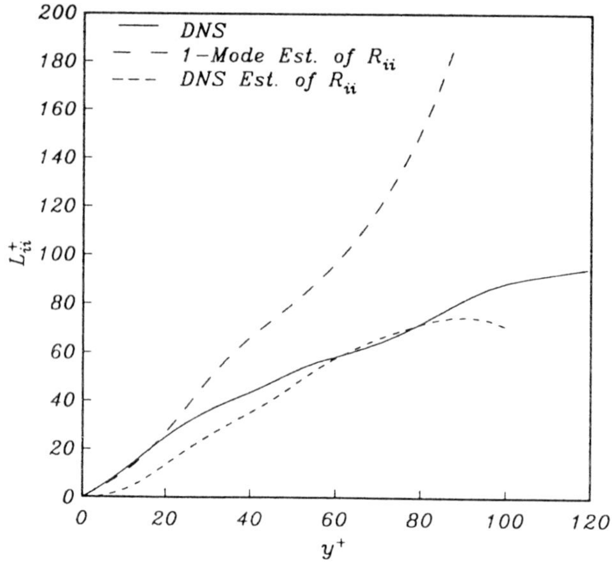

Figure 5: One POD mode reconstruction of length scale compared to that from the DNS results.

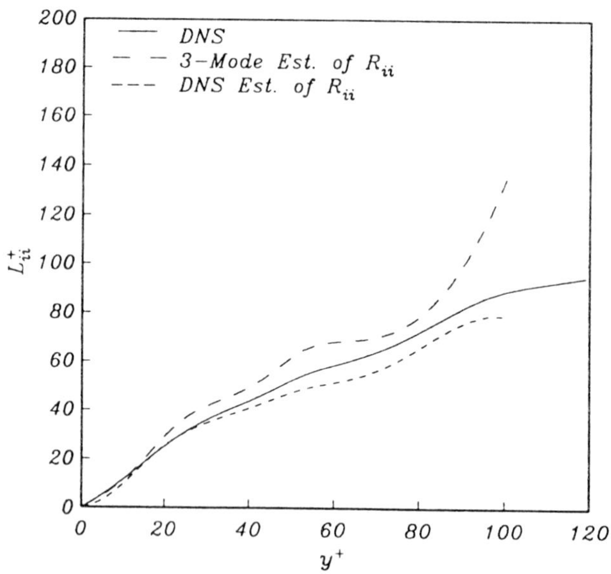

Figure 6: Three POD mode reconstruction of length scale compared to that from the DNS results.

ACKNOWLEDGEMENTS

One of us (MNG) would like to acknowledge the support of the NASA LaRC ASEE Summer Program and the organizers of the 1991 ICASE workshop on Instability, Transition and Turbulence.

REFERENCES

Dinavahi, S. and Zang, T.A., 1993 "Reynolds stress budgets in a transitional channel flow," *To be submitted for publication.*

Gatski, T.B. and Glauser, M.N. 1992 "Proper Orthogonal Decomposition Based Turbulence Modeling," *Instability, Transition, and Turbulence*, eds. M.Y. Hussaini, A. Kumar and C.L. Streett, Springer-Verlag, pp. 498-510.

W.K. George and H.J. Hussein, 1991 "Locally axisymmetric turbulence," *J. Fluid Mech.*, Vol 233, pp. 1-23.

N. Gilbert and L. Kleiser, 1991 "Turbulence model testing with the aid of direct numerical simulation results," *Eighth Symposium on Turbulent Shear Flows*, Technical University of Munich, Munich, Germany, September 9-11, pp. 26.1.1-26.1.6.

Lumley, J.L., 1967 "The structure of inhomogeneous turbulent flows," *Atmospheric Turbulence and Radio Wave Propagation*, (ed. A.M. Yaglom and V.I. Tatarski), Moscow: Nauka, pp. 166-178.

Sirovich, L. and Rodriguez, J.D., 1987 "Coherent structures and chaos: a model problem," *Physics Letters A*, vol. 120, pp. 211-214.

THE APPLICATION OF CLASSICAL POD AND SNAPSHOT POD IN A TURBULENT SHEAR LAYER WITH PERIODIC STRUCTURES*

D. HILBERG, W. LAZIK and H.E. FIEDLER
Hermann–Föttinger–Institut
Technische Universität Berlin, FRG

Abstract. The classical and snapshot proper–orthogonal–decomposition was applied to data taken in a one-stream mixing layer in a narrow channel. Due to this particular geometry the flow develops large periodic structures. POD-analysis of simultanously measured velocity components in spanwise direction identify as largest mode not only their periodic fraction, but also higher Fourier modes of the two-dimensional fluctuation. The energy content of the plane motion reaches values of about 90%. The amplitude of small three-dimensional vortices embedded in higher POD modes is correlated with the phase of the large structures, which indicates their influence on the entire turbulent motion.

Application of scalar snapshot POD on phase-averaged data of the entire flow field allows seperation into modes. The eigenvalues and eigenvectors show identical distribution for the u'– and v'–component. Comparision of streakline plots of the reconstructed velocity field from different number of modes with flow visualization exhibits that the largest physical structure is described by only the first two modes. This is also supported by calculation of the vorticity component in z-direction. The total energy content of the largest structure is approximately 60%.

1. Introduction

The classical POD as introduced by Lumley 1967, 1981 and by Sirovich 1989 and the snapshot-POD suggested by Sirovich & Kirby 1987, both based on the Karhunen–Loeve expansion (Loeve 1955), were applied to hotwire measurements taken in the turbulent region of a one-stream mixing layer with small sidewall distances B, where B is typically less 1/10 of the testsection length L. Due to this geometry this flow develops periodic structures of large scale corresponding to the test section length. The periodicity is caused by the feedback of coherent acoustic waves originating at the trailing edges of the two sidewalls, when a dominating structure leaves the test section. These waves travel upstream, where they trigger a new wave at the trailing edge of the mixing layer. Accordingly, the Fourier spectrum shows a strong peak with a periodic fraction of about 30%–45% of the total u'–value. Correlation measurements and flow visualisization in spanwise direction exhibit the two-dimensionality of the periodic motion (Hilberg & Fiedler 1989, Fiedler & Hilberg 1992).

* Supported by the Deutsche Forschungsgemeinschaft under Grant No. Fi 178/28

J. P. Bonnet and M. N. Glauser (eds.), Eddy Structure Identification in Free Turbulent Shear Flows, 251–259.

2. The classical and snapshot-POD method

The POD methods applied here are based on a scheme suggested by Glezer et al. 1989. Assuming u'-fluctuations to be measured simultaneously at M-positions in space and at N-points in time, the time series of the data is be combined in a matrix A as columns $a^{(i)}$

$$A = u'_{ij} \qquad i = 1,2,...,M;\ j = 1,2,...,N$$
$$a^{(i)} = (u'_{i1}, u'_{i2}, ..., u'_{iN}) \qquad i = 1,2,...,M.$$

For optimal representation of the time series an orthogonal basis Ξ is searched, for which the mean quadratic scalar-product

$$\chi^{(k)} = \sum_{i=1}^{M} (a^{(i)} \cdot \Xi^{(k)})^2 \qquad k = 1,2,...,M$$

is maximized. This leads to an eigenvalue problem

$$(A^T A)\ \Xi^{(k)} = \chi^{(k)}\ \Xi^{(k)} \qquad k = 1,2,...,M.$$

EISPACK routines were used for the numerical solution and decomposition of the signals into orthogonal POD modes of number M. The corresponding eigenvalues χ denote the energy content. The flow field corresponding to a POD-mode is calculated from

$$< u' >^{(k)} = a_t^{(k)}\ \Xi^{(k)}$$
$$a_t^{(k)} = A\ \Xi^{(k)}, \qquad k = 1,2,...,M$$

where the coefficients a_t are functions of time only. Summation over the flow-fields allows the reconstruction of the original velocity field.

This method is known as the classical POD. By exchanging rows and columns of the matrix A the snapshot-POD is obtained, where one column corresponds to one simultaneous 'snapshot' of the flow field. In this case then, the eigenvektors Ξ are functions of time and the coefficients a_t are functions of space.

3. Application of classical POD

The classical POD-method was applied to u'-time-series measured simultaneously with a rake of four hot-wires at different transverse positions, each 5 mm apart, in a narrow mixing layer with B = 30 mm and L = 1200 mm. The testsection with adjustable sidewalls is shown in figure 1. Typical dimensions are displayed in the sketch. Rake measurements were done at the streamwise positions x = 200, 400, 600, 800, 1000 and 1200 mm and at 41 different lateral positions with a resolution of $\Delta y = 10$ mm.

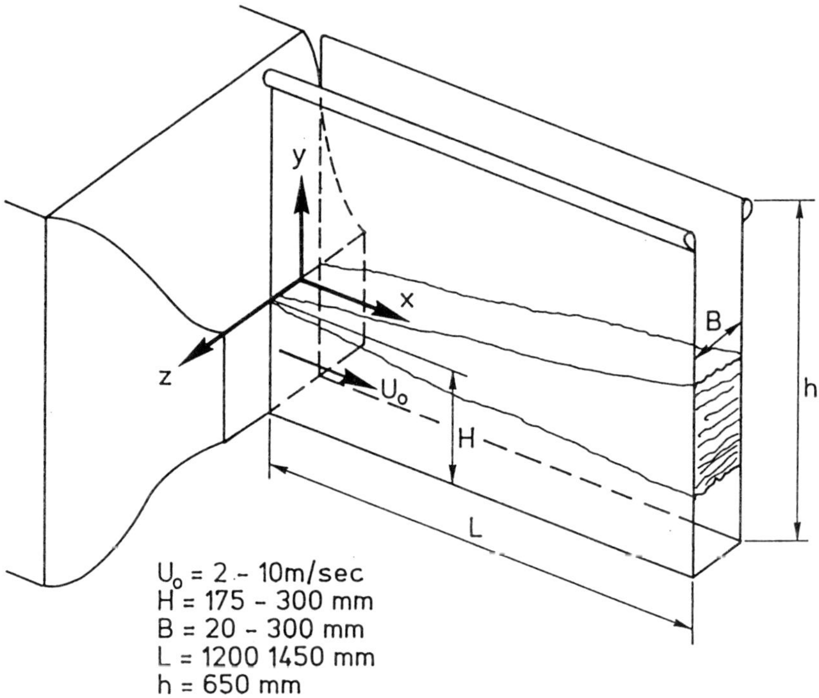

Fig. 1. Sketch of the testsection

Figure 2 shows the distribution of four eigenvalues over the y-axis at different streamwise positions. At all positions the normalized energy content of the first POD-mode is significantly higher than that of the other modes, reaching maximum values of 65%–90%. This indicates that more than just the periodic energy is captured in the first POD-mode. The y-position of the maximum is identical with the location of the largest perodic fraction.

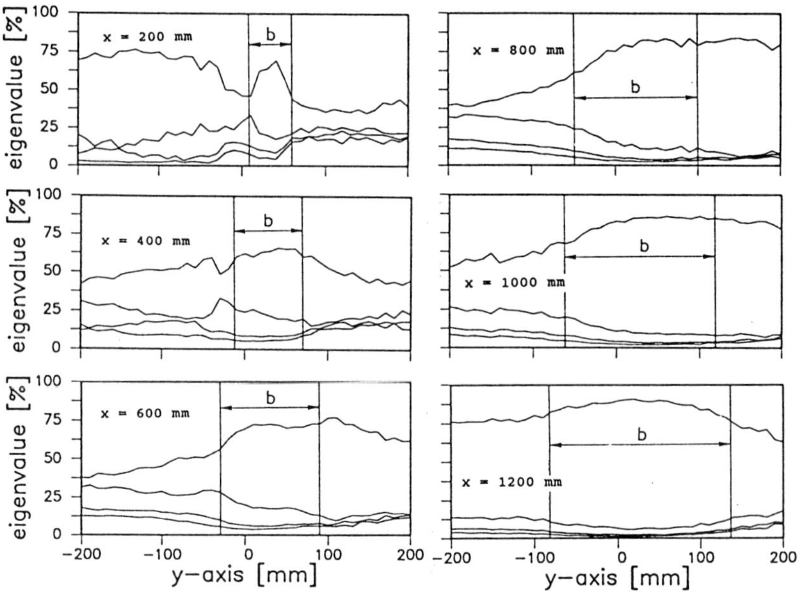

Fig. 2. Eigenvalues at different x-position as functions of y

At streamwise positions close to the splitterplate the first eigenvalue shows a second maximum outside the mixing layer at negativ y-positions. Closer inspection shows that the corresponding eigenvectors differ from those within the shear-layer. At positions further downstream this behaviour is no longer observed. The calculated components of the eigenvectors Ξ within the mixing layer have nearly equal distibution for each vector at all streamwise and lateral positions. An example is shown in fig. 3 for x = 1000 mm. Small differences of the components of respective eigenvektors at different lateral positions at x < 400 mm are vanishing further downstream, being independent of y. Particularly the components of the first eigenvector are identical also in spanwise direction, which points at the two-dimensionality of the large structures depicted by the first POD-mode. This is in agreement with correlation measurements of the periodic part of the turbulent motion. In relation to the corresponding eigenvalue the two-dimensional fraction of the turbulent fluctuations describes most of the turbulent motion within the narrow shear-layer.

Fig. 4 shows the coefficients a_t at x = 1000 mm and y = 40 mm, where the first eigenvalue has its maximum. The large periodic fraction is found only in the

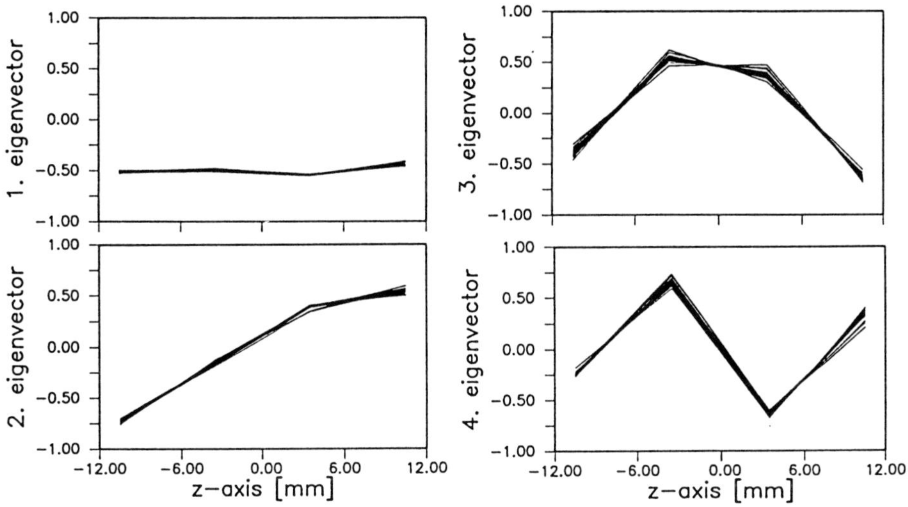

Fig. 3. Eigenvectors at x = 1000 mm

coefficient of the first mode. However, also higher Fourier-modes are embedded in this mode, indicating a two-dimensionalty of higher spectral fractions. Especially the small scale fluctuations near the maximum of the large scale motion suggest that the small three-dimensional vortices are partly stretched by the large structures, thereby becoming essentially two-dimensional.

The other POD-modes include only higher Fourier-modes. Obvioulsly, there is a correlation between the phase of the large fluctuation and the amplitude of the motion captured by the higer modes. This is clearly seen in the third and fourth mode, showing the influence of larger structures on the entire turbulent motion. Due to this phenomenon the mixing process in the shear-layer may be different from that in more or less stochastic turbulence. The large vortex rolls up layers of potential fluid from both sides, where the mixing between those is generated by small scale vortices. This is supported by measurements of intermittency, which shows considerably smaller values than those obtained in the wide shear layer.

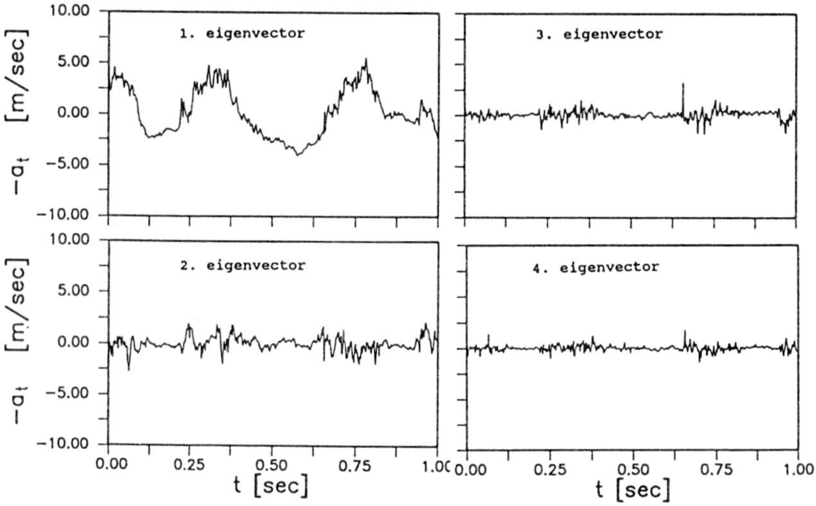

Fig. 4. Coefficients at x = 1000 mm

4. Application of snapshot-POD

For the application of snapshot-POD phase-averaged u- and v-data were taken with an X-wire probe in a field of 451 lateral and streamwise positions in the center plane between the sidewalls. The inherent periodicity of the flow was stabilized by weak forcing to simplify the eduction of data.

A total of 180 snapshots at equidistant phase positions were calculated. The components of corresponding eigenvectors of the u'- and v'-data show strong similarity. The eigenvectors of the first and second mode show the same periodicity of one wavelength, however, with a phase-shift of about 180 degrees (fig. 5). Corresponding normalized eigenvectors reach values of 30% . Higher modes have exponentially decreasing eigenvalues (fig. 6) and the components of the eigenvectors exhibit smaller wavelengths, decreasing by integer fractions. The first four modes alone describe about 80% of the total energy. According to the classical POD analysis, the energy containing modes of the snapshot-POD should be two-dimensional. The application of the snapshot- POD at other spanwise locations show, that the first four modes are independend of z.

Velocity- and streakline plots were calculated from the reconstructed flow-field with different numbers of modes. Thus, reconstruction from the first ten modes provides a reconstruction quality of 96%, according to a criterion by Sirovich & Kirby 1987. Comparison of streakline plots and flow-visualizisation enables the identification and association of physical patterns to POD-modes, showing in particular the largest physical structures to be contained in the first four modes (fig. 7).

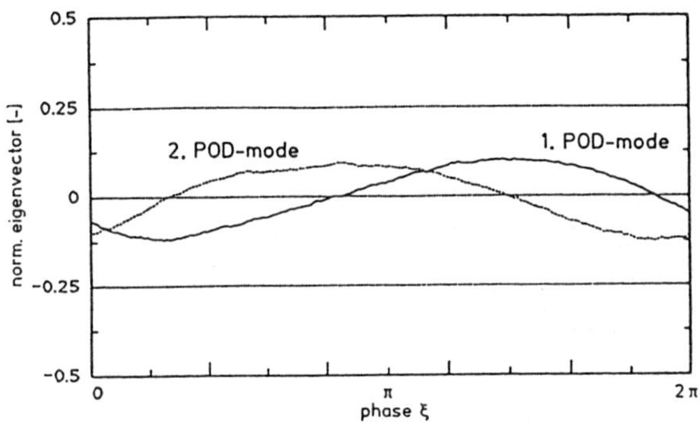

Fig. 5. First and second eigenvector

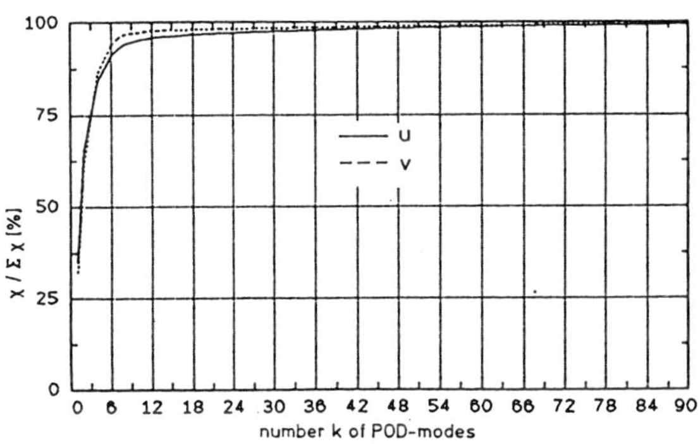

Fig. 6. Eigenvalue distribution

Calculation of the vorticity component $\omega_z \approx - \partial u / \partial y$ of the original and reconstructed velocity field by use of different number of POD-modes (k=2,4 and 180) shows that only 2 of 180 modes contain the largest fraction of coherent vorticity (fig. 8). Also streakline plots identify the largest visible structure to be represented by only the first two POD-modes.

Fig. 7. Streakline reconstruction and flow visualization

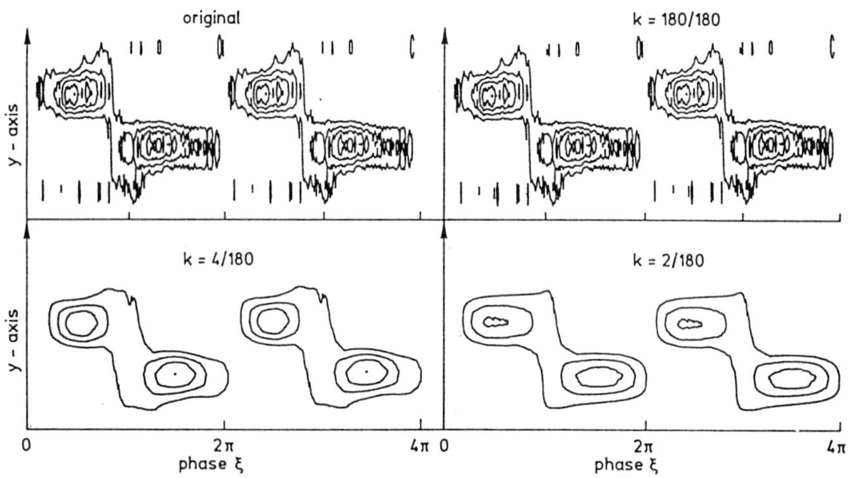

Fig. 8. Comparison of vorticity component ω_z of original and reconstructed velocity field at $x = L$ as function of phase (POD-modes: k=2,4 and 180)

5. Conclusions

Both, classical- and snapshot-POD was applied to data measured in a shear-layer with small sidewall distance. The classical POD showed that the major portion, up to 90%, of the turbulent energy within the shear-layer is typically of two-dimensional character. This fraction is descibed by the first POD-mode. This mode includes, however, not only the periodic part of the fluctuation referring to the significant peak in Fourier spectra, but also higher Fourier-modes. It was shown, that the coherent physical structures are described by only the first mode. In principle, those results may also be obtained by extracting coherent Fourier-modes by correlation calculations. However, the POD does not require pre-information about the correlation functions, so that the computation time is reduced. In contrast to methods requiring a given base, POD is an objective analysis (Lumley 1981).

Application of snapshot-POD on phase-averaged data permits unambiguous separation of modes, which are related to physical patterns. The first four modes contain about 80% of the energy. The reconstructed velocity field from 10 Modes out of 180 shows less than 4% overall deviation when compared to the original data. Comparison of streakline-plots with flow-visualization identifies the first four modes to be included in the typical structure.

References

H.E. Fiedler and D. Hilberg. Control of turbulent shear flows via stationary boundary conditions. *NATO Advances Research Workshop: The Global Geometry of Turbulence*, 1992.

A. Glezer, Z. Kadioglu, and A.J. Pearlstein. Development of an extended proper orthogonal decomposition and its application to a time periodically forced mixing layer. *Physics of Fluids A*, pages 1363–1373, 1989.

D. Hilberg and H.E. Fiedler. The spanwise confined mixing layer. In *Advances in Turbulence*, Fernholz and Fiedler, editors, vol. 2, pages 443–448. 1989.

M. Loeve. *Probabilty Theory*. van Nostrand, 1955.

J.L. Lumley. The structure of inhomogenous turbulent flows. In *Atmospheric Turbulence and Radio Wave Propagation*, Yaglom and Tatarski, editors. 1967.

J.L. Lumley. Coherent structures in turbulence. In *Transition and Turbulence*. 1981.

L. Sirovich. Chaotic dynamics of coherent structures. In *Physica D 37*, pages 126–145. 1989.

L. Sirovich and M. Kirby. Low dimensional procedure for the characterization of human faces. *J. Opt. Soc. Am.*, pages 519–525, 1987.

Free shear flow coherent structures in a low-dimensional eigenfunction space

Mojtaba Rajaee, Sture K.F. Karlsson and Lawrence Sirovich, Brown University, Center for Fluid Mechanics, Providence, R.I. 02912, U.S.A.

Large-scale coherent structures in the free shear layers have been the subject of many studies (see Ho & Huerre[1] for a comprehensive review) due to their important role in the mixing, entrainment and the laminar-turbulent transition processes.

An earlier work (Rajaee and Karlsson[2]) revealed that relatively few instability modes contain most of the fluctuation energy of our predominantly two-dimensional free shear flow. However, downstream of the first pairing event, the flow tends to become chaotic in a manner involving phase jitter and higher frequency fluctuations. Because of the phase jitter, difficulties arise in the use of Fourier methods. Use of the Karhunen-Loève expansion was found to overcome the deficiencies of Fourier methods in handling the more chaotic stages of the flow (Rajaee[3]). The basic idea is that spatial velocity correlations be orthogonally decomposed as a quantitative method of identifying coherent structures (Lumley[4]) which leads to an eigenfunction-eigenvalue problem, i.e.,

$$\int R_{ij}(x,x')\psi_j^n(x')dx' = \lambda^{(n)}\psi_i^{(n)}(x), \tag{1}$$

where the kernel, $\mathbf{R}$, is the two point correlation matrix

$$R_{ij}(x,x') = \overline{v_i(x)v_j(x')}, \tag{2}$$

and $\psi^{(n)}(x)$ are the eigenfunctions representing the coherent structures. We use the snapshot form (Sirovich[5]) of the K-L procedure for analyzing the hot-wire free shear measurements in order to describe the flow structures in terms of a low-dimensional basis set of empirical eigenfunctions. This basis set will then be used to develop a low-dimensional system of dynamical equations. The solution of this system of dynamical equations will be compared with the direct projections of flow realizations on this basis set of eigenfunctions. Procedures for developing a system of dynamical equations, employing the obtained eigenfunctions, for following the turbulent evolution of the flow have been presented (Sirovich[5,6]).

The experimental set up is the same as the one in the recent papers (Rajaee and Karlsson[2,7]). A closed return low speed wind tunnel is used to provide a free shear layer experimental facility. The mean velocity of the faster and slower

J. P. Bonnet and M. N. Glauser (eds.), Eddy Structure Identification in Free Turbulent Shear Flows, 261–267.

streams are set to 5.8 and 3.0 m/s, respectively. The velocity ratio, defined as $R = (u_1 - u_2)/(u_1 + u_2)$, is 0.32. Flow Reynolds number based on one half the sum of the free stream velocities and half the initial maximum shear thickness on the high speed side, $\delta_o = 1.25$ mm measured at $X = 2$ mm downstream of the splitter plate edge, is approximately 360. The initial turbulence level is 0.03% of the higher free stream velocity. As discussed in the previous paper, this free shear flow is predominantly two-dimensional.

Two dimensional excitation of the shear flow is achieved by a loud speaker. The speaker is driven by a computer generated signal consisting of two superimposed sine waves; the fundamental, which equals the unstable natural frequency, and its first subharmonic with a desired phase difference. The observed rms velocity fluctuation level at the trailing edge of the splitter plate was measured to be 0.4% of the average of the free stream velocities.

One experiment was made for the case of distinct pairing vortex interaction, i.e., corresponding to the initial phase angle difference of 270° between the subharmonic and the fundamental sine-wave components of the driving signal. The measurements were made between X = 10 to 35 cm and between Y = 6 to -6 cm in 1 mm intervals for both directions at the total of (251 × 121) locations. At each location 2/10 of a second of data were taken from the X-probe hot-wires as well as the subharmonic component of the driving signal at the sampling rate of 3060/sec, a multiple of 68 *Hz* which is the frequency of the subharmonic component of the driving signal. At each location data acquisition started at a reference time, i.e., the zero crossing with the positive slope on a cycle of the subharmonic component of the driving signal, and continued for 2/10 of a second, equivalent to more than 12 periods of the subharmonic component of the driving signal. Assuming flow stationarity in a time frame, phase locked with the driving signal, composite snapshots were produced using the data acquired in the above mentioned time frame (Rajaee[3]). Corresponding to each specific time, or phase, there is a composite snapshot of the flow field starting at t_o, the start of data collection at the formerly mentioned zero crossing time. These composite snapshots satisfy the two-dimensional continuity equation, a necessary condition to regard them as possible representations for an appropriate instantaneous flow. Figure 1 presents two of the composite snapshots. 540 composite snapshots, equivalent to the number of samples taken in 12 cycles of the subharmonic wave were considered for the eigenfunction-eigenvalue problem.

Solution of 540 × 540 matrix eigenvalue problem, through the method of snapshots, produced eigenvectors used in the calculations of time-independent eigenfunctions. The energies associated with different eigenfunctions are represented by their corresponding eigenvalues, λ. Table 1 lists the eigenvalues of the first 16 modes. To aid in the understanding of the coherent structures associated with different eigenfunctions we show these by means of color graphics. Figure 2 presents the vertical components of the first two eigenfunctions.

The dark red regions represent the maximum positive value, upward flow, and the

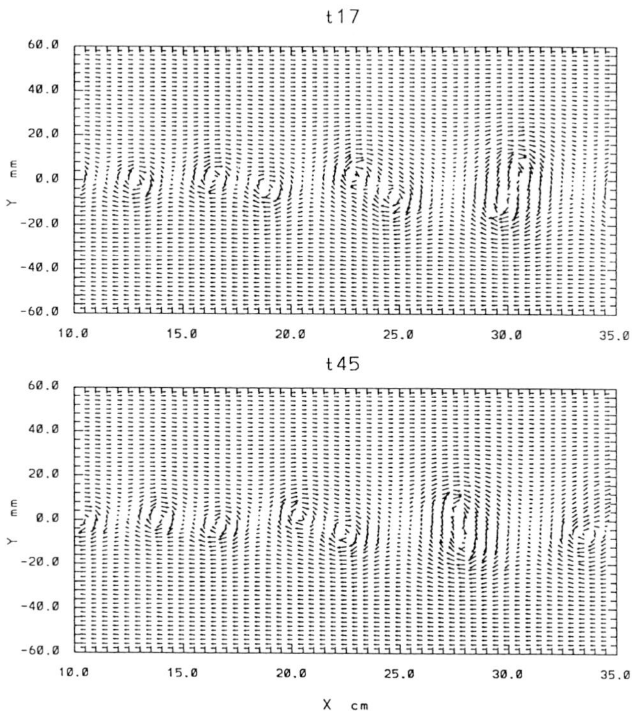

Figure 1: Two composite snapshots at t_{17} and t_{45}

Table 1: First 16 eigenvalues

i	λ_i	i	λ_i
1	0.2926	9	0.0031
2	0.2884	10	0.0029
3	0.1502	11	0.00134
4	0.1461	12	0.00131
5	0.0293	13	0.0011
6	0.0288	14	0.00107
7	0.0130	15	0.00105
8	0.0124	16	0.00101

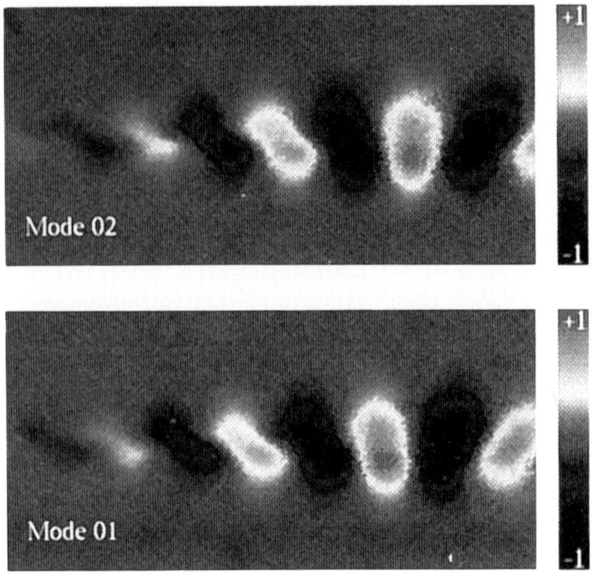

Figure 2: The mode 1 and 2 eigenfunctions, vertical component
Colour version in Colour Plates section, p. xxvii.

dark blue regions represent the maximum negative value, downward flow. They represent structures associated with the subharmonic instability wave, i.e., the projection of the shear flow velocity fluctuations on these eigenfunctions has a periodic sine-wave form corresponding to the subharmonic instability wave (Fig. 3(a)). The first two eigenfunctions are to a good approximation quarter wave translates of one another in the streamwise direction. Similar behavior is observed among the other pairs of modes, i.e., modes 3 and 4, etc. This is a consequence of near translational invariance in the streamwise direction. We note in Figs. 3(a) and 2 that mode 1 leads mode 2 by $\pi/2$ in time and lags it a quarter of wavelength in space.

Following the computations of a complete set of vector eigenfunctions, we can approximate the velocity field by its N-term expansion in this set of eigenfunctions, i.e.,

$$\mathbf{v}(x,y,t) \approx \overline{\mathbf{v}}(x,y) + \sum_{n=1}^{N} a_n(t)\Psi_n(x,y), \qquad (3)$$

where $\mathbf{v}$ refers to both velocity components, $\overline{\mathbf{v}}$ to the time-averaged mean velocity components, Ψ to the vector eigenfunctions and a_n to the time-dependent coefficients of the eigenfunctions in the expansion.

We substitute the N-term approximation of velocity field in the Navier-Stokes equations and project the N-S equations on the eigenfunctions, neglecting the insignificant fluctuating part of the viscous term, a reasonable assumption away from

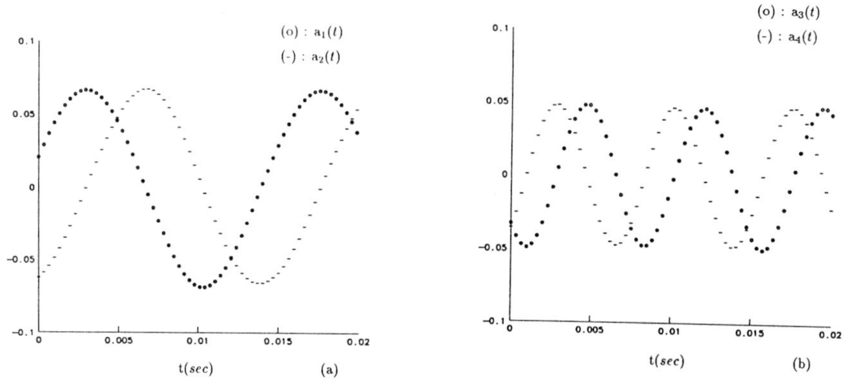

Figure 3: Projection of experimental velocity fluctuations on the eigenfunctions of : (a) modes 1 and 2, (b) modes 3 and 4

the wake of the splitter plate trailing edge. Knowing that the pressure fluctuations only come into effect at the cross stream boundaries of the flow region, they are also neglected. This procedure results in a system of nonlinear ordinary differential equations, i.e., the system of dynamical equations

$$\frac{da_k}{dt} + B^{kn} a_n + C^{knm} \left(a_n a_m - \overline{a_n^2} \right) = 0, \tag{4}$$

where

$$B^{kn} = \left(\psi_k^i, \left(\overline{\mathbf{v}} \cdot \nabla \psi_n^i + \Psi_n \cdot \nabla \overline{\mathbf{v}^i} \right) \right), \tag{5}$$

and

$$C^{knm} = \left(\psi_k^i, \left(\Psi_n \cdot \nabla \psi_m^i \right) \right). \tag{6}$$

The first sixteen modes, each with the energy percentage $\geq 0.1\%$ of the total two-dimensional fluctuation energy, are considered for the system of dynamical equations to include all the significant harmonics of the subharmonic instability wave. They represent more than 97.5% of the fluctuation energy. A system of sixteen equations with sixteen unknowns are solved using the experimental initial condition for the $a_n(t = 0)$, i.e., from the direct projection of the first snapshot, at t = 0, on the eigenfunctions. The fourth order Runge-Kutta scheme, for the explicit integration, is used with the running time average $\overline{a_n^2}$ term. The experimental mean flow velocity measurements are used in computing B^{kn}.

The computed $a_n(t)$ are compared with the direct projection results, for the first four modes, in Fig. 4. There is a second frequency, a low frequency modulation, observed in the simulation results which could be due to error in computing

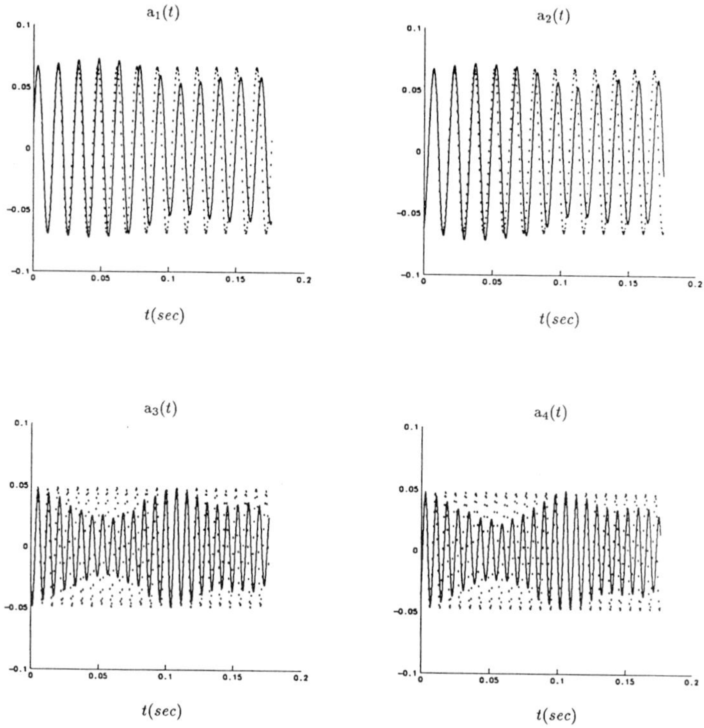

Figure 4: $a_1(t)$..........$a_4(t)$, (dotted line) from direct projection of snapshots on the eigenfunctions, (solid line) computed from dynamical equations, considering 16 modes

the constant coefficients of the dynamical equations, i.e., B^{kn} and C^{knm}. This is the consequence of the error in the numerical differencing scheme. Also, since the mean flow velocity is a time average, or ensemble average over the number of snapshots, i.e., 540 snapshots taken in less than 0.2 of a second, it could be possible that the mean flow estimate for the small transverse component includes some inaccuracy which could in turn cause an error in B^{kn}. Fig.4 shows that the simulation results, solid lines, obtained by solving the system of dynamical equations, compare reasonably well with the direct projections of snapshots on the eigenfunctions, the dotted lines. Especially for the first pair of the most energetic modes, with the sum of more than 58% of the fluctuation energy, the results from simulations and experiment are close. The good agreement of the solution of the system of dynamical equations with the direct projection results is a promising indication that the dynamical equations may be used to explore the turbulent evolution of the flow.

* This work was partly supported by DARPA under ONR N00014-86-K0754

References

[1] Ho C-M. and Huerre P., "Perturbed free shear layers," *Ann. Rev. Fluid Mech.* **16**, 365 (1984).

[2] Rajaee M. and Karlsson S.K.F., "On the Fourier space decomposition of free shear flow measurements and mode degeneration in the pairing process," *Phys. Fluids A* **4** (2), 321 (1992).

[3] Rajaee M., "Measurement and mathematical analysis of free shear flow," Ph.D. thesis, Brown University (December 1991).

[4] Lumley J.L., "Coherent Structures in Turbulence", in *Transition and Turbulence*, edited by by R.E. Meyer (Academic, New York), 215 (1981).

[5] Sirovich L., "Turbulence and the dynamics of coherent structures", *Q. Appl. Math.* **45/3**, 561 (1987).

[6] Sirovich L., "Chaotic dynamics of coherent structures," *Physica D* **37**, 126 (1989).

[7] Rajaee M. and Karlsson S.K.F., "Shear Flow Coherent Structures Via Karhunen-Loève Expansion", *Phys. Fluids A* **2**, 2249 (1990).

VII. Stochastic Estimation

STOCHASTIC ESTIMATION OF CONDITIONAL STRUCTURE

Ronald J. Adrian
Department of Theoretical and Applied Mechanics
University of Illinois
Urbana, Illinois 61801

1. Introduction

In a general sense, stochastic estimation is the approximation or estimation of a random variable in terms of some other random variables which are known. In the context of turbulent structure identification the estimated variable is typically the velocity field $\boldsymbol{u}' = \boldsymbol{u}(\boldsymbol{x}', t')$ for some domain of positions and times, and the known variables are associated with the occurrence of certain events at one or more points in the field. Their totality is referred to as the *event data vector*, $\boldsymbol{E}$. We shall denote the estimate of $\boldsymbol{u}'$ by $\hat{\mathbf{u}}'$.

The best mean square estimate of $\boldsymbol{u}'$ given the data $\boldsymbol{E}$ is the conditional average of $\boldsymbol{u}'$ given $\boldsymbol{E}$, $<\boldsymbol{u}'|\boldsymbol{E}>$. That is, of all possible estimates $\boldsymbol{u}' = \boldsymbol{F(E)}$, $<[u_i' - F_i(\boldsymbol{E})]^2>$ is a minimum when $F = <\boldsymbol{u}'|\boldsymbol{E}>$. In general, $<\boldsymbol{u}'|\boldsymbol{E}>$ is a non-linear function of $\boldsymbol{E}$, but under the condition that the elements of $\boldsymbol{u}'$ and $\boldsymbol{E}$ are joint normally distributed, it is well known that $<\boldsymbol{u}'|\boldsymbol{E}>$ is a linear function of $\boldsymbol{E}$ (Papoulis 1984). Often, this property is applied approximately to non-normal random variables by postulating an estimate of $\boldsymbol{u}'$ in the form of a linear combination of the event data. Then one speaks of *linear mean square estimation*. This is a well known subject, thoroughly discussed in the literature of stochastic theory (Papoulis 1984; Deutsch 1965). The mean square error of the linear estimate must be greater than or equal to that of the conditional average.

The application of linear mean square estimation to a turbulent field $\boldsymbol{u}'$ raises certain objections, in part because the linear approximation seems unlikely to be able to describe a strongly non-linear phenomenon, and in part because the joint normal property has not enjoyed much success as an approximation in turbulence. Another objection is that any estimate of $\boldsymbol{u}'$ in terms of event data $\boldsymbol{E}$ must become erroneous as the location $\boldsymbol{x}'$ moves away from the location of the event data, because $\boldsymbol{u}'$ will become uncorrelated with the event data for large spatial (or temporal) separations. However, this latter objection is easily answered because the conditional average $<\boldsymbol{u}'|\boldsymbol{E}>$ also has this same property, and we traditionally accept its consequences.

Instead of estimating $\boldsymbol{u}'$ as a function of $\boldsymbol{E}$, it is more straight forward conceptually to estimate $<\boldsymbol{u}'|\boldsymbol{E}>$, which can always be interpreted as random variable since it is a function of $\boldsymbol{E}$, and $\boldsymbol{E}$ is random. Then, it is natural to expand $<\boldsymbol{u}'|\boldsymbol{E}>$ in a Taylor series about $\boldsymbol{E}=\boldsymbol{0}$, and truncate at some level (Adrian 1977, 1979). When the series contains

J. P. Bonnet and M. N. Glauser (eds.), Eddy Structure Identification in Free Turbulent Shear Flows, 271–280.

only first order terms, we refer to this as the *linear stochastic estimation* (LSE) to distinguish it from linear mean square estimation. In fact, the linear stochastic estimate of $<\boldsymbol{u}'|\boldsymbol{E}>$ and the linear mean square estimate of $\boldsymbol{u}'$ are numerically equal, so the principal difference is one of interpretation. In particular, while the mean square error of the linear estimate of $\boldsymbol{u}'$ must be large when $\boldsymbol{u}'$ is uncorrelated with $\boldsymbol{E}$, due, for example, to large $|\mathbf{x}-\mathbf{x}'|$, the error of the linear stochastic estimate of $<\boldsymbol{u}'|\boldsymbol{E}>$ may be small because $<\boldsymbol{u}'|\boldsymbol{E}>$ also vanishes as $|\mathbf{x}-\mathbf{x}'|$ becomes large.

The equations for *linear* stochastic estimations of the *ith* component of $\boldsymbol{u}'$ are

$$\hat{u}_i' \quad = \quad \text{linear estimate of } <u_i'\,|\,\mathbf{E}> \quad = \quad \sum_{j=1}^{M} L_{ij}E_j, \tag{1}$$

where M is the number of event data, and $L_{ij} = L_{ij}(\mathbf{x},\mathbf{x}')$. The estimation coefficients L_{ij} are chosen so that the mean square error

$$<\left[<u_i'\,|\,\mathbf{E}> - \sum_j L_{ij}E_j\right]^2> \quad = \quad \text{minimum}, \quad i = 1,2,3 \quad \text{and} \quad j = 1,..N. \tag{2}$$

A necessary condition for minimization is the *orthogonally principle* which states that the errors $<u_i'|E> - \sum_j^M L_{ij}E_j$ are statistically orthogonal to the data:

$$<\left[<u_i'\,|\,\mathbf{E}> - \sum_j L_{ij}E_j\right]E_k> \quad = \quad 0, \quad i = 1,2,3 \quad j,k = 1,\ldots M. \tag{3}$$

Simple manipulation leads to the *Yule-Walker* equations, an $N \times N$ system of linear algebraic equations for the coefficients L_{ij} whose coefficients are the correlations between each event data and every other event data and the quantity being estimated,

$$\sum_{j=1}^{N} <E_jE_k> L_{ij} \quad = \quad <E_ku_i'>, \quad j,k = 1,\ldots,N \tag{4}$$

If the data consist of velocity vectors $\mathbf{u}_\alpha = \mathbf{u}(\mathbf{x}_\alpha, t_\alpha)$ at a set of space-time points $\{\mathbf{x}_\alpha, t_\alpha\}, \alpha = 1,..,N$, then $<E_jE_k>$ involve only the second-order, two-point space-time correlation $R_{jk}(\mathbf{x}_\alpha, \mathbf{x}_\beta, t_\alpha, t_\beta)$ and $<E_ku_i'>$ reduces to $R_{ki}(\mathbf{x}_\alpha, \mathbf{x}', t_\alpha, t')$. Thus, knowledge of $R_{jk}(\mathbf{x},\mathbf{x}',t,t')$ suffices to fully determine any linear estimate in terms of velocity data. Likewise, if the event vector contains derivatives of the velocity field, $(\partial u_i/\partial x_j)$ at points $\mathbf{x}_\alpha$, the resulting two-point correlations that appear in Eq. (4) contain correlations between velocities, velocities and derivatives and derivatives and derivatives. Since each type of correlation can be expressed in terms of R_{ij}, it follows that knowledge of $R_{jk}(\mathbf{x},\mathbf{x}',t,t')$ again suffices a fully determine the linear estimate of $\boldsymbol{u}'$ in terms of velocity data and derivatives of any order. This is a powerful result as it states that *very complicated conditional averages can be approximated once R_{ij} is known.*

The accuracy of LSE has been evaluated by direct comparison to experimentally measured conditional averages in a number of flows, such as grid turbulence, axisymmetric jets, wakes, plane shear layers and pipe flow (Adrian et al. 1989). Usually the agreement is surprisingly good, even for large values of the event data. To improve accuracy, it has been suggested that (1) be generalized to include higher order terms in $\boldsymbol{E}$ (Adrian 1977). Often, higher order terms have little effect for probable values of the events, but sometimes, they improve accuracy (Pesceli and Trulsen 1989; Brereton 1992). Interestingly, Brereton (1992) has also suggested a Laurent-type exposure which contains negative powers of the event data, and he shows improved estimation in some regions of the boundary layer.

2. Conditional eddies

The foregoing discussion shows how conditional averages can be linked to two-point correlation functions through the process of stochastic estimation. Thus, conditional averages and stochastic estimates can be considered to contain approximately the same information about the structure of the flow, and this information is given by a proper combination of the structure information contained in the two-point spatial correlation tensors and the event data. It is absolutely essential to distinguish conceptually between the structure of any particular conditional average and the structure of coherent elements of the flow, should they exist. There is no *a priori* guarantee that any conditional average (or, indeed, the structure extracted by any other pattern recognition procedure) *must* coincide with individual, random realizations of portions of the flow field. The conditional averages are well defined mathematical quantities, and they should be treated the same as the mean or correlation, which exist for any flow. In contrast, while some flows contain readily recognized coherent patterns, in the sense that similar patterns occur repeatedly, we have no rational basis for asserting that all turbulence flows contain such structures (although we may suspect this to be true). To avoid confusing conditional averages with realizations of the random field (and to avoid the problem of defining "coherent structure") the flow fields described by conditional averages have been referred to as "conditional eddies." There are many types of conditional eddies depending upon the conditional event. They always exist, and the same type of conditional eddy can be used to compare the structures of different flows. However, the relationship between conditional eddies and random realizations of the flow can only be established by empirical comparison. Thus, the interpretation of the structure found in a conditional eddy should be done by performing "calibration" experiments.

Conditional eddies are defined by the event. One of the simplest and best known events is the quadrant analysis method of Willmarth and Lu (1972) in which there are only four events corresponding to the four quadrants of the *u*-*v* plane. There are, however, infinitely many alternative choices. It seems clear that the definition of the conditional event for a particular application must be determined by the ultimate goals of the conditional structure analysis in that application. For example, it is well known that the equation governing the probability density function of the velocity at one point $f_1(\boldsymbol{v},\boldsymbol{x},t)$ contains factors which are two-point conditional averages given the velocity at the point $\boldsymbol{x}$ (Adrian 1977): $< \mathbf{u}(\mathbf{x}',t) \,|\, \mathbf{v} \le \mathbf{u}(\mathbf{x},t) < \mathbf{v} + \mathbf{dv} >$. Thus, if one is interested in closure approximations for the one-point *pdf* equation, it is natural to examine the structure of the one-point conditional eddies $<\boldsymbol{u}' \,|\boldsymbol{u}>$. If closure approximations at the two-point level are of interest, the two-point conditional eddy $<\boldsymbol{u}' \,|\boldsymbol{u}_1,\boldsymbol{u}_2>$ is relevant. Obviously, the latter, two-point conditional eddy will more faithfully reproduce the random field than the one-point eddy $<u'|u>$ because of the additional information contained in the second velocity, but for the purposes of one-point closures, the governing equation for f_1 does not require this information.

This approach to the selection of events seems more rational than simply searching for the condition that will extract the most faithful representation of the coherent structure because, after all, the best representation of a particular structure is given by the event that describes that structure completely, e.g., the structure itself. It is clear that what we really want is a good representation of the structure subject to certain constraints, and one method of defining those constraints is by recourse to governing equations which define the amount of information that is just adequate to close the theory.

In addition to one- and two-point events, conditional eddies have also been studied for multipoint events (Cole et al. 1992; Bagwell et al. 1992) and for events involving the velocity $u(x,t)$ and the local deformation tensor $d_{ij}(\mathbf{x},t) = \partial u_i(\mathbf{x},t)/\partial x_j$ (Adrian and Moin 1988; Ditter and Adrian 1992). The latter is referred to as a one-and-one-half point event in the sense that d_{ij} contains a subset of the information imbedded in the behavior of the flow at two neighboring points.

3. Structures found by stochastic estimation of conditional eddies

In isotropic turbulence the LSE of $<u'|u>$ is a vortex ring oriented axisymmetrically about the given velocity, $u(x,t)$ (Adrian 1979). Thus, specifying a vector direction in an isotropic field results in an axisymmetric field. Higher order estimates produce more complex ring structure, but do not significantly alter the basic structure for probable values of the velocity fluctuation (Tung and Adrian 1980). Although the flow is a ring on a scale of order many integral scales, Fig. 1, most of the flow around the ring is very weak, except in the vicinity of the conditional vector where the flow more nearly resembles a jet of fluid. Thus, observations of random isotropic turbulence are not likely to be able to see the weak outer structure of the ring and one expects to see long, thin jets at location where the velocity fluctuations are large. Axisymmetric structures have zero helicity, $u \cdot \omega = 0$, unless they are also spinning around their axis of symmetry. It is therefore of interest to impose the local deformation tensor as an additional condition. Ditter and Adrian (1992) have studied the conditional average $<u'|u,d>$ isotropic turbulence, and in u find a linear estimate of the form

$$< \mathbf{u}' \mid \mathbf{u}, \mathbf{d} > \quad = \quad \hat{u}_i(\mathbf{x}+\mathbf{r}) \quad = \quad \frac{R_{ji}(\mathbf{r})}{u^2} u_j(\mathbf{x}) - \frac{2\lambda^2}{15u^2}\{R_{ki,j}(\mathbf{r}) + 4R_{ji,k}(\mathbf{r})\} d_{jk}(\mathbf{x}). \tag{5}$$

The first term corresponds to the usual translational event, and the second gives the contribution of deformation. The Taylor microscale is λ and $u^2 =< u_i u_i > /3$. Setting $u = 0$, implies that x is a critical point of the flow. The local structures at such a point can be found by the methods described in Perry and Chong (1987), since the linear estimate is a solenoidal vector field. If the flow at x is purely rotated about the x_3-axis (with no translation), the streamlines and vortex lines are as shown in Fig. 2. In the vicinity of x the flow field is essentially a long, thin vortex tube whose length is of order of several longitudinal integral scales, L_{11}. As before, the strength of the flow at larger distances from x is small, making the total flow pattern difficult to observe when the eddy is immersed in an environment of random eddies. These considerations may explain the tendency of visualizations based on thresholded vorticity to produce vortex tubes.

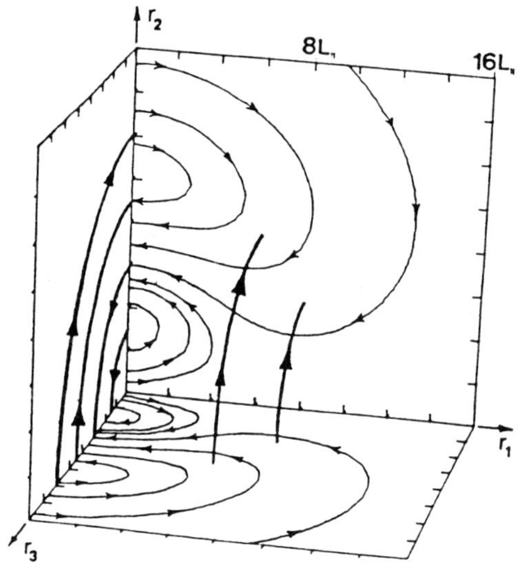

Fig. 1 Conditional eddy given $\mathbf{u}(\mathbf{x})=(u,0,0)$ in isotropic turbulence. Streamlines are shown in the light ink and vortex lines in heavy ink (Ditter and Adrian 1992).

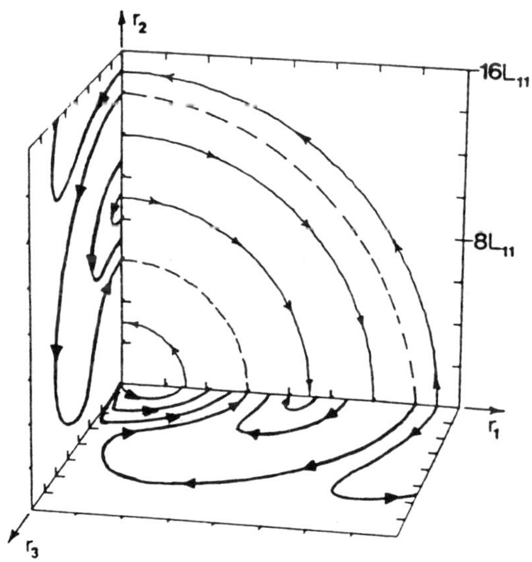

Fig. 2 Conditional eddy given pure rotation about the r_3-axis. Streamlines in r_1-r_2 plane and vortex lines in r_1-r_3 and r_2-r_3 planes demonstrate symmetry about r_3-axis (Ditter and Adrian 1992).

The conditional average $<u'|u,d>$ has also been stochastically estimated for homogeneous shear flow using R_{ij} computed from direct numerical simulation (Adrian and Moin 1988). In this case the conditional eddy has the form of a hairpin vortex inclined at 45° to the flow when the conditional vector corresponds to the value that contributes the largest amount to the mean Reynolds shear stress. If the mean shear flow is subtracted the hairpin vortex becomes a vortex ring.

Single-point conditional eddies have been estimated for turbulent channel flow using the two-point spatial correlation tensor computed from DNS simulations (Moin, Adrian and Kim 1987; Adrian, Moin and Moser 1987). The vector events were chosen to be the second- and fourth- quadrant *u-v* event that contribute most to the mean Reynolds shear stress by minimizing the product $uvf_1(u,v)$. Consider the Q2 event. Very close to the wall the structure is a low speed streak under the given vector with two counter-rotating streamwise vortices on either side of the given vector. Farther above the wall a bridge of vorticity forms between the streaks, resulting in a hairpin geometry. Figure 3 shows the contours of constant enstrophy of the LSE of $<u'|u>$ for $y^+ = 103$. The legs of the hairpin lay back along the wall with spacing of approximately 100 viscous wall units. The conditional vector is positioned between the legs and just under the head of the hairpin. Robinson et al. (1990) report observations of "canes" or one legged hairpins rather than symmetric hairpins. Such structures are obtained from stochastic estimation by letting the given velocity have a sufficiently strong *w*-component of velocity so that the streamwise vorticity associated with $\partial w/\partial y$ enhances the vorticity of one leg and cancels the vorticity of the other. Farther still from the wall, e.g. at the centerline of the channel, the one-point LSE becomes a vortex ring.

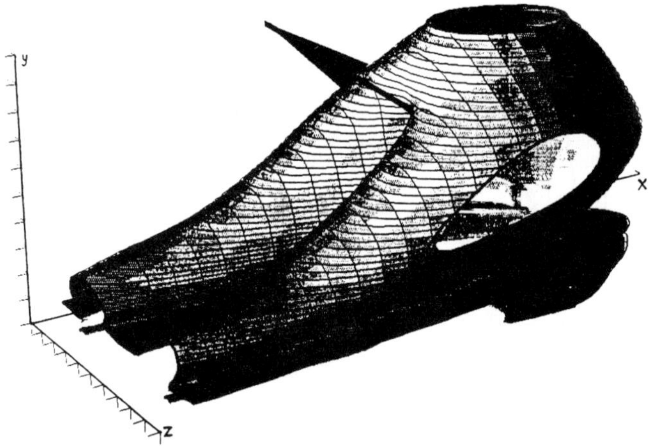

Fig. 3 Hairpin-like structure resulting from the $Q2$ event specified at y^+=103 in turbulent channel flow.

The one-point hairpins found by LSE possess many of the features observed in experiments and direct numerical simulations. However, they do not have the thin shear layer lying above the space between the legs that has been observed in computations and experiments. The shear layer is believed to be the part of the flow that creates the VITA signature. To obtain this element, it is necessary to combine a downstream Q2 vector with an upstream Q4 vector in close proximity. The Q2 event creates a hairpin as in Fig. 3, and the Q4 event creates a fast downward flow that collides with the upward motion from the Q2 event. The result is a stagnation point flow on a thin strip above the hairpin's legs. The fact that two events are required to represent the shear layer suggests that it should be considered as arising from the interaction of two separate conditional eddies that are colliding. It also suggests that closure of the *pdf* equation at the two-point level may be intrinsically superior to closure of the one-point *pdf*, because conditions and two-points may be essential to represent important flow interactions.

Single point estimation of the log-layer in the *x-y* plane of Re=50,000 pipe flow yields cross-stream vortex patterns that are consistant with a large hairpin vortex (Adrian 1990).

Single point estimation of a plane shear layer yields a cross-stream vortex, but not one that is particularly rich in detail. Multipoint estimation of the shear layer using points along streamwise line yields relatively detailed vector fields (Cole et al. 1992). Guezzenec (1989) has shown how one-point space-time estimates can be used to study evolution of the boundary layer.

Multipoint LSE of the form $<\mathbf{u}' | \mathbf{u}_1, \mathbf{u}_2, ..., \mathbf{u}_N>$ is best viewed as a means of stochastic interpolation or extrapolation. If the locations $\mathbf{x}_1, ... \mathbf{x}_N$ are grid points in a numerical algorithm, $<\mathbf{u}' | \mathbf{u}_1, ... \mathbf{u}_N>$ can be interpreted as the best estimate of the subgrid field given the resolved scale data (Adrian 1977, 1990). An effort is currently underway to use this approach to estimate the fluctuating pattern of subgrid scale wall shear stress in a large eddy simulation (Bagwell et al. 1993). The velocity vectors computed on the lowest grid plane are used to estimate the shear stress on the wall.

4. Detection of structures in random realizations by stochastic estimation

The greatest single virtue of LSE is that it makes it easy to estimate conditional averages for complex events. Even a relatively simple event like $\{\mathbf{u}, \mathbf{d}\}$ constitutes a twelve-dimensional space in which it would be almost impossible experimentally to find a reasonable number of realizations that satisfy the condition by conditional sampling. Once the LSE coefficients are found for a given flow and given type of event, the event can be changed quantitatively to cover the full range of probable values without recalculating the coefficients or re-sampling the data. This flexibility places more emphasis on the selection and definition of the events than is encountered in conditional averaging where the events must be constrained to have simple forms for the sake of feasibility. We have seen that certain events are indicated naturally by the theory, and it has been argued that the fidelity with which they represent the structure should be considered enough for the purpose of that particular theory. In this sense the problem of event definition is solved.

However, suppose we consider the classical coherent structure detection problem in which one attempts to find events which serve as accurate *detectors* of the presence of a coherent structure. How well can LSE address this problem? First, it should to pointed out that the simple one-point event $\{\mathbf{u}(\mathbf{x},t)\}$ is already significantly more specific that the quadrant detection methods which have been used extensively. Evenso, it may not be good enough if the flow contains structures that have the same velocity vector at several points in their volume. Then, the conditional average will average over all of this events indiscriminately. A simple example is the conditional average $< u(t') \mid u(t) >$ where $u(t) = \sin \omega t$. If $v < 1$ is the prescribed sampling level, $u(t) = v$ at two different times during any cycle, one with positive slope and one with negative slope. The conditional average averages over these two different events and predicts a cosine wave, e.g. a sine wave whose phase is the average of the two events. To define one event uniquely, more information must be specified. This can be done by adding the slope as an event, analogous to specifying **d**.

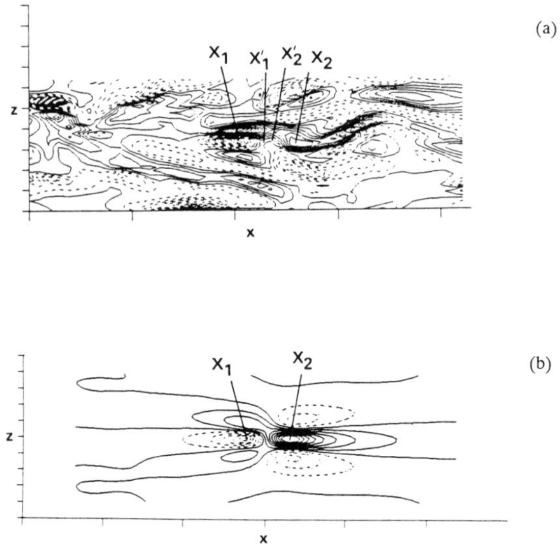

Fig. 4 (a) Contours of the instantaneous v-component in the x-z plane passing through $\mathbf{x}_1$ at y^+=30 in turbulent channel flow. (b) Linear estimate of v-component in the x-y plane passing through $\mathbf{x}_1$ given the $Q4$ event at $\mathbf{x}_1$ and the $Q2$ event at $\mathbf{x}_2$. Note how the pattern of alternating positive and negative velocities is made apparent (Adrian, Moin and Moser 1987).

In principle, a sufficiently detailed list of events will define a structure uniquely. The list is *sufficient* if the addition of more events results in negligible reduction of the mean square error. An example of this process is that of estimating $u(x')$ from uniformly spaced events $u(\Delta x), u(2\Delta x), u(3\Delta x)$. It can be shown that as the sample spacing decreases the mean square error decreases until it vanishes at the Nyquist sampling limit. In this case the linear estimation functions become the sinc function of the Nyquist sampling theorem.

Clearly, Nyquist sampling is too detailed for eddy structure identification, so we must constrain the set of given data by another principle. One approach which has been investigated to a limited extent is to identify points of interest in a random field and use the actual random values of the flow variables at those points as data for the LSE. The result is a smooth structure that is free from random background noise, and represents that average structure present when those events become. Figure 4 illustrates the procedure. Two points $\boldsymbol{x}_1$ and $\boldsymbol{x}_2$ were located by searching for the location of maximum in the fluctuating Reynolds shear stress on a y = constant plane of turbulent channel flow. The random field does not exhibit a readily recognized pattern, but the LSE using $\boldsymbol{u}(\boldsymbol{x}_1)$

and $\boldsymbol{u}(\boldsymbol{x}_2)$ shows a clear pattern of alternating signs in the v-component of velocity that corresponds to the average footprint of a Q4 event interacting with the back of a Q2 hairpin.

A new extension of this technique for filtering the coherent structure out of a more random pattern is to specify deformation data at the critical points of the vector field. This appears promising because the spacing of critical points is of the order of the eddy size so that data at these locations identify different eddies and their interactions.

Acknowledgement. This work is supported by AFOSR Grant 90-0169 and ONR Grant N000-14-90-J-1415.

References

Adrian, R. J., "Stochastic Estimation of Sub-Grid Scale Motions," *Appl. Mech. Rev.*, **43**, S214 (1990).

Adrian, R. J., "Linking Correlations and Structure: Stochastic Estimation and Conditional Averaging," in NEAR WALL TURBULENCE (Hemisphere Publ., New York, 1990) pp. 420–436.

Adrian, R. J., "Conditional Eddies in Isotropic Turbulence," *Phys. Fluids*, **22**, 2065 (1979).

Adrian, R. J., "On the Role of Conditional Averages in Turbulence Theory," TURBULENCE IN LIQUIDS (Science Press, Princeton, 1977).

Adrian, R. J., Jones, B. G., Chung, M. K., Hassan, Y., Nithanandan, C. K., and Tung, A. T. C., "Approximation of Turbulent Conditional Averages by Stochastic Estimation," *Phys. Fluids A*, **1**, 992 (1989).

Adrian, R. J., and Moin, P., "Stochastic Estimation of Organized Turbulent Structure: Homogeneous Shear Flow," *J. Fluid Mech.*, **190**, 531 (1988).

Adrian, R. J., Moin, P., and Moser, R. D., "Stochastic Estimation of Conditional Eddies in Channel Flow," Report CTR-S87, Center for Turbulence Research, NASA Ames/Stanford (1987).

Bagwell, T. G., Adrian, R. J., Moser, R. D., and Kim, J., "Improved Approximation of Wall Shear Stress Boundary Conditions for Large Eddy Simulations," in NEAR-WALL TURBULENT FLOWS (Elsevier Science, Amsterdam, 1993) pp. 265–275.

Brereton, G. J., "Stochastic Estimation as a Statistical Tool for Approximating Turbulent Conditional Averages," *Phys. Fluids A*, **4**, 2046–2054 (1992).

Cole, D. R., Glauser, M. N., and Guezennec, Y. G., "An Application of the Stochastic Estimation to the Jet Mixing Layers," *Phys. Fluids A*, **4**, 192 (1992).

Deutsch, R., ESTIMATION THEORY, Prentice-Hall, New York (1965) 269 pp.

Ditter, J. L., and Adrian, R. J., "Local Conditional Structure Defined by Kinematic Events in Isotropic Turbulence," manuscript under revision.

Guezennec, Y. G., "Stochastic Estimation of Coherent Structure in Turbulent Boundary Layers," *Phys. Fluids A*, **1**, 1054 (1989).

Moin, P., Adrian, R. J., and Kim, J., "Stochastic Estimation of Organized Structures in Turbulent Channel Flow," *Proc. 6th Turbulent Shear Flow Symp.*, Toulouse, pp. 16.9.1–16.9.8 (1987).

Papoulis, A., PROBABILITY, RANDOM VARIABLES AND STOCHASTIC THEORY, 2nd ed., (McGraw-Hill, New York 1984).

Pecseli, H. L., and Trulsen, J., "A Statistical Analysis of Numerically Simulated Plasma Turbulence, *Phys. Fluids B*, **1**, 1616–1636 (1989).

Perry, A. E., and Chong, M. S., "A Description of Eddying Motions and Flow Patterns Using Critical-Point Concepts," *Ann. Rev. Fluid Mech.*, **19**, 125–156 (1987).

Robinson, S. K., Kline, S. J., and Spalart, P. R., "Quasi-Coherent Structures in the Turbulent Boundary Layer: Part II. Verification and New Information from a Numerically Simulated Flat-Plate Layer, in NEAR WALL TURBULENCE (Hemisphere Publ., New York 1990) pp. 218–247.

Tung, T. C., and Adrian, R. J., "Higher-Order Estimates of Conditional Eddies in Isotropic Turbulence," *Phys. Fluids*, **23**, 1469 (1980).

Willmarth, W. W., and Lu, S. S., "Structure of the Reynolds Stress Near the Wall," *J. Fluid Mech.*, **55**, 65 (1972).

Stochastic Estimation of Multipoint Conditional Averages and their Spatio-Temporal Evolution[1]

T. J. Gieseke [2] and Y. G. Guezennec
The Ohio State University
Mechanical Engineering Department

Identifying coherent structures in turbulent flows has been successfully accomplished in the past using several techniques. Among them are flow visualization, proper orthogonal decomposition (POD), conditional averaging, pattern recognition, and stochastic estimation. Each of these methods has advantages, but the stochastic estimation is exceptional in its flexibility, objective nature, and ease of validation. As an illustrative problem and application of the stochastic estimation, a turbulent wake at Re_d=5000 is examined at x/d=100. Single and multiple point conditionally averaged structures will be estimated and a quantitative three-dimensional flow-reconstruction technique will be used to study the flow dynamics.

Three-dimensional structures in the turbulent wake of a two-dimensional body wake were first identified by Townsend (1956) and Grant (1957). After noting the structure in the transverse velocity correlations measured by Townsend, Grant extended Townsend's measurements to include spanwise and normal velocity component correlations. Based on the structure of the space-time correlation tensor, Grant proposed that the far-wake contains pairs of counter-rotating eddies, rotating in the plane of the wake with axes inclined with the mean shear. Supporting evidence for the existence of double-rollers was found by Lumley (1965), Payne (1966), and Payne and Lumley (1967) by using the correlation data of Grant to compute the first POD mode. The structure computed in this way was identical to that postulated by Grant with the exception that Payne and Lumley's eddies had a v component of velocity while Grant's original eddy model did not. To explain the absence of the normal velocity component in Grant's large eddies, Townsend (1970;1978) proposed that along with the rollers, the wake contains jets that eject turbulent fluid into the mean flow. Using the stochastic estimation procedure, single and two point estimates of conditionally averaged structures can be computed to be compared these flow patterns.

The description of far wake turbulence as a distribution of rollers and jets has

[1]Supported by Office of Naval Research under Grant N00014-90-J-1499.
[2]Supported by The OAI Space Grant Consortium.

J. P. Bonnet and M. N. Glauser (eds.), Eddy Structure Identification in Free Turbulent Shear Flows, 281–292.

been generally accepted as an accurate depiction. Many supporting studies have been conducted to determine more about the distribution of eddies and their role in the turbulent dynamics (Keffer, 1965; Barsoum, Kawall, Keffer, 1978; Townsend, 1978; Wlezien, 1981; Mumford, 1983; Ferrè, Giralt, 1989; Ferrè, Mumford, Savill, Giralt, 1990). Despite the extensive research in this field, no conclusive evidence has been found for a preferred distribution of eddies. Most studies indicate that they appear in groups and several have suggested that eddies causing entrainment of irrotational fluid are less important than those causing ejection of turbulent fluid.

The stochastic estimation procedure can be used to help answer questions about the distribution of large eddies and to estimate their temporal evolution. A technique called *pseudo-dynamic reconstruction* will be developed and used to reconstruct the time evolution of turbulent velocity fields from experimental data. Structures can be identified in the estimated velocity fields by visualization of physical quantities which will aid in identifying preferred distributions of eddies in the wake. In addition, the dynamics of these eddies can be inferred from the fields because the fields represent the actual (measured) evolution of far-wake turbulence.

DESCRIPTION OF EXPERIMENT

The experimental portion of this study was conducted in the subsonic wind-tunnel at in the Department of Mechanical Engineering at The Ohio State University. The wind-tunnel is an open-return type, the inlet of which consists of a bellmouth followed by a plastic-tube honey-comb, three steel wire screens, and a settling chamber. A 9 to 1 contraction leads to the test section. The test section has the internal dimensions 0.46m wide by 0.61m high by 3.66m long. A 0.635cm diameter steel cylinder was mounted horizontally in the center of the test section and was fitted with end-plates near the tunnel walls with a 10° angle to the mean flow. With the end-plates, the model has an aspect ratio L/d $\simeq$55. The coordinate system in the test section has its origin at the center of the cylindrical model. The streamwise direction is x, the normal direction is y, and z is the spanwise direction completing the orthonormal set.

Two precision traversing mechanisms are mounted on tracks on the top of the test section. The traversing mechanisms are powered by computer controlled stepper motors. The first traversing mechanism has the capability of motion in all three coordinate directions. The second traversing system, mounted 21cm downstream of the first, is capable of independent vertical motion only. The two mechanisms are mounted to the same carriage and therefore move together in the streamwise direction. Additional rigid probe supports are mounted through holes in the test section walls.

A Masscomp 55-S-05 computer with a 12 bit, 16 channel A/D interface and two D/A channels are used to control the traversing mechanisms and to acquire data from various instruments. Fourth order Butterworth analogue filters with a

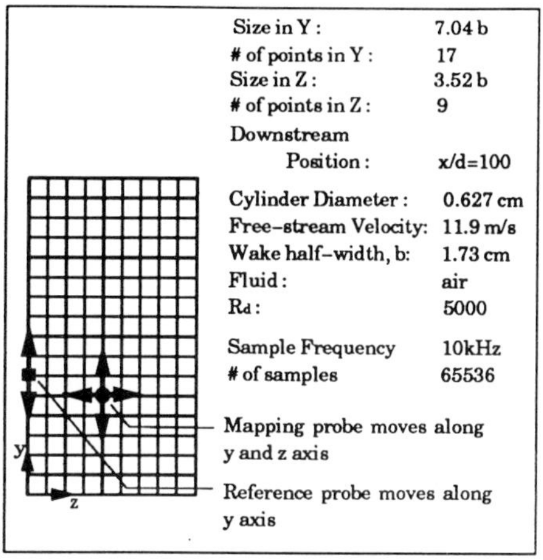

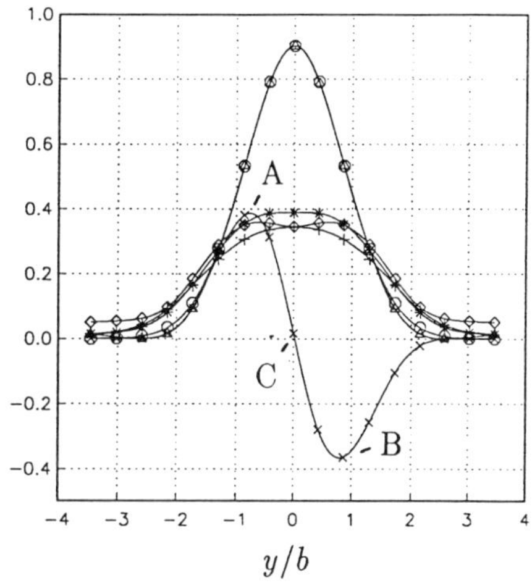

Figure 1: (left) Illustration of the measurement grid and probe movement with a summary of the experimental parameters.

Figure 2: (right) Statistics at x/d=100 and Re_d=5000:
○ Gaussian profile, $\triangle(1-U/U_\infty)(x/d)^{-1/2}$ (experimental), $\diamond(\sigma_u/U_\infty)(x/d)^{-1/2}$, $+(\sigma_u/U_\infty)(x/d)^{-1/2}$, $*(\sigma_w/U_\infty)(x/d)^{-1/2}$, $\times 100(\overline{uv}/U_\infty^2)(x/d)^{-1/2}$.

cut-off frequency tuned to 6.2kHz are used as anti-aliasing filters for the data acquisition system. Turbulence measurements were made using triple hot-wire probes connected to custom made constant temperature anemometers. Each anemometer has its own variable overheat setting, frequency compensation, and analogue linearizer. The two models of probes used were the TSI model 1295-20 and custom-made Auspex probes.

The first part of the experiment was designed to measure the space-time correlation tensor at x/d=100 and Re_d=5000. An Auspex triple-wire probe was mounted to each of the traversing mechanisms and they were traversed over a sampling grid, as shown in Figure 1. The grid was defined to span the width of the wake, $-3.52 \leq y/b \leq 3.52$, and an equivalent spanwise scale, $0.00 \leq z/b \leq 3.52$. The grid spacing was $0.44b$ which was sufficiently small to capture the spatial structure in the correlation function. The probes were positioned at each spatial separation on this grid. At each point, time histories of the anemometer signals were recorded. A total of 65536 samples were collected at 10kHz. These parameters were selected based on the convergence and the curvature of the autocorrelation function. Periodically during the course of the experiment, the hot-wires were calibrated against the freestream velocity to provide a function relating the anemometer output to cooling velocity. Using this calibration data, the anemometer outputs were converted to cooling velocities which were corrected for drifts in the ambient temperature and mean velocity. To determine the three components of velocity based on cooling velocity measurements, the procedure given by Gieseke (1990) and Gieseke and Guezennec (1992) was used.

For each probe separation, the temporal correlation between the velocity components was computed using an FFT approach. The original data was broken into 64 blocks of data which translates to calculating the correlation function at 1024 temporal separations. Owing to the flow symmetries, a total of four redundancies exist in the spatial correlation measurements. Therefore the correlation functions represent the average of 256 separate 1024 point measurements. The wake mean statistics were also computed from this data and are shown in Figure 2.

The second part of the experiment was designed to measure time traces of the three velocity components at three physical locations simultaneously. One Auspex probe was positioned at y/b=0.88, the other at y/b=0.0, and the TSI probe at y/b=-0.88. All three probes were at the same z location. The velocity histories were recorded and stored for use in the pseudo-dynamic reconstruction technique.

TURBULENT EVENTS

In order to identify coherent turbulent structures, some quantity must be defined that is indicative of their presence. From an experimental standpoint, this must be a measurable quantity, the realization of which during the course of an experiment should indicate the occurrence of the coherent structure of interest. Such a realization is termed an *event*.

The quantity or quantities that constitute an event can be determined based on physical arguments and consideration of the statistical properties of the turbulence. The structures that contribute most to the Reynolds stress are those that contribute to turbulence transport. To associate events with these structures, specific detection points were chosen corresponding to peaks in the Reynolds stress profile. These points are labeled A, B, and C in Figure 2.

Following the *Quadrant* technique used by Wallace, Eckelmann, and Brodkey (1972) and Willmarth and Lu (1973) in the turbulent boundary layer, the joint probability density functions (PDFs) of the velocity fluctuations were used to identify events at the points of interest. Figure 3 shows the joint PDF of u and v weighted by uv at point B. Two types of events can be clearly identified: a fourth quadrant event (which entrains irrotational fluid), and a second quadrant event (which ejects turbulent fluid away from the wake center). Similar events can be identified on the opposite side of the wake, where the entraining motions are found in the first quadrant and the ejecting motions are found in the third quadrant.

These events are based on very localized information and do not provide any information about the relative locations (in space or time) of various types of events. Events can also be defined by quantities simultaneously measured at two locations. To determine the importance of events that contribute to the Reynolds stress on both sides of the wake, a joint PFD of the velocity directions can be defined. Figure 4 shows the joint PDF of the flow angle defined by $\theta = tan^{-1}(v/u)$

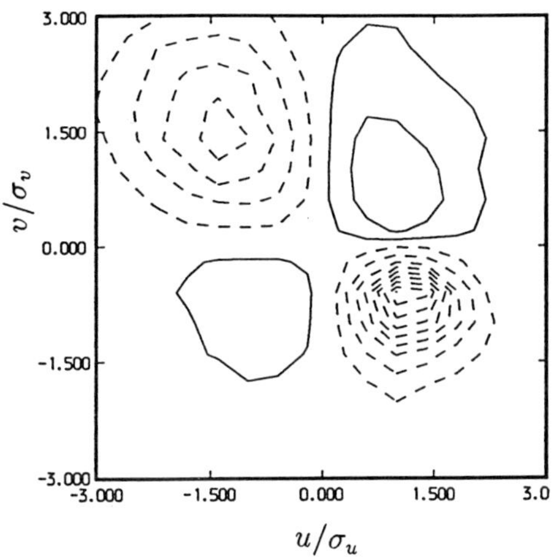

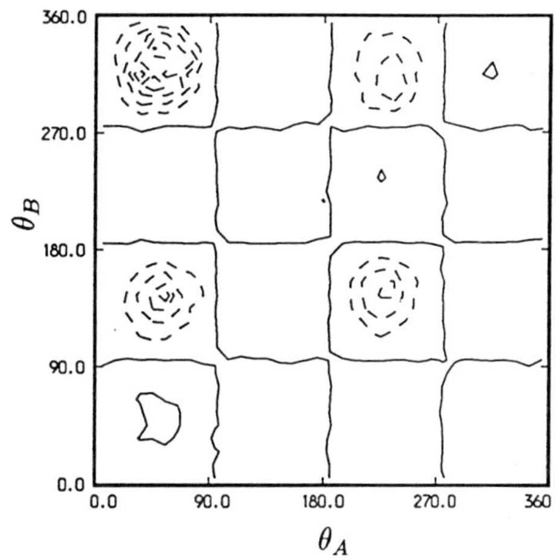

Figure 3: (left) Joint probability density function of u and v weighted by the Reynolds stress $P(u, v)uv$.

Figure 4: (right) Joint probability density function of the flow angles at points A and B weighted by the Reynolds stress at both points $P(\theta_A, \theta_B)uv_A uv_B$.

at points A and B weighted by the contributions to the Reynolds stress at each point. Peaks in the function $P(\theta_A, \theta_B)uv_A uv_B$ indicate what combinations of events are most likely to contribute to the Reynolds stress simultaneously on both sides of the wake. The cell associated with simultaneous entraining motions illustrates that these are important types of motions.

STOCHASTIC ESTIMATION OF COHERENT STRUCTURES

Conditional and phase averaging are the traditional approach to determining the flow structures associated with turbulent events. The stochastic estimation was originally proposed by Adrian (1976) as a way to estimate conditionally averaged structures without having to pre-select an averaging condition or event. This work was followed by many related studies testing and applying the theory (Adrian, 1979; Adrian et al. 1987; Guezennec, 1989; and others). The procedure can be summarized by the statement, *given velocities at some coarsely distributed points in space and/or time, the velocities at all other points in space and/or time can be estimated as functions of the known velocities.*

Single point conditionally averaged flow fields are computed with the stochastic estimation procedure by fixing the velocity at a single location and computing the velocities in a neighborhood around that point. The resulting field can be considered the most likely state of the field, in a least square error sense, given the prescribed condition. The corresponding conditionally averaged field is the average state of the field given that the velocity at the condition location is within a small neighborhood of a prescribed value. Despite the difference in definition, fields computed using the two methods have been found to be qualitatively similar (Adrian, 1987). The stochastic estimation approach has two major advantages over the conditional average approach; it can be extended from the single condition estimate to a multi-condition estimate very easily and the conditions used for the field reconstruction need not be fixed before the required experiment is conducted.

The procedure is build from the premise that given a set of N_s realizations of $(N + P)$ stochastic variables, some of the variables $(\mathbf{y}^1, \cdots, \mathbf{y}^p)$ can be estimated by functions of the others $(\mathbf{x}_1, \cdots, \mathbf{x}_n)$ having the form,

$$\mathbf{y}^l \approx F^l(\mathbf{x}_1, \cdots, \mathbf{x}_n) = a^l_j f^l_j(\mathbf{x}_1, \cdots, \mathbf{x}_n) \tag{1}$$

The basis functions f^l_j and the number of them can be chosen arbitrarily. The constants a^l_j must be determined such that the function F^l predicts $\mathbf{y}^l$ as best as is possible based on the available statistical sample.

The error for a given variable $\mathbf{y}^l$ is defined as the sum of the square of the difference between F^l and $\mathbf{y}^l$,

$$\epsilon^{l\,2}_T =< \left(y^l - a^l_j f^l_j\right)^2 > \tag{2}$$

where <> denotes averaging over all realizations. This error can be minimized with respect to each of the coefficients by taking the partial derivatives of $\epsilon^{l\,2}_T$ with respect to each a^l_j and setting the result equal to zero,

$$\frac{\partial \epsilon^{l\,2}_T}{\partial a^l_j} =< \left(-2y^l f^l_j + 2a^l_k f^l_k f^l_j\right) >= 0 \tag{3}$$

Sets of simultaneous equations are generated that can be solved for the coefficients,

$$a^l_k f^l_k f^l_j = y^l f^l_j \tag{4}$$

The functions F^l can be expanded as polynomials,

$$F^l(\mathbf{x}_1, \cdots, \mathbf{x}_n) \approx a^l_0 + a^l_1 \mathbf{x}_1 + \cdots a^l_n \mathbf{x}_n \tag{5}$$

Substituting these definitions for f_j into Equation 4 and defining correlation matrices $L_{jk} =< x_k x_j >$ and $D_{jl} =< y^l x_j >$, the simultaneous equations can be expressed as a matrix problem

$$\mathbf{AL} = \mathbf{D} \tag{6}$$

This problem can be solved for all of the coefficients $\mathbf{A}$ so that the variables $\mathbf{y}$ can be predicted based on values of the variables $\mathbf{x}$ using a matrix multiplication,

$$\mathbf{y} \approx \mathbf{Ax} \tag{7}$$

The procedure can be used to estimate the coherent structures associated with the turbulent events defined in the preceding section by considering space and time dependent perturbation velocities $\mathbf{u}(\vec{r},t)$ as the variables that are to be estimated. The simplest application of the stochastic estimation is the estimated single-point conditional average. For this calculation, the three components of velocity are prescribed at a single point in space. This velocity vector is called the *condition* and noted $\mathbf{u}_c$. Near this condition, the estimated flow field $\mathbf{u}_e(\vec{r},\tau)$ is the most likely state of the field given the condition. Here τ represents the time delay between the occurrence of the condition and the time when the velocities are to be estimated. The field can be considered the coherent structure associated with the condition and is reconstructed using the formulas

$$\begin{aligned}
u_e(\vec{r},\tau) &= a_{11}(\vec{r},\tau)\,u_c + a_{12}(\vec{r},\tau)\,v_c + a_{13}(\vec{r},\tau)\,w_c \\
v_e(\vec{r},\tau) &= a_{21}(\vec{r},\tau)\,u_c + a_{22}(\vec{r},\tau)\,v_c + a_{23}(\vec{r},\tau)\,w_c \\
w_e(\vec{r},\tau) &= a_{31}(\vec{r},\tau)\,u_c + a_{32}(\vec{r},\tau)\,v_c + a_{33}(\vec{r},\tau)\,w_c
\end{aligned} \tag{8}$$

or more concisely by

$$\mathbf{u}_e(\vec{r},\tau) = \mathbf{A}(\vec{r},\tau)\,\mathbf{u}_c \tag{9}$$

Here, the estimated variable vector is $\mathbf{y}(\vec{r},\tau) = [u_e, v_e, w_e]^T(\vec{r},\tau)$ and the conditions are $\mathbf{x} = [u_c, v_c, w_c]^T$. In these expressions, the regression coefficients have been presented as functions of space and temporal separation. This extension of the procedure simply requires that the regression analysis be conducted for each point in space and time offset where the reconstruction formulas are to be applied. To construct the matrix of coefficients $\mathbf{A}(\vec{r},\tau)$, the two-point space-time correlation tensor is required between the velocities at the condition location and each location at which the reconstructed velocities are needed. These correlations are available for the turbulent wake as measured in the experimental portion of this study.

An example estimated perturbation velocity field associated with a fourth quadrant event at point B is shown in Figure 5. The velocities shown are the u and v components in the y-t plane in which the event condition was fixed. The flow is from left to right. The information obtained from this single-point estimate is very localized, suggesting that multiple point estimates would be more informative.

Extending the reconstruction technique to estimate the most likely velocity field given the velocities at two or more points in space requires the redefinition of $\mathbf{u}_c$ and computation of the required coefficients. For a reconstruction based on conditions at points A and B, $\mathbf{u}_c$ is defined as

$$\mathbf{u}_c = [u_A, v_A, w_A, u_B, v_B, w_B]^T \tag{10}$$

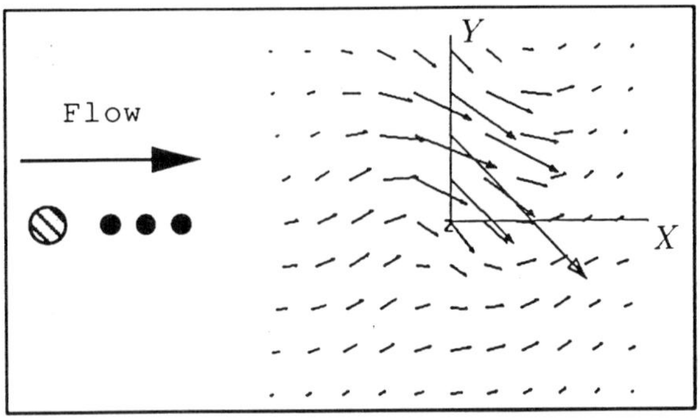

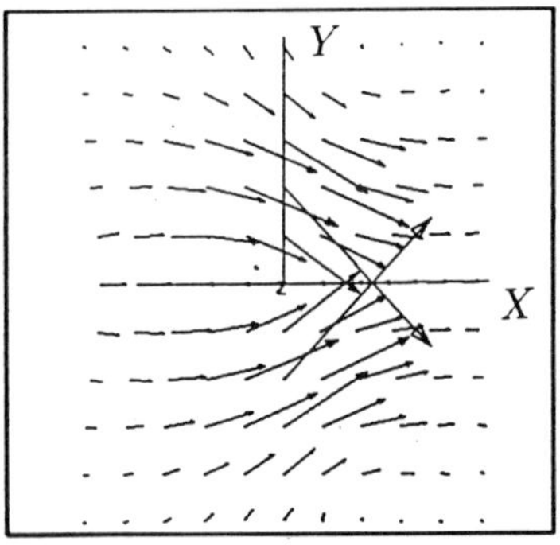

Figure 5: (left) A two-dimensional slice in the x-y plane of the estimated velocity field based on a Q4 event at point B.

Figure 6: (right) A two-dimensional slice in the x-y plane of the estimated velocity field based on a simultaneous Q4 event at point B and a Q1 event at point A.

The matrix of coefficients can be computed from the space-time correlation tensor using Equation 6.

The estimated field for the event associated with simultaneous entrainment on both sides of the wake is shown in Figures 6, 7, and 8. The two point estimation reconstructs the flow across the entire width of the wake much better than the single point estimation of Figure 5. The three-dimensional topology is illustrated in Figures 7 and 8 using an iso-surface of constant perturbation enstrophy, represented as a solid object, superposed on a selected planar cut through the field showing the velocity vectors. The structure suggested by this analysis is a pair of counter-rotating eddies, inclined approximately 45° in the direction of the mean shear.

PSEUDO-DYNAMIC RECONSTRUCTION

In an effort to estimate the dynamics of the turbulent motions and the interactions between neighboring events, the stochastic estimation procedure was extended to the pseudo-dynamic reconstruction technique. In this technique, time histories of conditions are measured using a coarsely distributed array of hot-wire probes. Within that history, short time series of simultaneously measured conditions are considered at each of the probe locations. The condition vector of Equation 9 is defined as

$$\mathbf{u}_c = [u_1, v_1, w_1, u_2, \cdots, u_3, \cdots]^T \tag{11}$$

Each condition velocity in this vector is given at the location of one of the hot-wire probes and at some time within the short time segments. The velocity field in the y-z plane is estimated based on these measured conditions, the regression equations of Equation 6, and the space-time correlation tensor (illustrated in Figure 9). The condition histories are then advanced in the time record thus defining a new set of

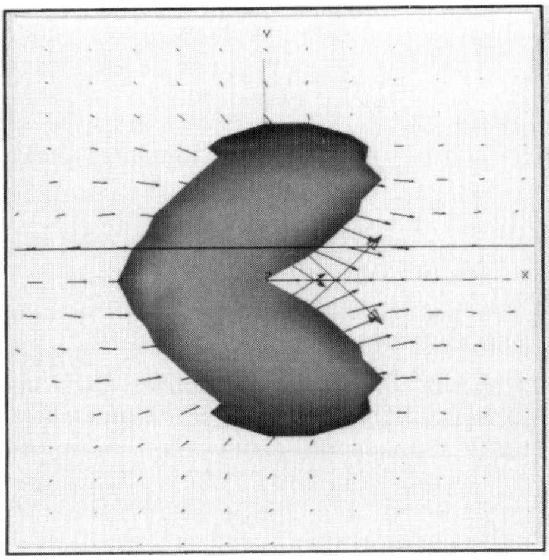
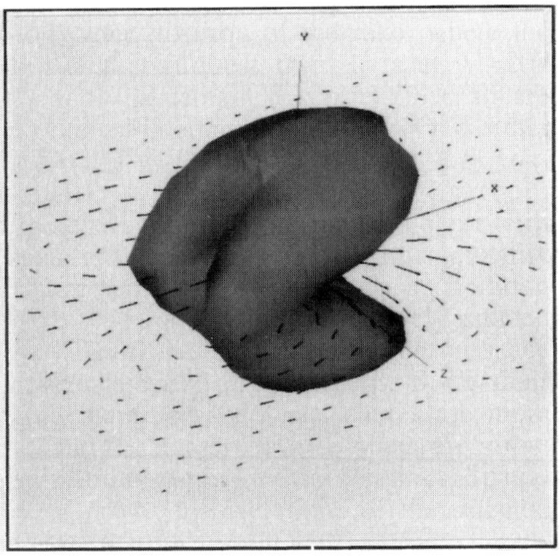

Figure 7: (left) A side view of a surface of constant vorticity magnitude with a x-y plane of velocity vectors for the Q4-Q1 event

Figure 8: (right) As above but viewed at an oblique view from above with an x-t plane of vectors.

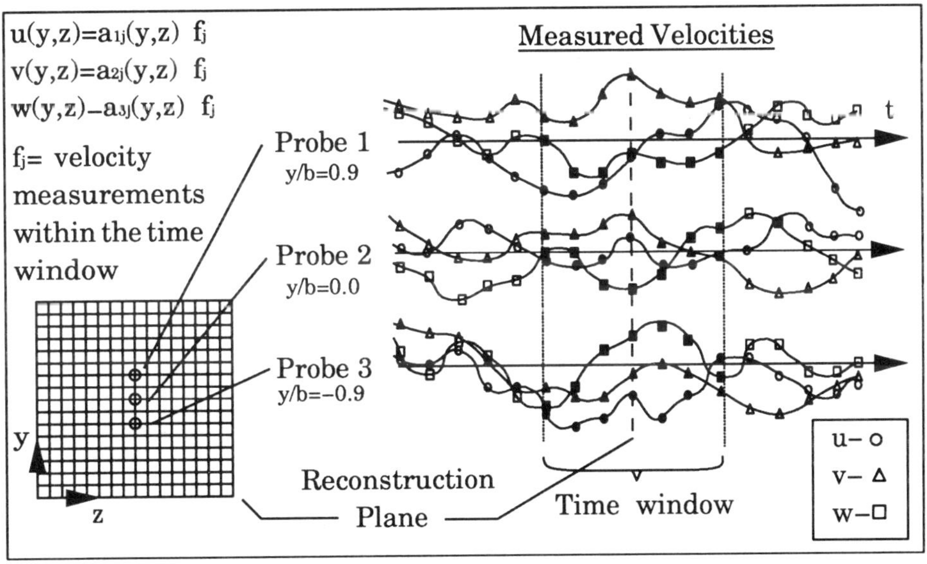

Figure 9: Illustration of the pseudo-dynamic reconstruction technique to reconstruct a single y-z plane of velocity vectors. The process is used to construct a three-dimensional velocity field by repeatedly advancing the time record one step and estimating additional planes of velocities.

conditions to estimate another plane of velocities. This process is repeated until a complete time history of reconstructed two-dimensional velocity fields is obtained. By panning a short time window over the data set, the dynamics of the turbulence can be approximated.

Figure 10 contains views of one instant of a reconstruction made using simultaneously measured velocities at points A, B, and C. This illustrates the complex distribution of eddy structures. The streamwise direction is to the right so the sequence of events is from right to left. At the far right of the figures, a set of double roller eddies are observed which are nearly identical to those shown in Figure 7. They are strongly symmetric about the wake centerline, appear in relative isolation, and have a fairly large streamwise scale. Immediately to the left is an interaction between a symmetric entraining event and a strong ejection on the bottom of the wake. A similar combination of events is found slightly upstream on the opposite side of the wake. In contrast to the entraining motions, these front-like structures do not appear in symmetric configurations and are not clearly associated with rollers. Still further upstream, a combination of strong jet motions can be seen. No clear association with rollers is observed and the motions are anti-symmetric. Just off the left edge of the figure are wavy motions with strong span-wise components of velocity.

The sequence of events shown in these figures are specific to this selected time segment. There do not appear to be any clearly observable patterns of streamwise distributions of structure in the remainder of the data. However, the results of Figure 4 can be better understood by considering the cross-stream distribution of structures in this sample data set. Entrainments tend to be symmetric across the wake center. In this sample, ejections appear isolated on one side of the wake or on both sides with a phase delay. It should be noted that in other sample data sets strong symmetric ejection motions were also found. In all of the data studied, the entrainment motions are associated with rollers while ejection motions generally are not. Jets tend to occur with very strong spanwise components of velocity and in more chaotic turbulent regions.

CONCLUSION

Coherent structures can be identified using the stochastic estimation technique as has been illustrated here and in many other studies. With those structures established, a new technique, the pseudo-dynamic reconstruction, was developed to estimate the evolution of three-dimensional velocity fields from experimental data. This new technique has several very useful properties. The most important is that three-dimensional velocity fields can be estimated from several point measurements. This data can then be analyzed using many techniques. The most direct is to visualize the data to look for flow structures and repeated flow patterns. The very strongly three-dimensional nature of turbulent motions is preserved by this type of reconstruction and visualization. In addition, this type of visualiza-

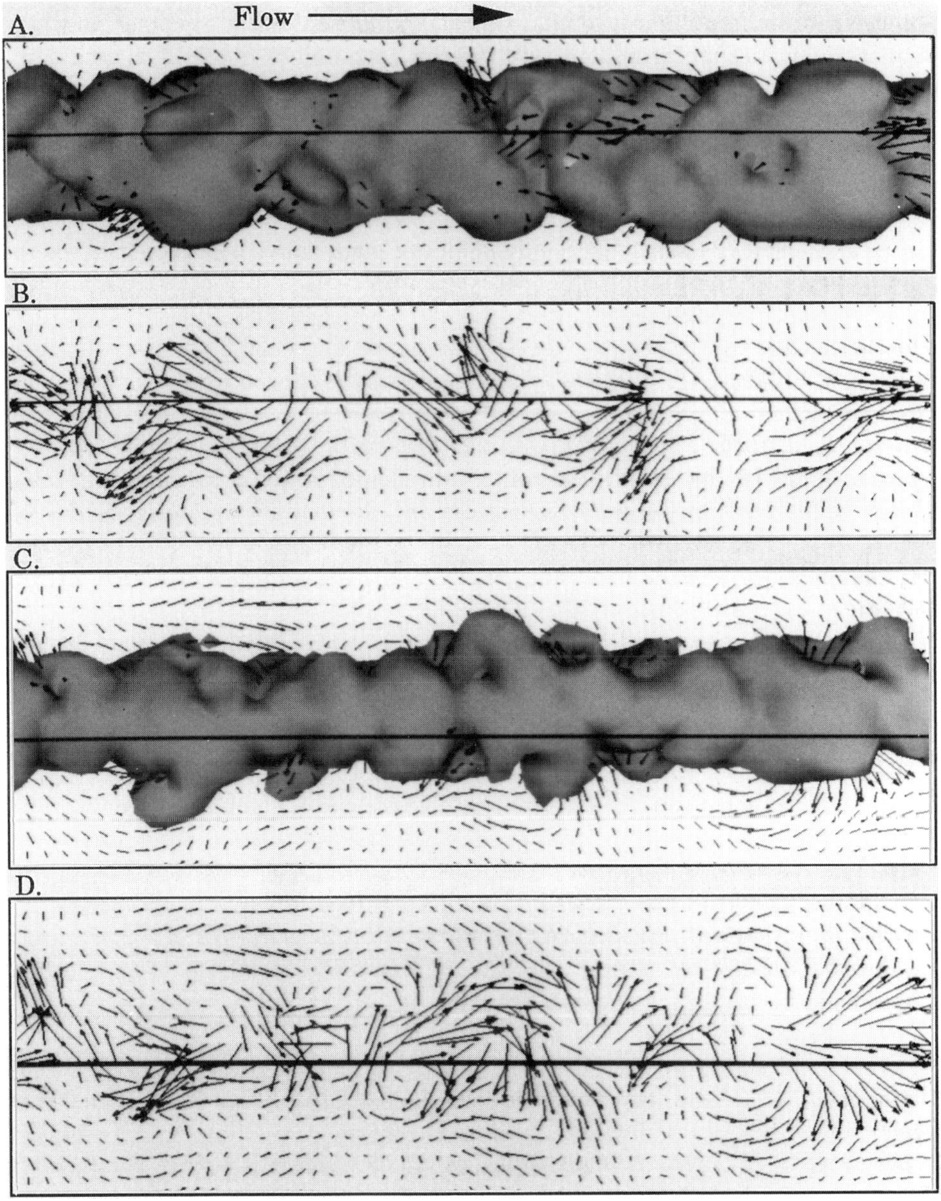

Figure 10: A short time history of a reconstructed field: A. side view of an iso-surface of constant vorticity magnitude with velocity vectors. B. side view of the velocity vectors only. C. top view of the same reconstructed field, iso-surface and velocities. D. top view with only the velocities shown.

tion overcomes the ambiguity found in normal flow visualization results as to what physical quantities are correlated with the concentration of smoke, dye, etc. Here the visualized quantities are relevant physical quantities so the observed structures can be interpreted literally and identifying structures is much more direct.

Using the pseudo-dynamic reconstruction to observe flow dynamics can be a very useful application of the technique. No clearly observable patterns were visualized in these data sets for the wake flow; however, it is clear that this technique can be successful in identifying the dynamics of many other flow problems where the turbulent activity has a more repetitive nature.

REFERENCES

[1]Adrian, R. J. 1976. *Proceedings of the 4th Symposium on Turbulence, Univ. of Missouri-Rolla.*

[2]Adrian, R. J. 1979. *Physics of Fluids*, **22**, 2065-2070.

[3]Adrian, R. J., Chung, M. K., Hassan, Yassin, Jones, B. G., Nithianandan, C. K. and Tung, A. T. C. 1987. *Sixth Symposium on Turbulent Shear Flows, Toulouse, France, Sept. 7-9, 1987.*

[4]Barsoum, M. L., Kawall, J. G., and Keffer, J. F. 1978. *Physics of Fluids*, **21**(2), 157-161.

[5]Ferrè, J. A., Mumford, J. C., Savill, A. M., and Giralt, F. (1990). *J. Fluid Mech.*, **210**, 371-414.

[6]Ferrè, J. A. and Giralt, F. (1989). *J. Fluid Mech.*, **198**, 65-78.

[7]Ferrè, J. A. and Giralt, F. (1989). *J. Fluid Mech.*, **198**, 27-64.

[8]Gieseke, T. J. 1990. M.S. Thesis, The Ohio State Univ.

[9]Gieseke, T. J. and Guezennec, Y. G. 1992. Accepted for publication in *Experiments in Fluids.*

[10]Grant, H. L. 1957. *J. Fluid Mech.*, **3**, 149-190.

[11]Guezennec, Y. G. 1989. *Physics of Fluids*, **1**(6), 1054-1060.

[14]Keffer, J. F. 1965. *J. Fluid Mech.*, **22** *part 1*, 135-159.

[15]Lumley, J. 1965. *Proc. Intern. Colloquium, Moscow.*

[16]Mumford, J. C. 1983. *J. Fluid Mech.*, **137**, 447-456.

[17]Payne, F. R. 1966. *PhD. Thesis, Penn State Univ.*

[18]Payne, F. R. and Lumley, J. L. 1967. *Physics of Fluids Supplement*, S194-S196.

[19]Townsend, A. A. 1956, The Structure of Turbulent Shear Flow. Cambridge Univ. Press.

[20]Townsend, A. A. 1970. *J. Fluid Mech.*, **41**, 13-33.

[21]Townsend, A. A. 1978. *J. Fluid Mech.*, **95**, 515-537.

[22]Wallace, J. M., Eckelmann, H., and Brodkey, R. S. 1972. *J. Fluid Mech.*, **54**, 39-48.

[23]Willmarth, W. W. and Lu, S. S. 1972. *J. Fluid Mech.*, **50**, 65-92.

[24]Wlezien, R. W. 1981. PhD. Thesis, I.I.T.

VIII. Combined or Comparative Approaches

THE PROPER ORTHOGONAL DECOMPOSITION, WAVELETS AND MODAL APPROACHES TO THE DYNAMICS OF COHERENT STRUCTURES*

Gal Berkooz, Juan Elezgaray, Philip Holmes, John Lumley, Andrew Poje

Cornell University
Ithaca, NY 14853, USA

ABSTRACT

We expand the velocity field in the vicinity of the wall in a turbulent boundary layer in empirical eigenfunctions obtained from experiment (Herzog, 1986). This is the Proper Orthogonal Decomposition, or Karhunen-Loève expansion. This approach can be applied equally well to other flows, but requires the eigenfunctions, which in most cases have not been measured or calculated. Some progress has been made in predicting the eigenfunctions for other flows, using non-linear stability analyses of two types (Poje & Lumley, 1991; Leibovich, 1991). In the boundary layer, the eigenfunctions describe streamwise rolls. We are ultimately interested in the dynamics of these rolls.

Coherent structures have also been educed in turbulent flows by a method called linear stochastic estimation, notably by Adrian (1979) and co-workers. We show below that there is a close relation between the eigenfunctions educed in this way and those obtained from the proper orthogonal decomposition.

So far our decompositions have used Fourier modes in the homogeneous spanwise and streamwise directions. These distributed functions, while energy-optimal on average, are not well adapted to reproduce spatially localized events such as bursts. In particular, one of the points on which the work of Aubry *et al* was criticised, was that the structures produced were not localized in space like the structures observed in the turbulent boundary layer. Wavelet bases, having a localized support, seem much more appropriate. We are currently studying (in collaboration with J. Elezgaray and I. Moroz) applications of wavelets to the study of coherent structures arising from the Kuramoto-Sivashinsky equation (Berkooz *et al*, 1992). This study is preliminary to a more involved study of the more complex turbulent boundary layer, permitting us to gain experience with the application of wavelets to the study of modal dynamics in chaotic systems. We have successfully modeled the wavelet-wavelet interactions, and have produced a low-

* Prepared for presentation at International Union of Theoretical and Applied Mechanics Symposium "Eddy structure identification in free turbulent shear flows," Poitiers, France, 12-14 October 1992. Supported in part by: the U. S. Air Force Office of Scientific Research, The U. S. Office of Naval Research (Mechanics Branch and Physical Oceanography Program), and the U. S. National Science Foundation (program in Physical Oceanography).

J. P. Bonnet and M. N. Glauser (eds.), Eddy Structure Identification in Free Turbulent Shear Flows, 295–309.

dimensional wavelet model that reproduces some of the behavior of the Kuramoto-Sivashinsky equation.

1. Linear stochastic estimation.

In this section we comment on the connection between the POD and linear stochastic estimation, as applied by Adrian and coworkers in Adrian (1979), Adrian and Moin (1988), Adrian, Moin and Moser (1987) and Moin, Adrian and Kim (1987). Suppose one wanted to find the conditional probability density function (CPDF) of $(u(x)|u(x'))$. For the sake of simplicity of the exposition, we limit ourselves to single point scalar events. Suppose, moreover, that one seeks an estimate linear in $u(x')$; this is called a linear stochastic estimate. We are seeking $A(x, x')$ such that $A(x, x')u(x')$ will be the estimate for $u(x)$, and we want to find $A(x, x')$ such that:

$$\min_{V(x, x')} \langle | u(x) - V(x, x')u(x') |^2 \rangle = \langle | u(x) - A(x, x') u(x') |^2 \rangle \tag{1}$$

is achieved by $A(x, x')$. We use the classical calculus of variations (cf. Adrian, 1979), which leads directly to:

$$A(x,x') = \langle u(x)u^*(x')\rangle \bullet \langle u(x')u^*(x')\rangle^{-1} \tag{2}$$

In (2) the two point correlation tensor $R(x,x') = \langle u(x)u(x')\rangle$ appears with some normalization. Using classical results for the POD, we can write

$$A(x,x') = \frac{\Sigma_{i-1}^{\infty} \lambda_i \phi_i(x)\phi_i^*(x')}{\Sigma_{i-1}^{\infty} \lambda_i |\phi_i(x')|^2} = \sum_{i=1}^{\infty} \phi_i(x) f_i(x') , \tag{3}$$

where $f_i(x') = \lambda_i \phi_i^*(x')/\Sigma_{j=1}^{\infty} \lambda_j |\phi_j(x')|^2$. We may interpret $f_i(x')$ as the relative contribution of ϕ_i to $u(x')$ on the average.

It is remarkable that we get exactly the same result from a simplified PDF model based on the POD. Here we assume that the coefficients a_i in a modal expansion $u(x, t) = \Sigma a_i(t)\phi_i(x)$ are jointly normal and independent, with zero mean and variance λ_i. Let us compute the estimator $\langle u(x)|u(x')\rangle$. Since we have an expression for the PDF we can compute this explicitly. Recall from probability theory that if the x_i are independent and normal with zero mean and variance σ_i^2 for $i = 1,..., m$ then

$$\langle x_i \mid \sum_{j=1}^{m} x_j = C \rangle = \frac{\sigma_i^2 C}{\Sigma_{j=1}^{m} \sigma_j^2} \quad . \tag{4}$$

(See the formula for the conditional expectation of joint normal variables in Feller, 1957). Using (4), we have

$$\langle a_i\phi_i(x') \mid \sum_{j=1}^{\infty} a_j\phi_j(x') = u(x')\rangle = \frac{\lambda_i |\phi_i(x')|^2 u(x')}{\Sigma_{j=1}^{\infty} \lambda_j |\phi_j(x')|^2} \tag{5}$$

which gives

$$<u(x) \mid u(x')> = \frac{\Sigma_{i=1}^{\infty}\lambda_i|\phi_i(x')|^2\, u(x')\phi_i(x)/\phi_i(x')}{\Sigma_{j=1}^{\infty}\lambda_j \mid \phi_i(x') \mid^2}$$

$$= \frac{\Sigma_{i=1}^{\infty} \lambda_i\, \phi_i^*(x')\phi_i(x)u(x')}{\Sigma_{j=1}^{\infty}\lambda_j\, |\phi_j(x')|^2} \tag{6}$$

This is exactly the same result obtained from linear stochastic estimation: (3).

2. Wavelets

All the flows of interest to us have one or more homogeneous directions. We are accustomed to use in these directions the Fourier transform, which is the homogeneous equivalent of the POD. However, the Fourier transform is not nearly so appropriate in the homogeneous case as is the POD in the inhomogeneous case. This is because the Fourier modes are not confined to a neighborhood, but extend to infinity without attenuation. All disturbances in a fluid, and coherent structures in particular, are localized. There is therefore considerable motivation to find another representation that is more appropriate.

In Tennekes & Lumley (1972) it was suggested that a more appropriate quantity would be the energy surrounding a wavenumber κ, say from κ/a to $a\kappa$, where $a = 1.62$. In physical space, this packet with appropriate phase relations is confined to a region, essentially dropping to zero in about $2\pi/\kappa$ from the origin. Tennekes & Lumley called these "eddies", but they are an example of what are now called wavelets.

While wavelets appear to make more physical sense, we might worry because we would be discarding the optimality of the Fourier representation; would convergence be much slower, so that we would need many more terms, or would we lose considerable energy if we used the same number of terms? A main result of a recent paper (Berkooz et al, 1992) is that very little energy is lost when using a wavelet basis instead of a Fourier basis. Although wavelets are physically appealing, it would also be nice to have reassurance from calculations that physical behavior would be preserved in a wavelet representation. To set our minds at rest on this point, Berkooz et al (1992) also display a *relatively* low-dimensional wavelet model of the Kuramoto-Sivashinsky equation that shows dynamical behavior similar to the full equation.

Without getting involved with mathematical details, an orthonormal wavelet basis is constructed by starting with a function, say $\psi(x)$, similar to the eddy suggested by Tennekes & Lumley (1972). From this, construct a set $\psi_{j,0}(x) = \psi(x2^j)$, $j = 1, 2, \ldots$. Each of these is shrunk affinely, but is geometrically similar to the original function. Now consider the translates of $\psi_{j,0}$: $\psi_{j,k}(x) = \psi_{j,0}(x-k2^j)$, $k = 1, 2 \ldots$.

Berkooz *et al* (1992) consider a periodic, homogeneous stochastic process. It is then obvious that the POD decomposition becomes identical to the Fourier decomposition. Now if, for a given ε, we need $N(\varepsilon)$ POD modes in order to satisfy

$$\left|\sum_{i\geq 1}<|a_i(t)|^2> - \sum_{i=1}^{N(\varepsilon)}<|a_i(t)|^2>\right| \leq \varepsilon \tag{7}$$

then, if the $\{b_i\}$ are the coefficients in a wavelet basis, we get for some constant C, depending only on the process (and not on ε):

$$\left|\sum_{i\geq 1}<|a_i(t)|^2> - \sum_{i=1}^{\mathcal{N}(\varepsilon)}<|b_i(t)|^2>\right| \leq C\varepsilon \tag{8}$$

for some $\mathcal{N}(\varepsilon) > N(\varepsilon)$ slightly bigger than $N(\varepsilon)$ (the precise statement is given in Berkooz *et al* 1992).

Numerical results

Berkooz *et al* (1992) wished to apply these ideas to a simple situation. The three-dimensional, three component Navier Stokes equations are too complicated for a first effort.

The one-dimensional, scalar Kuramoto-Sivashinsky (K-S) equation appears in a variety of contexts, such as quasi-planar fronts, chemical turbulence, etc . It shares some properties with Burgers' equation and the Navier Stokes equations, but is much easier to deal with. Up to some rescaling, this equation can be written as:

$$u_t + u_{xx} + u_{xxxx} + u_x u = 0, \qquad 0 \leq x \leq L \tag{9}$$

u periodic on [0, L], where L, the length of the spatial domain, is the only free parameter in the problem.

Although the dynamical behavior for small values of L is fairly well understood (see Hyman & Nicolaenko, 1986 for an overview), many open questions remain concerning the limit $L \to \infty$ (Zaleski, 1989; Pomeau *et al*, 1984). As can be seen from the numerical simulations, for $L > 30$, a chaotic regime involving both space and time disorder occurs (see Figure 1, where we plot a space-time representation of a typical solution, $L=400$, $0 \leq t \leq 100$).

In order to check the estimate (8), we compare the energy resolved by a given number of modes using either a Fourier (POD) or wavelet basis. Note that, to compare the Fourier and wavelet bases, all the translates in the wavelet basis must be considered. Here are some results (see Fig.2):

No. of modes	wavelets (m = 6)	wavelets (m = 8)	Fourier
64 (j = 6)	70.84%	71.5%	72.2%
96 (j = 6,5)	79.1%	79.43%	83.3%
127(0≤j≤6)	84.1%	84.9%	89.7%
255(0≤j≤7)	99.9%	99.9%	99.9%

The scale j=6 which captures most of the energy on the average, corresponds to a characteristic length $2^{-6}L \approx q_m^{-1}$, which is also the length scale associated with the most unstable wavelength q_m. In agreement with the general shape of the energy spectrum, the scales $0 \leq j \leq 5$ are shown to capture more energy than the scales in the dissipative range ($j \geq 7$). The above figures show that (for sufficiently smooth splines) the wavelet projection captures almost the same amount of energy as the Fourier (= POD) decomposition (within 5%).

These results prompted us to conclude that from an average energy point of view wavelets are a reasonable candidate for a modal decomposition of the K-S equation. The localized nature of the wavelets may give us a unique view of the spatial attributes of the coherent structures. We outline our approach. We conjecture the existence of a dynamically relevant length scale L_C such that interactions between physical regions of distance greater than L_C are dynamically insignificant (a dynamical St. Venant principle). Determining the validity of this conjecture is part of our study. We use this conjecture to remove terms in a wavelet-Galerkin projection that correspond to interactions between regions of distance greater than L_C. To study whether the dynamics of coherent structures are indeed locally determined we construct truncations corresponding to a small box size L_B in the larger box of size L.

We need to address the role of unresolved physical space (i.e.- modes located outside the box of size L_B). It is obvious that the Dirichlet type of boundary condition imposed will create a boundary layer which will affect the dynamics, especially in small boxes which are of interest to us. There are two plausible approaches to remove this effect. One approach uses a stochastic boundary condition (which is hard to implement numerically and treat analytrically). The other approach appeals to the conjecture on the existence of L_C. One takes a box of size L_B greater that $2L_C$ so that one can periodize the small model using resolved relatively distant modes instead of unresolved ones. We opted for the second approach.

We present some preliminary results of the integration of one such model. We resolved a box of size $L_B = 50$ (this is 1/8 of the original box). Figure 3 shows the spatio-temporal evolution of the full system, with a Fourier basis. Figure 4 shows the spatio-temporal evolution of a (rescaled) model with $L_C = 50\times3/8$, which is in excellent qualitative agreement with the dynamics of the full system. If L_C is too small, after a long initial transient, the system eventually settles down to a periodic oscillatory state, in which no interaction between the localized structures is observed. This can be avoided by an increase in L_C, which adds significant non-linear interactions between relatively distant physical space locations. It might also be avoided by a stochastic boundary condition.

3. Eduction of coherent structures.

Ideally, one would like to apply the POD approach to a wide range of flows where coherent structures are known to play an important role in the dynamics. The POD procedure, however, requires the two-point velocity autocorrelation tensor as input, thus necessitating complete documentation of the flow before the analysis can proceed. For flows with very high Reynolds numbers or complicated geometries this can be prohibitively expensive given current computational and experimental capablities. In this section we will describe an analytic procedure for extracting basis functions (structures) which approximate those given by the POD but which requires much less *a priori* statistical information about the flow.

The method presented is based on energy stability considerations put forth by Lumley (1971). First, the instantaneous flow field is decomposed into three components in order to isolate the large scale structures. Evolution equations can then be written for the coherent velocity field and the coherent kinetic energy. A procedure can then be formalized to search for the structures which maximize the instantaneous growth rate of coherent energy, the rationale being that the structures which on average have the largest growth rates will compare well with the structures which contribute the most to the average turbulent kinetic energy (POD eigenfunctions).

As an example we consider turbulent channel flow assumed statistically homogeneous in both the downstream (x_1) and crosstream (x_3) directions. In order to extract spatial structures from the total velocity field, we avoid traditional Reynolds averaging and instead decompose the instantaneous velocity field into three components: the spatial mean (U), the coherent field ($\mathbf{v}$) and the incoherent background turbulence ($\mathbf{u}'$).

$$u_i(\mathbf{x}, t) = U(x_2) + v_i(\mathbf{x}, t) + u_i'(\mathbf{x}, t) \tag{10}$$

The spatial mean is an average over the x_1 - x_3 plane; we will indicate it by [...]. We introduce a second averaging procedure, denoted by $<...>$, which eliminates the small-scale turbulence while leaving the coherent field intact.

$$<u_i(\mathbf{x}, t)> = U(x_2) + v_i(\mathbf{x}, t) \tag{11}$$

Practically this can be accomplished in several ways (Reynolds & Hussain, 1972; Liu, 1988; Brereton & Kodal ,1992; Berkooz, 1991b). We will refer to this average as a phase average. For our purposes here it is sufficient that the phase average and space average commute, and that the cross correlations are negligible.

$$<u'_i v_j> = [u'_i v_j] = 0 \tag{12}$$

Given these averaging procedures, we can manipulate the Navier-Stokes equations to arrive at evolution equations for the coherent velocity field.

$$Dv_i/Dt + v_j v_{i,j} = -\pi_{,i} + \nu v_{i,jj} + \tau_{ij,j} - S_{ij} v_j + [v_i v_j]_{,j} \tag{13}$$

where D/Dt denotes the mean convective derivative, S_{ij} the mean rate of strain, and ν the kinematic viscosity. τ_{ij} represents the rectified effects of the small scale fluctuations on the coherent field and is defined by

$$\tau_{ij} = [u'_i u'_j] - <u'_i u'_j> \tag{14}$$

This can be thought of as a perturbed Reynolds stress, which is unknown and will ultimately require modeling. In the limit of a completely random turbulence containing no structure (ie $<...> = 0$) this quantity is equal to the usual Reynolds stress. In the case when the turbulence is completely structured so that $<...> = [...]$, τ_{ij} is identically zero.

We now follow classical energy method stability analysis for the coherent field. First, the growth rate of the volume averaged coherent energy E is defined as a functional of the coherent velocity field.

$$\lambda(\mathbf{v}, \mathrm{U}, \nu\, \tau) = (1/2E)dE/dt \tag{15}$$

Integration by parts and continuity are used to eliminate the nonlinear convective and pressure terms. We seek the solenoidal velocity field which maximizes λ. Application of the calculus of variations then gives the Euler equations for the maximizing **v** field in the form of an eigenvalue relation.

$$\lambda v_i + S_{ij}v_j = -\pi_{,i} + \nu v_{i,jj} + \tau_{ij,j}$$

$$v_{i,i} = 0 \tag{16}$$

We consider coherent fields which are periodic in the homogeneous directions. This allows a decompositon into poloidal and toroidal components which satisfy continuity exactly (Joseph , 1973)

$$v_1 = \psi_{,12} - v_{,3}; \quad v_2 = -(\psi_{,11} + \psi_{,33}); \quad v_3 = \psi_{,23} + v_{,1} \tag{17}$$

The two scalar functions are then expanded in normal modes in the streamwise and spanwise directions.

$$\psi(\mathbf{x}) = \psi(x_2)\exp\{i(k_1x_1 + k_3x_3)\}$$

$$v(\mathbf{x}) = v(x_2)\exp\{i(k_1x_1 + k3x_3)\} \tag{18}$$

Substituting the above into equation (16) and eliminating the pressure π results in two coupled equations, forming a differential eigenvalue problem.

In order to procede we need to specify a mean velocity field and a model for the unknown stress terms.

Closure Models

We have investigated two different models for the unknown stress terms appearing in the eigenvalue relation. It should be noted that, modulo the modeled terms, equation (16) is linear in the coherent velocities, providing an inexpensive means of determining basis functions. This linearity is an essential advantage of the method and for this reason we will constrain any stress model to be both linear and homogeneous in the **v** field insuring that the governing equation remains a regular eigenvalue problem. Tensorially this requires

$$\tau_{ij} - \tau_{kk}\delta_{ij}/3 = \Gamma_{ijkl}(v_{k,l} + v_{l,k}) \tag{19}$$

The nature of the averaging procedure implies that the scales of the coherent field and the background turbulence are different. Assuming that the background turbulence evolves on much shorter time and length scales then the structures, it seems plausible that a Newtonian stress-strain relationship like that for the molecular stresses will provide the basis for a model. We set

$$\tau_{ij} - \tau_{kk}\delta_{ij}/3 = \nu_t(v_{ij} + v_{j,i}) \tag{20}$$

Due to the inhomogenity of the turbulence in the wall normal direction, we specify ν_t as a function of x_2 corresponding to experimentally determined values of the traditional eddy viscosity. We will refer to this basic model as the isotropic eddy viscosity model.

Using the basic stress model and an analytic expression for the fully turbulent mean profile (Reynolds & Tiederman, 1967) we have solved the resulting equations numerically. Figure 5 shows comparisons between the calculated eigenvectors and the POD results of Moin & Moser (1989) obtained from a numerical data base. Although there are qualitative similiarities in the shape of the structures, the modes predicted by the stability method fall off much more rapidly away from the wall than do the POD functions. The eigenvalue spectrum clearly shows that the stability analysis favors modes which have a much higher wavelength than the maximum energy modes of the POD.

Although there may be a number of reasons for this discrepancy, we choose to first examine more closely the closure model. Since both the POD analysis and the stability method favor modes which are infinitely long in the streamwise direction, we exmine our equations setting $k_1 = 0$. We find that the isotropic eddy viscosity model creates no coupling between the different components of the coherent velocity. When there is no streamwise variation of the coherent field the only coupling terms in the equations are those multiplying the mean gradient. For realistic mean profiles, regions of high shear are confined to thin regions near the wall, and the structures predicted may be expected to fall off as quickly as the shear.

We now seek to develop a stress model that allows for some anisotropy in the eddy viscosity, and thus couples the component equations through the stress terms. We begin with the evolution equation for the Reynolds stresses, where we are obliged to model a number of terms to obtain a closed system. We use standard second-order turbulence models. We model the pressure-strain correlation by a return-to-isotropy term and an isotropization-of-production term (Naot et al, 1970); we use an isotropic dissipation. We assume the stresses are in local equilibrium: $D[u_i u_j]/Dt = 0$. This reduces the evolution equation for the Reynolds stress to an algebraic expression.

Now we set up a perturbation expansion in terms of mean field quantities, taking the coherent field as an order ε perturbation to the spatial mean. On physical grounds, we argue that the perturbed stress field is due entirely to the presence of the structures and consequently we restrict the model to include only production due directly to coherent velocity gradients. This is in agreement with a cascade analogy for the complete flow: the coherent structures are fed energy directly by the mean gradients while the small scale turbulence is in turn fed by gradients of the coherent field. If we identify the 0th order stresses with an eddy viscosity tensor, then the closure model can be written as

$$\tau_{ij} - \tau_{kk}\delta_{ij}/3 = -\nu_{ik}s_{kj} \tag{21}$$

where the tensor viscosity has the following structure in this specific case: $\nu_{13} = \nu_{23} = 0$, $\nu_{33} = \nu_{22}$.

Despite the absence of mean production terms, this model is still a major improvement over the isotropic eddy viscosity formulation. In the simple model the effects of the mean field have been neglected entirely. Here we have allowed for modulation of the perturbation stresses by the mean field through the 0th order stresses appearing in the production terms. Also we have unconstrained the model in an important way since the tensorial form of the eddy viscosity allows the pricipal axes of the stress tensor to be unaligned with the axes of the rate of strain. This is more realistic considering the three-dimensionality of the coherent field. This model leads to the expected cross coupling of the equations through the stress terms.

Figure 6 shows eigensolutions for several values of k_3. The results compare well with the POD eigenvalues, especially for wavenumbers at or below the peak in the POD spectrum. The improvement with decreasing wavenumber is expected given the modeling considerations. The separation of scales between the background turbulence and the coherent structures increases as the wave number decreases adding to the expected accuracy of the stress model. The comparison of the two models indicates significant improvements in the results given by the anisotropic eddy viscosity form. The energy method procedure with the more refined closure model appears capable of extracting structures which closely approximate those given by the POD at least at the energy containing scales of motion.

Despite the general improvement, it is still clear that more needs to be done. From Figures 5 and 6, it is evident that the eigen spectrum produced by solution of Equation (10),while improved by the use of the anisotropic closure model, still predicts structures with maximum growth rate that are a factor of 2 smaller than those containing the most energy (as given by the POD). We next considered the effect on the spatial mean velocity field of the growing coherent perturbation.

Interaction with the mean

At this point we consider the role of the mean velocity in the two methods. The POD structures are derived from solutions to the non-linear Navier-Stokes equations which allow for complicated interaction between the different scales of motion. The structures evolve in a mean velocity field that is changing due to the presence of the structures themselves. Conditionally averaged mean profiles clearly show the evolution of the local shear in the presence of coherent structures (see Figure 7, 8). We see that structures, in the relatively long period before bursting, act to erode the shear that they see. The POD eigenfunctions are given by averages of contributions from different mean profiles.

The stability method on the other hand does not allow for any interaction between the mean and the coherent field. The mean flow is imposed and the resulting structures are calculated. The mean profiles we have used are time averages which mask any contribution from the coherent field. As such the stability analysis predicts that the highest growth modes are those which can best extract energy from the time averaged mean shear which is concentrated in the small near-wall region. Since the structures have an aspect ratio of about 1, the narrow region of high imposed shear leads to a peak in the eigen spectrum at a large wave number.

To allow the mean field to evolve under the influence of the coherent field, we follow Liu (1988) and write time evolution equations for the energy density of the coherent field. We allow the mean profile to depend on the coherent velocity as it does in reality. We expect that equilibrium solutions for the energy density as a function of cross stream wave number will approximate the average energy content as given by trhe POD spectrum.

We assume that the coherent field is given by the eigenvalues of the stability problem, but we now allow them to vary in time

$$v_i(\mathbf{x}, t) = A(t)\psi_i(\mathbf{x}) \tag{22}$$

$$<u_iu_j>(\mathbf{x}, t) = E(t)B_{ij}(\mathbf{x}) \tag{23}$$

where $\psi_1 = v \exp\{ikx_3\}$, $\psi_2 = ik\psi \exp\{ikx_3\}$, $\psi_3 = -D\psi \exp\{ikx_3\}$. By examining the evolution equation for $r_{ij} = <u_iu_j> - [u_iu_j]$, the forcing terms are of the form $<u_iu_j>v_{i,j}$. Consequently, we assume the perturbation stresses also to be a product:

$$r_{ij}(\mathbf{x}, t) = A(t)E(t)R_{ij}(\mathbf{x}) \tag{26}$$

Since we have used an eddy viscosity in obtaining the coherent forms we further assume that:

$$R_{ij}(\mathbf{x}) = -\nu_{ik}(\psi_{j,k} + \psi_{k,j}) \tag{27}$$

where $\nu_{11} = \nu_{22} = \nu_{33} = \nu_\tau$, $\nu_{12} = \nu_{21} = \beta_\tau$, $\nu_{13} = \nu_{23} = 0$. All that remains is to model the mean profile. For this we adopt the quasi-steady model used in Aubry *et al* (1988). This allows the mean to respond to growing structures providing the necessary feed-back to the evolving modes. Using the friction velocity,u_τ and the channel half height, a, the scaled equation for the mean gradient is:

$$\partial U(x_2, t)/\partial x_2 = Re_\tau\{<u_1u_2> + <v_1v_2> - x_2\} \tag{28}$$

The rate of dissipation of turbulent energy is given by a simple model adopted from second order closures schemes.

$$D\varepsilon/Dt = c_1(\varepsilon/k)<u_iu_k>U_{i,k} - c_2\varepsilon^2/k + \text{Transport} \tag{29}$$

Substituting these various models into the energy equations results in a set of three, coupled ODEs for the temporal evolution of the energies and dissipation.

In order to evaluate the integrals appearing in these equations, we need to assume the spatial form of the averaged turbulence quantities B_{ij} and the dissipation $D(x_2)$. For this simple model we have assumed that while the intensity of the turbulence varies its spatial dependence remains unchanged. We use experimental data for fully developed turbulent channel flow to determine both B and D. The coherent structures are found by energy stability analysis, as described above.

Conclusions

We find that A^2_{equil} is at approx $k = 7.5$, which is where the POD spectrum peaks (figure 9).

It seems plausible that the equilibrium value of the coherent energy density can be related to the time averaged energy (POD) in a natural way. This is certainly not true for the time derivative of the coherent energy (the eigenvalues of the energy stability problem). We make the following remarks:

- Small scale structures grow rapidly in the region of high shear. They quickly reduce the local shear and decay by energy loss to small scale fluctuations.

 There exists an optimal mode shape which best balances energy extraction from local shear and loss to background turbulence.

 The equilibrium solutions of the modeled, single mode temporal problem compare favorably with the POD spectrum.

Bibliography.

Adrian, R.J. 1979. Conditional eddies in isotropic turbulence. *Physics of Fluids,* 22(11):2065-70.

Adrian, R.J. and Moin P. 1988. Stochastic estimation of organized turbulent structure: homogeneous shear flow. *Journal of Fluid Mechanics*, 190:531-59.

Adrian, R.J., Moin, P. and Moser, R.D. 1987. Stochastic estimation of conditional eddies in turbulent channel flow. In *CTR, Proceedings of the Summer Program 1987* .

Aubry, N., Holmes, P., Lumley, J. L. and Stone, E. 1988. The dynamics of coherent structures in the wall region of a turbulent boundary layer. J. Fluid Mech. 192: 115-173.

Berkooz, G. 1991b. *Turbulence, coherent structures and low dimensional models.* Ph.D. Thesis, Cornell University.

Berkooz, G., Elezgaray, J. & Holmes, P. 1992. Coherent structures in random media and wavelets. *Physica D.* in press.

Berkooz, G., Holmes, P. & Lumley, J. L. 1991. Intermittent dynamics in simple models of the turbulent wall layer. *J. Fluid Mech.* 230: 75-95.

Brereton, G. J. & Kodal, A. 1992. A frequency domain filtering technique for triple decomposition of unsteady turbulent flow. *J. Fluids Engineering*. To appear.

Cantwell, B. J. 1981. Organized motions in turbulent flow. *Ann. Rev. Fluid Mech.* 13: 457-515.

Clavin, P. 1985 *Prog. Energy Comb. Sci.* 11: 1.

Hyman, J. M. & Nicolaenko, B. 1986. *Physica D* 18: 113.

Joseph, D. D. 1973. *Stability of Fluid Flow.* Berlin etc.: Springer.

Liu, J. T. C. 1988. Contributions to the understanding of large scale coherent structures in developing free turbulent shear flows. *Advances in Applied Mechanics* 26: 183-309.

Lumley, J. L. 1971. Some comments on the energy method. In *Developments in Mechanics 6*, L. H. N. Lee and A. H. Szewczyk, eds. Notre Dame, IN: N. D. Press.

Moin, P. and Moser, R.D. 1989. Characteristic-eddy decomposition of turbulence in a channel. *Journal of Fluid Mechanics*,200:471-509.

Moin, P., Adrian, R.J. and Kim, J. Stochastic estimation of organized structures in turbulent channel flow. In 6th *Turbulence Shear Flow Symposium,* Toulouse, France.

Pomeau, Y., Pumir, A. & Pelce, P. 1984. *J. Stat. Phys* 37: 39.

Reynolds, W. C. & Hussain, A. K. M. F. 1972. The mechanics of an organized wave in turbulent shear flow. Part 3. Theoretical models and comparisons with experiment. *J. Fluid Mech.* 54: 263-287.

Reynolds, W. C. and Tiederman, W. G. 1967. Stability of turbulent channel flow with application to Malkus' theory. *J. Fluid Mech.*, 27: 253-272.

Tennekes, H. and Lumley, J.L. 1972. *A First Course in Turbulence*. Cambridge, MA: The MIT Press.

Zaleski, S. 1989. *Physica D* 34: 427

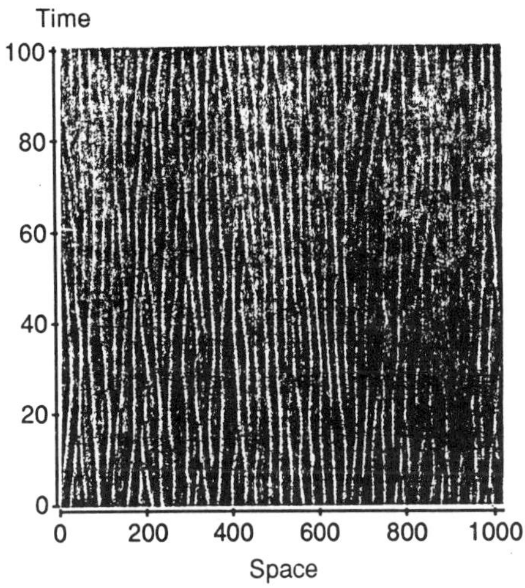

Figure 1. A typical solution of the periodic Kuramoto-Sivashinski equation (L = 400).

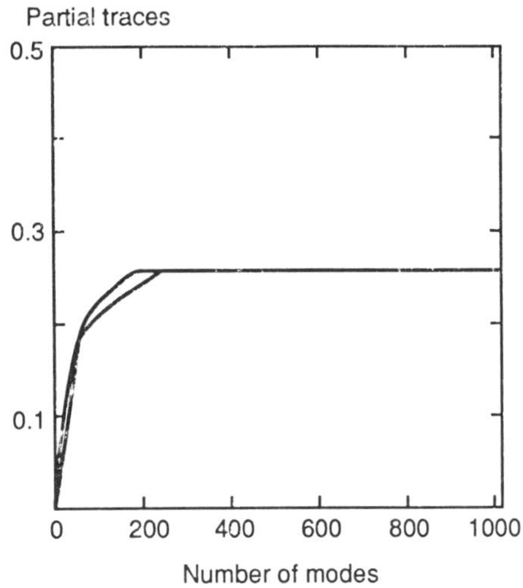

Figure 2. The energy resolved by the Fourier (POD) basis (upper curve) and the wavelet basis (lower curve), as functions of the number of modes.

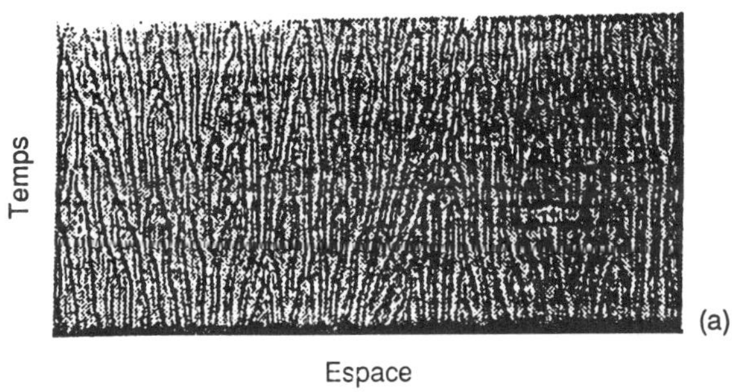

Figure 3. Full simulation of the Kuramoto-Sivashinski equation (Fourier basis).

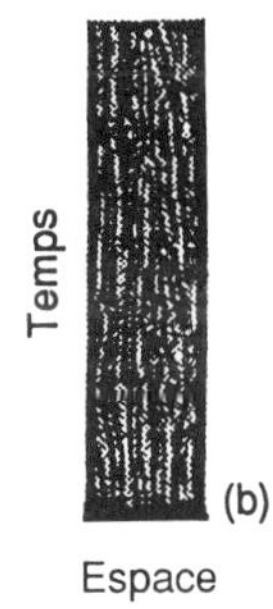

Figure 4. Wavelet simulation of the K-equation (4 mode) with $L_B = 50$ (rescaled

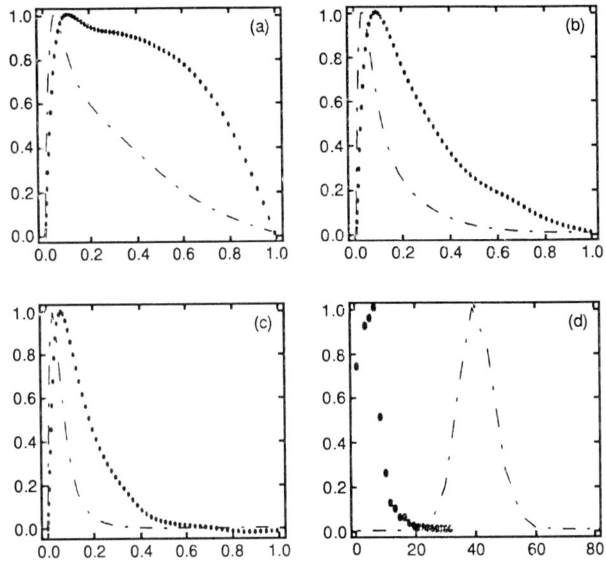

Figure 5. Isotropic eddy viscosity model. ••••• POD; ----- . ------ . ------- . , isotropic model; (a) $k_3 = 3.00$, (b) $k_3 = 6.00$, $k_3 = 12.00$, (d) spectrum.

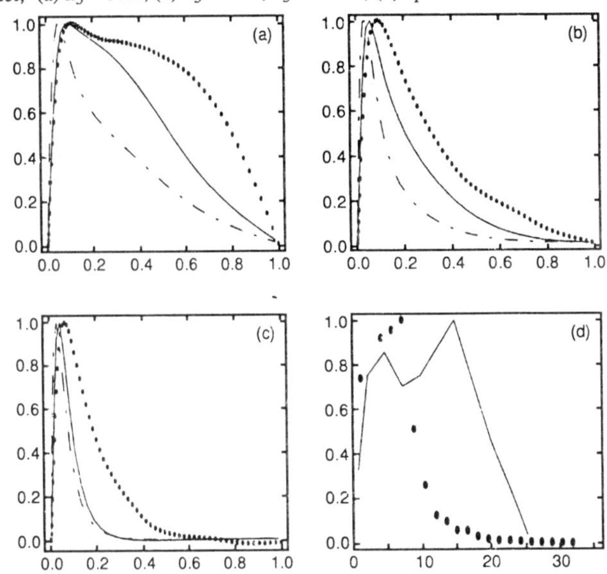

Figure 6. Anisotropic eddy viscosity model: •••••, POD; ------ . ------ . ------ . , isotropic model; -------------------- , anisotropic model; (a) $k_3 = 3.00$, (b) $k_3 = 6.00$, (c) $k_3 = 12.00$, (d) spectrum.

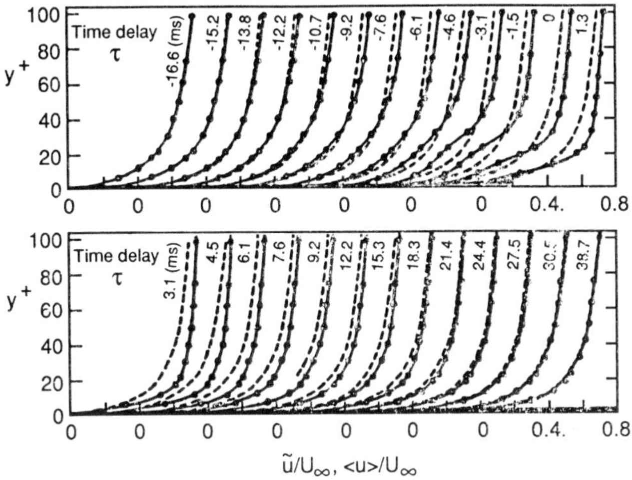

Figure 7. Conditionally averaged velocity profiles before and after a burst, showing the influence on the local shear of the coherent structures (Blackwelder & Kaplan, 1976).

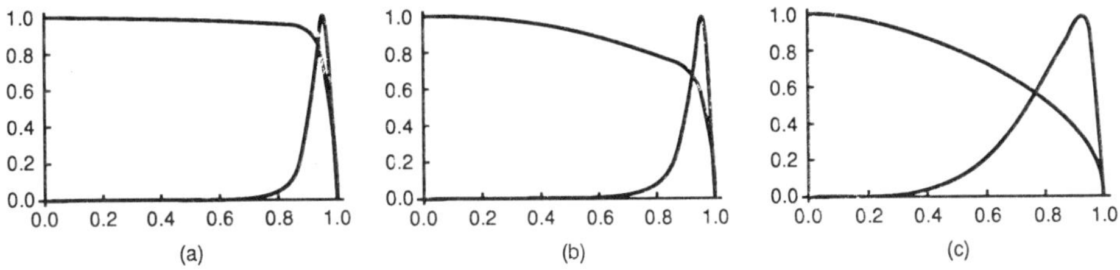

Figure 8. The eigenfunctions generated by the anisotropic eddy viscosity model with various mean velocity profiles.

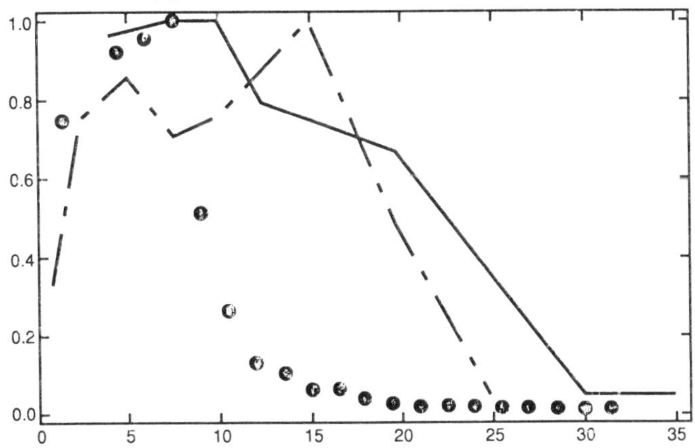

Figure 9. ●●●●●●●● , the POD eigenvalue spectrum; --------- · --------- · --------- · , the eigenvalue spectrum generated by the anisotropic model without mean flow interaction; ----- ---------------, the eigenvalue spectrum generated with mean flow interaction.

A comparison of different analytical techniques for identifying structures in turbulence

N. K.-R. KEVLAHAN[1], J. C. R. HUNT[1,2]
AND J. C. VASSILICOS[1]

[1]DAMTP, University of Cambridge, U.K.
[2]Meteorological Office, Bracknell, U.K.

Vortical structures play an important role in the kinematics and dynamics of turbulence, but in order to understand this role we require techniques to identify and classify them. Proper Orthogonal Decomposition (POD), conditional sampling with ensemble statistics, and conditional sampling with conditional statistics are applied to a simple test function and the results are compared to determine the strengths and weaknesses of each approach. The second method gives the closest approximation to the test signal and is the easiest to use, although it is sensitive to the choice of conditions. None of these techniques can give much insight into the dynamics of turbulence, or into the organisation of eddies with complex, fine-scale structure.

New methods for investigating complex (self-similar) structures based on fractal and wavelet analyses are presented. Methods of distinguishing between locally (accumulating) and globally (fractal) self-similar structures are suggested.

1. Introduction

There has been much interest recently in the identification of characteristic structures in turbulence as a way of understanding its dynamical properties such as intermittency, the transfer of energy between length scales and the dispersion of contaminants. Experiments and Direct Numerical Simulations (DNS) have provided evidence that turbulence is not random, but contains structures with complex internal organisation (eg. Vincent & Meneguzzi 1991). Statistical approaches to turbulence are capable of predicting some quantities averaged over many realisations of a turbulent flow, but they are not able to provide insight into dynamical quantities that are directly related to the flow structures of a given realisation. In order to be effective the structural approach requires mathematical tools capable of unambiguously identifying complex turbulence structures according to appropriate kinematical and dynamical criteria. Once different sorts of structures have been identified it is possible to determine which of them play important roles in the dynamics of turbulence.

Many mathematical techniques have been used to identify and describe structures, eg. the Fourier transform, Karhunen-Loeve orthogonal decomposition (Lumley 1967, Aubry *et al.* 1988), functional/pattern recognition (Mumford 1982), non-

J. P. Bonnet and M. N. Glauser (eds.), Eddy Structure Identification in Free Turbulent Shear Flows, 311–324.

functional conditional sampling (Hussain 1986), kinematic classification (Chen *et al.* 1990, Kevlahan 1992, Wray & Hunt 1990), and, more recently, fractals (Vassilicos & Hunt 1991) and the wavelet transform (Brasseur & Wang 1992, Farge 1992, Hunt *et al.* 1993). Each of these techniques breaks down the flow in different ways and places a different emphasis on the various aspects of turbulence (usually velocity, pressure, and their gradients). The question then arises: which methods are best able to detect, classify and describe eddy structures which may contain significant complexity? This paper has two parts: the first is concerned with the identification and eduction of large coherent structures. The second presents and discusses subtler tools capable of identifying small scale complex eddy structure.

We evaluate three common eduction techniques for coherent structures, namely Proper Orthogonal Decomposition (POD), conditional sampling with ensemble statistics, and conditional sampling with conditional statistics by applying each of these techniques to a simple test function that is asymmetric and whose energy is not contained in a single mode (i.e. $f(x) = (a + xb) \exp -(x^2)$). Then we assess the strengths and weaknesses of each approach. The results show that these techniques can lead to quite ambiguous and erroneous inferences as to the structure of eddies.

In the second part of this paper we address the problem of identifying and analysing eddies with significant internal complexity; none of the traditional approaches described above are capable of giving this sort of information. By 'complex structure' we mean eddies with a self-similar or singular internal structure. The usual Fourier transform can provide some limited information about the existence of such singular regions or structures in the flow. A non-integer power law energy spectrum (eg. Kolmogorov's $E(k) \propto k^{-5/3}$ for the inertial range of isotropic turbulence) implies that the velocity field must contain singularities 'worse' than discontinuities (Moffatt 1984, Hunt & Vassilicos 1991, Zeldovich & Sokolov 1985). These singular structures could be cusps, accumulating discontinuities or oscillations (eg. functions of the form $f(x) = x^s \sin(x^{-t})$), fractal functions (eg. the non-differentiable Weierstrass function $f(x) = \sum_{k=1}^{\infty} \lambda^{(s-2)k} \sin(\lambda^k x)$), or some more complex superposition of these. Both the accumulating and fractal functions are self-similar, however the accumulating functions are self-similar only about a specific point while fractals (characterised by a non-integer Hausdorff dimension) are self-similar everywhere. Accumulating and fractal structures may be considered globally and locally self-similar respectively. Turbulence models have been developed using both these types of structures as building blocks (eg. the fractal β-model of Frisch *et al.* (1978) and the strained spiral vortex model of Lundgren (1982)).

To detect whether or not the structure of eddies actually corresponds to that assumed by these models requires a method for unambiguously determining and distinguishing between global and local self-similarity. The differences between these classes of self-similar structures are investigated by applying Fourier, wavelet and fractal methods to an accumulating test function such as $f(x) = \sin 2\pi/x^t$ (which is locally self-similar) and a related function constructed from the random superposition of its Fourier components (which is globally self-similar). Finally, general analytical methods of distinguishing between locally and globally self-similar structures are proposed.

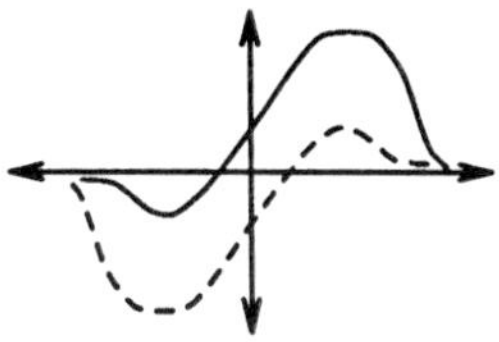

FIGURE 1. Typical eddy structure from the test function $u(x) = (a + bx)\exp(-x^2)$.

2. Comparison of methods for large eddy eduction

In this section we investigate the relative strengths and weaknesses of methods for eduction based on unconditional, or ensemble, statistics compared to those based on conditional sampling by applying each of them in turn to a simple one-dimensional test function.

Methods based on invariants of the deformation tensor and/or pressure (e.g. Wray & Hunt 1990, Chen *et al.* 1990) are in a different class, since unlike the above approaches they classify each point of the flow in a given realisation and do not extract whole eddy structures. Invariant/pressure methods do, however, have the advantage of being equally applicable to both large and small scale coherent structures. The invariant/pressure approach also requires more sophisticated and physically realistic test signals for evaluation, and for this and the previous reasons these methods are not included in the comparison.

The test function $u(x)$ we use is a simple sum of the Gaussian profile and its first derivative, both having random amplitudes, viz:

$$u(x) = (a + bx)\exp(-x^2) \tag{2.1}$$

where a and b are normally distributed random variables with zero mean. This is representative of the velocity profile across a flow (eg. a free shear flow) in which there are different kinds of eddies that are asymmetrical about the centre line $x = 0$ (see figure 1). If the eddies are uncorrelated with each other a and b are also uncorrelated. It is assumed that the variances σ_a^2, σ_b^2 of a and b are of similar magnitude. Each of the structure eduction techniques will be judged on its ability to pick out these typical eddies with the least amount of subjective manipulation. This is a more demanding task than in many flows, where the dynamics leads to a selection of a few dominant modes or types of eddy (Hussain 1988).

2.1. *Proper Orthogonal Decomposition*

POD decomposes a signal onto a basis of non-specified functions chosen to represent the energy of the signal in the fewest number of modes. This method is completely objective (no possibility of subjective bias or control in the process), and is optimally efficient. The attempt to find the structure that is best correlated (in terms of energy) with the random turbulence field reduces to a problem of maximisation (Aubry *et al.* 1988). The calculus of variations reduces this maximisation problem to the solution of a Fredholm integral equation of the first kind

$$\int R_{ij}(x,x')\phi_j(x')\,dx' = \lambda\phi_i(x), \tag{2.2}$$

where $R_{ij}(x,x')$ is the autocorrelation tensor and $\phi_i(x)$ are the eigenfunctions, or modes (which are often interpreted as independent eddy structures). Then the signal $u_i(x)$ may be written as the sum of the eigenfunctions $\phi_i^{(n)}(x)$ with random amplitudes $a^{(n)}$

$$u_i(x) = \sum_{n=1}^{\infty} a^{(n)} \phi_i^{(n)}(x). \tag{2.3}$$

The turbulent kinetic energy is the sum over the n eigenvalues

$$E = \int \langle u_i u_i \rangle \, dx = \sum_{n=1}^{\infty} \lambda^{(n)}, \tag{2.4}$$

because of the orthogonality of the eigenfunctions and the statistical independence of the amplitudes, i.e.

$$\int \phi^{(n)} \phi^{(m)} \, dx = \delta_{nm} \quad \text{and} \quad \langle a^{(m)} a^{(n)} \rangle = \delta_{mn} \lambda^m, \tag{2.5}$$

where $\langle\rangle$ denotes the ensemble average. In the case of discrete data, solving equation (2.2) reduces to a matrix inversion problem. Note that although (2.3) converges optimally fast, some signals require many POD modes to capture a significant portion of the energy. POD is only useful for those signals where the energy can be captured in a few modes, or where the coefficients $a^{(n)}$ decrease abruptly after a certain mode.

The correlation function of the test signal (2.1) is

$$R(x,x') = (\sigma_a^2 + \sigma_b^2 x x') \exp(-x^2 - x'^2), \tag{2.6}$$

where a and b are taken to be random uncorrelated variables with standard deviations σ_a and σ_b. For this simple correlation function equation (2.2) can be solved analytically by the separable kernel method (Arfken 1985, p. 872) to give

$$u(x) = a^{(1)} \phi^{(1)}(x) + a^{(2)} \phi^{(2)}(x), \quad a^{(1)} = a^{(2)} = \sqrt{\pi}(\sigma_a^2 + \sigma_b^2) \tag{2.7}$$

where the two modes are symmetric and anti-symmetric,

$$\phi^{(1)} = \sqrt{2}\pi^{-1/4} x \exp(-x^2), \quad \phi^{(2)} = \pi^{-1/4} \exp(-x^2). \tag{2.8}$$

Note that neither $\phi^{(1)}$ nor $\phi^{(2)}$ describes the asymmetrical form of the general eddy (see figure 1). It is evidently misleading to interpret the largest mode, $\phi^{(1)}$, as a typical eddy of the flow. The problem of interpreting the first mode arises here because the energy of the signal is fairly evenly distributed between the two largest modes. If the majority of the energy is not captured by the first mode it is more appropriate to consider the truncated series $\sum_{n=1}^{N} a^{(n)} \phi_i^n$ for which an increase of N by 1 results in little increase in energy, but for which a decrease of N by 1 removes a large portion of the energy. The first N modes together are then interpreted as the typical flow structure. It is important to remember that signals are possible for which even this method of interpreting the POD is not suitable (eg. the complex eddies studied in §3).

2.2. *Conditional sampling with ensemble statistics*

The goal in conditional sampling is to estimate $\mathbf{u}(\mathbf{x},t)$ given that the velocity at $(\mathbf{x}',t)$ assumes a specific value $\mathbf{u}(\mathbf{x}',t)$. It can be shown (see Adrian *et al.* 1988) that the best estimate of the profile $\mathbf{u}(\mathbf{x}, t)$, which is defined as $\hat{\mathbf{u}}(\mathbf{x})$, is the *conditional average*

$$\hat{\mathbf{u}}(\mathbf{x}) = \langle \mathbf{u}(\mathbf{x}, t) | \mathbf{u}(\mathbf{x}', t) \rangle. \tag{2.9}$$

The conditional average (2.9) can be approximated in terms of second-order (linear) and third-order (nonlinear) two-point spatial correlations by,

$$\hat{u}_i(\mathbf{x}) \sim A_{ij}(\mathbf{x}', \mathbf{x}) u_j(\mathbf{x}') + B_{ijk}(\mathbf{x}', \mathbf{x}) u_j(\mathbf{x}') u_k(\mathbf{x}') + O(u^3(\mathbf{x}')), \tag{2.10}$$

where A_{ij} and B_{ijk} are found by requiring that the mean-square errors $e_i = \langle (\hat{u}_i(\mathbf{x}) - u_i(\mathbf{x}))^2 \rangle$ be minima. Setting the derivatives of e_i with respect to A_{ij} and B_{ijk} equal to zero and carrying out the indicated averages gives the following equations for the components of A_{ij} and B_{ijk} for inhomogeneous, anisotropic turbulence

$$A_{ij}(\mathbf{x}', \mathbf{x}) \langle u_j u_l \rangle (\mathbf{x}') + B_{ijk}(\mathbf{x}', \mathbf{x}) \langle u_j u_k u_l \rangle (\mathbf{x}') = \langle u_l(\mathbf{x}') u_i(\mathbf{x}) \rangle \tag{2.11}$$

$$A_{ij}(\mathbf{x}', \mathbf{x}) \langle u_j u_l u_m \rangle (\mathbf{x}') + B_{ijk}(\mathbf{x}', \mathbf{x}) \langle u_j u_k u_l u_m \rangle (\mathbf{x}') = \langle u_l(\mathbf{x}') u_m(\mathbf{x}') u_i(\mathbf{x}) \rangle. \tag{2.12}$$

In the case of homogeneous, isotropic turbulence these relations simplify to

$$A_{ij}(\mathbf{x}', \mathbf{x}) = \langle u_j(\mathbf{x}') u_i(\mathbf{x}) \rangle / u^2, \tag{2.13}$$

$$B_{ijk}(\mathbf{x}', \mathbf{x}) = 3 \langle u_j(\mathbf{x}') u_k(\mathbf{x}') u_i(\mathbf{x}) \rangle / 2 K u^4 \tag{2.14}$$

where $u^2 = 1/3 \langle u_i u_i \rangle$, and K is the kurtosis. Note that these correlations are *unconditional*, and they are not selected for particular 'eddies'; they are ensemble statistics. The Linear Stochastic Estimation (LSE) for $\mathbf{u}(\mathbf{x}, t)$ retains only the first term on the right hand side of equation (2.10).

From (2.10), (2.11), and (2.12) the LSE for our test function is

$$u(x) = \frac{R(x, x')}{\langle u^2(x') \rangle} u(x'), \tag{2.15}$$

which, using (2.6), becomes

$$\hat{u}(x) = \frac{(\overline{\sigma_a^2} + \overline{\sigma_b^2} x x') \exp(-x^2 + x'^2)}{(\sigma_a^2 + \sigma_b^2 x'^2)} u(x'). \tag{2.16}$$

This purely statistical estimate may be combined with conditional sampling to determine the form of all eddies whose maximum amplitude is negative, i.e. $u_{mx} < 0$. Conditional sampling governs how the reference point (in space and time) is chosen. In general, conditions are placed on both the location of the reference point, x', and the value of the velocity at that point, $u(x')$. If the location of the maximum (x_{mx}) is taken as the reference point x' then (2.16) shows that if $u_{mx} < 0$ the point $u = 0$ lies at $x = -\sigma_a^2/(\sigma_b^2 x')$. Therefore the location of the 'zero point' always has the opposite sign to that of the location of u_{mx}. This is a significant improvement over POD: the conditional eddy correctly represents the qualitative characteristics of the typical physical eddy as shown in figure 1.

How robust is the conditional eddy to changes in the sampling criteria? Consider a purely conditional approach (not based on the correlation tensor). If the conditional average is based on the criterion $x_{mx} > 0$ then the conditional eddy becomes

$$\langle u(x)^2 | x_{mx} > 0\rangle = \sigma_b^2 x \exp(-x^2). \tag{2.17}$$

This conditional eddy has lost all information about the symmetrical part of the velocity field (and half the energy)! If we now add an additional criterion $u(0) > 0$ the conditional eddy becomes

$$\langle u(x)^2 | x_{mx} > 0, u(0) > 0\rangle = (\sigma_a^2 + \sigma_b^2 x) \exp(-x^2). \tag{2.18}$$

This is again a very different structure from the one obtained by conditioning only on $x_{mx} > 0$. These examples show how sensitive the LSE method can be to the conditions chosen: slight changes in conditions can lead to the identification of completely different eddies.

2.3. *Conditional sampling and conditional statistics*

The Coherent Structures method (Hussain 1986) is an interactive and iterative form of conditional sampling. In this approach one first defines the signal, its nature and the relevant length scales of the coherent structure for there to be a conditional event. The location of the structure is adjusted until an optimal mean structure pattern is obtained. If the spread of structures is too great one reclassifies, introducing more structure classes. This process continues until an optimal mean structure pattern with minimum spread is obtained.

Consider our test signal (2.1). Initially we may decide that there is only one coherent structure, and then filter the signal to remove white noise. On examining the signal we could then split the structures into two classes based on the sign of the maximum velocity: $u_{mx} > 0$ and $u_{mx} < 0$. This classification would still allow too great a spread of structures so we would reclassify further into four groups based on the sign and position of the maximum velocity: $u_{mx} > 0, x_{mx} > 0$, $u_{mx} > 0, x_{mx} < 0$, $u_{mx} < 0, x_{mx} > 0$, $u_{mx} < 0, x_{mx} < 0$. These structure classes can be further divided by the magnitude of the maximum velocity: e.g. $1/2 < u_{mx} < 1$, $1 < u_{mx} < 3/2$, $3/2 < u_{mx} < 2$, $2 < u_{mx} < 5/2$, $5/2 < u_{mx}$. Finally, by this iterative, subjective process we have ended up with twenty structure classes. This example shows that the coherent structures approach can require a large amount of manipulation to obtain representative structures, but also permits a great deal of fine control and sensitivity. An advantage of this approach is that, unlike LSE, it is able to separate out the incoherent noise of the signal.

3. Small scale complex eddy structures

How can small scale structures with complex internal organisation (e.g. the accumulating oscillatory signal from a cut through a spiral vortex) be identified? The continuous spectrum and accumulating structure of these eddies makes them difficult to represent in a few terms of a series (their energy is not concentrated in the first few POD modes). Conditional sampling is also inappropriate since it is difficult to extrapolate on such a highly oscillatory function. New tools are definitely needed to identify and analyse eddies with complex fine scale structure.

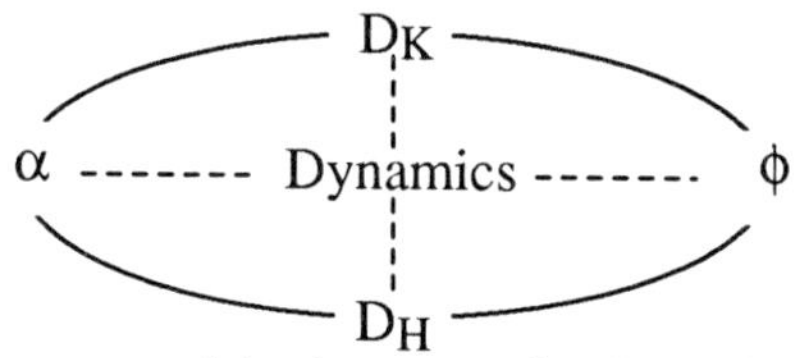

FIGURE 2. The relation between quantities important for the analysis of the fine-scale structure of turbulence: the exponent of the Fourier energy spectrum α, Fourier phase spectrum $\phi(k)$, Kolmogorov capacity D_K, and Hausdorff dimension D_H. The solid lines indicate primarily fixed relationships, while the dotted lines indicate that the relationship is determined by the dynamics of the turbulence.

In this section we introduce some new tools for the analysis of complex eddy structures and apply them to two test signals: a $\sin(2\pi/x)$ 'spiral' which is an oscillatory locally self-similar function, and the Fourier phase scrambled version which is a globally self-similar fractal function. We use two new tools: the box-counting algorithm and the wavelet transform, and one old tool: the Fourier transform. The problem we address here is how to determine whether turbulence has a locally self-similar ('spiral') structure, or a globally self-similar fractal structure. The long term goal of the analysis of complex eddy structures is to find the relation between the quantities α (the exponent of the self-similar Fourier energy spectrum), $\phi(k)$ (the Fourier phase spectrum), D_K (the Kolmogorov capacity, a measure of self-similarity), D_H (the Hausdorff dimension, a non-integer Hausdorff dimension defines a fractal) and the dynamics and geometry of the flow (see figure 2).

The box-counting algorithm gives a number (the Kolmogorov capacity) which characterises the self-similarity of a signal. A non-integer Kolmogorov capacity means that the signal is in some way self similar, however both locally and globally self similar functions will have a non-integer dimension. The Kolmogorov capacity alone cannot tell us whether the turbulence is fractal or spiral.

The general form of our accumulating 'spiral' test function

$$f(x) = x^s \sin(2\pi x^{-t}) \tag{3.1}$$

is locally self-similar; its zero crossings form a set of points with Hausdorff dimension $D'_H = 0$, but a non-trivial Kolmogorov capacity $D'_K = t/(t+1)$ (Hunt *et al.* 1993). Using the method of stationary phase, one finds that for large k the energy and phase spectra of (3.1) are given by (Hunt *et al.* 1993)

$$E(k) \propto k^{-2p} \tag{3.2}$$

$$2p = (2s+1)(1-D'_K)+1 \tag{3.3}$$

$$\phi(k) \propto k^{D'_K} \tag{3.4}$$

provided that $1 < 2p < 2$, $-t < s < 1$, and $t > 0$.

Orey (1970) showed that Gaussian sample functions with an energy spectrum $E(k) = k^{-2p}$ are fractal and that their zero crossings have a Hausdorff dimension

$$D'_H = \frac{1}{2}(3-2p). \tag{3.5}$$

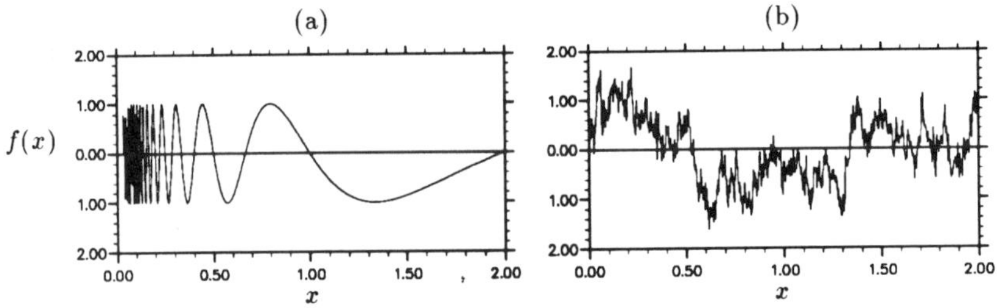

FIGURE 3. (a) Original, and (b) scrambled versions of the accumulating function $f(x) = \sin(2\pi/x)$.

Notice that the two functions just mentioned differ in three respects: (i) $D'_H = 0$ for (3.1), $D'_H \neq 0$ in (3.5), (ii) they have different relations between p and respectively D'_K and D'_H, (iii) they have different phase spectra—the phase spectrum of a Gaussian random function is random. A question which then arises is whether the *phase spectrum* can discriminate between local and global self-similarity.

To study the relation between an ordered phased spectrum $\phi(k)$ and the type of self-similar structure we took the Fourier transform of the accumulating function (3.1) with $s = 0$ and $t = 1$, scrambled the phases and then transformed back to physical space. This procedure destroys the spatial organisation of the original signal and produces a function that is quantitatively and qualitatively indistinguishable from a fractal in the Hausdorff sense. The Kolmogorov capacity of the signal was significantly altered by this operation! In fact it is then given by the equation (3.5) for Gaussian random functions. This property appears to be generally true for power-law spectra with random phases (Osborne & Provenzale 1989), but may not hold for practical reasons related to the scrambling procedure when $s \neq 0$ in the initial signal. By considering both the scrambled and original versions of the signal we will try to find analytical ways capable of distinguishing between such localised and global self-similar structures that have the same energy spectra or Kolmogorov capacities, but different phase spectra and Hausdorff dimensions.

3.1. *Box counting analysis and Kolmogorov capacity*

The Kolmogorov capacity (D_K) of a curve is calculated by finding the slope on a log-log graph of the plot of the number of boxes required to cover the curve as a function of the size of the box. The range of length-scales which need to be resolved to calculate D_K is in certain cases much smaller than that required to obtain a good estimate of the exponent of the energy spectrum (Vassilicos & Hunt 1991). Thus, the Kolmogorov capacity is a more practical measure of self-similarity than the energy spectrum. Although D_K alone cannot distinguish between the original and scrambled signals, the overall shapes of the curves do differ significantly. In the case of the scrambled signal the transition from a slope of -1.5 (the true dimension of the function) to 0 (caused by the finite number of points in the representation of the function) is very abrupt, while in the case of the original signal the transition is much more gradual (see figure 4). This difference is due to the fact that in the original signal the points representing the function are distributed evenly, but the increasingly smaller scales emerge only as one nears the origin. The scrambled signal

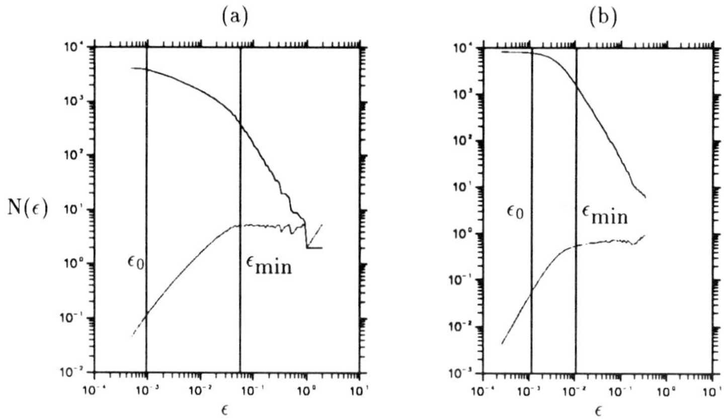

FIGURE 4. Calculating the box-counting dimension. (a) Spiral function $\sin(2\pi/x)$, lower curve is $\times\epsilon^{1.5}$. (b) Scrambled spiral, lower curve is $\times\epsilon^{1.75}$. Note the much more gradual transition between ϵ_{min} (the lower limit of the power-law section of the curve) and ϵ_0 (the minimum spacing between points in the signal) for the locally self-similar spiral.

is self-similar at every location, so the box-counting algorithm suddenly saturates everywhere at a particular length-scale whereas in the original signal the saturation occurs at different length-scales at different positions, thus smoothing the transition of the box-counting curve. This result may give a practical way of distinguishing between locally and globally self-similar structures.

The on-off function generated by the zero-crossings, i.e. the function that jumps from 0 to 1 and from 1 to 0 at the zero-crossings, but remains constant elsewhere, has a Fourier power spectrum $\Gamma(k) \sim k^{-2\rho}$, $2\rho = 2 - D'_K$ (Vassilicos & Hunt 1991), whether the zero-crossings are of a fractal or a spiral signal. Two such different signals with the same value of D'_K will have the same ρ, but different p. In the case of Gaussian sample functions, $p = 2\rho - 1/2$ (see (3.5)); in the case of spiral functions, such as (3.1), the relation between p and ρ is different. A good understanding of this relation may again lead to new practical ways of detecting whether $D'_H = 0$.

3.2. *Wavelet analysis*

A more systematic tool for obtaining local information about spectral quantities is the recently developed wavelet transform (Grossman & Morlet 1984, Farge 1992),

$$\tilde{f}(\lambda, x_0) = \lambda^{-\frac{1}{r}} \int_{-\infty}^{\infty} f(x)\psi^*(\frac{x - x_0}{\lambda})\, dx. \qquad (3.6)$$

The wavelet transform allows one to investigate the structure of the signal $f(x)$ at different scales λ and different positions x_0 simultaneously. Because we are interested in what are essentially local fractal dimensions and fractal dimensions are related to energy spectra the wavelet transform would seem to be a good way of determining whether a flow contains locally self-similar accumulating functions or global fractal structures. In this section we use the wavelet analysis to develop another way of distinguishing between our locally and globally self-similar test functions.

Using the Morlet wavelet,

$$\psi(x) = e^{ik_g x} e^{-\frac{|x|^2}{2}} \tag{3.7}$$

the squared modulus of the wavelet transform of the $\sin(2\pi x^{-t})$ spiral is

$$|\tilde{f}(\lambda, x_0)|^2 \sim \lambda^{1/r} \lambda^{-((2s+1)(D'_K-1)+1)} \exp\left[-\frac{(x_0 - (t\lambda)^{1-D'_K})^2}{2\lambda^2}\right], \tag{3.8}$$

where λ is the length-scale and x_0 is the location. The wavelet transform of the spiral is thus peaked around $x_0 = (t\lambda)^{1-D'_K}$. By contrast, the wavelet transform of the scrambled spiral shows no spatial organisation, although, of course, it reflects the overall distribution of the energy of the signal between different length-scales (see figure 5). This difference is brought out most clearly by looking at the local wavelet energy spectra at various locations. The local wavelet energy spectrum is obtained from the modulus of the wavelet transform by taking a slice through length-scale space at a particular location. The difference between the scrambled and spiral test functions is obvious in figure 5. The scrambled spiral has essentially the same local energy spectrum everywhere (on average $\propto k^{-((2s+1)(D'_K-1)+1)}$ where D'_K is the capacity of the original spiral), but the spiral test function has different ranges of $k^{-((2s+1)(D'_K-1)+1)}$ spectrum depending on the location. Looking at the way the local energy spectrum varies from position to position may provide another way of determining whether turbulence has a 'spiral' or fractal structure.

It has been claimed that branching in the wavelet transform indicates the signal is fractal (e.g. Arnéodo *et al.* 1989), however this is not necessarily the case. If a signal holds only one singularity of the type (3.1) and is totally regular elsewhere, then (3.8) shows that the wavelet transform will point towards $x_0 = 0$ since most of the power $|\tilde{f}(\lambda, x_0)|^2$ is concentrated around a curve $x_0 \sim \lambda^{1-D'_K}$ in the $x_0 - \lambda$ plane of the wavelet transform. The exponents t and s or D'_K and s characterising this singularity can be extracted from the examination of that curve and of the scaling of the wavelet transform along that curve. Problems arise when the signal carries an unknown distribution of singularities of the type (3.1), thereby producing a plethora of intertwined inverse branchings which may not be easily distinguishable from the branching pattern obtained when a fractal or multifractal signal is wavelet transformed (or even when random noise is transformed). It is then not a trivial task to untangle the local scaling exponents from the wavelet transform of a complex signal, let alone interpret them. A simple example of the way this sort of fractal-like branching may arise is shown in figure 6.

Note also that one must be careful in applying real wavelet transforms to determine scaling exponents (related to the Hausdorff dimension). Vergassola *et al.* 1991 have shown that spurious scaling exponents can be obtained from the Mexican Hat wavelet transform of a fractional Brownian signal. The same conclusion is reached, for other reasons, on the basis of a different example: the spiral accumulation function $\sin 2\pi x^{-t}$. From (3.8) it can be seen that the closer x_0 is to 0, the centre of the accumulation, the smaller the range over which $|\tilde{f}|^2$ is dominated by the power law $\lambda^{1/r}\lambda^{((2s+1)(D'_k-1)+1)}$ rather than by the Gaussian filter. A calculation of the scaling over the wrong range will give misleading results. Furthermore, even in the correct

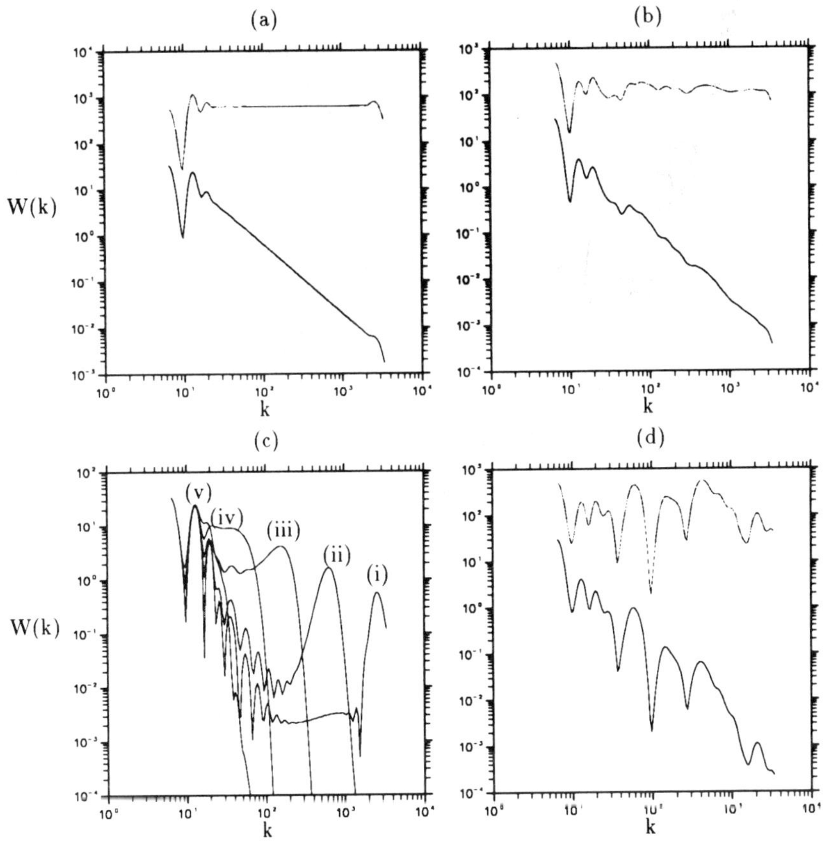

FIGURE 5. Distinguishing between locally and globally self-similar signals using the wavelet transform. (a) Wavelet energy spectrum averaged over all locations for the spiral (upper curve is $\times k^{1.5}$). (b) Wavelet energy spectrum averaged over all locations for the scrambled spiral (upper curve is $\times k^{1.5}$). (c) Local wavelet spectra $\times k^{1.5}$ of the spiral at locations (i) 0.05, (ii) 0.1, (iii) 0.2, (iv) 0.4, (v) 0.8. The flat regions indicate the locations of $k^{-1.5}$ spectrum. (d) Typical local wavelet spectrum of the scrambled spiral at location $x = 0.4$ (upper curve is $\times k^{1.5}$).

range, the detection of a power law does not prove that the signal is fractal; it can evidently also be a spiral accumulation. These two examples show the difficulty of determining the precise self-similar nature of a signal.

4. Conclusions

By applying the POD, conditional sampling with conditional statistics (Linear Stochastic Estimation), and conditional sampling with conditional statistics (Coherent Structures) to a test signal we have clarified the strengths and weaknesses of each approach. First, the physical interpretation of POD is unclear: the first eigenmode will *not* be a typical flow structure unless it contains most of the energy. Secondly,

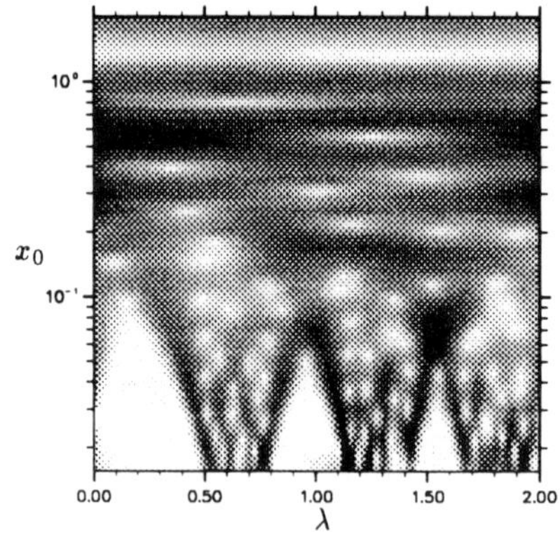

FIGURE 6. Branching pattern in the modulus of the Morlet wavelet transform produced by a random superposition of four $\sin(2\pi/x)$ spirals.

conditional sampling can give a qualitatively accurate physical representation of the flow structures, but the resulting eddy may be sensitive to the precise conditions chosen. A combination of ensemble statistics with conditions determining the location (in space and time) of the reference point gives physically realistic results with a minimum of manipulation. Thirdly, the Coherent Structures method requires significant manipulation to obtain representative structures, although the interactive nature of this approach also makes it the most flexible and least liable to misinterpretation. The Coherent Structures approach also gives a way of separating out the incoherent noise of the signal.

Approaches based on invariants of the deformation tensor and pressure can give information about the relative number and location of regions with particular dynamics, but say nothing about their shape or internal structure. No clear way of interpreting invariant plots exists yet!

None of the above methods are incapable of analysing or identifying the local fine-scale structure of the flow.

The new techniques of fractal analysis and the wavelet transform have been applied to the problem of distinguishing between globally ('fractals') and locally ('spirals') self-similar signals. The Kolmogorov capacity gives one parameter for describing the average degree of convolutedness within a coherent structure where the fields are self similar over a range of length scales that may be quite small compared to those necessary for conventional Fourier methods. Alone among the methods examined, the wavelet transform is able to provide a simple functional description in terms of *local* length scales and position of the typically complex distribution of velocity and vorticity within turbulence structures. Three methods of distinguishing between fractals and spirals have been suggested: the first is based on the box counting algorithm (the variation of the range over which a Kolmogorov capacity is clearly defined), the second exploits the difference between the Fourier power spectrum of the original

$\phi(k)$ random	$\phi(k)$ non-random
$D_H = D_K = \frac{1}{2}(5-\alpha)$	D_H, D_K, α - no fixed relation
- no energy cascade	- allows for cascade
- no structures	- allows for structures

TABLE 1. The differences between signals with random and non-random phase spectra.

function and the on-off function obtained from the zero-crossings, and the third uses the variation of the local wavelet energy spectrum with position.

The long term goal of the development of new techniques to analyse complex structure is to clarify the relationship between the Fourier energy spectrum (exponent α), the Fourier phase spectrum ($\phi(k)$), the Hausdorff dimension (D_H), and the Kolmogorov capacity (D_K). This knowledge will have far-reaching implications for a deeper understanding of the dynamics of turbulence. A major part of this investigation is a study of the differences between signals with a non-random $\phi(k)$ ('spirals') and those with a random $\phi(k)$ ('fractals'). The phase spectrum of turbulence largely determines its dynamics and structure. For example, if the phases of a DNS are randomised the characteristic small-scale vortex tubes vanish. The nature of the energy cascade also depends crucially on the phase spectrum; in a velocity field with random phases there can be no net cascade of energy. The phase spectrum also changes the relationship between D_K and D_H. In a $\sin 1/x$ spiral $D_K = 1.5$ and $D_H = 1$, while if the phases of the spiral are scrambled $D_K = D_H = 1.75$. The fixed relation between D_H and α given by Orey's formula (3.5) holds for a Gaussian (random phase) signal. It is interesting to note that the Kolmogorov cascade in turbulence implies a non-random $\phi(k)$ which in turn means that D_H in turbulence is probably not given by (3.5), i.e. if $\alpha = 5/3$ then $D_H \neq 2.67$ in three-dimensional turbulence (contrast with results of Procaccia *et al.* 1991). Some differences between signals with random and non-random phases are summarised in table 1.

This investigation has helped to quantify the differences between the many methods of analysing a turbulent flow into turbulence structures, and has suggested optimum structure types for elucidating the dynamics of turbulence. The results obtained will give greater power and precision to the structural approach to understanding turbulence.

NKRK and JCV have been supported by British Gas during this work.

REFERENCES

ADRIAN, R.J. & MOIN, P. 1988 Stochastic estimation of organised turbulent structure: homogeneous shear flow. *J. Fluid Mech.*, **190**, 531–559.

ARFKEN, G. 1985 *Mathematical methods for physicists.* Orlando: Academic Press.

ARNÉODO, A., ARGOUL, F., BACRY, E., ELEZGARAY, J., FREYSZ, E., GRASSEAU, G., MUZY, J.F., & POULIGNY, B. 1989 Wavelet transform of fractals. In *Wavelets and some of their applications, proceedings of the conference held in Marseille-Luminy, June 1989.*

AUBRY, N., HOLMES, P., LUMLEY, J.L. & STONE, E. 1988 The dynamics of coherent structures in the wall region of a turbulent boundary layer. *J. Fluid Mech.* **192**, 115–173.

BRASSEUR, J.G. & WANG, Q. 1992 Structural evolution of intermittency and anisotropy at different scales analysed using three-dimensional wavelet transforms. *Phys. Fluids* A. **4**(11), 2538–54.

CHEN, J.H., CHONG, M.S., SORIA, J., SONDERGAARD, R., PERRY, A.E., ROGERS, M., MOSER, R., & CANTWELL, B.J. 1990 A study of the topology of dissipating motions in direct numerical simulations of time-developing compressible and incompressible mixing layers. In *Studying turbulence using numerical simulation databases - III, Proceedings of the 1990 summer program.* Stanford: CTR.

FALCONER, K. 1990 *Fractal geometry—mathematical foundations and applications.* John Wiley & Sons.

FARGE, M. 1992 The wavelet transform. In *Ann. Rev. Fluid. Mech.* **23**.

FRISCH, U., SULEM, P.-L. & NELKIN, M. 1978 A simple dynamical model of intermittent fully developed turbulence. *J. Fluid Mech.* **87**(4), 719–736.

GROSSMAN, A. & MORLET, J. 1984 Decomposition of Hardy functions into square integrable wavelets of constant shape. *SIAM J. Math. Anal.* **15**, 723–736.

HUNT, J.C.R., KEVLAHAN, N.K.-R., VASSILICOS, J.C. & FARGE, M. 1993 Wavelets, fractals and Fourier transforms: detection and analysis of structure. To appear in: *Wavelets, fractals and Fourier transforms: new developments and new applications, proceedings of the IMA conference at Newnham College, Cambridge, December 16–18, 1990.*

HUNT, J.C.R. & J.C. VASSILICOS. 1991 Kolmogorov's contributions to the physical and geometrical understanding of small-scale turbulence and recent developments. *Proc. R. Soc. Lond. A.*, **434**, 183–210.

HUSSAIN, A.K.M.F. 1986 Coherent structures and turbulence. *J. Fluid Mech.*, **173**, 303–356.

LUMLEY, J.L. 1967 The structure of inhomogeneous turbulent flows. In *Proceedings of the international colloquium on the fine structure of the atmosphere and its influence on radio wave propagation, Moscow, June 15-22, 1965*, eds. A.K. Yaglom & U.I. Tatarsky. Nauka: Moscow, 166–176.

LUNDGREN, T.S. 1982 Strained spiral vortex model for turbulent fine structure. *Phys. Fluids.* **25**(12), 2193–2203.

MOFFATT, H.K. 1984 Simple topological aspects of turbulent vorticity dynamics. In *Turbulence and chaotic phenomena in fluids* ed. T. Tatsumi. Elsevier, 223.

MUMFORD, J.C. 1982 The structure of the large eddies in fully developed turbulent shear flows. Part 1. The plane jet. *J. Fluid Mech.*, **118**, 241–268.

OREY, S. 1970 Gaussian sample functions and the Hausdorff dimension of a level crossing. *Z. Wahrscheinlichkeitstheorie verw. Geb.* **15**, 249.

OSBORNE, A.R. & PROVENZALE, A. 1989 Finite correlation dimension for stochastic systems with power-law spectra. *Physica D*, 357–381.

PROCACCIA, I., BRANDENBURG, A., JENSEN, M. & VINCENT, A. 1991 The fractal dimension of iso-vorticity structures in 3-dimensional turbulence. NORDITA preprint (submitted to *Europhys. Lett.*).

VASSILICOS, J.C. & HUNT, J.C.R. 1991 Fractal dimensions and the spectra of interfaces with applications to turbulence. *Proc. R. Soc. Lond.* A. **435**, 505–534.

VERGASSOLA, M. , BENZI, R., BIFERALE, L. & PISARENKO, D. 1991 Wavelet analysis of a Gaussian Kolmogorov signal. *Proceedings of the USA-French workshop on 'Wavelets and turbulence''.*

VINCENT, A. & MENEGUZZI, M. 1991 The spatial structure and statistical properties of homogeneous turbulence. *J. Fluid Mech.* **225**, 1–20.

ZEL'DOVICH, YA. B. & SOKOLOV, D.D. 1985 Fractals, similarity, intermediate asymptotics. *Sov. Phys. Usp.* **28**(7), 608–616.

AN EXAMINATION OF THE AXISYMMETRIC JET MIXING LAYER USING COHERENT STRUCTURE DETECTION TECHNIQUES

L. S. Ukeiley, D. R. Cole and M. N. Glauser
Department of Mechanical and Aeronautical Engineering
Clarkson University
Potsdam, NY 13699

ABSTRACT

The Proper Orthogonal Decomposition (POD) [12] and Linear Stochastic Estimation (LSE) [1] are used to identify structure in the axisymmetric jet mixing layer. Cole et al [4] found that the POD produces a better global representation of the structures, whereas the LSE yields better results on a local scale (relative to the integral length scale of the flow). In this paper we will briefly discuss the application of each method, then focus on a novel technique which employs the strengths of each. This complementary technique is composed of three main steps. First, the eigenfunctions and eigenvalues are obtained from direct application of the POD to the two-point spectral tensor (see Glauser et al [7], [8] and [10]). Second, the LSE is applied to the cross-correlation tensor and multipoint estimates of the random vector field are computed as described by Cole et al [3]. Third, the eigenfunctions obtained from step one are projected onto the estimated velocity field obtained from step two to obtain estimated random coefficients. The estimated random coefficients are then used in conjunction with the POD eigenfunctions to reconstruct the random velocity field. A qualitative comparison between the first POD mode representation of the estimated random velocity field and that obtained utilizing the full field indicates that the two are remarkably similar. In order to quantitatively assess the technique, the root mean square (RMS) energies are computed and compared for both cases. The RMS kinetic energy captured using the first POD mode of the estimated field is very close to that obtained from the first POD mode of the unestimated original field. These results show that the complementary technique, which combines LSE and POD, allows one to obtain time dependent information from the POD while greatly reducing the amount of instantaneous data required. That is, it is not necessary to have the instantaneous data at all points in space simultaneously, but only at a few select positions.

J. P. Bonnet and M. N. Glauser (eds.), Eddy Structure Identification in Free Turbulent Shear Flows, 325–336.

INTRODUCTION

Proper Orthogonal Decomposition

The POD was first proposed by Lumley [12] in 1967, as a mathematically unbiased technique for extracting structures from turbulent flows. Lumley proposed that the coherent structure be that structure which has the largest mean square projection on the velocity field. This maximization leads to an integral eigenvalue problem which has as its kernel the cross-correlation tensor, $R_{ij}(\vec{x}, \vec{x}', t, t')$. This technique has become much more popular with the development of more powerful computers and advanced data acquisition equipment. For example, Moin and Moser [14] applied the POD to the full correlation tensor which was generated numerically from simulations of turbulent channel flow and Delville et al [5] applied it to an experimental data base obtained in a mixing layer. For a recent comprehensive review of the POD see Berkooz et al 1993 [2].

In this study, only time and the radial direction are examined. We recognize that coherent structures are 3 dimensional in nature and that it can be quite misleading to extract information from 2 dimensional slices. However, in this paper the thrust is not to extract physics, but to demonstrate the usefulness of the complementary technique. The POD reduces to the harmonic orthogonal decomposition in time since the jet flow is stationary, hence the eigenfunctions used in time are Fourier Modes. Only the radial direction (r), which is strongly inhomogeneous, is examined with the POD. A more complete analysis of this data, including streamwise and azimuthal dependence, has been reported by Glauser et al[10].

Under these conditions the spectral tensor may be defined by the following equation,

$$S_{ij}(r, r', f, \overline{\theta}, \overline{z}) = \int R_{ij}(r, r', \tau, \overline{\theta}, \overline{z}) e^{-i(2\pi f \tau)} d\tau, \tag{1}$$

where f denotes frequency, τ is the separation in time and $\overline{\theta}$ and $\overline{z}$ represent the azimuthal and streamwise location where the correlation tensor was measured. In this formulation, S_{ij} becomes the kernel in the integral eigenvalue problem which is written as:

$$\int S_{ij}(r, r', f, \overline{\theta}, \overline{z}) \psi_j^{(n)}(r', f, \overline{\theta}, \overline{z}) r' dr' = \lambda^{(n)}(f) \psi_i^{(n)}(r, f, \overline{\theta}, \overline{z}). \tag{2}$$

The ψ's and $\lambda^{(n)}(f)$ are the frequency dependent eigenfunctions and eigenspectra respectively. The Fourier Transform of the velocity can be reconstructed in terms of the ψ's as;

$$\hat{u}_i(r, f) = \sum_{n=1}^{\infty} a_n(f) \psi_i^{(n)}(r, f), \tag{3}$$

where

$$a_n(f) = \int \hat{u}_i(r, f) \psi_i^{(n)}(r, f) r dr. \tag{4}$$

$S_{ij}(r, r', f, \overline{\theta}, \overline{z})$ is then obtained from experimental measurements and used in equation 2 to obtain the eigenvalues and eigenfunctions. These eigenfunctions and eigenvalues are then utilized to reconstruct the original random velocity field and the RMS kinetic energy. The numerical approximation consists of replacing the integral in equation 2 by an appropriate quadrature rule (in this study a trapezoidal rule) as detailed by Glauser et al[7].

Stochastic Estimation

In 1975, Adrian [1] proposed that stochastic estimation could be applied to unconditional correlation data. This method uses what Adrian calls a "conditional eddy". This eddy is a *candidate* structure used to detect, within certain limits, other structures of similar type. Stochastic estimation uses the conditional information specified about the flow at one or more locations in conjunction with its statistical properties to estimate the information at the remaining locations.

Adrian [1] studied conditional flow structures in isotropic turbulence by computing estimates of the velocity $u(x+r,t)$ given that the velocity at (x,t) assumes some specified value $u(x,t)$. He found that this simple flow, when sampled in a statistical sense, shows the existence of organized structures. He used a second order stochastic estimation technique, but concluded that first order (linear) stochastic estimation (LSE) would have resulted in nearly identical estimates. This indicates that the second order term contributed little to the overall estimate. Tung and Adrian [16] studied the influence of the third and fourth order terms on the estimate as well as the second order term. Their results confirmed the insignificance of the higher order terms on the overall estimate. Moin, Adrian and Kim [13] applied stochastic estimation, in order to approximate conditional vector fields, to a numerically simulated channel flow. One of their most interesting results, but only discussed briefly, was that they found good agreement between the LSE and Lumley's characteristic eddy. This was further examined by Moser [15] but comparisons were difficult because of the ambiguity in domain selection for the application of the POD (i.e., different subdomains in the boundary layer).

In general, the conditional average is defined as

$$\overline{g(u)|E} = expected \;\; value \;\; of \;\; g(u) \tag{5}$$

provided that the event E, the detector of coherent structure, occurs. The properties of these coherent structures are not known beforehand, therefore it is difficult to select reliable unambiguous and unbiased detector events. Adrian [1] suggested choosing

$$g(u) = u(x') \tag{6}$$

and

$$E = e \leq u(x) < e + de. \tag{7}$$

This confines the velocity vector to a small interval between e and $e+de$ where e is any arbitrary vector. This can be expressed as

$$\tilde{u}(x') = \overline{u(x')|u(x)} \tag{8}$$

which can be approximated, using a Taylor series expansion about $u(x)=0$, as

$$\tilde{u}_i(x') = A_{ij}(x')u_j(x) + B_{ijk}(x')u_j(x)u_k(x) + C_{ijkl}(x')u_j(x)u_k(x)u_l(x) + \cdots \tag{9}$$

Values for the estimation coefficients, $A_{ij}(x')$, $B_{ijk}(x')$, $C_{ijkl}(x')$ are selected such that the mean square error is minimized, ie.,

$$e_i = \overline{[\tilde{u}_i(x') - \overline{u_i(x')|u(x)}]^2} \tag{10}$$

for $i = 1, 2, 3$. This minimization requires that

$$\frac{\partial e_i}{\partial A_{ij}(x')} = \frac{\partial e_i}{\partial B_{ijk}(x')} = \frac{\partial e_i}{\partial C_{ijkl}(x')} = 0 \tag{11}$$

which results in the following equation

$$\begin{aligned} &\overline{u_j(x)u_k(x)}A_{ik}(x') + \overline{u_j(x)u_k(x)u_l(x)}B_{ikl}(x') \\ &+\overline{u_j(x)u_k(x)u_l(x)u_m(x)}C_{iklm}(x') = \overline{u_j(x)u_i(x')}. \end{aligned} \tag{12}$$

Tung and Adrian [16] have shown that linear stochastic estimation produces reasonable qualitative estimates and little is to be gained by using second order or higher. By applying linear stochastic estimation only the first term on the right hand side of equation 9 is retained. As a result, equation 9 becomes

$$\tilde{u}_i(x') = A_{ij}(x')u_j(x) \tag{13}$$

This also reduces equation 12 to:

$$\overline{u_j(x)u_k(x)}A_{ik}(x') = \overline{u_j(x)u_i(x')} \tag{14}$$

where $\overline{u_j(x)u_k(x)}$ is the Reynolds stress tensor and $\overline{u_j(x)u_i(x')}$ is the two-point correlation tensor.

Linear stochastic estimation is a useful tool for the identification of coherent structures in the axisymmetric jet mixing layer as has been demonstrated by Cole et al [3]. For the u,v jet data ($u = u_1$ and $v = u_2$), the matrices that result from the expansion of equation 14 for a two probe estimate are:

First System:

$$\begin{bmatrix} \overline{u_{ref_1}^2} & \overline{u_{ref_1}v_{ref_1}} & \overline{u_{ref_1}u_{ref_2}} & \overline{u_{ref_1}v_{ref_2}} \\ \overline{v_{ref_1}u_{ref_1}} & \overline{v_{ref_1}^2} & \overline{v_{ref_1}u_{ref_2}} & \overline{v_{ref_1}v_{ref_2}} \\ \overline{u_{ref_2}u_{ref_1}} & \overline{u_{ref_2}v_{ref_1}} & \overline{u_{ref_2}^2} & \overline{u_{ref_2}v_{ref_2}} \\ \overline{v_{ref_2}u_{ref_1}} & \overline{v_{ref_2}v_{ref_1}} & \overline{v_{ref_2}u_{ref_2}} & \overline{v_{ref_2}^2} \end{bmatrix} \begin{bmatrix} A_{11p}^{ref_1} \\ A_{12p}^{ref_1} \\ A_{11p}^{ref_2} \\ A_{12p}^{ref_2} \end{bmatrix} = \begin{bmatrix} \overline{u_{ref_1}u_p} \\ \overline{u_{ref_1}v_p} \\ \overline{u_{ref_2}u_p} \\ \overline{u_{ref_2}v_p} \end{bmatrix} \tag{15}$$

Second System:

$$\begin{bmatrix} \overline{u^2_{ref_1}} & \overline{u_{ref_1}v_{ref_1}} & \overline{u_{ref_1}u_{ref_2}} & \overline{u_{ref_1}v_{ref_2}} \\ \overline{v_{ref_1}u_{ref_1}} & \overline{v^2_{ref_1}} & \overline{v_{ref_1}u_{ref_2}} & \overline{v_{ref_1}v_{ref_2}} \\ \overline{u_{ref_2}u_{ref_1}} & \overline{u_{ref_2}v_{ref_1}} & \overline{u^2_{ref_2}} & \overline{u_{ref_2}v_{ref_2}} \\ \overline{v_{ref_2}u_{ref_1}} & \overline{v_{ref_2}v_{ref_1}} & \overline{v_{ref_2}u_{ref_2}} & \overline{v^2_{ref_2}} \end{bmatrix} \begin{bmatrix} A^{ref_1}_{21p} \\ A^{ref_1}_{22p} \\ A^{ref_2}_{21p} \\ A^{ref_2}_{22p} \end{bmatrix} = \begin{bmatrix} \overline{v_{ref_1}u_p} \\ \overline{v_{ref_1}v_p} \\ \overline{v_{ref_2}u_p} \\ \overline{v_{ref_2}v_p} \end{bmatrix} \tag{16}$$

where ref_1 and ref_2 refer to reference probes 1 and 2 respectively, and p refers to the probe number. It should be noted, that for these systems of equations, only the two-point space-time correlation data is utilized. These systems **are not** a function of the condition being investigated. The estimated velocity components for the two probe reference case can then be found from the expansion of equation 13 ie.,

$$\tilde{u}_p = A^{ref_1}_{11p} uc_{ref_1} + A^{ref_1}_{12p} vc_{ref_1} + A^{ref_2}_{11p} uc_{ref_2} + A^{ref_2}_{12p} vc_{ref_2} \tag{17}$$

and

$$\tilde{v}_p = A^{ref_1}_{21p} uc_{ref_1} + A^{ref_1}_{22p} vc_{ref_1} + A^{ref_2}_{21p} uc_{ref_2} + A^{ref_2}_{22p} vc_{ref_2}. \tag{18}$$

It is in these estimated velocity equations that the condition selected plays a role (i.e., through uc_{ref_1}, uc_{ref_2}, vc_{ref_1} and vc_{ref_2}). A single probe estimate is obtained by merely setting all terms containing $ref_2 = 0$. Without much trouble this system can easily be expanded to include estimates for all of the probes. This should result in the estimated velocities being exactly the same as the actual velocities. This property can then be used as a check.

Complementary Technique

Mathematically the stochastic estimates of the random coefficients are calculated from:

$$a^{est}_n(f) = \int \hat{u}^{est}_i(r,f)\psi^{(n)*}_i(r,f)rdr \tag{19}$$

where $\hat{u}^{est}_i(r,f)$ is either a single or multipoint (in this study, 2 point) linear stochastic estimate of the velocity field (from the time Fourier Transform of equations 17 and 18) and $\psi^{(n)*}_i(r,f)$ is obtained from the original POD eigenvalue problem. Note the similarity to equation 4, here however the actual velocity field is replaced by that estimated from the 2 point linear stochastic estimate. The estimated streamwise or radial velocity can then be reproduced in Fourier space by

$$\hat{u}^{est}_i(r,f) = \sum_{n=1}^{\infty} a^{est}_n(f)\psi^{(n)}_i(r,f) \tag{20}$$

and then inverse transformed to obtain $u^{est}_i(r,t)$. Comparisons are then made between the reconstructed estimated velocity field as described by equation 20 and those obtained using the full instantaneous velocity data as given by equation 3.

EXPERIMENT

The experiment conducted to obtain the data in the axisymmetric jet was first reported by Glauser and George [7] and Glauser et al [8]. The jet had an exit diameter (D) of 0.098 m with a centerline exit velocity of 20 m/s. The Reynolds number based on exit diameter was 110,000 with a 0.35% turbulence intensity in the core region. The data was collected by two rakes each containing 4 "X" wire probes. The rakes were placed at z/D=3. In the original experiment the rakes were traversed through 25 azimuthal locations, however in this study only the azimuthal position of $\theta = 0$ is examined. The hot wires were spaced by approximately 10.9 mm, making the total distance spanned 76.2 mm, which is twice the vorticity thickness. The sensing wires used were 5 μm in diameter and had a sensing length of 1.2 mm. In order to achieve the ensemble averages necessary to calculate the correlation tensor 300 blocks of 1024 samples were collected. The data was low pass filtered at 800 Hz while being sampled at 2000 Hz. The data acquisition system was based around a 15 bit, 16 channel A/D converter with simultaneous sample and hold capability.

RESULTS

The results from the application of the two techniques (independently) have shown that they both provide a rational basis for detection of the large scale events in this flow field. This can be seen quantitatively in tables 1a and 1b by the rapid convergence of the energy as well as the turbulent shear stress. The values given are the percentage of the total of the particular quantity.

	$\overline{u^2}$	$\overline{v^2}$	$\overline{uv}$
1st Mode	35.3	43.8	68.8
2nd Mode	57.1	56.5	89.3
3rd Mode	68.7	69.8	95.5

Table 1a: Proper Orthogonal Decomposition, percent of total.

Probe location(s) (r/D)	$\overline{u^2}$	$\overline{v^2}$	$\overline{uv}$
0.35	27.4	23.5	35.5
0.46	30.4	35.0	49.6
0.57	28.2	45.2	43.6
0.35, 0.57	53.6	62.0	70.0
0.35, 0.68	42.3	54.7	60.2
0.35, 0.46, 0.57	70.3	75.6	85.9

Table 1b: Stochastic Estimation, percent of total.

Vector plots are used here to qualitatively examine the flow. Figure 1 presents the original velocity field as seen in a frame of reference moving at 12 m/s over a time duration equivalent to 0.073 seconds. Vector plots of the contribution from the first and summation of the first three POD modes, obtained from direct application of the POD are presented in Figure 2. A one POD mode reconstruction exhibits most of the large scale features observed in the original velocity field although it does not have the same spatial extent. The spatial information is filled in when more modes are included, as seen in Figure 2b (consistent with the findings of previous studies e.g. Ukeiley et al [17] and Delville [6]).

Figures 3 and 4 show two-point estimates obtained from application of the LSE. In Figure 3 the instantaneous data located at r/D =0.35 and 0.57 has been used as the condition (case 1), while in Figure 4 the instantaneous data located at r/D =0.35 and 0.68 has been used (case 2). In both cases the instantaneous data across the mixing layer at the remaining points has been estimated. Figures 3 and 4 show that these estimates are quite similar and compare quite well with the original field shown in Figure 1. From these figures it can be seen that the results from application of the two techniques are remarkably similar. While neither technique has a good representation of the outer regions, it appears that the POD produces a better representation globally, whereas the LSE appears to produce (on the order of the integral length scale) a better representation locally.

The results of the application of the complementary technique (Application of equations 19 and 20 using the estimated data presented in Figures 3 and 4) are presented in Figures 5 and 6. Remember that only a small percentage of the original instantaneous data has been used. The results from the first POD mode are remarkably similar to the first POD mode representation of the full field as presented in Figure 2 (i.e., computed using the **full** measured instantaneous velocity field presented in Figure 1). This suggests that this type of an approach can be used to verify or check low dimensional dynamical systems models for the POD coefficients (for the first POD mode) which have been developed for this flowfield by Glauser et al [9]. It should also be noted that more of the detail is recovered away from the center of the jet mixing layer from the first POD mode reconstruction of the stochastically estimated field than the original estimated field contains itself. This can be seen by comparing Figures 5a and 6a to Figures 3 and 4 respectively. Evidently, the first POD eigenfunction, used in the projection, contains a significant amount of knowledge of the velocity field (in the averaged mean square sense) and hence provides the additional information. As was observed earlier, more spatial information is filled in when additional POD modes are included as can be seen in Figures 5b and 6b. Note however that the 3 POD mode representations of the stochastically estimated fields differ slightly more from the 3 POD mode representation of the original field when compared to the 1 POD representations (i.e., the results differ more as additional POD modes are used). This is not unexpected since the sum of all the terms in equation 20 will result in the estimated field being recovered and not the actual field.

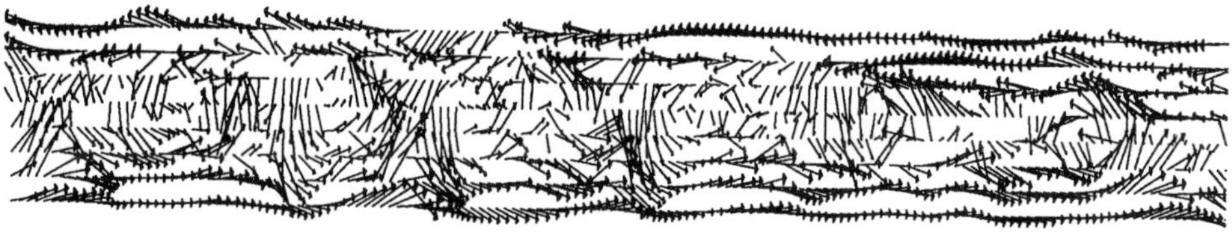

Figure 1: Instantaneous velocity vector plots of the jet seen in a frame of reference moving at $U_c = 12$ m/s.

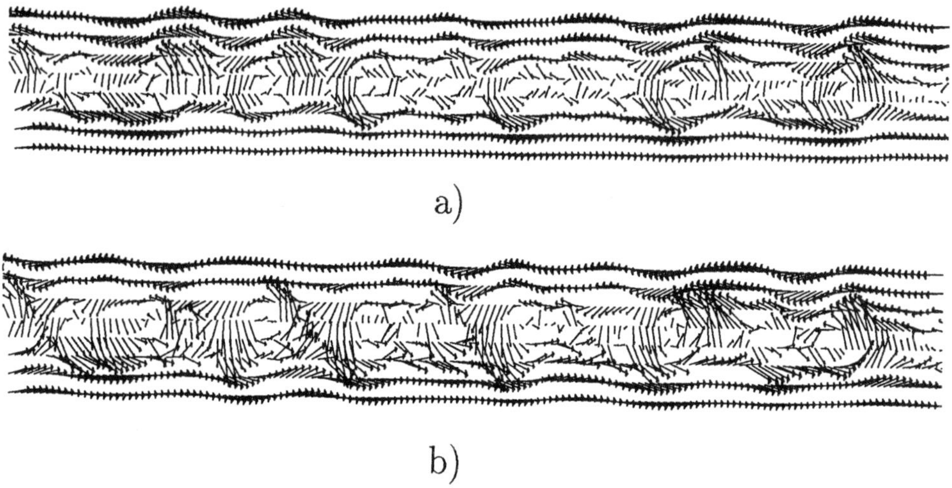

Figure 2: a) Reconstruction using 1 POD mode. b)Reconstruction using summation of first three POD modes

These results are summed up quantitatively in Figure 7, which shows RMS energy comparisons between the original (direct application of the POD) and the complementary approach using the stochastic estimate of case 1 (case 2 followed the same trends). The dotted line depicts the two-point stochastic estimates. The total energy is depicted by the solid line. Clearly the complementary technique, indicated by the dashed line, captures most of the RMS energy in a single POD mode as the direct application of the POD, the long dashed line, which is computed using the *original* flow field. Although not shown here, the comparisons differ more as additional POD modes are used, consistent with the results from the instantaneous data.

CONCLUSIONS

Both techniques have demonstrated the qualitative ability to reconstruct the large scale events in the instantaneous flow field. The techniques have been combined, in a novel fashion utilizing the global nature of the POD and the local nature of the LSE.

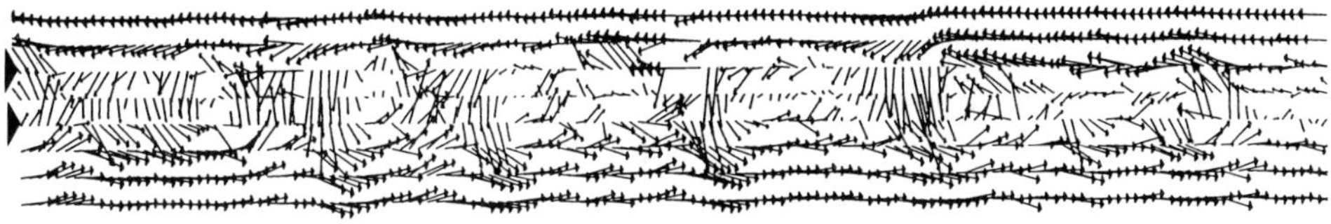

Figure 3: Stochastic Estimated field of the jet using the probes indicated by the arrows as reference (case 1).

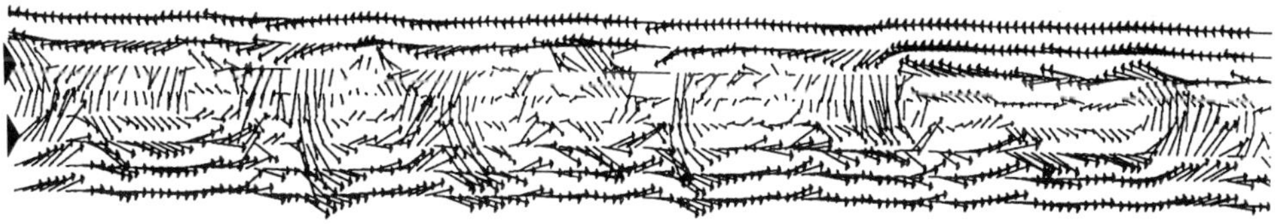

Figure 4: Stochastic Estimated field of the jet using the probes indicated by the arrows as reference (case 2).

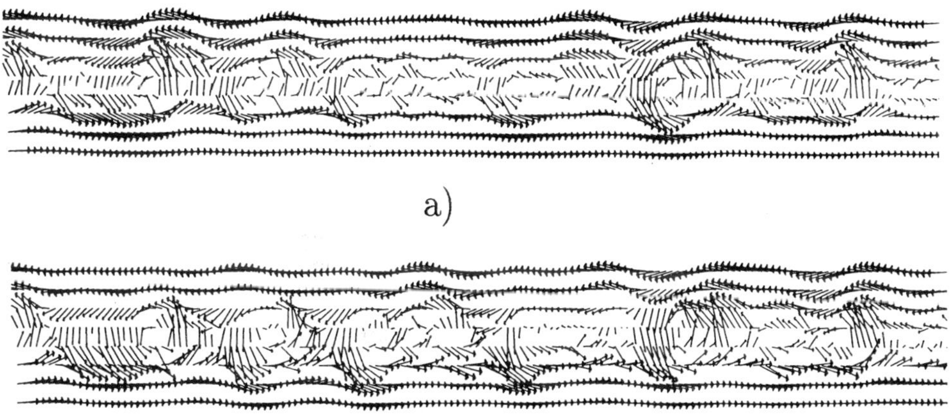

a)

b)

Figure 5: COMPLEMENTARY TECHNIQUE: (case 1) a) Reconstruction using 1 POD mode of the stochastically estimated field. b) Reconstruction using summation of first three POD modes of the stochastically estimated field.

a)

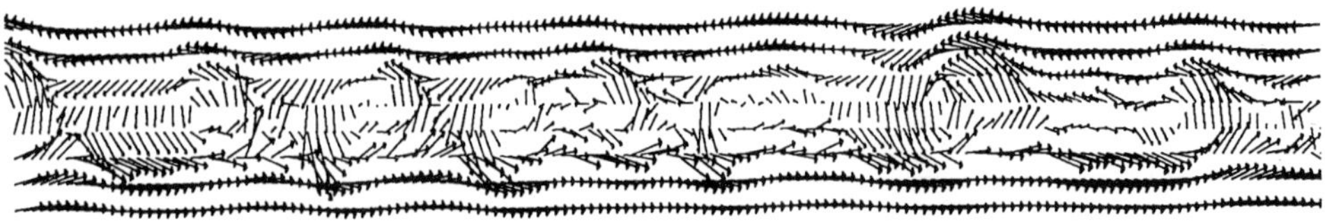

b)

Figure 6: COMPLEMENTARY TECHNIQUE: (case 2) a) Reconstruction using 1 POD mode of the stochastically estimated field. b) Reconstruction using summation of first three POD modes of the stochastically estimated field.

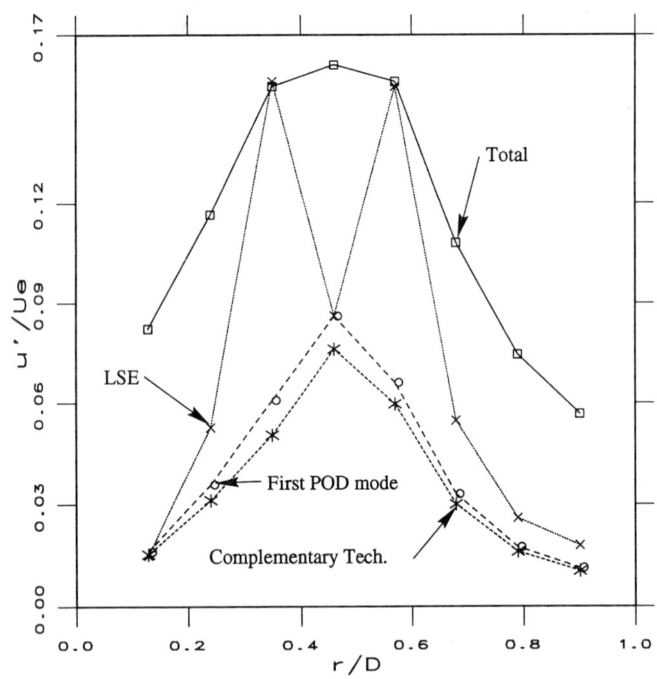

Figure 7: RMS comparisons for (case 1)

The POD eigenfunctions (not estimated) are projected onto a two-point LSE estimated instantaneous field, and a 1 POD mode representation was computed. These were then compared to a 1 POD mode representation of the original instantaneous field. We see for a two probe estimate the comparison of the first POD mode representations is remarkable. Most importantly, this technique retains the phase information of the POD modes from the instantaneous signal obtained at a "few" select positions. (i.e., Knowledge of the instantaneous field at all positions is not required.) The use of root mean square plots quantifies how well the complementary technique works. We see that the root mean square energy captured using a 1 POD mode of the estimated field is very close to that which is obtained from a 1 POD mode of the original field.

One can postulate the possibility of extending the LSE over the azimuthal direction (similar to the application of the LSE by Guezennec and Gieseke [11] in the wake of a cylinder) to obtain the instantaneous flow field for all radial and azimuthal positions. It would then be possible to project the eigenfunctions of Glauser et al [10] onto this estimated field and obtain the time evolution of the so called random coefficients of the POD. These then can be used to verify or check dynamical systems model for the coefficients which have been developed for this flowfield by Glauser et al [9]. In summary the complementary technique allows one to obtain the time dependent information from the POD while greatly reducing the amount of instantaneous data necessary.

References

[1] Adrian, R.J. (1975) "On the Role of Conditional Averages in Turbulence Theory". In *Proc. 4th Biennal Symp. on Turbulence in Liquids*, Princeton, N.J.

[2] Berkooz, G., Holmes, P., and Lumley, J.L., (1993), "The Proper Orthogonal Decomposition in the Analysis of Turbulent Flows". *Annu. Rev. Fluid Mech.*, **25**: 539.

[3] Cole, D.R., Glauser, M.N. and Guezennec, Y.G., (1992), "An Application of Stochastic Estimation to the Jet Mixing Layer". *Physics of Fluids A.*,**4**(1): 192.

[4] Cole, D.R., Ukeiley, L.S. and Glauser, M.N., (1991) "A Comparison of Coherent Structure Detection Techniques in the Axisymmetric Jet Mixing Layer", Bulletin of the American Physical Society, **36**, 10: 2699

[5] Delville, J., Bellin, S. and Bonnet, J.P. (1989) "Use of the Proper Orthogonal Decomposition a Plane Turbulent Mixing Layer". in *Turbulence and Coherent Structures* (eds. O. Metais and M. Lesieur), Kluwer Academic Publishers: 75.

[6] Delville, J. (1992) "Use of Stochastic Quantities for the Characterization of Structures in Shear Layers,"*Proceedings of the IUTAM Symposium on Eddy Structure Identification in Free Turbulent Shear Flows*, Poitiers France.

[7] Glauser, M.N., Leib, S.J. and George, W.K. (1987) "Coherent Structures in the Axisymmetric Jet Mixing Layer". *Turbulent Shear Flows 5*, (eds. F. Durst et al), Springer-Verlag: 134.

[8] Glauser, M.N. and George, W.K. (1987) "An Orthogonal Decomposition of the Axisymmetric Jet Mixing Layer Utilizing Cross-Wire Measurements". *Proceeding Sixth Symposium on Turbulent Shear Flows*, Toulouse, France.

[9] Glauser, M., Zheng, Z., and Doering, C.,(1992), "A Low Dimensional Description of the Axisymmetric Jet Mixing Layer Utilizing the Proper Orthogonal Decomposition", *Submitted to Physics of Fluids A.*

[10] Glauser, M.N., Zheng, X and George, W.K. (1992) "An Analysis of the Turbulent Axisymmetric Jet Mixing Layer". *Submitted to J. Fluid Mech.*

[11] Guezennec, Y. G. and Gieseke, T. J. (1992) "Stochastic Estimation of Multipoint Conditional Averages and their Spatio-Temporal Evolution," *Proceedings of the IUTAM Symposium on Eddy Structure Identification in Free Turbulent Shear Flows*, Poitiers France.

[12] Lumley, J.L. (1967) "The Structure of Inhomogeneous Turbulent Flows". *Atm. Turb. and Radio Wave Prop.*. (eds. Yaglom and Tatarsky) Nauka, Moscow: 166.

[13] Moin, P., Adrian, R.J. and Kim, J., (1987) "Stochastic estimation of organized structures in turbulent channel flow," *Proceedings, Sixth Symposium on Turbulent Shear Flows*, Toulouse, France.

[14] Moin, P. and Moser, R.D., (1989) "Characteristic-eddy decomposition of turbulence in a channel," *J. Fluid Mech.* **200**: 471.

[15] Moser, R.D., (1988) "Statistical analysis of near-wall structures in turbulent channel flow," *Proceedings, Symposium on Near Wall Turbulence*, Dubrovnik, Yugoslavia.

[16] Tung, T.C. and Adrian, R.J. (1980) "Higher-Order Estimates of Conditional Eddies in Isotropic Turbulence", *Phys. Fluids* **23**: 1469.

[17] Ukeiley, L., Glauser, M. and Wick, D. (1993) "Coherent Structure Downstream Evolution in a Lobed Mixer," To appear *AIAA Journal.*

IX. Topological Concepts

TOPOLOGY OF FLOW PATTERNS IN VORTEX MOTIONS AND TURBULENCE

A.E. Perry and M.S. Chong
Mechanical and Manufacturing Engineering Department
University of Melbourne
Parkville, Victoria 3052, Australia

1 SUMMARY

Some applications of critical point theory are shown for the description and identification of eddying motions in turbulence and in vortex shedding. This includes both large scale and fine scale motions. Difficulties in the interpretations of flow topology are outlined with some examples.

2 INTRODUCTION

In turbulence research the term *eddying motions* has long been used without any precise definition in mind and means of identifying and classifying these motions have been arbitrary and ad hoc. It is only in recent times that rigorous mathematical techniques and concepts adopted from topology have been applied to flow fields and eddying motions. In the past we have been forced to examine fields described only with smoke or dye or be content with single point measurements from a small number of probes. However with the aid of massive computers aided data acquisition systems applied to experiments with scanning probes and rakes and new optical techniques and with full direct numerical simulations (DNS) of the Navier-Stokes equations, entire flow fields for a series of time frames can be generated. In the case of simulations one has the choice of many quantities for describing the field and its eddying motions such as velocity, vorticity, pressure, pressure gradient and the various components of the velocity-gradient tensor. With such an explosion of information one asks, *What do we do with all these numbers? What quantities do we plot? What will give us the best insight into the physical processes occurring?* The answer to some of these questions will be addressed here in the form of case studies.

3 FLOW VISUALISATION AND VELOCITY VECTOR FIELDS

The most basic and primitive indicators of flow patterns are smoke and dye. If judicially introduced at places where vorticity is generated (e.g. at solid boundaries in the case of incompressible uniform density fluids) then such dye will follow the

J. P. Bonnet and M. N. Glauser (eds.), Eddy Structure Identification in Free Turbulent Shear Flows, 339–361.

vorticity since vorticity like dye or heat is *locked* in the fluid. This of course is true only if we can neglect viscous diffusion and this will be true in the initial roll-up of vortices and in the early stages of development of a jet, wake or boundary layer. Dye then is useful in indicating the position and geometry of vortex sheets but there are inumerable counter examples where dye is introduced incorrectly, giving erroneous interpretation of the flow field.

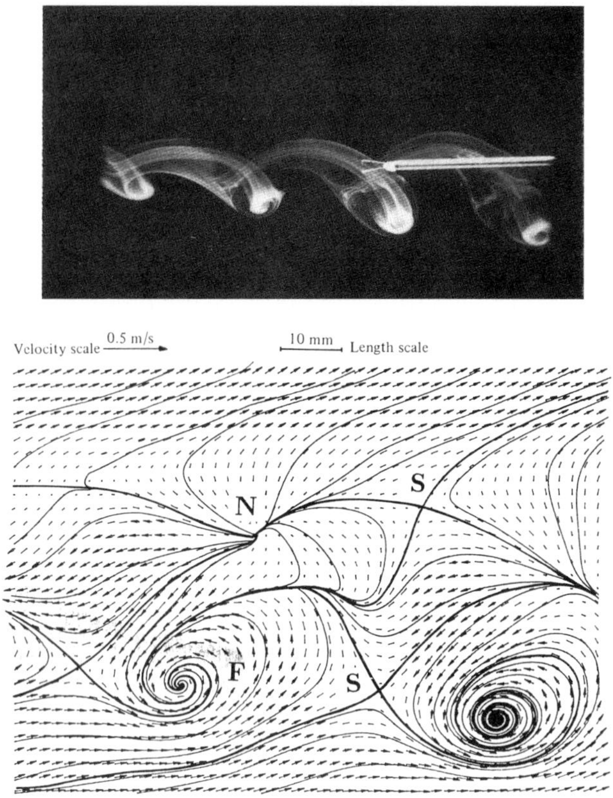

(b)

Figure 1. (a) Externally illuminated single-sided wake pattern passing hot wire probe. (b) Typical instantaneous (phase-averaged) velocity vector field for smoke pattern above. F =focus, S =saddle, N =node. Measurements made with flying hot wires. After Perry & Chong (1987).

One of the successful application of dye or smoke was in the study of coflowing jets and wakes. Perry & Lim (1978), Perry, Lim & Chong (1980) and Perry & Tan (1984) were able to relate smoke patterns with phase-sampled velocity vector fields for periodically perturbed coflowing wakes and jets. Provided an appropriate convection velocity was chosen large streamwise lengths of the instantaneous streamline patterns can be related to the smoke patterns (see figures 1(a) and 1(b)). The topology of the velocity fields and instantaneous streamline patterns can be described qualitatively in terms of critical points as labeled in the figure. These are points in the flow field where the velocity is zero and the streamline slope is indeterminate. These critical points have been identified and classified below.

4 CRITICAL POINT CLASSIFICATION USING PHASE-PLANE TECHNIQUES

The velocity field u_i about a critical point can be expanded in terms of the space coordinate x_j using a Taylor series. Asymptotically close to the critical point, we can neglect the higher order terms giving

$$\dot{x}_i = u_i = A_{ij}x_j \tag{1}$$

By integrating this equation with respect to time we obtain solution trajectories in x_i space. This correspond to instantaneous streamlines for the given A_{ij}. If the eigenvalues of the velocity gradient tensor A_{ij} are all real, then it can be shown that there will exists 3 planes (which are not necessarily orthogonal) which will contain solution trajectories (osculating planes or eigenvector planes). If the eigenvalues of A_{ij} are complex then there will be only one plane which contain solution trajectories and in that plane we have a focus. In each of these planes we can use simple phase plane analysis, i.e.

$$\dot{\mathbf{x}} = \mathbf{u} = \mathbf{F}\mathbf{x} \tag{2}$$

where $\mathbf{F}$ is a 2×2 matrix. The phase-plane portrait of the critical points can be classified by the two invariants, p and q, of $\mathbf{F}$, where p and q are given respectively by

$$p = -trace\mathbf{F} \tag{3}$$

$$q = det\mathbf{F} \tag{4}$$

The classification of all possible points can be mapped out on the so-called $p-q$ chart as shown in figure 2 where the classification of nodes, saddles and foci are shown. The eigenvectors shown may be orthogonal or nonorthogonal. Nonorthogonality can occur if there is vorticity present.

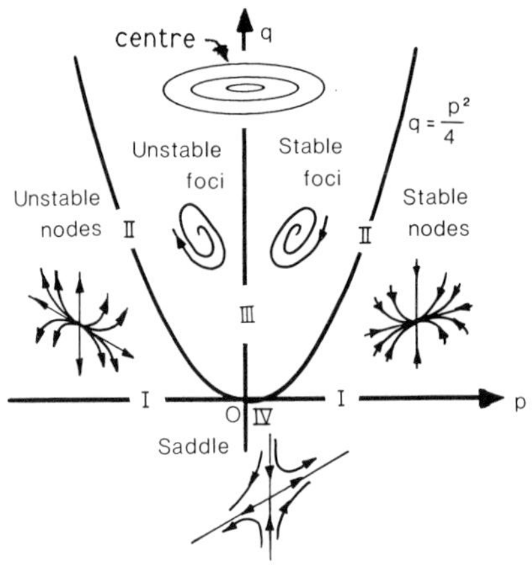

Figure 2. Classification of critical points on the $p - q$ chart. Critical points on the boundarries I, II and III and the origin IV are degenerate. In particular III corresponds to a *centre* (two-dimensional flow).

The basic patterns of nodes, foci and saddles can be related to phenomena such as vortex stretching or contraction and the stretching of vortex sheets. This has led to the construction of simplified *vortex skeletons* which describe the patterns in a simple and compact way. Just as Prandtl could describe the essential features of flow about a finite span wing with a vortex skeleton consisting of a bound vortex and two trailing vortices, so also can simple coflowing jets and wakes be described in terms of vortex skeletons. For example the vortex skeleton for the flow pattern shown in figure 1 is shown in figures 3(a) and (b). By using the Biot-Savart law, the computed flow field in the plane of symmetry for the vortex skeleton is shown in figure 3(c) and it can be seen to agree topologically (correct distribution of saddles, nodes and foci) with the experimental flow pattern given in figure 1(b). Even three-dimensional separation patterns have been described in terms of vortex skeletons where the important transport properties of the velocity field can be constructed using the Biot-Savart law as was shown by Perry & Hornung (1984).

Features of turbulent spots and the statistical features of fully developed wall turbulence can also be described in terms of representative vortex skeletons (e.g. see Perry & Chong, 1982) which serve the same purpose as the representative eddies of Townsend (1976). This has led to the idea that perhaps vorticity might be a much better quantity to study than velocity fields. For a start, the vorticity field is Galilean invariant whereas the velocity field and its associated distribution of critical points depends on the velocity of the observer. It is only when a Galilean frame of reference can be found which gives an almost steady velocity field that we have a velocity pattern with some useful physical meaning. Transport processes and other properties can then be correctly interpreted. For example, large streamwise lengths of a *Kármán* vortex street at moderate Reynolds numbers can yield velocity fields which are steady with a correctly chosen Galilean frame of reference as was shown by Perry, Lim & Chong (1982). Dye introduced at the source of the vorticity generation tends to align with the eigenvectors of the saddle points and accumulate at the centre points which are the centroids of vorticity. On the other hand, there are patterns, e.g. jets in cross-flow examined by Kelso (1991) where there is no single Galilean frame which can be chosen to see all the eddies at once. Here the eddies are accelerating and for a given frame of reference only local parts of the pattern become steady and eddies appear as foci and then only for a short time (see figure 4).

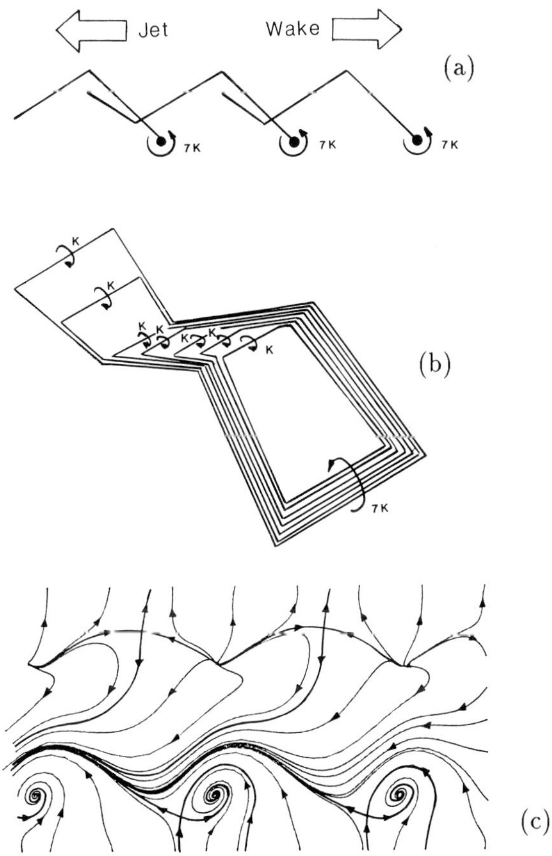

Figure 3. (a) Side view of vortex skeleton for single-sided structure. (b) Oblique view of a typical cell. K denotes a unit of circulation. (c) Computed velocity field using the Biot-Savart law (as seen by an observer moving with the eddies).

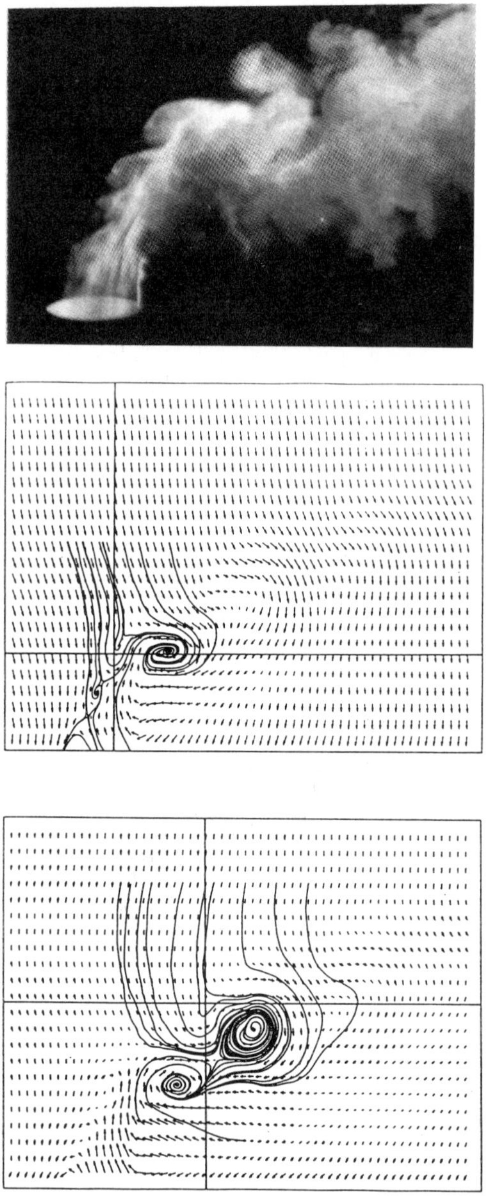

(a)

(b)

(c)

Figure 4. (a) Externally illuminated of jet in cross-flow. (b) & (c) Instantaneous phase-averaged velocity field and streamline pattern at the same phase as seen by an observer moving at two different velocities. Flying hot-wire measurements. After Kelso (1990).

In many of these flow fields taken along planes of symmetry, limit cycles occur in the vortices shed from nominally two-dimensional bluff bodies (e.g. see Steiner & Perry, 1987). A possible explanation is that this may be due to spanwise waviness in the roll-up of the vortex sheets. An example of this is the Sullivan vortex as illustrated in figure 5 (see Sullivan, 1959).

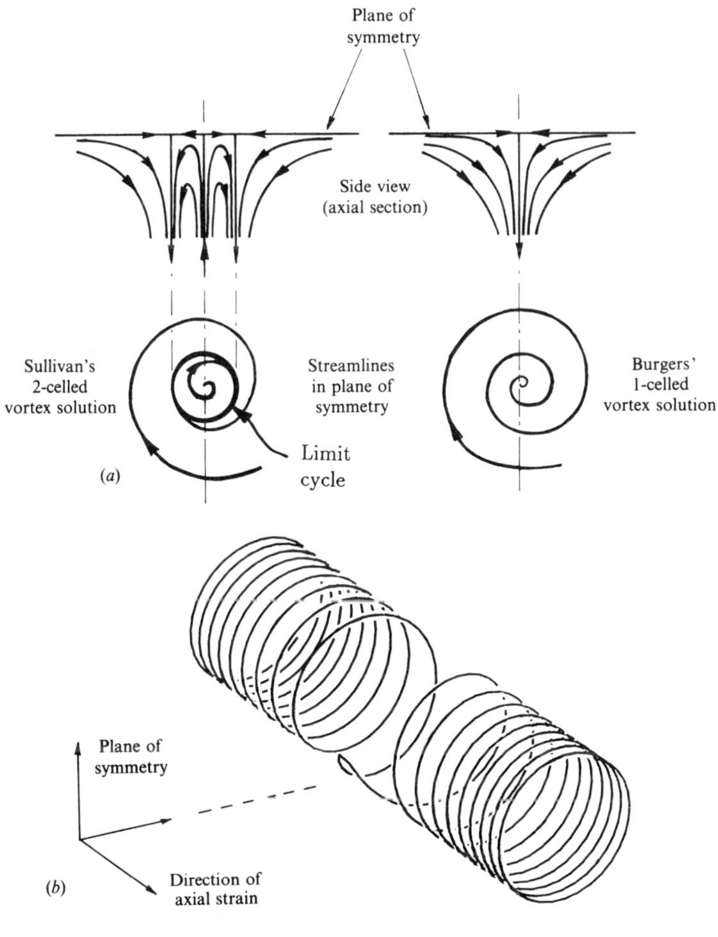

Figure 5. (a) Vortex flows for Sullivan's two-celled vortex solution and Burgers' one-celled solution. (b) Typical vortex line for the Sullivan vortex. After Perry & Steiner (1987).

Perry & Steiner (1987) calculated the integrated vorticity field and one typical vortex line is shown plotted. One can see that on one side of the plane of symmetry the vortex line is a right-handed helix and on the other side it is a left-handed helix.

Such vortex line windings have been mentioned by Professor Hussain in his keynote lecture presented in this conference. This vortex field feature is consistent with the roll-up of vortex sheets from nominally two-dimensional bluff bodies but with slight spanwise waviness in the vorticity field. It is conjectured that this waviness then becomes magnified as the vortex is formed.

Vorticity vector fields and associated plots of vortex lines have their own set of critical points which have the same basic classification as the velocity field. However, computed vorticity fields tend to be extremely complex which is somewhat dissapointing. However, since the induced velocity field a short distance away from a vortex filament is insensitive to detailed complexities, perhaps some form of spatial filtering might be permitted to aid in forming a clearer but essentially correct interpretation. This at least would be valid for the large scale motions. Such fields might give more insights if used in conjunction with the *grad p* field (i.e ∇ p), where p is pressure, as was done by Chen, Cantwell & Mansour (1989). One very appealing property of *grad p* field is that it has the features of an irrotational velocity vector field, i.e. it has zero curl and therefore all associated critical points have real eigenvalues and all eigenvectors are orthogonal. Thus the integrated field, i.e. the lines-of-force field tends to be simple and may reveal features about eddying motions in a more apparent way. Also, such a field is Galilean invariant and being a force field makes it an excellent quantity for forming physical interpretations.

Another salient feature of flow patterns besides critical points are bifurcation lines (Perry & Hornung, 1984)) or asymptotic trajectories. In the case of velocity fields these are streamlines to which other streamlines asymptote exponentially. These occur in a variety of free shear flows and in wall turbulence. Figure 6 shows some examples. If the bifurcation line is closed, then we have a limit cycle as shown in figures 6(d) and (e).

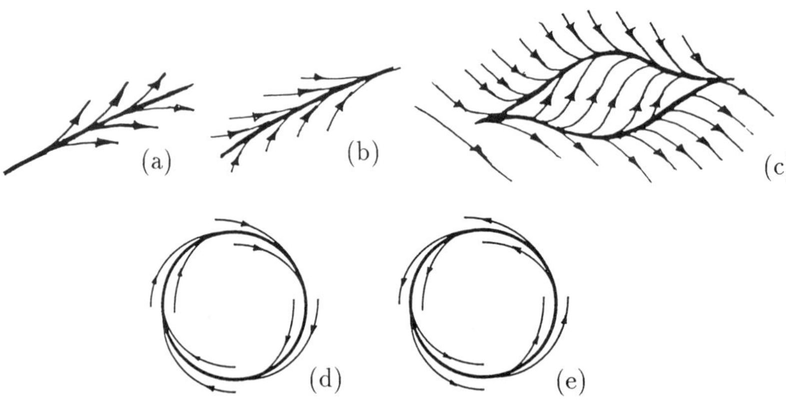

Figure 6. (a) Positive bifurcation lines. (b) Negative bifurcation lines. (c) A combination of bifurcation lines produced by a vortex close to a surface. (d) Stable limit cycle. (e) Unstable limit cycle.

5 EXAMPLES OF MISINTERPRETATIONS

In order to know fully the topology of a flow pattern about a critical point it is of great importance to have full three-dimensional information. In some work it has been tempting to simply look at *sectional-streamline* patterns in a plane through a critical point and make physical interpretations on this basis. This is fraught with danger. A sectional-streamline is defined as the integrated vector field of the vectors at the sectioning plane resolved onto the sectioning plane. Figure 7(a) shows an example of a stretched vortex described by a critical point with complex eigenvalues. There is really only one plane which contain solution trajectories (i.e. the $x_1 - x_2$ plane, $\theta = 0$). This is a plane of symmetry and the stable focus in that plane indicates correctly that we have vortex stretching. As θ is increased, the sectioning plane invariants move down the $p - q$ chart as is shown in figure 7(b) and there are cases where we have an unstable focus, i.e a focus where trajectories spiral out, which could be incorrectly interpreted as a vortex undergoing contraction. Such problems might occur in particle-image velocimetry. Thus, it is only on planes of symmetry and other eigenvector planes that sectioning streamline patterns can be safely interpreted.

Another example of misinterpretation occurs with the Burgers vortex. This is simply a vortex with a Gaussian distribution of vorticity undergoing axisymmetric stretching in the direction of the vortex axis (Burgers, 1948). A sectioning plane at the plane of symmetry would show a stable focus. If however the sectioning streamline pattern on a plane tilted at 60° to the vortex axis is plotted, an unstable focus occurs at the origin surrounded by a limit cycle as shown in figure 8 and could be misinterpreted as a Sullivan vortex discussed earlier. This again illustrates that if one is off a plane of symmetry, interpretations from sectional streamlines are most dangerous.

6 NO-SLIP CRITICAL POINTS AND APPLICATION OF ISOCLINES

A critical point is a point where the velocity is zero. More importantly it is a point where the streamline slope is indeterminate. This is particularly relevant to no-slip boundaries since all points at a no-slip boundary have zero velocity. At the no-slip boundary we have to use the concept of limiting streamlines or trajectories. These are trajectories with directions obtained by approaching the boundary from above and extrapolating to the boundary. Such trajectories correspond to skin friction lines and they are orthogonal to vortex lines at the boundary (see Lighthill, 1963). By Taylor series expanding about a critical point on the boundary, the lowest order term is quadratic in order that the no-slip condition on the boundary is satisfied, i.e.

$$\overset{\circ}{x}_i = \frac{\dot{x}_i}{x_3} = \frac{u_i}{x_3} = B_{i,j}x_j \tag{5}$$

where x_3 is the coordinate normal to the boundary (and the orthogonal coordinates x_1 and x_2 define the no-slip boundary). Hence $u_i \rightarrow 0$ as $x_3 \rightarrow 0$ but u_i/x_3 remains finite for for finite x_1 and x_2 as $x_3 \rightarrow 0$. $B_{i,j}$ is the appropriate tensor to use in place

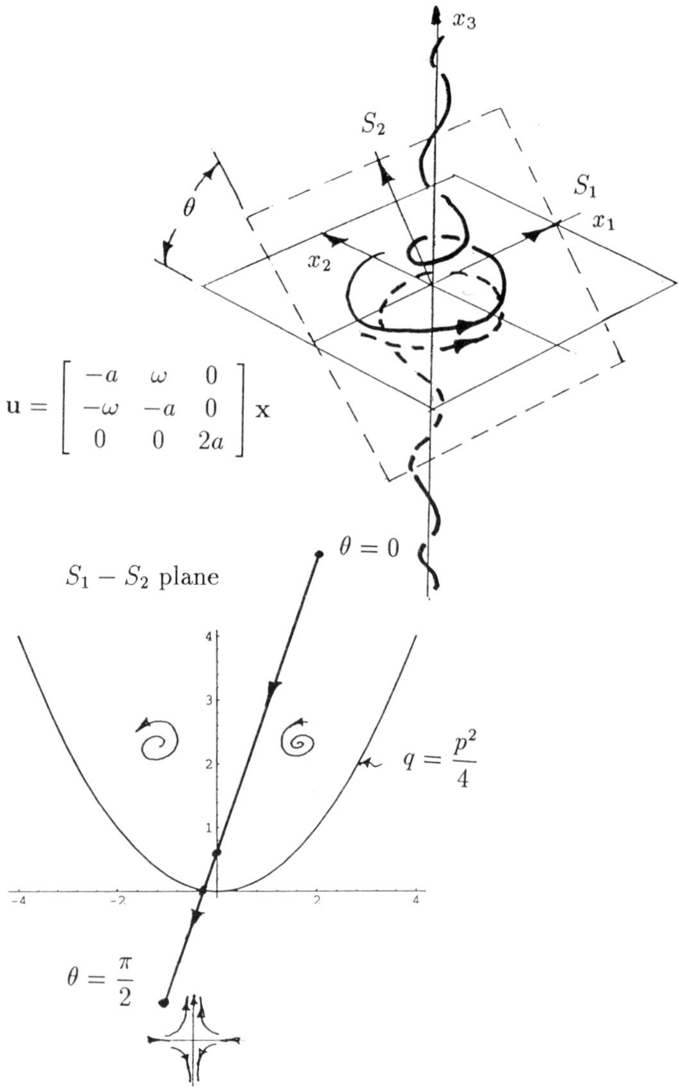

Figure 7. (a) Definition of sectioning planes. (b) Variation of sectioning plane invariants as θ increases.

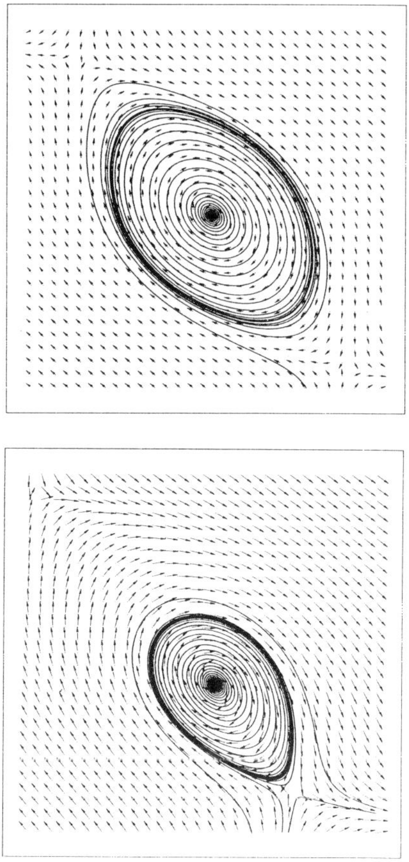

(a)

(b)

Figure 8. (a) Sectional velocity field and integrated streamline pattern around a Burgers vortex cut by a sectioning plane through the origin $x_3 = 0$, where x_3 is in the direction of the vorticity. The planes intersect at 60°. (b) The sectioning plane now intersects the vortex away from the origin, i.e. $x_3 \neq 0$. After Steiner & Perry (1987).

of A_{ij} given earlier for *free-slip* critical points and the same rules apply regarding the properties and classification of critical points. The quantity $\overset{\circ}{x}_i$ could be thought of as a derivative with respect to a transformed time given by $d\tau = x_3 dt$, where t is time. On a no-slip boundary, a critical point in velocity is also a critical point in vorticity but this is not true elsewhere.

A useful concept from phase-plane analysis of nonlinear dynamical systems is the *isocline*. This is a curve along which the trajectory slope is constant. Figure 9 shows two isoclines (with trajectory slope $M = \infty$ and trajectory slope $M = 0$)

intersecting. The intersecting point must be a critical point since the streamline slope at this point is indeterminate. In the Taylor series expansion solution of the Navier-Stokes equations given in Perry & Chong (1986), in the early stages of development some three-dimensional separation patterns were synthesized by constructing higher order polynomials which gave sets of isoclines as shown in figure 10(a). By shifting the two ellipses of fixed size and shape relative to each other, the three standard separation patterns (the simple U-separation, the owl-face of the first kind and the owl-face of the second kind) can be generated as shown respectively in figures 10(b), (c) and (d). The flow patterns which occur above the surface have also been computed and are discussed in Perry & Chong (1986).

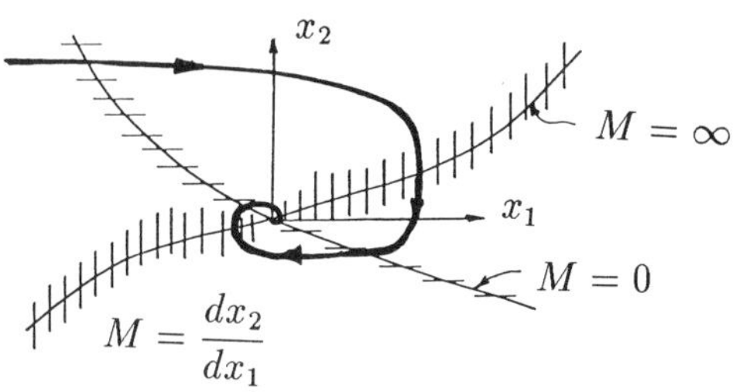

Figure 9. Critical point located by two intersecting isoclines.

The authors have often wondered what the limiting instantaneous streamlines would look like in a turbulent boundary layer. Figure 11(a) shows such a pattern computed using the DNS data of Spalart(1988) for a zero pressure gradient boundary layer. One can see a spectacular display of bifurcation lines. The question is : *do critical points occur?* At such critical points there would exist a streamline at a finite angle to the wall (all limiting streamlines are parallel to the wall except at critical points). Perry & Spalart (unpublished) attempted to locate these critical points. It is very time consuming to seek such critical points, particularly at low Reynolds numbers, but using the isocline method these were located rapidly. At low Reynolds numbers, such points were rare but as the Reynolds number increased from $R_\theta = 670$ to $R_\theta = 1400$, the number of points increased. This is completely consistent with the attached hypothesis of Townsend (1976) as further developed by Perry, Henbest & Chong (1986). With increasing Reynolds number, the turbulence intensity (based on local mean velocity) at a fixed $x_3 u_\tau / \nu$ in the logarithmic region increases and so flow reversal and critical points at the wall are inevitable. Herre u_τ is the friction velocity and ν is the kinematic viscosity. Figures 11(b) show critical points which are made clearer by a blow-up of the localized region shown in figure 11(a).

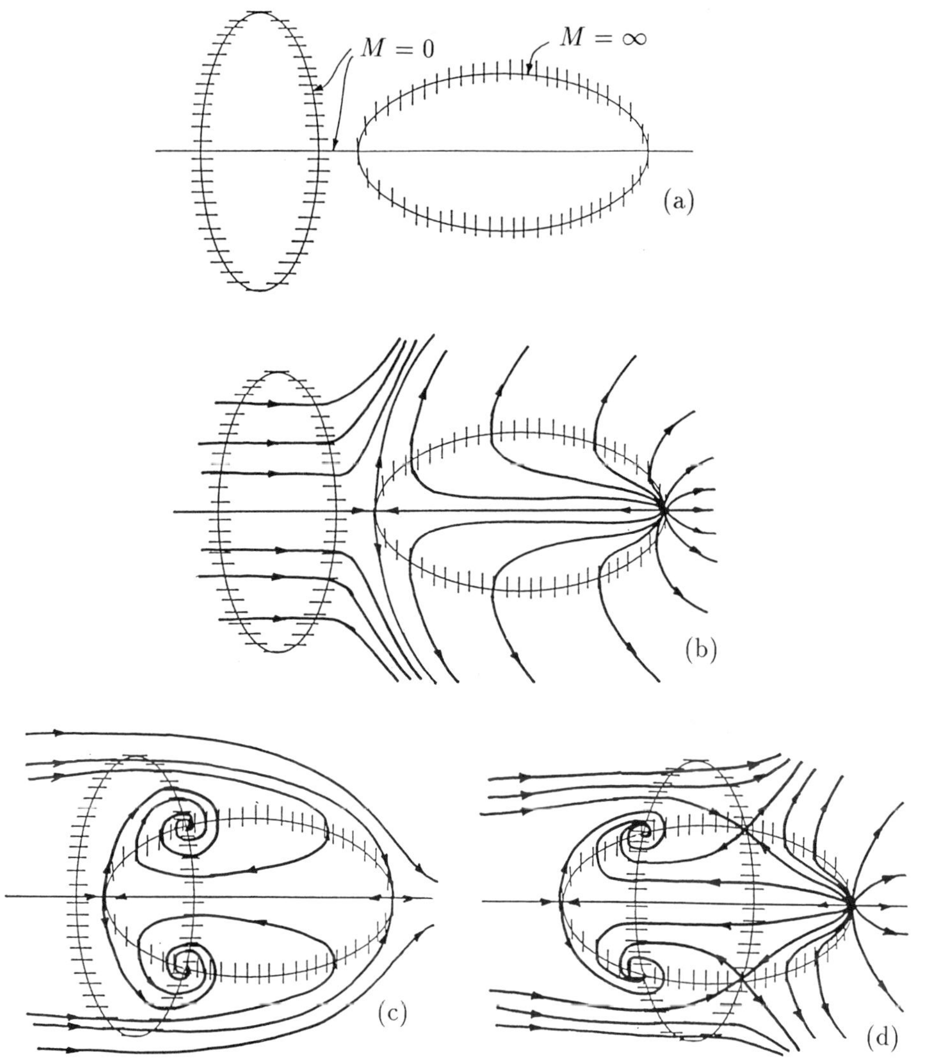

Figure 10. (a) Isoclines (ellipses and a straight horizontal line) used to generate simple U-separation, owl-face of the first kind and owl-face of the second kind shown respectively in (b), (c) and (d).

These surface pattern features are undoubtedly vortical structures which extend away from the boundary as shown in figure 12 where some sectional instantaneous streamlines have been plotted. In some recent work by Spalart & Coleman (unpublished), the surface trajectories for an adverse pressure gradient boundary layer were computed and are shown in figure 13. Here the critical points are most obvious. There is even the suggestion of owl-faces. The important point which the authors wish to make is that a turbulent boundary layer should be looked upon as a forest of three-dimensional unsteady separations.

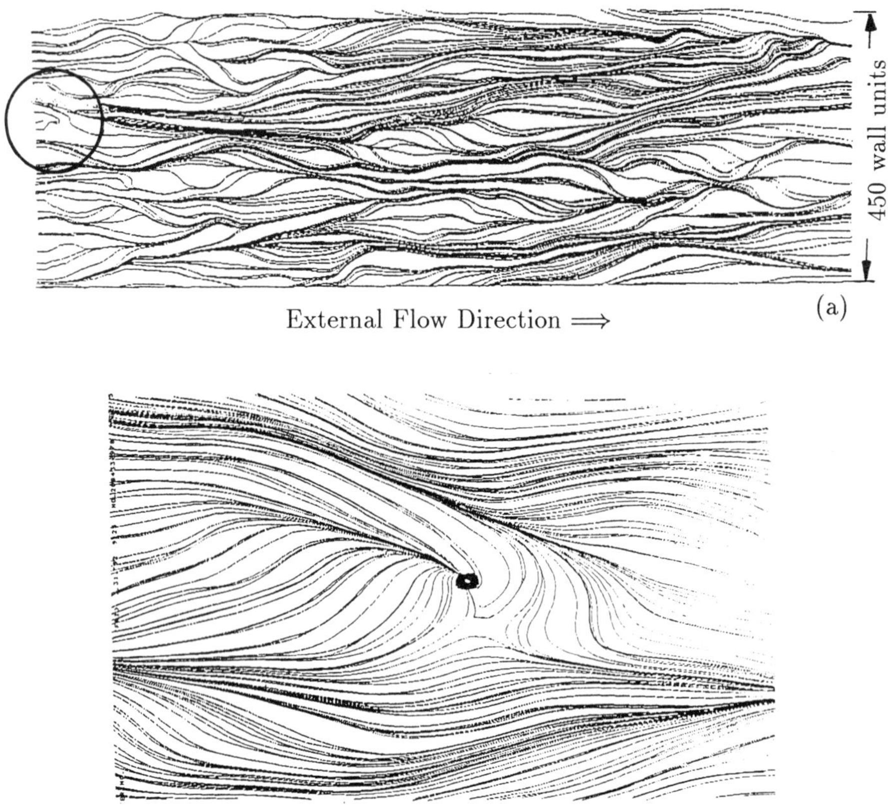

Figure 11. (a) Instantaneous streamlines in a zero pressure gradient turbulent boundary, $R_\theta = 670$, from Perry & Spalart (unpublished). (b) Blow-up of circled region shown in (a).

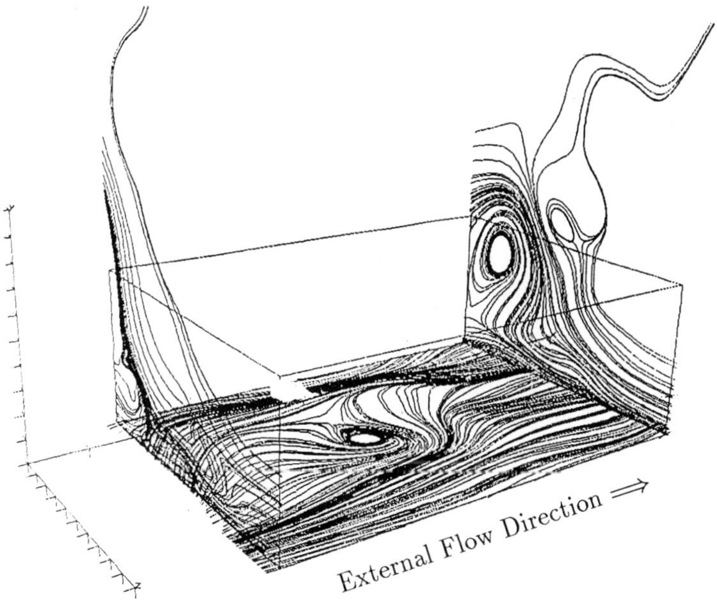

Figure 12. Surface streamlines and vertical plane sectional streamlines, from Spalart & Perry (1987) (unpublished).

Figure 13. Surface streamlines for adverse pressure gradient boundary layers, from Spalart & Coleman (private communication). From simulations given in Spalart & Watmuff (1992).

7 CLASSIFICATION BASED ON THREE INVARIANTS AND APPLICATION TO TURBULENT MOTIONS

Critical points have been classified by Perry & Fairlie (1974) using the concept of eigenvector planes and simple phase-plane analysis. However, an alternative approach has been given in a paper by Chong, Perry & Cantwell (1990). Here, three invariants of the velocity gradient tensor are used for the general classification of critical points and the various restrictions for compressible and incompressible flows away from and at no-slip boundaries are derived. This analysis shows how important it is to have full three-dimensional information about a flow field before correct topological classification can be made and this reinforces what was stated earlier in section 3.

The three-invariant topological classification can be applied to every point in the flow field. If an observer moves in a non-rotating frame with any particle in the field, then the flow surrounding the particle can be described in terms of the nine components of the velocity gradient tensor A_{ij}. The velocity gradient tensor may be broken up into a symmetric and an antisymmetric part, i.e. $A_{ij} = \partial u_i/\partial x_j = S_{ij}+W_{i,j}$ where $S_{ij} = (\partial u_i/\partial x_j + \partial u_j/\partial x_i)/2$ and $W_{ij} = (\partial u_i/\partial x_j - \partial u_j/\partial x_i)/2$ are the rate-of-strain and rate-of-rotation tensors respectively. The eigenvalues of A_{ij} satisfy the characteristic equation

$$\lambda^3 + P\lambda^2 + Q\lambda + R = 0 \tag{6}$$

where the matrix invariants are:

$$P = -(A_{11} + A_{22} + A_{33}) = -\text{trace}[A] = -S_{ii} \tag{7}$$

$$\begin{aligned} Q &= \begin{vmatrix} A_{11} & A_{12} \\ A_{21} & A_{22} \end{vmatrix} + \begin{vmatrix} A_{11} & A_{13} \\ A_{31} & A_{33} \end{vmatrix} + \begin{vmatrix} A_{22} & A_{23} \\ A_{32} & A_{33} \end{vmatrix} \\ &= \frac{1}{2}\left[P^2 - \text{trace}[A^2]\right] \\ &= \frac{1}{2}[P^2 - S_{ij}S_{ji} - W_{ij}W_{ji}] \end{aligned} \tag{8}$$

and

$$\begin{aligned} R &= -\begin{vmatrix} A_{11} & A_{12} & A_{13} \\ A_{21} & A_{22} & A_{23} \\ A_{31} & A_{32} & A_{33} \end{vmatrix} = -\det[A] \\ &= \frac{1}{3}(-P^3 + 3PQ - \text{trace}[A^3]) \\ &= \frac{1}{3}(-P^3 + 3PQ - S_{ij}S_{jk}S_{ki} - 3W_{ij}W_{jk}S_{ki}). \end{aligned} \tag{9}$$

The three invariants P, Q and R form a space which is divided into various topological classification and these are Galilean invariant. There are two surfaces S_{1a} and S_{1b}, given respectively by

$$\frac{1}{3}P(Q-\frac{2}{9}P^2)-\frac{2}{27}(-3Q+P^2)^{\frac{3}{2}}-R = 0 \qquad (10)$$

$$\frac{1}{3}P(Q-\frac{2}{9}P^2)+\frac{2}{27}(-3Q+P^2)^{\frac{3}{2}}-R = 0 \qquad (11)$$

which join along a *ridge* and divide the $P-Q-R$ space as shown in figure 14. All points *below* the surface are for real eigenvalues of A_{ij} while points *above* the surface belong to complex eigenvalues of A_{ij} (two complex eigenvalues and one real eigenvalue).

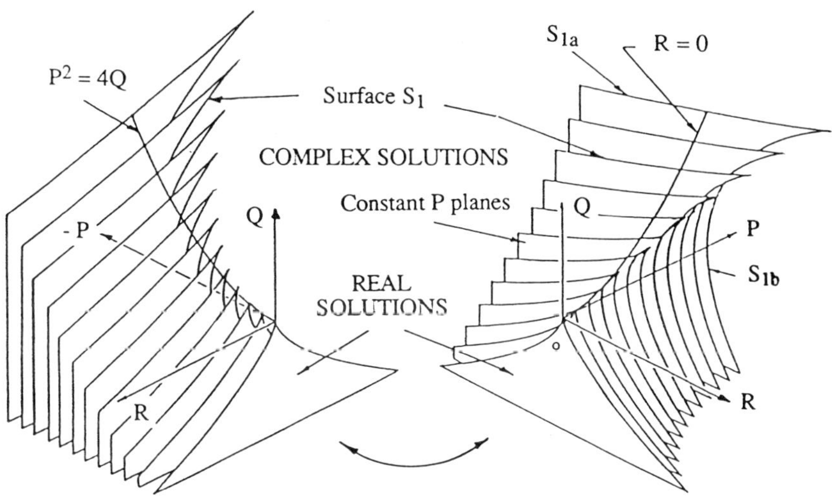

Figure 14. $P-Q-R$ space showing surface S_{1a} and S_{1b} which divides real solution from complex solutions. Diagram has been split along the $Q-R$ plane and opened up for clarity.

This diagram serves as an excellent means of defining a vortex core. A region of vorticity need not necessarily be part of a vortex. *When does such a region qualify as being part of a vortex?* It is asserted here that a particle belongs to a vortex core if its velocity gradient tensor A_{ij} has invariants such that it is above the surface shown in the $P-Q-R$ space. This is effectively the same definition used by Cantwell (1978) and later by Vollmers (1983) and Dallman (1983). Any such point will have a flow pattern with a plane containing trajectories in the form of a focus. Perhaps subconsciously this is what we have all meant by the term *eddying motions*. For some inexplicable reason to the authors, some workers oppose this definition.

For incompressible flow, $P = 0$ and so we are confined completely to the $Q - R$ plane and figure 15 from Soria et al. (1992) shows the topological classification on this plane. Below the *tent-like* curves, the eigenvalues are real and the eigenvectors are not orthogonal if the flow has vorticity. The classifications shown are (reading from top right corner counter-clockwise) unstable focus/contracting, stable focus/stretching, stable node/saddle/saddle and unstable node/saddle/saddle.

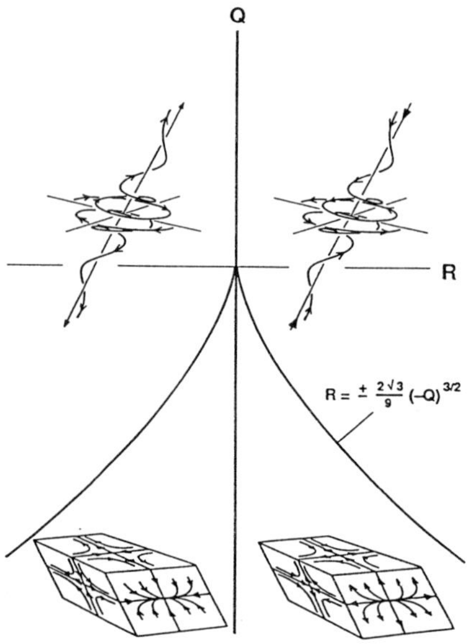

Figure 15. Three-dimensional topologies in the $Q - R(P = 0)$ plane. From Soria et al. (1992).

From these basic quantities other quantities can be derived, e.g. the invariants of the rate of strain tensor, P_s, Q_s and R_s, can be formed from S_{ij}. The mechanical dissipation for kinetic energy due to viscous friction is

$$\phi = 2\nu S_{ij}S_{ij} = -4\nu Q_s \tag{12}$$

Another important quantity is Q_w which is the second invariant of the rate of rotation tensor W_{ij}. This is non zero and is given by

$$Q_w = \frac{1}{2}W_{ij}W_{ji} = \frac{1}{2}W_{ij}W_{ij} \tag{13}$$

This is proportional to the enstrophy density.

At two recent *Center of Turbulence Research* summer programs, DNS data for plane mixing layers and other shear layers were examined. The second and third invariants for A_{ij}, S_{ij} and W_{ij} were calculated using all nine components of the velocity gradient tensor for every grid point in the flow field. Scatter diagrams of Q versus R, Q_s versus R_s and $-Q_s$ versus Q_w were produced (see Chen, et al. , 1990 and Soria, et al., 1992). Some typical results are shown in figure 16 for a plane mixing layer with Reynolds number of 3000 based on velocity difference and vorticity thickness.

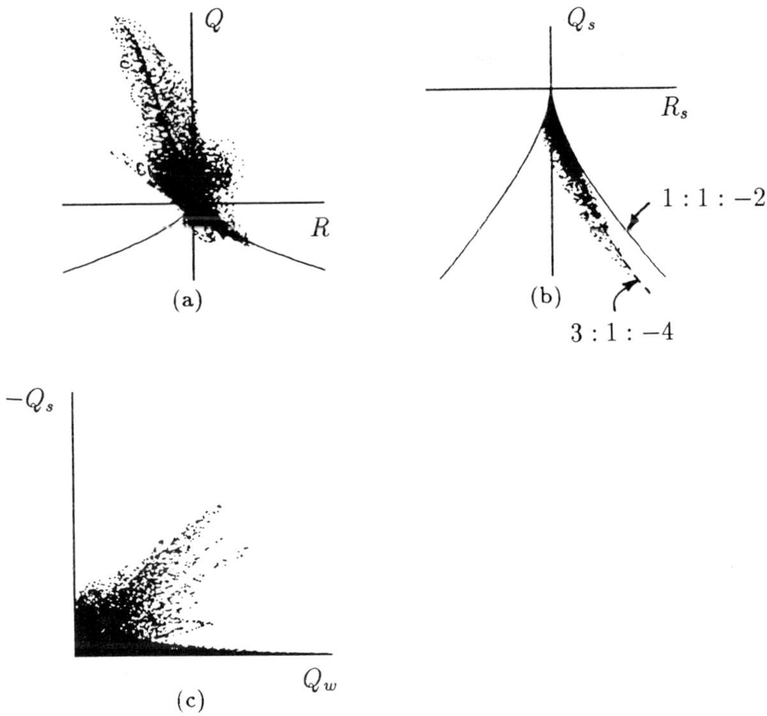

Figure 16. Scatter plots of (a) Q vs. R, (b) Q_s vs. R_s, and (c) $-Q_s$ vs. Q_w for incompressible mixing layer. From Soria et al. (1992)

Large positive values of Q indicate large enstrophy density which dominates over the rate of strain effect. This is because $-S_{ij}S_{ji} = S_{ij}Sij$ is always negative ($= Q_s$) and $W_{ij}W_{ji} = -W_{ij}W_{ij}$ and is always positive ($= Q_w$). One can see from figure 16(a) that such points are most likely to have the topology of stable focus/stretching. Motions which have large $-Q_s$ have high dissipation and tend to lie along curves on the Q_s versus R_s plot as shown. Such curves are scaled vertically from the boundary curve and denote curves of constant ratio of the principle rates of strain $\alpha : \beta : \gamma$. The boundary curve is $1 : 1 : -2$ whereas the data shown is $3 : 1 : -4$. However, from the work of Soria et al. (1992), this result was not found to be universal.

The plot of $-Q_s$ versus R_s is most interesting as shown in figure 16(c). Figure

17 shows the physical interpretation for various regions of the plot. It can be seen from figure 16(c) that a large number of points appear to belong to vortex sheets but there are very large number of points which belong to vortex tubes. Figure 18 shows a most intriguing result of Chen, et al. (1990) for a plane mixing layer. This flow is slightly compressible but indicates that most of the motion belonging to vortex sheets (i.e. a high enstrophy density also corresponds with a high dissipation).

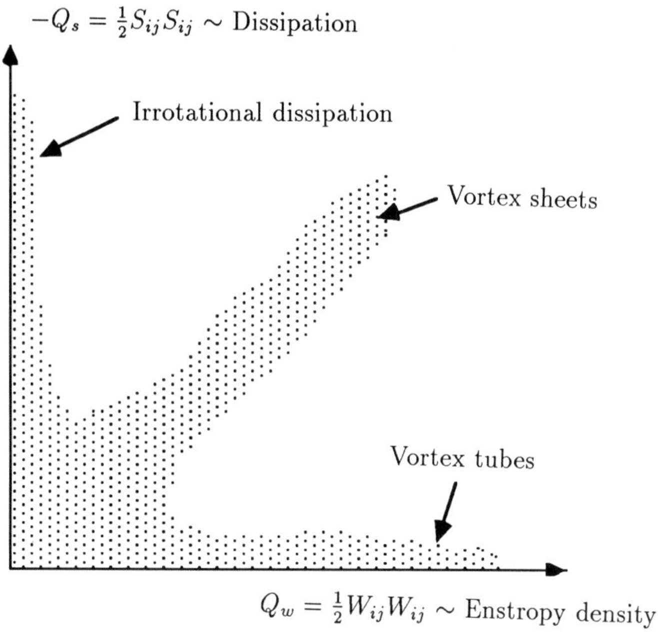

Figure 17. Physical interpretation of various regions in the $-Q_s$ vs. Q_w plot.

Figure 19 shows a plot of $-Q_s$ versus Q_w from some preliminary work on turbulent boundary layers using the DNS data of Spalart. Although the plot has poor resolution (the figure is a blow-up from another plot) it indicates that most high velocity gradient motion are sheet like. All these intriguing results should be pursued in future work.

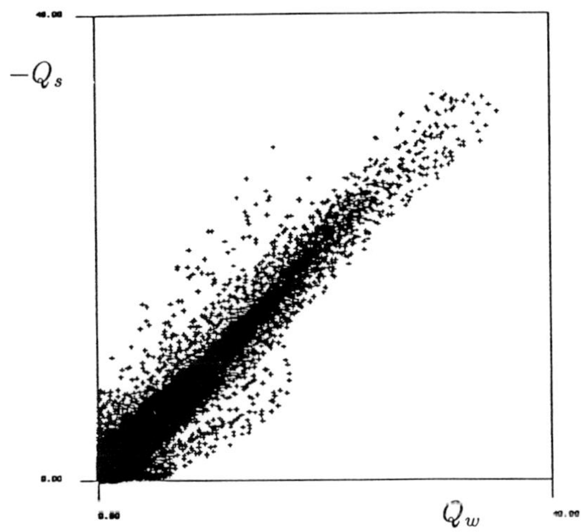

Figure 18. Plot of $-Q_s$ vs. Q_w for compressible mixing layer computed by Chen (1990).

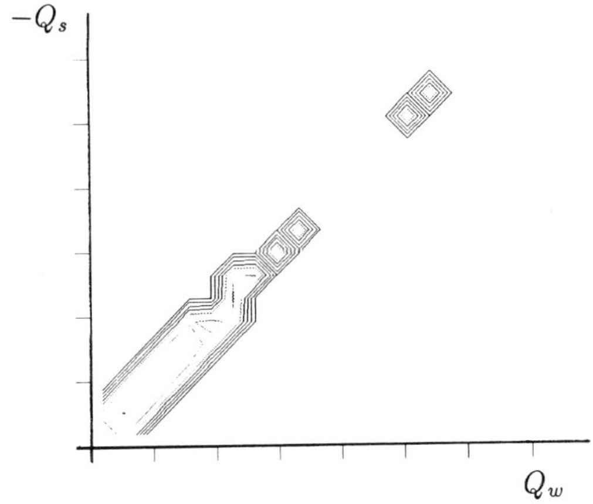

Figure 19. Plot of $-Q_s$ versus Q_w for turbulent boundary layer using DNS data of Spalart.

8 CONCLUSIONS AND DISCUSSIONS

In the study of the topology of flow patterns and eddying motions, instantaneous streamlines have found considerable use even though they are not Galilean invariant and appear to lose physical meaning if the flow is unsteady. However, cases have been identified where they are of some use in decribing the large scale motions of turbulence and the vortex shedding processes. Much of their use stems from the fact that they are the easiest to measure and are the most accurate feature to compute.

The usefulness of other vector fields needs to be explored. Now that we have DNS data almost any quantity can be mapped out and studied. Unfortunately this situation does not hold at the moment with experimental data. With DNS data, not only can the large scale topology be examined but also the fine scale dissipating motions. The concept of critical points for velocity fields can be applied to every point in the flow field by chosing a non-rotating observer fixed to a particle at the chosen point. The local instantaneous streamline patterns give a geometric meaning to the concepts of the symmetric rate of strain tensor and the anti-symmetric rate of rotation tensor. By definition, the observer does not need to be specified. Applying this to all points of a DNS data set with scatter diagrams of the tensor invariants has shown some intriguing results which need to be pursued. No matter what vector field is chosen the same critical point concepts outlined here will be needed for describing and understanding what we are looking at and this after all is what our job as researchers is all about.

REFERENCES

BURGERS, J.M. 1948. A mathematical model illustrating the theory of turbulence. Adv. in Applied Mechanics. **1**, 171-199.

CANTWELL, B.J. 1978. Cohererent turbulent structures as critical points in unsteady flow. Arch. Mech. Strosow. (archives of Mechanics), **31**, 707-721.

CHEN, J., CANTWELL, B. & MANSOUR, N. 1989 The topology and vortex dynamics of a three-dimensional plane compressible wake. Proceedings of the Tenth Australasian Fluid Mechanics Conference, University of Melbourne, Australia.

CHEN, J.H., CHONG, M.S., SORIA, J. SONDERGAARD, R., PERRY, A.E., ROGERS, M., MOSER, R. & CANTWELL, B.J. 1990. A study of the topology of dissipating motions in direct numerical simulations of time-developing compressible and incompressible mixing layers. Proceedings of Center for Turbulence Research, CTR-S90.

CHONG, M.S., PERRY, A.E. & CANTWELL, B.J. 1990. A general classification of three-dimensional flow fields. Physics of Fluids **A.2(5)**, 765-777.

DALLMANN, U. 1983. Topological structures of three-dimensional flow separations. DFVLR Rep. IB 221-82-A07. Gottingen. West Germany.

KELSO, R.M. 1990. A study of free shear flows near rigid boundaries. Ph D. thesis, Mechanical Engineering Department, University of Melbourne.

LIGHTHILL, M.J. 1963. Attachment and separation in three-dimensional flow. In *Laminar Boundary Layers*, ed. L. Rosenhead, 72-82. Oxford: Oxford University Press.

PERRY, A.E. & FAIRLIE, B.D. 1974. Critical points in flow patterns. Adv. in Geophysics, **18B**, 299-315.

PERRY, A.E. & LIM, T.T. 1978. Coherent structures in coflowing jets and wakes. J. Fluid Mech., **88**, 451-463.

PERRY, A.E., LIM, T.T. & CHONG, M.S. 1980. The instantaneous velocity fields of coherent structures in coflowing jets and wakes. J. Fluid Mech., **101**, 243-256.

PERRY, A.E. & CHONG, M.S. 1982. On the mechanism of wall turbulence. J. Fluid Mech. **119**, 173-217.

PERRY, A.E. & TAN, D.K.M. 1984. Simple three-dimensional vortex motions in coflowing jets and wakes. J. Fluid Mech., **141**, 197-231.

PERRY, A.E. & HORNUNG, H.G. 1984. Some aspects of three-dimensional separation. Part II. Vortex skeletons. Z. Flugwiss. Weltraumforsch., **8**, 155-160.

PERRY, A.E. & CHONG, M.S. 1986. A series-expansion study of the Navier-Stokes equations with applications to three-dimensional separation aptterns. J. Fluid Mech., **173**, 173-217.

PERRY, A.E. & STEINER, T.R. 1987. Large-scale vortex structures in turbulent wakes behind bluff bodies. Part 1. J. Fluid Mech., **174**, 233-270.

PERRY, A.E. & CHONG, M.S. 1987. A study of eddying motions and flow patterns using critical point concepts. Annual Review of Fluid Mech., bf 19, 125-155.

SORIA, J., CHONG, M.S., SONDERGAARD, R., PERRY, A.E. & CANTWELL, B. 1992. Topology of time developing incompressible mixing layers. Proceeding of Center of Turbulence Research, CTR-SS92.

SPALART, P.R. 1988. Direct simulation of a turbulent boundary layer up to $R_\theta = 1410$. J. Fluid Mech., **187**, 61-98.

SPALART, P.R. & WATMUFF J.H. 1992. Experimental and numerical study of turbulent boundary layers with pressure gradients. J. Fluid Mech., **249**, 337-371.

STEINER, T.R. & PERRY, A.E. 1987. Large-scale vortex structures in turbulent wakes behind bluff bodies. Part 2. J. Fluid Mech., **173**, 271-298.

SULLIVAN, R.D. 1959. A two-celled solution of the Navier-stokes equations. Journal of the Aerospace Sciences, **26**, 767-768.

TOWNSEND, A.A. 1976. The Structure of Turbulent Shear Flow. 2nd edn. Cambridge University Press.

VOLLMERS, H. 1983. Separation and vortical-type flow around a prolate spheroid. Evaluation of relevant parameters. AGARD Symposium on Aerodyn. of Vortical Type Flow in Three-Dimensions. Rotterdam. AGARD-CP 342 14.1-14.14.

INSTABILITY PATTERNS, WAKES AND TOPOLOGICAL LIMIT SETS

Robert M. Kiehn
Physics Department, University of Houston, Houston TX 77004

Abstract

Many hydrodynamic instability patterns can be put into correspondence with a subset of characteristic surfaces of tangential discontinuities. These topological limits sets to systems of hyperbolic PDE's are locally unstable, but a certain subset associated with minimal surfaces are globally stabilized, persistent and non-dissipative. Sections of these surfaces are the spiral scrolls so often observed in hydrodynamic wakes. This method of wake production does not depend explicitly upon viscosity.

Introduction

From the topological point of view it is remarkable how often flow instabilities and wakes take on one or another of two basic scroll patterns. The first scroll pattern is epitomized by the Kelvin-Helmholtz instability (Figure 1a) and the second scroll pattern is epitomized by the Raleigh-Taylor instability (Figure 1b).

Figure 1a. The pattern of the Kelvin-Helmholtz instability.

Figure 1b. The pattern of the Rayleigh-Taylor Instability.

The repeated occurrence of these two patterns (one similar to a Cornu Spiral and the other similar to a Mushroom Spiral) often deformed but still recognizable and persistent, even in dissipative media, suggests that a basic simple underlying topological principle is responsible for their creation. The mushroom pattern is of particular interest to this author, who long ago was fascinated by the long lived ionized ring that persists in

J. P. Bonnet and M. N. Glauser (eds.), Eddy Structure Identification in Free Turbulent Shear Flows, 363–378.

the mushroom cloud of an atomic explosion. Although the mushroom pattern appears in many diverse physical systems (in the Frank-Reed source of crystal growth, in the scroll patterns generated in excitable systems, in the generation of the wake behind an aircraft, ...), no simple functional description of the mushroom pattern is given in the literature known to this author. Classical geometric analysis applied to the equations of hydrodynamics has failed to give a satisfactory description of these persistent structures, so often observed in many different situations.

Viscosity vs. Compressibility

Conventional analysis of wake production and boundary layer formation subsumes that the dominant physical effect is governed by viscous creation (and destruction) of vorticity. To quote Batchelor, "The term wake is commonly applied to the whole region of non-zero vorticity on the downstream side of a body in an otherwise uniform stream of fluid." The conventional perspective presumes that the wake is approximated as a tangential discontinuity, defined as a vortex sheet. The viscous creation of the vortex sheet is not precisely defined, but once the sheet is formed, its evolution is presumed to be governed by an integral form of the Biot-Savart law, known as the Rott-Birchoff equation. These assumptions have been shown to lead to the production of discontinuities in finite time. Asymptotic spiral type solutions in the vicinity of the singularity have been investigated both analytically and numerically [Kaden, Rott, Kambe, Pullin, Cayflisch, Kransky].

The viscous process essentially is one of diffusion, and is governed by a non-hyperbolic system of PDE's. It is most certainly applicable to the decay of wake phenomena. However, this point of view, which has its domain of applicability in the infinitely far field behind a body, is often at odds with experiments of the near field, which indicate that wake features persist with sharp definition for long periods of time, over many characteristic lengths, without diffusive blurring, and at high Reynolds numbers where viscosity effects are minimal. To quote Browand [Browand, 1986] "At the present time, there is no theoretical framework to describe the structural features of high Reynolds number shear flows."

In this article an alternate approach to the creation of wakes is proposed. The basic physical mechanism for wake production is assumed to be associated with the fact that all real fluids have a finite speed of sound, hence a finite compressibility. Therefore, there can exist domains in every (perhaps slightly) compressible fluid where the system of PDE's describing the fluid evolution is hyperbolic, and not diffusively parabolic or elliptic. Hyperbolic domains have the feature that they can be associated with topological limit sets upon which the solutions to the PDE's are not unique. Therefore topological discontinuities are admissible in such dynamical systems. Such discontinuities are of two types: shock waves, where C^0 differentiability is lost for the component of velocity normal to the discontinuity surface (but the tangential components of velocity are continuous) and tangential discontinuities, where C^0 differentiability is lost for the tangential components of the velocity, but the normal component is continuous. It is this set of limit points associated with the tangential discontinuities that can be put into correspondence with the two basic instability patterns described above. Recall that propagating discontinuities are associated with singular solutions to wave equations. In fact, the very definition of a wave, according to Hadamard, is a propagating discontinuity. To recap, the point of view taken in this article is that the *creation* of a wake is to be associated with a discontinuous process in a hyperbolic domain, while the *decay* of a wake is associated with a diffusive process in an elliptic domain.

This alternate topological approach to the problem of wake creation is independent from viscosity, and gains credence from the fact that not only is a mechanism offered for the creation of the tangential discontinuity, but also closed form solutions to the vector fields that describe the aforementioned instability patterns can be obtained. An equivalent result is not known in the literature familiar to this author. Not only does the topological point of view give new insight into how wakes may be generated, but also points out how such wakes possibly may be controlled. As described below, the resultant wake creation phenomena is closely related to the problems of diffraction and interference in electromagnetism. In fact, it may be said that from the topological point of view the near field hydrodynamic wake is a diffraction pattern caused by the physical obstruction. Hence, it is plausible that phase interference mechanisms may be used to control wakes. At high Reynolds numbers the effect of viscous diffusion is to smear out or thicken the tangential discontinuity.

Spiral Space Curves

The clue that wave fronts, representing propagating tangential discontinuities, are the basis for the two basic spiral instability patterns in fluid dynamics came to this author during a study of Cartan's methods of differential topology as applied to the production of defects and topological torsion in dynamical systems [Kiehn 1990, 1991, 1992]. Cartan's ideas start from a generalization of the Frenet-Serret concept of the "moving frame". The Frenet analysis indicates that a space curve, conventionally defined by a position vector, **R**(t), relative to a specific coordinate system, is completely characterized by its arc length, s, its Frenet curvature, $\kappa(s)$, and its Frenet torsion, $\tau(s)$. [Struik 1956] All curves having the same $\kappa(s)$ and $\tau(s)$ are congruent, hence are independent from the choice of coordinates used to define the original position vector. The variables $\{s,\kappa,\tau\}$ are the intrinsic variables of a space curve.

The null set of a function $f(s,\kappa,\tau)$ defines a surface in the space of the intrinsic coordinates, and the intersection of two such surfaces defines a curve in the space of intrinsic coordinates that has a preimage in the space of initial variables, $\{x,y,z\}$. Consider "plane" curves given by the relations

$$f_1(s,\tau,\kappa) = \tau = 0 \qquad \text{and} \qquad f_2(s,\tau,\kappa) = \kappa - g(s) = 0 \,. \tag{1}(2)$$

For $g(s) = 0$, the intrinsic curve has zero curvature and zero torsion. It is a straight line in $\{x,y,z\}$ space, too. For $g(s) = +1$, the curve in $\{s,\kappa,\tau\}$ space is a straight line displaced from the origin to the right, parallel to the s axis and in the $\tau = 0$ plane. The preimage of this intrinsic curve with constant curvature and zero torsion in $\{x,y,z\}$ space is a circle.

For $g(s) = 1/s$ in the right half plane, the resulting space curve is the logarithmic spiral in the space $\{x,y,z)$. For $g(s) = s$, a straight line in the intrinsic coordinate space, $\{s,\kappa,\tau\}$, the resulting $\{x,y,z\}$ image is the Cornu spiral. These facts have been known for more than 100 years to differential geometers.

However, a simple sequence is to be recognized:

$$\ldots \kappa = g(s) = s^{-1}, \quad \kappa = s^{0}, \quad \kappa = s^{1} \ldots$$

The question arises: what are the characteristic shapes of intrinsic space curves for which the curvature is proportional to an arbitrary power of the arc length, $\kappa = g(s) = s^n$, for all positive and negative integers (or rational fractions)? Through the power of the PC these questions may be answered quickly by integrating the equations,

$$dx/ds = u = \sin(Q(s)) \quad \text{and} \quad dy/ds = v = \cos(Q(s)), \qquad (3)\ (4)$$

where Q is some "phase" function of s. The vector with components, u and v, is a unit tangent vector to a curve with arc length, s. The Frenet curvature is given by $\kappa = dQ/ds$. Hence if Q is $s^{n+1}/n+1$, then the Frenet curvature is $\kappa = s^n$. The results of the numerical integrations are presented in Figure 2 for $n = 1$ and $n = 2$. A most remarkable result is that the Cornu spiral of Figure 2a is the deformable equivalent for all odd-integer $n > 0$, and the Mushroom spiral of Figure 2b is the deformable equivalent for all even-integer $n > 0$.

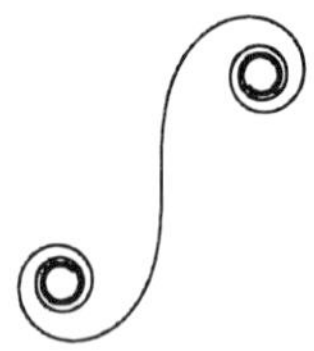

Figure 2a. The Cornu spiral is a harmonic function for which the phase function, $Q = s^2$.

Figure 2b. The Mushroom spiral is the harmonic function for which the phase function, $Q = s^3$.

Not only has the missing analytic description of the mushroom spiral been found, but also a *raison d'être* has been established for the universality of the two spiral patterns. They belong to the even and odd classes of arc length exponents describing plane curves in terms of the formula, $\kappa = s^n$. The Cornu spiral is the first odd harmonic function that maps the infinite interval into a bounded region of the plane, and the mushroom spiral is the first even function that maps the infinite interval into a bounded region of the plane. Note that the zeroth harmonic function maps the infinite interval into the bounded region of the plane, but the trajectory is unique in that it is not only bounded but it is closed; i.e., a circle. A similar sequence can be generated for the half-integer exponents with $n = +3/2$, $7/2$, $11/2$... giving the Mushroom spirals and $5/2, 9/2, 13/2$... giving the Cornu spirals. In numeric and experimental studies of certain shear flows, the Cornu spiral appears to dominate the motion in the longitudinal direction, while the Mushroom spiral appears in the transverse direction. Periodic patterns can be obtained by examining phase functions of various forms. For example, the Kelvin Helmholtz instability of Figure 1a is homeomorphic to the choice $Q(s) = 1/\cos^2(s)$. Similarly, the Rayleigh-Taylor instability of Figure 1b is homeomorphic to the case $Q(s) = \tan(s)/\cos(s)$.

Figure 1a'. The pattern of the Kelvin-Helmholtz instability. $Q(s) = 1/\cos^2(s)$.

Figure 1b'. The pattern of the Rayleigh-Taylor Instability. $Q(s) = \tan(s)/\cos(s)$.

The Kelvin-Helmoltz instability function is highly singular at points where $\cos(s) = 0$. To study the effect of this singularity, consider a modification of the phase function, such that $Q(s) = 1/(a^2 + \cos^2(s))$. The constant term a represents the radius of a finite "core" or

$a^2 = 0.167$ $a^2 = 0.067$ $a^2 = 0.050$

Figure 3. Ying Yang patterns $Q(s) = 1/(a^2 + \cos^2(s))$

hole. The solutions to (3-4) for such a phase function are extraordinary. Above a certain minimum core size the wave pattern exhibits no spiral shapes or self-intersections. At a value of $a^2 = .167$, the wave pattern describes a curve which exhibits the first double tangency, in the shape of a Ying-Yang diagram. As the core radius is decreased further, another size is reached when the double tangency again occurs, but now the pattern contains a dipole, or circulation pair. The process can be continued to generate a sequence of such double tangencies, ultimately generating the spiral patterns of Figure 1a. In the interior of each of the double tangencies is a circulation pair, about which the curve describing the pattern winds, first in one direction, then in the other direction. At each new double tangency the winding number about each doublet center increases by one. A description of this effect is given in Figure 3, where the first few double tangencies are displayed. The interpretation and application of this spin-pair production mechanism has not been developed.

Torsion vs. Curvature

The full Frenet equations in three dimensions are given by the expressions,

$$\begin{aligned} d\mathbf{t}/ds &= \quad\quad +\kappa\,\mathbf{n} \\ d\mathbf{n}/ds &= -\kappa\,\mathbf{t} \quad\quad + \tau\,\mathbf{b} \\ d\mathbf{b}/ds &= \quad\quad -\tau\,\mathbf{n}\ . \end{aligned}$$

The vectors {**t, n, b**} form an orthogonal moving frame at the point p along the space curve. The variables s, κ, τ, are the arc length, curvature and torsion, respectively.

There are two extreme situations. In the first extremal case the torsion is negligible, and the space curve is confined to the plane described by the vectorst and **n**. In classical mechanics this result is interpreted as the conservation of angular momentum. The unit binormal maintains a constant direction ($d\mathbf{b}/ds \approx 0$) along the angular momentum vector. Space curves are generated by a parametric displacement in the direction of the tangent vector. The tangent and the normal vectors form a harmonically conjugate pair if the curvature is a constant, but if the curvature is exactly zero, then the directions of the vectors **n** and **b** are indeterminate.

However, suppose that the curvature, κ, is small but finite at some initial condition, such that the normal and the binormal field can be specified. Now, however, assert that the torsion of the curve is not negligible. Then the normal and binormal form a harmonically conjugate pair. To first order, the unit tangent vector is an invariant ($d\mathbf{t}/ds \approx 0$). A curve can be generated in the (**n**, **b**) plane, parametrized by a displacement,ν, along **b**. If the torsion, $\tau(\nu)$, is one of the functions described above, then spiral shapes for the curve in the (**n**, **b**) plane are to be expected. Hence there are two extremes, the first extreme where the torsion tends to zero, and the space curves are dominated by curvature, and can be embedded in scrolled surfaces which locally approximate a cylinder with generators along the binormal. The second extreme is where the space curves are dominated by torsion, and are sections of a scrolled surface with local "cylindrical" generators along the tangent field. In both cases, the tangent plane to the surface is constructed (only approximately) in terms of the unit tangent and the binormal. In the first extreme, angular momentum is conserved, $d\mathbf{b}/ds \approx 0$. In the second extreme, $d\mathbf{t}/ds \approx 0$. As derived below, the curvature, κ, for an Eulerian flow is proportional to the component of the pressure gradient along the normal field for a single surface. In certain situations, it would appear that the free wake consists of a thin double layer.

From experimental photographs, it would appear that the Kelvin-Helmholtz instability is related to characteristic spirals along the tangent line (and is curvature dominated) and the Rayleigh-Taylor instability is related to characteristic spirals along the binormal line (and is torsion dominated). The latter case is that associated with the rollup of the wake behind and aircraft in flight, schematically described in Figure 4.

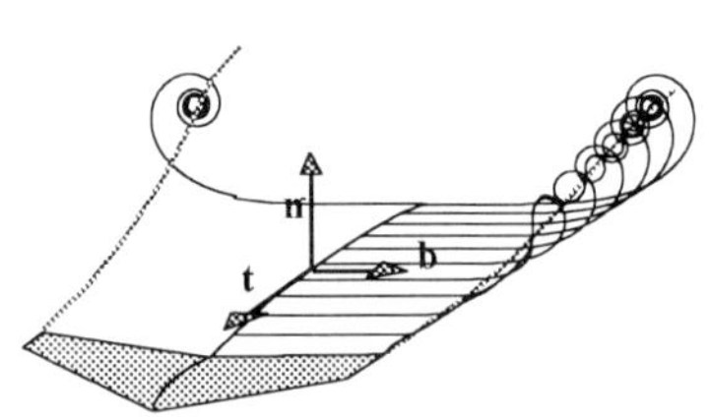

Figure 4. Torsion dominated wake. Characteristic Spirals along the binormal.

(The curvature dominated case is described in Figure 8 .)

Wakes as Characteristic Minimal Surfaces

A fundamental observation is that these spiral space curves are universally among the "characteristic" solutions to hyperbolic systems of PDE's, of which the Euler equation of two dimensional compressible flow,

$$A\phi_{\eta\eta} + 2B\phi_{\eta\xi} + C\,\phi_{\xi\xi} = D\,,$$

is a prime example. Following Landau [Landau, 1959, p.380], the characteristic surfaces contain embedded curves which are given by solutions to ordinary differential equations,

$$d\eta/d\xi = [B \pm (B^2 - AC)^{1/2}\,]/C\,.$$

Recall that characteristic surfaces represent point sets upon which the solutions to the PDE's are not unique; i.e., the characteristic surfaces include surfaces of tangential discontinuities (on which the normal components of the vector field are continuous, but the tangential components are not continuous), and shocks (on which the normal components are discontinuous, but the tangential components are continuous). Shocks permit the flow of mass across the discontinuity and permit pressure discontinuities, tangential discontinuities do not. Shocks can be dissipative and involve entropy change; tangential discontinuities are not dissipative. Shocks are generally stable, tangential discontinuities are generally unstable. Landau states that tangential discontinuities are to be ignored, for they will lead to turbulence. However, a certain subset of characteristic surfaces, which always are locally unstable, may be globally stabilized. It is these special characteristic surfaces, those associated with minimal surfaces, that are of that interest to this article.

Define a position vector $\{u,v,w\}$ to a point on a surface in terms of the characteristic coordinates, with parametrization, σ, as

$$u - d\eta/d\sigma = A(\sigma)\sin(Q(\sigma)),$$
$$v = d\xi/d\sigma = A(\sigma)\cos(Q(\sigma)),$$
$$\text{and} \qquad w = F(u,v) = f(\sigma). \qquad (5)$$

If $F(u,v)$ satisfies the equation

$$(1 + F_v{}^2)F_{uu} - 2\,F_uF_vF_{uv} + (1 + F_u{}^2)F_{vv} = 0 \qquad (6)$$

then $F(u,v)$ defines a minimal surface. There is one solution of (6) which is unique in that it is the only *harmonic* minimal surface; it is the solution $F(u,v) = \tan^{-1}(u/v) = Q(\sigma)$, or the right helicoid [Osserman, 1983]. It is this special subset of characteristic surfaces that is to be associated with the spiral solutions generated by Equations (3-4). Working backwards, assert that the surface of characteristics is a minimal surface; then to generate the ordinary differential equations (3-4), the minimal surface must be the right helicoid. More precisely, the fundamental result is that of the infinite number of surfaces of tangential discontinuities, there is a special subset whose sections yield spiral space curves that are solutions of Equations (3-4). It turns out that this special subset can be related to the unique harmonically generated minimal surface, the right helicoid.

As mentioned above, all such surfaces of tangential discontinuities are locally unstable [Landau, 1959 p.114], for they are associated with a hyperbolic domains. In fact Landau claims that these surfaces are the precursors of turbulence. However, the special subset of locally unstable surfaces of tangential discontinuities (in fact those which contain the spiral characteristic lines described above, and which can be associated with minimal surfaces) can, like soap films, exhibit domains of global stability [Barbossa, 1976]. In the domain of global stability, the locally unstable surfaces create the persistent wake patterns observed at moderate Reynolds numbers. When the Reynolds number exceeds a certain value such that the global stability of the minimal surface is lost, then these special surfaces of tangential discontinuity lose their global stability and the flow becomes turbulent. It is the thesis of this article that it is these globally stabilized surfaces of tangential discontinuities, which are not diffusive, but which are persist and observable, which are those surfaces that generate the distorted but persistent mushroom patterns in the Von Karman wake, or the Lanchester tip vortices in the wake of an aircraft in flight, or the Kelvin-Helmholtz instability pattern of a shear layer.

Such patterns are reproduced in a qualitative way by appropriate adaptations of the phase function Q(s). For example, in Figure 5a, the phase function used to generate the Kelvin-Helmoltz instability of Figure 1a has been modified to include an amplitude factor $A(s) = .1\ s^{3/2}$. The exponent of the amplitude factor is related to the ratio of the relative scales between successive spiral singularities. The physical interpretation of this mathematical observation has not been resolved, but it points to an experimental property that could be measured. If the phase function used to describe the Rayleigh-Taylor

Figure 5a $Q(s) = 0.1(s)^{3/2} / \cos^2(s)$

Figure 5b. Von Karman wake

instability of Figure 1b is modified to include a translation term proportional to cos(s), then the spiral wave pattern of Figure 5b is produced. It is apparent that this modification of the phase function is related to the staggered Von Karman wake.

Further modifications of the Von Karman wake of Figure 5b to include a decaying minimal core radius produces the complex spiral pattern of Figure 6. If the phase factor includes a linear and a quadratic term in arc length s, as well as the Kelvin-Helmholtz

Figure 6. Von Karman Wake past a cylinder.

term, then the generated pattern imitates the observed flow pattern generated around a sharp edge. See Figure 7.

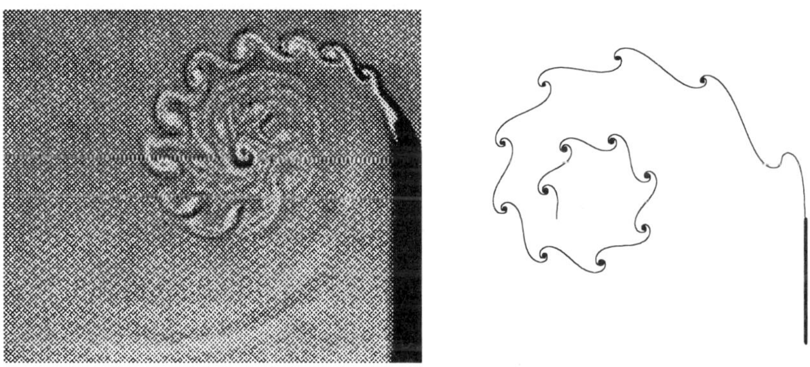

Figure 7. Wake past a sharp edge

The fact that the spiral characteristic lines are contained in the only harmonic minimal surface (generated by the function w = arctan Q(s)), and their obvious connection with the Cornu Spiral, demonstrates that the production of the Kelvin-Helmholtz instability pattern is a first order (odd) diffraction problem. It seems natural to suggest that the Rayleigh-Taylor instability pattern is a second order (even) diffraction problem. Being topological limit sets, these sets of tangential discontinuities can create extensions of material boundaries into the bulk fluid. Typically, in the region of a "sharp" edge, a material boundary will be extended by a topological limit set of tangential discontinuities in the form of a scroll pattern, independent from any viscous, diffusive effect. During the transient accelerations from rest to a state of constant velocity, the flow about sharp edges will become hyperbolic, and (in section) lines of tangential discontinuity will grow into the bulk fluid. These surfaces (in section, lines) of topological limit sets form extensions of the material boundary, and will evolve into the spiral shapes described above. As mentioned above, tangential discontinuities also can be generated at points where the pressure thermodynamically permits a change of phase, or cavitation, to take place. For example, in potential flow around a cylinder, the kinetic energy density at the top of the cylinder is four times the kinetic energy density at infinity. At points along the streamline in the neighborhood of the cylindrical surface, between the leading stagnation point and the top of the cylinder, and where the kinetic energy density exceeds one third the pressure energy density at infinity, Bernoulli's law would indicate a domain of "zero" or "negative" pressure. Physically, it would be expected that a surface of tangential discontinuity (separation) would be created when the Bernoulli pressure goes to zero.

The characteristic surface is a domain where more than one solution to a set of PDE's can exist. In the hydrodynamic case, such a surface can be used to divide a domain where the flow in the nearby neighborhood is represented by a non-harmonic potential function ($\text{div}\,\mathbf{V} = \nabla^2\phi \neq 0$, $\text{curl}\,\mathbf{V} = 0$) from a nearby neighborhood where the flow is represented by a non-harmonic stream function ($\text{div}\,\mathbf{V} = 0$, $\text{curl}\,\mathbf{V} = \nabla^2\psi \neq 0$). In the first case, the fluid is compressible, but without vorticity, and in the second case the fluid is incompressible, but with vorticity. A classical wake could be modeled as a thin layer bounded on either side by such a surface of tangential discontinuity. On the surfaces of boundary limit points, both functions are harmonic, but admit discontinuities. The interior of the thin layer would be diffusive, and would exhibit a evolutionary thickening. The tangential discontinuity would not permit mass flow between the two regions. In the classic analysis of wakes, the viscous-vorticity point of view ($\text{div}\,\mathbf{V} = 0$, $\text{curl}\,\mathbf{V} = \nabla^2\psi \neq 0$) has dominated the literature. In this article, the alternate point of view based upon compressibility ($\text{div}\,\mathbf{V} = \nabla^2\phi \neq 0$, $\text{curl}\,\mathbf{V} = 0$) is assumed to be dominant, for it gives not only a plausible physical theory for describing the production of the tangential discontinuity, but also leads the way to finding close form generators for the wake patterns created by such tangential discontinuities.

Integrability and Topological Defects

The two extremes of wake patterns mentioned above indicate that the vector field tangent to the spiral characteristics is (to first order) two dimensional. More precisely, the observation is that these characteristic lines (not necessarily the flow field) are integrable in the sense of Frobenious. Either the tangent field is integrable, or the binormal field is integrable. It is of some interest to see how these two extremes for the characteristics are related to hydrodynamic flows. To this end, a useful procedure is obtained by making a transformation to a space-time coordinate system involving the intrinsic properties of the particle paths (in this case streamlines). That is, consider the velocity field, $\mathbf{V}$, which may be composed of a unit tangent field, $\mathbf{t}$, and a normalization factor U.

$$\mathbf{V} = U\,\mathbf{t}.$$

Assume that the unit tangent field is not explicitly time dependent (such that the particle paths are along streamlines), but the normalization factor might be time dependent, U = U(x,y,z,t).

This representation for a vector field is very convenient, for if a unit tangent field can be found that satisfies boundary conditions at infinity and has zero normal component along a material boundary, U(x,y,z,t) = 0, then the zero set of the product function, $\mathbf{V} = U(x,y,z,t)\,\mathbf{t}(x,y,z)$, can be used to define no-slip boundary conditions. The unit tangent field, **t**, is permitted to have its own singularities (zeros), which are different from the zero sets of the normalization function, U. On the surface U(x,y,z,t) = 0, the velocity field vanishes identically.

The equations for the divergence and curl of the vector field become

$$\mathbf{V} = U\,\mathbf{t}, \qquad \mathbf{V}\bullet\mathbf{V}/2 = U^2/2,$$
$$\text{div}\,\mathbf{V} = \text{grad}U \bullet \mathbf{t} + U\,\text{div}\,\mathbf{t}, \qquad \text{curl}\,\mathbf{V} = \text{grad}U \times \mathbf{t} + U\,\text{curl}\,\mathbf{t},$$
$$\mathbf{V} \bullet \text{curl}\,\mathbf{V} = U^2\,\mathbf{t} \bullet \text{curl}\,\mathbf{t},$$
$$\mathbf{V} \times \text{curl}\,\mathbf{V} = U\text{grad}U - U\,(\mathbf{t} \bullet \text{grad}U)\,\mathbf{t} + U^2\,\mathbf{t} \times \text{curl}\,\mathbf{t},$$

The divergence in the nearby vicinity of the set U(x,y,z,t) = 0 is related to gradU• **t**, and the curl is equal to gradU **x t**. Note that curl **V** always resides in the surface U(x,y,z,t) = 0, (i.e., the vorticity is in the tangent plane of the surface whose normal is given by grad U). The divergence, div **V**, is not zero unless **t** resides in the surface represented by equation, U(x,y,z,t) = 0.

The non-exact differential, ds, of arc length is defined such that

$$d\mathbf{R} = \mathbf{t}\,ds = \mathbf{V}\,dt.$$

Assuming locally a euclidean metric, it follows that ds = U dt. The differential dt is exact; the differential ds is not. Closed loop integrals for dt vanish, but closed loop integrals for ds do not. From the Frenet theory,

$$d\mathbf{t}/ds = \kappa\,\mathbf{n} = (\partial\mathbf{t}/\partial t + \partial\mathbf{t}/\partial x^{\upsilon}\, dx^{\upsilon}/dt)(1/U) = (1/U)\partial\mathbf{t}/\partial t + \text{curl}\,\mathbf{t} \times \mathbf{t}. \tag{7}$$

where κ is the curvature, **n** is the normal vector, orthogonal to the unit tangent field. Curl **t** is the Darboux vector ,

$$\text{curl}\,\mathbf{t} = \tau\,\mathbf{t} + \kappa\,\mathbf{b}, \tag{8}$$

given in terms of the binormal, **b**, of the Frenet frame, and the torsion, τ, of the space curve, as well as the curvature, κ, and the unit tangent field, **t**.

The Frenet theory becomes useful in a hydrodynamic representation when it is realized the convective derivative may be expressed in terms of the Frenet derivative:

$$\begin{aligned} d\mathbf{V}/ds &= (d\mathbf{V}/dt)dt/ds \\ &= (\partial\mathbf{V}/\partial t + \partial\mathbf{V}/\partial x^\nu\, dx^\nu/dt)(1/U) \\ &= (\partial\mathbf{V}/\partial t + \mathbf{V}\bullet \text{grad}\,\mathbf{V})(1/U) \\ &= (\partial\mathbf{V}/\partial t + \text{grad}(U^2/2) - \mathbf{V}\,\mathbf{x}\,\text{curl}\,\mathbf{V})(1/U) \\ &= (dU/ds)\,\mathbf{t} + U\,\text{curl}\,\mathbf{t}\,\mathbf{x}\,\mathbf{t} + \partial\mathbf{t}/\partial t \\ &= (dU/ds)\,\mathbf{t} + U\kappa\,\mathbf{n} + \partial\mathbf{t}/\partial t\,. \end{aligned}$$

The Euler equations of hydrodynamics therefore have an intrinsic representation as

$$(dU/ds)\,\mathbf{t} + U\,d\mathbf{t}/ds = (dU/ds)\,\mathbf{t} + U\kappa\,\mathbf{n} + \partial\mathbf{t}/\partial t = -\,\text{grad}\,P/(\rho U) + \mathbf{f}/(\rho U), \qquad (9)$$

or,

$$[(\rho U)dU/ds + \mathbf{t}\bullet \text{grad}\,P]\,\mathbf{t} + [(\rho U^2\kappa + \mathbf{n}\bullet \text{grad}P]\,\mathbf{n} = \mathbf{f} - [\mathbf{b}\bullet \text{grad}P]\,\mathbf{b}. \qquad (10)$$

If the external forces were identically zero, it would follow that the square bracket factors must vanish everywhere, and the curvature or bending becomes

$$\kappa = -\,\mathbf{n}\bullet \text{grad}P\,/\,\rho U^2\,. \qquad (11)$$

Pressure gradients in the direction of the binormal cannot be sustained without a component of external force in he same direction. In other words, the Euler equations (with no external forces) constrain the flow such that the torsion is null and the curvature is given by (11)

Equation (10) is predicated upon the assumption that the velocity vector field can be represented by a single parameter group of evolution, such that $d\mathbf{R} - \mathbf{V}dt = 0$. Suppose, however that this is only true on an "equilibrium singular" surface, in the sense that off this surface, $d\mathbf{R} - \mathbf{V}dt = \mathbf{e}$, where $\mathbf{e}$ is a small vector of fluctuations. The external forces, $\mathbf{f}$, then could be attributed to fluctuations. In any case rewrite (10) as:

$$A(s)\,\mathbf{t} + B(s)\,\mathbf{n} = (A^2 + B^2)^{1/2}\,d\mathbf{P}/ds = \mathbf{f} - C(s)\,\mathbf{b}. \qquad (12)$$

The zero set of the norm of this vector defines the equilibrium singular surface. Construct the derivative of this constraint equation with respect to s to yield,

$$(dA/ds - \kappa B)\,\mathbf{t} + (dB/ds + \kappa A - \tau C)\,\mathbf{n} + (dC/ds + \tau B)\,\mathbf{b} = d\mathbf{f}/ds. \qquad (13)$$

First hypothecate the vector $\mathbf{f}$ in the direction of the binormal is vanishingly small. Then, subject to the assumptions imposed above, $C = 0$, and $dC/ds = 0$, and the torsion,τ, must vanish:

$$\tau = \mathbf{t} \bullet \text{curl } \mathbf{t} = \mathbf{V} \bullet \text{curl } \mathbf{V}/U^2 = 0. \tag{14}$$

This condition is equivalent to the Frobenius integrability condition. Note that the integrability condition is equivalent to the vanishing of the helicity density, $\mathbf{V} \bullet \text{curl } \mathbf{V} = 0$, for the flow. The flow field is of Pfaff dimension 2 or less. The coefficients A and B are *harmonic* in the sense that,

$$dA/ds - \kappa B) = 0,$$
$$(dB/ds + \kappa A) = 0 . \tag{15}$$

In fact, divide (15) by $(A^2 + B^2)^{1/2}$ to form a unit tangent vector with the same format as equations (3-4); spiral solutions in the direction of the tangent field are to be expected for the curvature dominated characteristics of such a surface. Figure 8 is a schematic of the curvature dominated case.

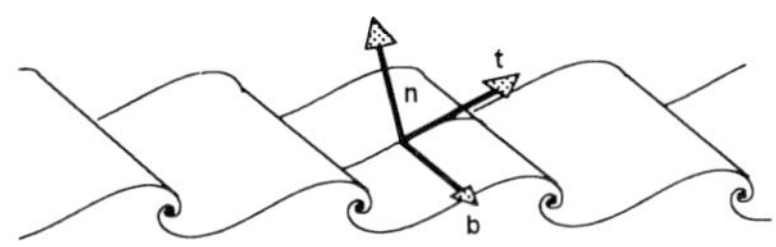

Figure 8. Curvature Dominated Wake Characteristic spirals along the tangent vector

Suppose that **f** is not identically zero, but d**f**/ds is approximated by zero. Then the "power" equation given by (13) can be examined in the limit of $\kappa \approx 0$. This would be the case of torsion dominated evolution as discussed above. For torsion dominated flows, the pressure gradient is either perpendicular to the normal, or zero. The spiral characteristic lines are in the direction of the binormal. A schematic representation of the torsion dominated case was presented in Figure 4. Note that the velocity field is NOT integrable in the Frobenius sense, for a torsion dominated wake. The flow field is of Pfaff dimension 3 or higher.

The topological defects (characteristic limit sets) described above are associated with flow fields of Pfaff dimension 2 (vector fields that satisfy the Frobenius integrability theorem) and Pfaff dimension 3. Other topological defects for flow fields of Pfaff dimension higher than 2 yield more complicated wakes. For example, in a rotating frame of reference it is possible to find closed form solutions to the Navier-Stokes equations which are of Pfaff dimension 3. Such vector fields have finite torsion, τ, and exhibit re-entrant flow patterns (contained within a swirling bubble) above a critical flow parameter. The topological defect associated with the transition to a re-entrant flow is not a dislocation hole, but can be put into correspondence with when the surface of zero torsion changes from a hyperboloid of one sheet into a hyperboloid of two sheets. The surface of zero torsion becomes *disconnected*. The analytic results, which mimic the "vortex" bursting problem, are presented in [Kiehn, 1991]. Note that in this problem the sheet of zero torsion is a point set (a surface, not flat) upon which the Frobenius integrability conditions are true.

Lagrangian Wakes and Dislocation Defects

When a harmonic potential flow or a harmonic stream function is used to describe a fluid flow, the symmetry of the Eulerian streamlines created about obstacles gives no indication of a wake. However, if it is assumed possible to synchronize globally and transversely a set of "points" in a fluid, then the kinematic evolution of such a set will give an indication of a wake. In this section these ideas will be examined. Potential flows (irrotational vector fields) are always sychronizable over the domain of uniqueness. Stream function flows (as solenoidal vector fields constructed from two independent variables at most, in 3 dimensions) are also globally synchronizable over the domain of uniqueness. However, the combination of a potential flow and a solenoidal flow in 3D is not necessarily integrable

The Frobenius integrability theorem can be used to determine if a given vector field admits globally a set of points that can be synchronized. That is, if a set of initial points lies on a smooth surface everywhere transversal to the flow field, then after an interval of evolution (time increment) the initial set of points can still be connected by a smooth surface in the final state if and only if the vector field satisfies the Frobenius condition. This fact implies that the Pfaff dimension of the vector field never exceeds 2 over the domain of interest, and that there exists a global submersion of this domain to a space of two dimensions! The smooth surface of synchronous points may be grossly distorted, and this distortion is indicative of different topological features of the flow and its wake.

For example, consider the two dimensional potential flow about a cylindrical rod. Start a set of synchronized particles at very far distances upstream from the obstacle. Assert that at the initial position the synchronized set of points resides on a surface that is orthogonal to the streamlines, much like a plane wave front in optics. Now forward integrate the kinematic equations for these points using the velocity potential for the streamline flow. As the "wave front" advances to the obstacle, the front retains its shape as a plane which is transversal and almost orthogonal to the streamlines. However as the "wave front" reaches the obstacle, the forward stagnation point greatly deforms the shape of the "synchronized" set of points that make up the "wave front". After the "wave front" of synchronous points has passed over the obstacle, a noticeable near field wake is observed downstream, a wake pattern that is not at all apparent from the format of the streamlines, which do form a pattern which is symmetric with respect to time inversion. The Eulerian description of the flow pattern is symmetric, but the Lagrangian description, with a choice of initial conditions, is not.

At very far distances to the rear of the obstacle, the "wave front" for synchronous points again looks as if were a plane surface almost perpendicular to the flow lines. Different times correspond to uniformly spaced "almost plane waves". Only very close inspection (on scales smaller that the obstacle size) reveals that in the neighborhood of the positive x-axis the set of equal time surfaces has "sharp" bend on the downstream side of the obstacle. The equal time surface does not cross the x-axis downstream from the obstacle, but it does cross the x axis on the upstream side. See Figure 9a. On measurement scales much larger than the radius of the cylindrical obstruction, the wave fronts form a lattice with only a "local" defect in the downstream side of the obstacle. This defect (on these very large scales) probably would be ignored observationally.

On the other hand, if the potential flow that generated the wake of Figure 9a is modified to include some circulation, but no vorticity, then the "wave front" pattern exhibits an apparent "global" defect as it passes over the obstacle. On scales much larger than the radius of the cylindrical obstacle, the "plane waves" that approximate the synchronous particles have a "dislocation" defect in the downstream wake. This topological defect, associated with the circulation of the flow about the obstacle, has nothing to do with vorticity! It would not be ignored observationally. See Figure 9b. An extra line, or defect, appears to be inserted into the large scale pattern. The circulation integral around the contour that is a section of the closed surface composed of two equal time faces is zero, but the contribution from the loop around the circular cylinder is balanced by the contribution along the dislocation defect.

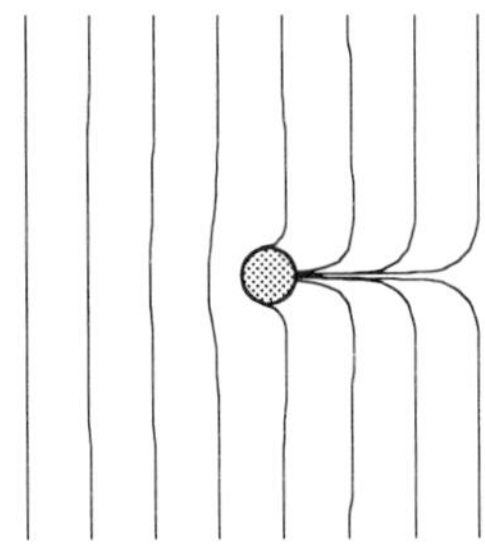

Figure 9a Lagrangian wake, no circulation

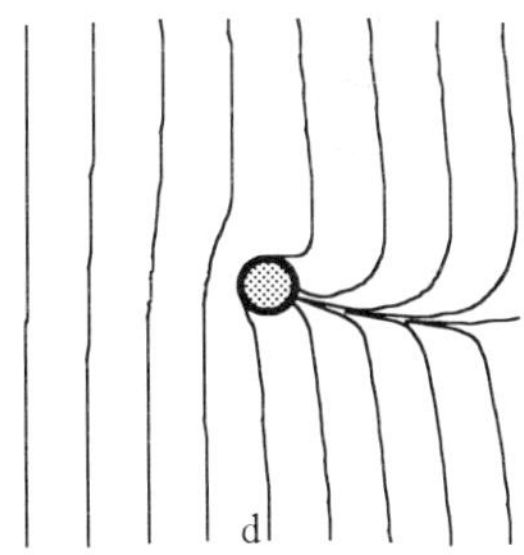

Figure 9b. Lagrangian wake with circulation. Note the zebra stripe dislocation defect.

This work was supported in part by the ISSO at the University of Houston.

REFERENCES

J. L. Barbossa and M. do Carmo (1976), Amer. Journ. Math. **98**, 515

G. K. Batchelor (1967), *Fluid Dynamics*, (Cambridge University Press) p.354.

H. K. Browand (1986), Physica D **118** 173

R. E. Cayflish (1991), in "Vortex Dynamics and Vortex Methods", C.R. Anderson and C. Greengard, editors, (Am. Math. Soc., Providence, RI) p. 67

L.D. Landau & E.M. Lifschitz (1959), *Fluid Mechanics*, Addison-Wesley, Reading Mass

J. Hadamard (1952) *Lectures on Cauch's Problem in Linear Partial Differential Equations* (Dover, N.Y.) p.21

H.Kaden (1931), Ing. Arch. 2, p.140

T. Kambe (1989), Physica D **37** p.403

R.M.Kiehn (1990), "Topological Torsion, Pfaff Dimension and Coherent Structures" in *Topological Fluid Mechanics*, H. K. Moffatt and A. Tsinober, editors, (Cambridge University Press), p. 225.

R. M. Kiehn (1991), "Compact Dissipative Flow Structures with Topological Coherence Embedded in Eulerian Environments" in *The Generation of Large Scale Structures in Continuous Media* , (Singapore World Press).

R. M. Kiehn (1992), "Topological Defects, Coherent Structures, and Turbulence in Terms of Cartan's theory of Differential Topology" in *Developments in Theoretical and Applied Mechanics*, SECTAM XVI Conference Proceedings, B.N.Antar,R. Engels, A.A.Prinaris, T.H.Molden. editors (University of Tennessee Space Institute, TN)

R. Krasny (1991), in *Vortex Dynamics and Vortex Methods*, C.R. Anderson and C. Greengard, editors, (Am. Math. Soc., Providence, RI) p. 385

R. Osserman (1986) A Survey of Minimal Surfaces (Dover, N.Y.) p.18

D. I. Pullin (1989), in *Mathematical Aspects of Vortex Dynamics,* R. E. Cayflisch, editors (SIAM)

N. Rott (1956), J. Fluid Mech. 1 p.111

D. Struik (1961), *Differential Geometry*, (Addison Wesley, Reading, Mass)

IDENTIFICATION AND CLASSIFICATION OF TOPOLOGICAL STRUCTURES IN FREE SHEAR FLOWS

J. Soria[1] **and B.J. Cantwell**[2]
[1]Commonwealth Scientific and Industrial Research Organisation - DBCE, Highett, Vic 3190, Australia
[2]Aeronautics and Astronautics Department, Stanford University, Stanford, CA 94305, USA

ABSTRACT
This paper describes a method for identifying and visualising the three-dimensional geometry of focal (*vortex*) structures in complex flows. The method is based primarily on the classification of the local topology as it is identified from the values of the velocity gradient tensor invariants. The identification of the local topology is reference frame invariant. Therefore, focal (*vortex*) structures can be unambiguously identified in these flows. A novel flow visualisation method is introduced whereby focal structures are rendered using a solid model view of the local topology. This new approach is applied to the identification of focal structures in three-dimensional plane mixing layer and plane wake flows.

1 INTRODUCTION

The recent success in direct numerical simulation (DNS) of three-dimensional free shear flows at moderate Reynolds number has made available large data sets of transitional and turbulent flow fields. Topological methodology has been identified as a useful tool in the study of the geometry of the kinematics of these unsteady flows. Recent examples are Chen *et al.* 1990 and Sondergaard *et al.* 1991, who have applied this methodology to the study of the local topology of small scale motions in turbulent compressible and incompressible shear flows. Soria *et al.* 1992, have studied the topology of the dissipating motions in turbulent mixing layers with different initial conditions.

Although the local flow patterns are described using critical point theory (Chong *et al.* 1990), one does not have to rely on stationary points in the flow to apply this topological identification and classification methodology to the entire flow field. If an observer moves in a non-inertial, non-rotating frame with any particle in the flow field, then it is possible to characterise the local topology of the kinematics at each

J. P. Bonnet and M. N. Glauser (eds.), Eddy Structure Identification in Free Turbulent Shear Flows, 379–390.

point in the flow by the invariants of the velocity gradient tensor at that point in the flow.

Note, that since the velocity gradient tensor is reference frame invariant, the local topological characterisation of the flow field is independent of any particular observer. Note also, that the velocity gradient tensor is not the only quantity that can be used to classify the geometry of the flow; other quantities such as the vorticity or pressure gradient can also be used in the topological analysis. Chen 1991 has considered the topology of the pressure gradient field in a compressible mixing layer. The relative usefulness of using a particular quantity as the input to this type of analysis remains to be explored. This paper considers exclusively the characterisation of the geometry of the kinematics as derived from the velocity gradient tensor for an incompressible fluid. The approach is illustrated by its application to free shear flows.

2 THE LOCAL TOPOLOGY OF THREE-DIMENSIONAL INCOMPRESSIBLE FLOWS

Let the velocity gradient tensor

$$a_{ij} = \frac{\partial u_i}{\partial x_j}(\mathbf{x}) \tag{1}$$

be divided into a symmetric and an anti-symmetric part. The symmetric part is known as the rate-of-strain tensor and is given by

$$S_{ij} = \frac{1}{2}(\frac{\partial u_i}{\partial x_j} + \frac{\partial u_j}{\partial x_i}), \tag{2}$$

while the anti-symmetric part is known as the rate-of-rotation tensor, and it is given by

$$W_{ij} = \frac{1}{2}(\frac{\partial u_i}{\partial x_j} - \frac{\partial u_j}{\partial x_i}), \tag{3}$$

a_{ij}, S_{ij} and W_{ij} are second order tensors. The second order tensor, $\mathbf{a}(\mathbf{x})$ has eigenvalues λ_1, λ_2 and λ_3 and corresponding eigenvectors e_1, e_2 and e_3, given by the eigenvalue problem

$$(\mathbf{a} - \lambda\mathbf{I})\mathbf{e} = 0. \tag{4}$$

The eigenvalues can be determined by solving the corresponding characteristic equation

$$\det[\mathbf{a} - \lambda\mathbf{I}] = 0, \tag{5}$$

which, for a second order tensor yields

$$\lambda^3 + P\lambda^2 + Q\lambda + R = 0. \tag{6}$$

The invariants P, Q and R characterising the $a_{ij}(\mathbf{x})$ tensor are

$$P = -S_{ii}, \tag{7}$$

$$Q = \frac{1}{2}(P^2 - S_{ij}S_{ji} - W_{ij}W_{ji}), \tag{8}$$

$$R = \frac{1}{3}(-P^3 + 3PQ - S_{ij}S_{jk}S_{ki} - 3W_{ij}W_{jk}S_{ki}). \tag{9}$$

Solutions of the characteristic equation (6) determine the local, linearised flow patterns. It can be shown that in the (P, Q, R) space of the tensor invariants the surface which divides characteristic equations with three real solutions for the eigenvalues from characteristic equations with one real and two complex solutions is

$$27R^2 + (4P^3 - 18PQ)R + (4Q^3 - P^2Q^2) = 0. \tag{10}$$

A detailed discussion of the properties of this surface and a guide to the various possible flow geometries and their terminology which can occur in the entire (P, Q, R) space (includes compressible flows) is given in Chong *et al.* 1990. At each point $\mathbf{x}$ in a general compressible flow field the values of the triplet (P, Q, R) specify the local topology of the fluid motion.

For an incompressible fluid, $P = 0$ and the (P, Q, R) space is reduced to the (Q, R) plane with $P = 0$. Thus equations (7) - (9) reduce to

$$Q = -\frac{1}{2}(S_{ij}S_{ji} + W_{ij}W_{ji}), \tag{11}$$

$$R = -\frac{1}{3}(S_{ij}S_{jk}S_{ki} + 3W_{ij}W_{jk}S_{ki}). \tag{12}$$

Figure 1 shows the two-dimensional (Q, R) space with the four possible non-degenerate topologies. The tent-like curve indicated in Figure 1 is part of the surface defined by equation (10) when $P = 0$. Below this curve the eigenvalues of the velocity gradient tensor are real, while above it the velocity gradient tensor has one real eigenvalue and a complex conjugate pair of eigenvalues.

Strain-rate dominated regions in the flow field are areas which have local topologies denoted by either stable-node/saddle/saddle (SN/S/S - below the tent with $R < 0$) or unstable-node/saddle/saddle (USN/S/S - below the tent with $R > 0$) according to the terminology of Chong *et al.* 1990. Vortex-like structures are a conglomerate of local regions which have as their local topology stable-focus/stretching (SF/S - above the tent with $R < 0$) or unstable-focus/contracting (UF/C - above the

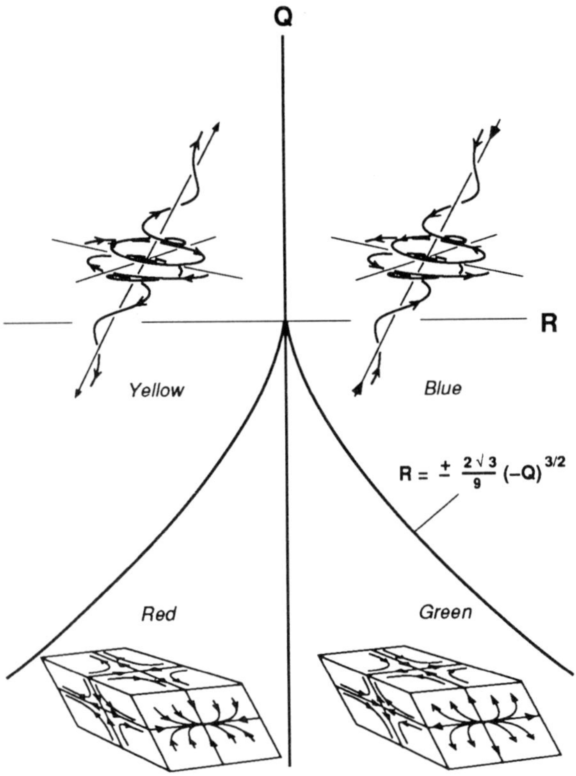

Figure 1. Identification of local non-degenerate flow topologies in the plane $P = 0$.

tent with $R > 0$).

In the compressible flow considered by Chen *et al.* 1990 only two additional topologies were found to be vortex-like in nature for this more general flow, namely stable-focus/contracting (SF/C) and unstable-focus/stretching (UF/S). Colour coding the different topologies has been found to be the most appropriate method for the visualisation of the geometry of the different topological structures which exist in the flow. Figure 1 includes the different colours which have been assigned to the four possible non-degenerate incompressible fluid flow topologies.

In the degenerate case when $R = 0$ which corresponds to two-dimensional flow, regions which are focal in nature can be established from the second invariant alone. Closer inspection of the second invariant as given by equation (11), shows that (i) $S_{ij}S_{ji}$ gives a measure of the local strength of the irrotational stretching and is a positive quantity because S_{ij} is a symmetric tensor and (ii) $W_{ij}W_{ji}$ gives a measure of the local strength of the vorticity (enstrophy) and is negative because W_{ij} is an anti-symmetric tensor. Thus, in this restricted case when $R = 0$ vortex-like regions exist only if the strength of the local vorticity dominates over the strength of the irrotational stretching. In the more general case when $R \neq 0$ this condition is no longer applicable and the third invariant given by equation (12) must also be considered.

Implicit in this definition of regions which are vortex-like (*focal*) in nature is the concept that whether a region of vorticity appears as a vortex structure or not depends also on its environment, i.e. it depends on the local strain field induced by motions outside the region of interest as pointed out by Chong *et al.* 1990. Also, it is important to again emphasise that this definition of a vortex structure is independent of the observer, since its characterisation depends entirely on the properties of the velocity gradient tensor a_{ij}. Therefore, the classification and identification of focal regions in the flow is unambiguous and accurately calculable from the invariants of the velocity gradient tensor.

3 APPLICATION OF LOCAL TOPOLOGICAL CLASSIFICATION TO THE IDENTIFICATION OF FOCAL (VORTEX) STRUCTURES

The topological classification and identification just described can be used as a flow visualisation tool for large DNS data bases. It is quite difficult, if not impossible to load a complete DNS flow field for one instant in time (e.g. more than 2 million data points) into the computer for three-dimensional flow visualisation. The computation of instantaneous streamlines (which are not reference frame invariant and depend on an observer) and the graphical manipulation is a computationally intensive task. Contrast with this, the visualisation of the flow field using one characteristic (e.g. one colour code) per data point. Although this still represents a large amount of data, it is manageable and it is possible to visualise interactively the three-dimensional geometry of the kinematics.

The identification of the geometry of reference frame invariant structures in free shear flow by identifying the local topology at each point in the flow is illustrated by applying it to two numerically computed data bases: one a temporal-developing plane mixing layer and the other a temporal-developing plane wake. In both cases the emphasis is on the identification of focal structures in the flow.

3.1 Focal Structures in a Temporal-developing Mixing Layer

A data base of a time-developing plane mixing layer computed by Rogers and Moser 1990 was used for the visualisation of the topologies and the identification of focal structures in this type of flow. The data presented is for a time frame which corresponds to a Re = 3000, where Re is based on the velocity difference and vorticity thickness. At this time frame, the flow has undergone pairing and post-mixing transition. The initial conditions to the simulation consisted of a mean velocity profile (a function of y, the cross-stream direction only) plus the corresponding most unstable two-dimensional fundamental eigenfunction plus the sub-harmonic of this fundamental plus streamwise vorticity distributed uniformly in the streamwise direction. The z-direction is the spanwise direction and is assumed periodic. The initial Re was 500. The direct simulation of the Navier-Stokes equations was a fully de-aliased spectral calculation. Further details on this simulation are described in Rogers and Moser 1992.

Figure 2 shows the cross-sectional topology visualisation of this three-dimensional time-developing plane mixing layer. Only topologies with $|\omega| > 0.1|\omega|_{max}$ are shown. Included in this figure are equally spaced contour lines for $|\omega| \geq 0.5|\omega|_{max}$. All the regions which are yellow and blue in Figure 2 correspond to focal regions. Note that a significant amount of this flow field has $|\omega| < 0.5|\omega|_{max}$ and is still part of a focus. The topological map shown in Figures 2 (a),(b) shows two different xy-planes. Figure 2(b) shows streamwise structures which are topologically focal/stretching in nature. In fact, these regions of the flow are dominated by rotation. They contain the largest vorticity magnitude and very little strain as identified by Chen *et al.* 1990, and correspond to streamwise stretching vortices. Figures 2(c)-(e) show the topologies in yz-planes of this flow. Figure 2(c) corresponds to a cross-sectional cut through the middle of the streamwise stretching vortices. The topological visualisation of these structure shows them to have an elliptical cross-sectional geometry at this yz-plane. In contrast, the cross-sectional geometry of these structures is circular towards their extremities (Figure 2(e)).

Figure 3 shows a three-dimensional flow visualisation of regions in the flow which are focal in nature by showing iso-surfaces with $|\omega| = 0.3|\omega|_{max}$. This visualisation clearly shows the dominant focal structures in this flow which are identified as streamwise vortices. It is curious to note the helical nature of these structures at their extremities. At these locations there is an intertwining of topologies which are SF/S and UF/C. The spiralling nature of the streamwise vortices for this particular flow has been the subject of a recent study by Lopez and Bulbeck 1992.

3.2 Focal Structures in a Temporal-developing Wake

Regions with focal structures in plane wake flows were identified by considering the data base of a direct numerical calculation of a time-developing plane wake at a Re = 500, where Re is based on the velocity deficit and the displacement thickness. The initial conditions for this calculation were a Gaussian mean velocity profile with the corresponding most unstable two-dimensional mode plus an oblique mode at 45° with the same streamwise wavenumber as the two-dimensional mode. The numerical calculation was performed by solving the incompressible vorticity equations using periodic boundary conditions in the streamwise and spanwise spatial directions and an infinite domain in the cross-stream direction. A pseudo-spectral numerical code developed by Rogers and Moser 1990 was adapted for this three-dimensional time-developing plane wake calculation. Details of the pseudo-spectral numerical method are described in Spalart *et al.* 1991. The numerical calculation was performed with $64 \times 128 \times 64$ Fourier/Jacobi modes.

Figure 4(a)-(c) shows the topology map of the wake in xy-planes at a certain time frame in its evolution. Flow is left to right in the positive x-direction. The topology of the flow is only shown for $|\omega| > 0.1|\omega|_{max}$. Equally spaced contours of vorticity magnitude for each of the xy-planes are also shown for $|\omega| \geq 0.5|\omega|_{max}$. Close correspondence is observed between the regions of high $|\omega|$ and the topological regions which are focal in nature. Figure 4(c) shows regions of focal topology which can be identified as the classical cross-stream rollers of a wake and some focal regions which are SF/S which are elongated and tilted in the streamwise direction. Figure 4(b) also shows regions of SF/S which can be identified as part of the cross-stream rollers. A Large focal structure which is elongated and tilted in the streamwise direction is also identified in Figure 4(b). Figures 4(d)-(f) show the corresponding topology map for yz-planes, clearly showing streamwise oriented focal structures. Figures 4(d) and (f) also show clearly that (i) the cross-stream vortex structures contain significant regions of UF/C topology and (ii) that regions with the largest vorticity magnitude are associated with the streamwise vortex structures which are topologically focal in nature.

Much information can be deduced about the geometry of focal structures in this wake flow from planer topology maps. However, all this information and more about the vortex structures in this flow is even better summarised and visualised by considering a three-dimensional solid model of the focal structures. This was achieved by an iso-surface visualisation of regions of focal topology with $|\omega| = 0.4|\omega|_{max}$.

On a graphics computer the iso-surface visualisation is easily rotated and hence, provides a great deal of geometric insight into the complex three-dimensional nature of the focal structures in this flow. Some of the views of this visualisation are shown in Figure 5. The cross-stream rollers and the streamwise vortices are clearly identified in this visualisation. Curiously the streamwise vortices have a helical geometry in the region close to the cross-stream vortices. The streamwise vortices consist of topologies SF/S with the core consisting of UF/C. The helical region of these vortices is made up of intertwining regions of SF/S and UF/C similar to the helical regions of the streamwise vortices identified in the plane mixing layer. The cross-stream focal structures are clearly identified between a pair of the streamwise vortical structures, but only a remanence of them exist near the region of the streamwise vortical structures.

4 CONCLUDING REMARKS

An unambiguous and reference frame invariant method for the identification of focal (vortex-like) regions has been presented. The method is based on the local topology classification determined from the calculation of the velocity gradient tensor invariants. The method has been illustrated in incompressible free shear flows, although, its applicability is more general and also encompasses compressible fluid flows.

A novel method of visualisation of the flow geometry has been introduced. A rendition of the geometry of focal structures in any flow can be generated by a solid model view of regions with a focal topology. These focal regions can be determined exactly from the values of the velocity gradient tensor invariants. This method of visualising the geometry of focal structures, and hence, the identification of vortex structures in a complex three-dimensional flow has been illustrated by considering a three-dimensional plane mixing layer and plane wake flow. Cross-stream focal structures and streamwise oriented focal structures have been identified using this methodology. In both flows the streamwise vortex structures are found to have an interesting helical geometry.

Acknowledgments

Support for J. Soria under a CSIRO Post-doctoral Fellowship for some of this work is acknowledged. We are indebted to Drs. M. Rogers and R. Moser for their help in using their mixing layer data base and their invaluable assistance with the wake simulation. We would also like to acknowledge support from ONR Grant N00014-90-J-1976.

5 REFERENCES

CHEN, J., CHONG, M., SORIA, J., SONDERGARRD, R., PERRY, A.E., ROGERS, M., MOSER, R. & CANTWELL, B. 1990 A study of the topology of dissipating motions in direct numerical simulations of time-developing compressible and incompressible mixing layers. CTR Report SS90, Stanford University, Stanford, CA.

CHEN, J.H. 1991 The topology of the compressible mixing layer. *Bull. Am. Phys. Soc.*, **36(10)**, 2615.

CHONG, M., PERRY, A.E. & CANTWELL, B. 1990 A general classification of three-dimensional flow fields. *Phys. Fluids A*, **2(5)**, 765 - 777.

LOPEZ, J.M. & BULBECK, C.J. 1992 Behaviour of streamwise rib vortices in a three-dimensional mixing layer. CTR Report SS92, Stanford University, Stanford,

CA.

ROGERS, M. & MOSER, M. 1990 Mixing transition and the cascade to small scales in a plane mixing layer. Proc of the IUTAM Sym. on Stirring and Mixing. UCSD, LaJolla, CA.

ROGERS, M.M. & MOSER, R.D. 1992 Spanwise scale selection in plane mixing layers. submitted for publication to *J. Fluid Mech.*.

SONDERGAARD, R., CHEN, J., SORIA, J. & CANTWELL , B. 1991 Local topology of small scale motions in turbulent shear flows. 8th Sym. on Turbulent Shear Flows, Munich, Germany.

SORIA, J., CHONG, M.S., SONDERGAARD, R., PERRY, A.E. & CANTWELL, B.J. 1992 A study of the fine scale motions of incompressible time-developing mixing layers. CTR Report SS92, Stanford University, Stanford, CA.

SPALART, P.R., MOSER, R.D. & MOSER, M.M. 1991 Spectral methods for the Navier-Stokes equations with one infinite and two periodic directions. *J. Comp. Phys.*, **96**, 297 - 324.

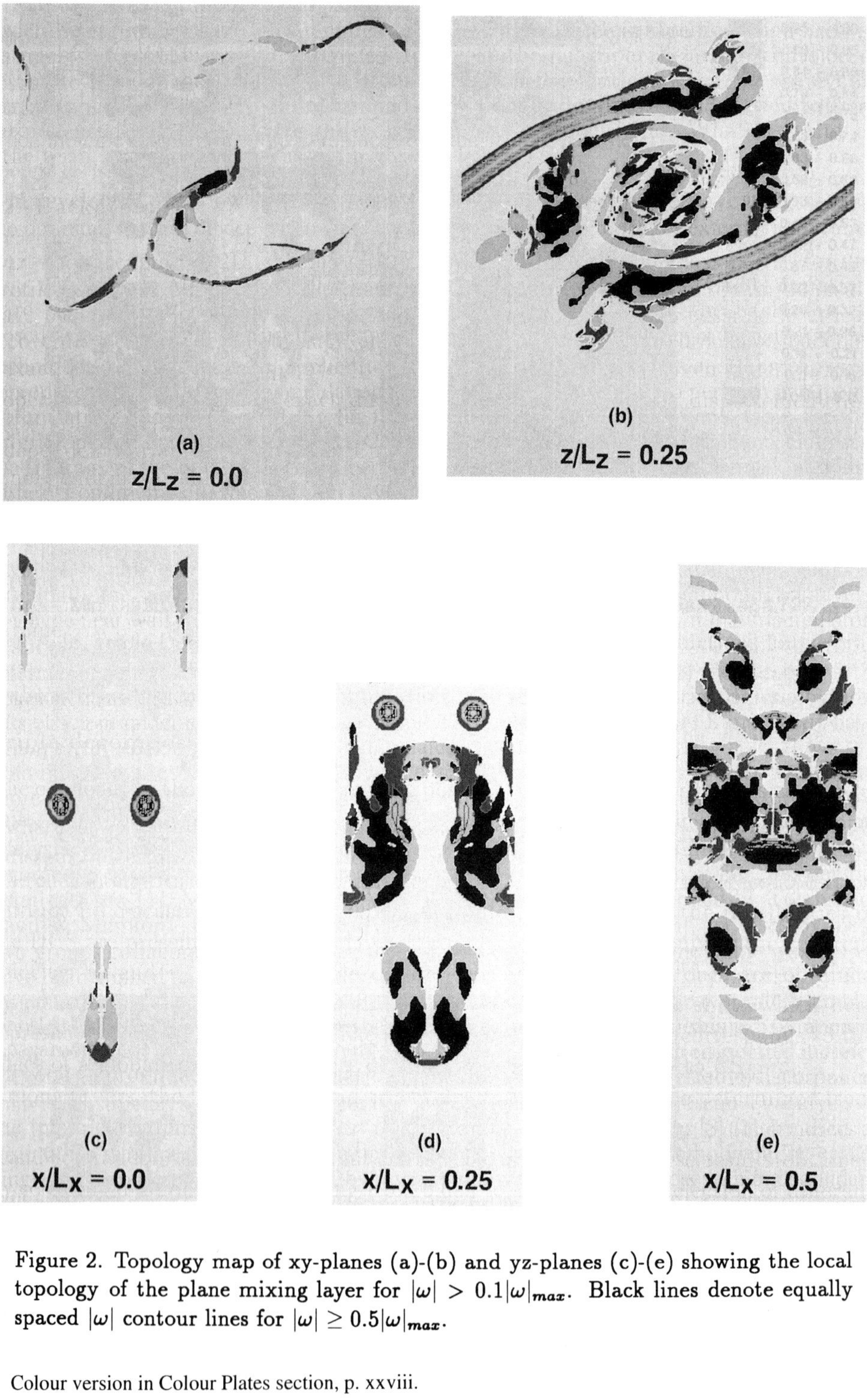

Figure 2. Topology map of xy-planes (a)-(b) and yz-planes (c)-(e) showing the local topology of the plane mixing layer for $|\omega| > 0.1|\omega|_{max}$. Black lines denote equally spaced $|\omega|$ contour lines for $|\omega| \geq 0.5|\omega|_{max}$.

Colour version in Colour Plates section, p. xxviii.

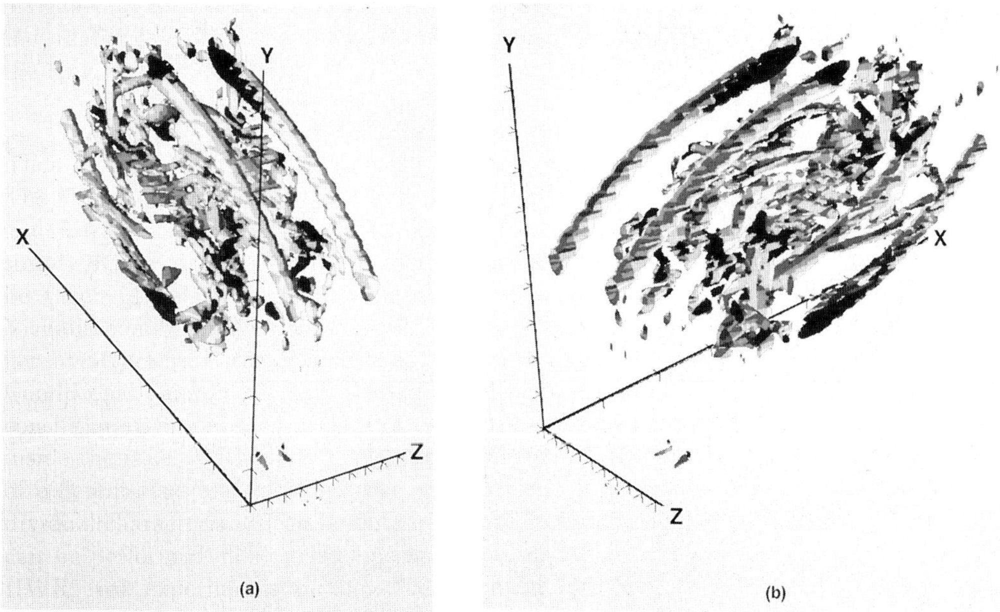

Figure 3. Visualisation of SF/S and UF/C focal (vortical) structures in the plane mixing layer shown as iso-surfaces with $|\omega| = 0.3|\omega|_{max}$.

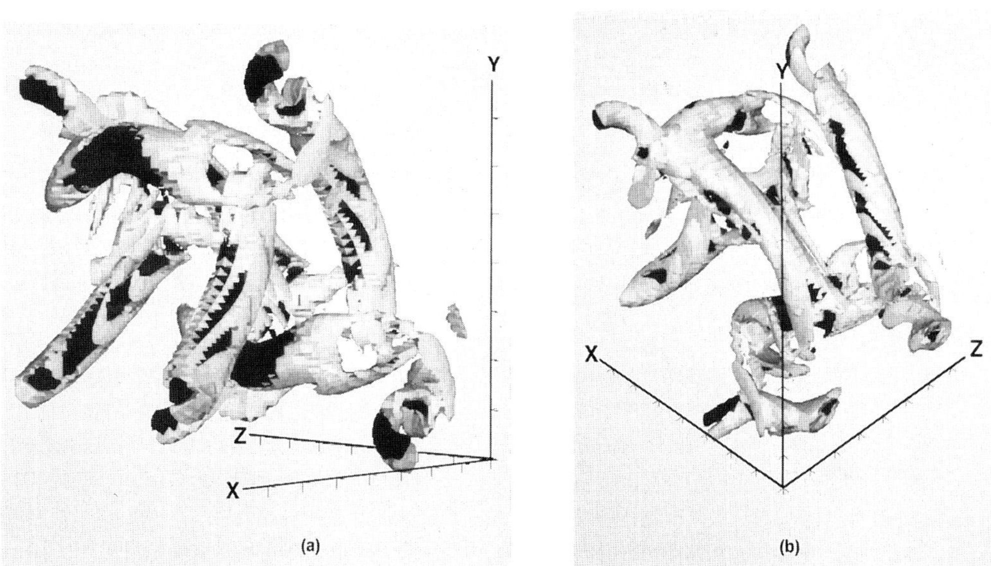

Figure 5. Visualisation of SF/S and UF/C focal (vortical) structures in the plane wake shown as iso-surfaces with $|\omega| = 0.4|\omega|_{max}$.

Colour version in Colour Plates section, p. xxix.

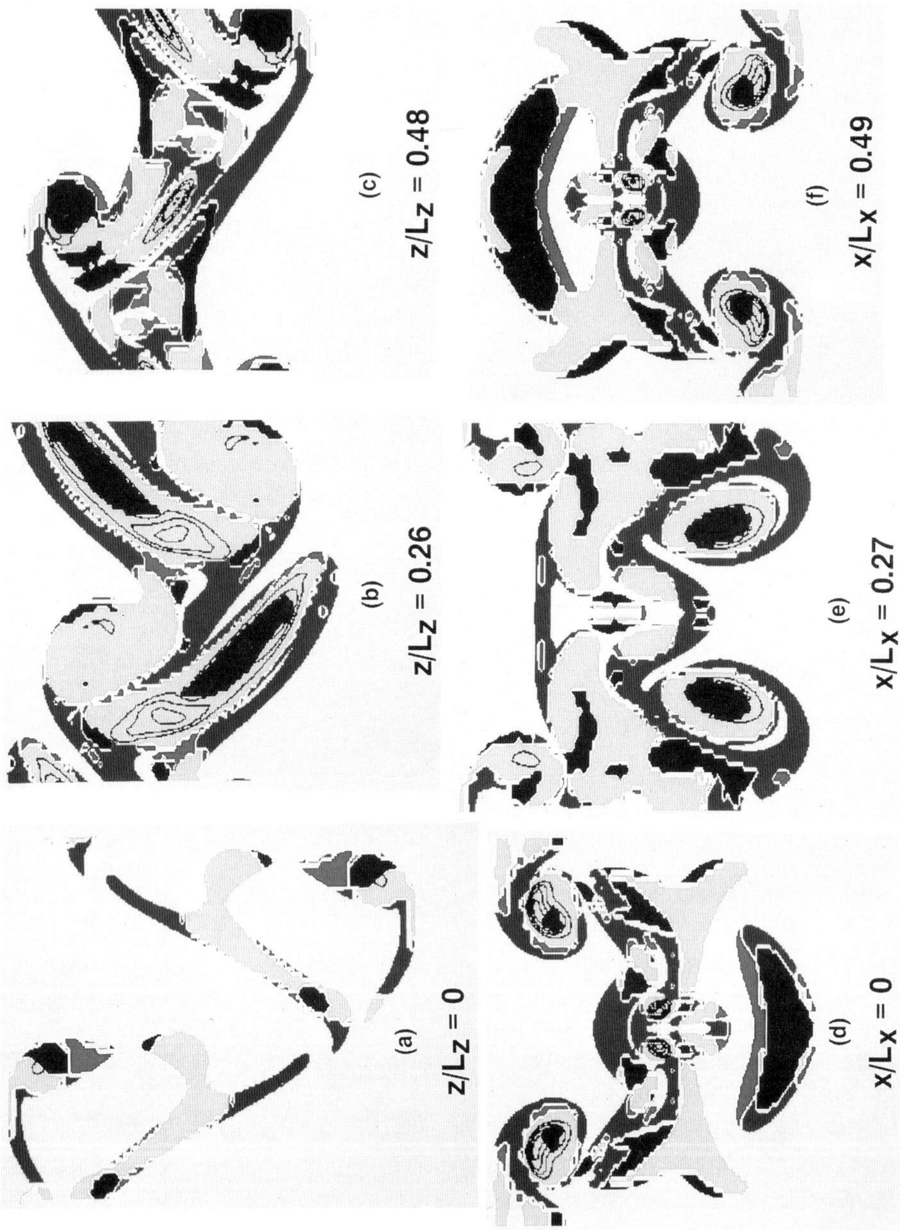

Figure 4. Topology map of xy-planes (a)-(c) and yz-planes (d)-(f) showing the local topology of the plane wake for $|\omega| > 0.1|\omega|_{max}$. Black lines denote equally spaced $|\omega|$ contour lines for $|\omega| \geq 0.5|\omega|_{max}$.

Colour version in Colour Plates section, p. xxx.

X. Full Field Methods

VISUALIZATION AND IDENTIFICATION OF VORTEX STRUCTURES IN STRATIFIED WAKES

Yuli D. Chashechkin
Institute for Problems in Mechanics R.A.S.
prospect Vernadskogo 101, Moscow 117526, Russia

Isolated eddies and coherent structures (vortex arrays) effect the drag and lift acting upon obstacles in a flow, energy and contaminants transfer, flow structure and stability. They are widely studied in physical, geophysical and technological hydromechanics [1]. Modern computer techniques together with conventional dyeing, markers and tracers are used for visualization and vortex measuring [2]. Fine features of vortex interactions are studied by numerical simulation [3]. The critical point concept is employed to describe quantitatively the topology of the vortex field [4].

It is shown experimentally that even weak continuous stratification which brings about vorticity and energy transfer to the internal gravity wave field and wake influences the fine structure and dynamics of localized free vortices (laminar vortex rings [5]).

The aim of this paper is to visualize and identify eddies in the wake of $2D$ and $3D$ bluff bodies in continuously stratified liquids. Fine flow features due to stratification are identified, structural elements characterizing different flow forms are determined, critical conditions for restructuring the flow pattern summarized in a flow regime diagram are found.

The system of equations describing the flow of a deep viscous isothermal continuously stratified liquid includes equations of momentum, mass and salinity conservation, state equation and appropriate boundary conditions (non-slipping for velocity and non-permeability for salinity, as well as attenuation of all perturbations at infinity).

$$\frac{d(\rho\vec{v})}{dt} = -\nabla p + \rho\vec{g} + \rho\nu\Delta\vec{v}$$

$$\frac{d\rho}{dt} = 0 \quad \frac{d(\rho S)}{dt} = \nabla(\rho k \nabla S)$$

$$\rho = \rho(p, S); \quad \rho_0(z) = \rho_{00}\exp\{-z/\Lambda\}$$

$$\vec{v}_\infty = \{U, 0, 0\}; \quad \vec{v}|_\Sigma = 0 \quad \frac{\partial S}{\partial n}\Big|_\Sigma = 0$$

They contain as dimensional parameters as density ρ and its gradient $d\rho/dz$ (which can be characterized by reference density ρ_{00} and buoyancy scale $\Lambda = |d\ln\rho/dz|^{-1}$, velocity u, pressure p, gravity acceleration g, salinity S, kinematic viscosity ν, diffusion coefficient k, obstacle size d. The intrinsic time scale here is buoyancy period (frequency) $T_b = 2\pi/N = 2\pi\sqrt{\Lambda/g}$. There are several length scales in the problem.

J. P. Bonnet and M. N. Glauser (eds.), Eddy Structure Identification in Free Turbulent Shear Flows, 393–403.

As independent (basic or fundamental) length scales it is natural to choose scales characterizing the geometry and main structural features of the flow near an obstacle. The buoyancy scale Λ and body size (diameter d) are external scales in the problem. Dynamic length scales are dependent on the body velocity. They characterize the length of attached internal waves $\lambda_i = 2\pi\lambda = 2\pi U/N$, velocity $\delta_u = \nu/U$ and density $\delta_\rho = k/U$ boundary layer thickness.

A boundary layer splitting effect due to "discerntion" of medium, i.e. the difference of kinetic coefficients ($\nu \neq k$) is clearly evident in the boundary flow on a sloping plane in a stratified liquid at rest, as well as in other types of flow in which interruption of molecular flux occurs. Conventional dynamic parameters that are Reynolds number $\mathrm{Re} = Ud/\nu = d/\delta_u$; Froude number $\mathrm{Fr} = U/Nd = \lambda/d$; Pekle number $\mathrm{Pe} = Ud/k = d/\delta_\rho$ length ratio $C = \Lambda/d$ are ratios of basic scales [7].

For weak stratification and "discerntion" of media the values of basic scales are essentially different $\Lambda \gg d \gg \delta_u \gg \delta_\rho$; $\Lambda \gg \lambda \gg \delta_u$; because for aqueous solutions $\nu \gg k$. The countable number of secondary (derivative) scales characterizes the sizes and variability scales of secondary flow elements (shear layers, high gradient sheets, waves, vortices, ...). Flow structural elements can be divided into two groups: uni- and bi(multi)-scale. Spatial distributions of all parameters are similar in uni-scale flow elements, e.g. in internal waves. The scales of the spatial variability of fast (or smooth) parameters (e.g. velocity) are greater then the scales of "slow" (sharp) variables (e.g. density) in the multiscales elements of flow such as vortices. The countable set of derivative scales include translation ones (independent of kinematic properties, e.g. ($L = U^2/g$; $L^* = \sqrt{\lambda\Lambda}$ ...), viscous ($L_\nu = \sqrt[3]{\lambda\Lambda\delta_u} = \sqrt[3]{g\nu}/N$, $l_u = \sqrt[3]{\nu^2/g} = \sqrt[3]{\delta_u^2/\Lambda}$), concentrated ($l_s = \sqrt[3]{k^2/g}$, $l_k = \sqrt{k/N}$) and mixed $L_m = \sqrt[4]{\nu k/N^2}$ [10].The modal structure of harmonic internal waves [8] and the mode of vortex shedding [9] are characterized by the viscous wave scale $L_\nu = \sqrt[3]{g\nu}/N$, dissipative-gravity waves [10] or stationary boundary flow induced by diffusion at impermeable surface [11] are characterized by the mixed-length scale $L_m = \sqrt[4]{k\nu/N^2}$. The latter two scales are independent of the body velocity. The set of scales describes the existence, the possibility to identify and relative stability of the structural elements of the flow (isolated vortices, vortex arrays, shear layers, waves, etc.) It allows one to determine the Eigen-form of flow and find the parameter range of its existence. There from the conditions of similarity of different kinds of scale describing different Eigen-forms, i.e. the condition of scales of different origin synchronism with strong interaction or mutual transformation of different structure it follows that the boundaries of regime maps in the double logarithmic scales are straight line sections. It follows that the experimental technique should provide for registering fields of different natures with high spatial and temporal resolutions.

The experiments have been carried out in several lab tanks ($0,7 \times 0,25 \times 0,7\ \mathrm{m}^3$; $1,5 \times 0,4 \times 0,4\ \mathrm{m}^3$; $2,4 \times 0,4 \times 0,6\ \mathrm{m}^3$; $9 \times 0,6 \times 0,6\ \mathrm{m}^3$). The tanks were filled with linearly stratified brine. Bodies of different sizes and shapes (spheres, discs, cylinders) were towed at a constant velocity. Different modifications of the schlieren technique namely, the vertical slit — the Foucault knife, the slit-thread in focus, the horizontal slit the regular grating in focus producing a natural colour shadow picture (rainbow method) [12]. In addition, markers, tracers, dye and electrolytic precipitation [13] were used. The different optical methods complement each other and make it possible to create a more adequate model of flow. Dynamic characteristics of probes due to their strong interaction with the stratified environment were used to analysis of the data [13].

From the results of many observations and probe measurements in the flow near obstacles, the following structural elements were found: up-stream disturbances including dissipative and non-stationary gravity waves, attached internal waves, density and velocity boundary layers and split density and velocity wakes.

The structural elements manifest themselves in a black and white replica of the colour schlieren image of the flow past the towing sphere (see Fig. 1) The points of isopleth condensation on the front part of the body separate the blocking area from the transient internal waves and the free stream forming boundary layers on the body surface. The high gradient density boundary layer (DBL) causes isopleth condensation near the sphere poles. The separated DBL forms the envelope of the density wake. The dark and light strips outside the wake are the result of light refraction in the internal wake field. The calculation of transient internal wakes is given in [14]. The forms of the phase surfaces of attached internal waves are described by $X^2 = (Y^2 + Z^2)(Z^{-2} - 1)$ past a $3D$ body and by $X^2 + Z^2 = 1$ past a $2D$ body, here A-phase, $\{X; Y; Z\} = \dfrac{N}{AU}\{x, y, z\}$. The internal wake length in $\lambda_0 = 2\pi U/N$.

Phase surfaces curve due to the Dopler effect inside the velocity wake and do not penetrate through the high gradient envelope inside the density wake (see Fig. 2). To evaluate the envelope parameters, one can use the calculation of the boundary flow on a sloping plane at an angle of $\alpha°$ to the horizon in a liquid at rest. In this case, the effect of the decomposition of fields manifest in the splitting of the scales of their spatial variability.

The main parts of the first terms of the exact solution for salinity S and velocity u [6] are

$$S = 2\sqrt{kt}\cos\alpha\, i\,\mathrm{erfc}\{\eta/\sqrt{2kt}\}/\Lambda$$

$$u = 4g\sqrt[3]{kt}\sin 2\alpha\{i^3\,\mathrm{erfc}(\eta/\sqrt{2\nu t}) - i^3\,\mathrm{erfc}(\eta/\sqrt{2kt})\}/\Lambda(\nu - k)$$

where $i\ \mathrm{erfc}(x)$ is the repeated integral of the complementary error function, η — is the distance from the plane. It can be seen that the ratio of squares of spatial variability scales of these monotonous functions does not depend on time and in proportional to the Schmidt number (for brine $R_u = \delta_u/\delta_\rho = \sqrt{\mathrm{Sc}} = \sqrt{\nu/k} =$ 26,5).

The results of the measurements of the scales (by the displacement of the vertical marker and photometry of the shadow image) in the area of laminar boundary layers separation from sphere well agrees with the above evaluation. These ratios are $R_u = 20(\mathrm{Re} = 116)$; $R_u = 22(\mathrm{Re} = 148)$; $R_u = 23(\mathrm{Re} = 164)$.

A special method has made it possible to measure both the thickness of the thin envelope and the density gradient in the split laminary wake behind the cylinder [15]. Depending on the Reynolds number it thickness was 0,06–0,5 mm, the original density gradient reinforced by 15–160 times, the ratios of scale variability by 10–140 times somewhat exceeded the evaluation [6], which may be due to non-linear effects, because an increase of the gradient is also observed at the laminar vortex boundary [5].

Discontinuity surfaces essentially effect the form, topology, symmetry, stability of the flow past the body and generation of the vortices. Their effect is most pronounced in the laminary regime where the wake behind a $3D$ body has a prismatic form with enough distinct angles formed by the intersection of two horizontal and two vertical high gradient layers [16]. At the smallest values of the Reynolds number, the density wake is compressed from the sides and stretched along the vertical. When velocity of the body increases the vertical and horizontal sizes of the wake are equalized and the cross section of the wake transforms into a square. The minimal vertical size being attained at $0{,}05 < F < 0,3$ when the section of the wake is a horizontal strip. In subsequent regimes the vertical wake size starts to increase due to the vorticity accumulation in the rear part of the body.

All vortex structures can be divided into a number of groups depending on their dimension ($2D$ and $3D$), form, size, orientation, the intensity of interaction with the stratified environment, stucture of the boundary, life time, degeneration character. In the wakes behind $2D$ and $3D$ bodies one observed compact vortex elements (puffs, bubbles, rings, loops, columns, whirls), alongated elements (threads, loops, "horse-shoes", "staples") and their combinations. Vortices were observed both inside and outside the $2D$ density wake (submerged and soaring vortices). The maximal vortex size is determined by the wake thickness and the attached wave length $L \sim O(d)$; $L \sim \lambda = UT_b$, the minimal size — by the thickness of high gradient interface or value of the ultimate scale $l_u = \sqrt[3]{\nu^2/g}$.

The abundance of forms demonstrates the diversity of mechanisms generating compact vortices in the wake due to the instability and whirling of the vertical (side) and horizontal (upper/lower) parts of the wake envelope, accumulation and shedding of vorticity from the circular flow area in the rear part of the body, reconnection of vortex lines in the turning points of the meandering wake [17], instabilities of the attached waves ("soaring" vortices behind $2D$ bodies [9]), as a result of the interaction of waves with the wake or waves with vortices, interaction and fractioning of vortices. Many kinds of vortices are contoured by a high gradient envelope. When moving they radiate internal waves coincident in structure with those attached behind a solid body [14] which allow them to be identified against the background of intensive small scale disturbances inside the degenerating wake and translational velocity ($u = \lambda/T_b$) to be determined. Vertical vortex columns (whirls) have diffuse side boundaries and are located between horizontal high gradient boundaries of the density wake.

The following criteria were used to identify vortices:

- existence of domain with (almost) closed streamlines where streaklines of (tracks, dye) or high gradient interfaces form a typical helix of the laminar vortex core;
- the presence of high gradient vortex envelopes;
- captured mass transfer;
- generation of attached internal waves;
- retention of topological features when changing visualization methods if the time for their finding exceeds that which in necessary for identification (one cycle period).

Some vortex types have been identified in the wake behind a horizontally moving sphere: vertical vortex columns inside the laminary wake (as result of the developing of instability of vertical side envelopes (see Fig. 3); small-scale vortices located in the interlayers, wave-vortex billows positioned in the areas of contact of crests and troughs of the attached internal waves with the density wake (see Fig. 4); rear vortices attached to the body; shedding laminar vortices or vortex puffs, compact vortices in areas where the meandering wake makes turns (see Fig. 5). Longest living of them are vertical vortex columns (internal whirls) weakly interacting with the background stratification. But all the other types of vortices or turbulent flows in a continuously stratified liquid transform into this form of the movement.

A map of flow regimes (see Fig. 6) constructed in a double logarithmic scale shows parameter areas where a laminar prismatic wake, vertical vortices (columns) inside the prismatic wake (I), vortex billows (II), rear vortices with a sharp boundary (III), regular vortex axis-symmetrical blobs in the wake ($m = d/L_\nu > 1$, IV), regular vortices in the flow turning area ($M_\nu = d/L_\nu < 1$, V) are observed. The boundaries between the regimes in the logarithmic scale are straight line sections, for three first $\mathrm{Re}^2 \sim \mathrm{Fr}^{-3}$, for the lower boundary of regimes IV, V $\mathrm{Re}^2 \sim \mathrm{Fr}^{-1}$ (the dependence $\mathrm{Re}^2 \cdot \mathrm{Fr}^3 \sim$ const corresponds the proportionality of two combinational scales $L_*^2 \sim \lambda_c^2$, $L_* = U^3/N^2\nu$, $\lambda_c = \sqrt{d\lambda}$. The relation ($\mathrm{Re}^2\mathrm{Fr} \sim$ const) or condition of proportionality of combinational $(\lambda d/\delta_u^2)^3 \sim$ const and $\lambda_c \sim \delta_u$ scales correspond to the dependence.

A relatively large number and the regular location of simultaneously observed vortex elements allow them to be treated as a unique structure and be characterized by the Strouhal number $\mathrm{Sh} = \mathrm{fd/U}$, where f is the shedding frequency. Its value was determined by measuring the period of vortex shedding ($\mathrm{Sh} = \mathrm{d/U\Delta t}$), as well as by the drift velocity u_* and the distance between structure elements Δl ($\mathrm{Sh} = \mathrm{d(1 - u_*/U)/\Delta l}$). A dependence of the log of the Strouhal number on the log of the Reynoolds number is presented on Fig. 7. Data for a homogeneous fluid are approximated by the line $\mathrm{Sh} = 6,5 \times 10^{-4}\mathrm{Re}^{0.95}$. Individual points correspond to Strouhal number values in stratified liquid. Averaged Strouhal number for given sphere diameter and buoyancy period (see symbols on Fig. 7) are drawn by solid horizontal lines. Unlike the case of a homogeneous fluid Strouhal number value is a step-like function of Reynolds numbers (as well as Froude numbers and on the ratio Re/Fr). The influence of stratification in a considered parameters range is more weak than that of body size. The mean value of the Reynolds number increses in a step way proportionally to the sphere diameter value ($\mathrm{Sh} = 0.25$ for $d = 1$ cm, $T_b = 4.5$ s, $\mathrm{Sh} = 0.99$ for $d = 4$ cm, $T_b = 4.5$ s). For a fixed body diameter variations of Strouhal number for fluids with different buoyancy period are not so significant ($d = 4$ cm $\mathrm{Sh} = 0.92$ when $T_b = 3.59$ s and $\mathrm{Sh} = 0.99$ when $T_b = 8.0$ s).

All the set of experimental data is collected in a single universal dependence $\mathrm{Sh} = 5 \cdot 10^{-4}(\mathrm{Re^2C/Fr^2})^{0.37} = 5 \cdot 10^{-4}(\mathrm{d/l_u})^{1.1}\cdot$; $\mathrm{Re^2C/Fr^2} = (\mathrm{d/l_u})^3$; $l_u = \sqrt[3]{\nu^2/g}$ (see Fig. 8).

The topological structure of the density wake past a sphere depends on the angle of the path with the horizon. As the slope increases the upper part of the density wake envelope deflexes greater than lower one, they approaches each other and eventually merges forming a trough– "opening" downstream density valley. In the wide part of density wake far from the body a pair of vortex columns can occur. The density envelope acted upon by bouyancy forces looses stability and breaks into a sequence of salt fingers due to Rayleigh-Taylor instability.

A more complicated structure has a flow behind a $2D$ body where, together with eddies submerged in the wake, "soaring" or suspended vortices are observed appearing during the development of instability of internal wave in the free flow [9]. Wakes of the suspended vortices also create discontinuities (interfaces) in the density gradient distribution. Depending on experimental conditions, isolated suspended vortices (Fig. 8), a sequence of turbulized suspended vortices located in the depression of the varicose density wake (Fig. 9) and a system of vortex bubbles (Fig. 10) inside the density wake were observed. In all regimes, the density wake is contoured with high gradient interfaces which can merge into a single central layer behind a thin body ($d/L_\nu < 1$). All in all, 14 types of flow have been identified which are separated by straight line sections in a logarithm plot of regimes [9].

The data presented and results of other works [3, 5, 7, 9, 17] show that the Eigen-forms are structural stable elements of flows. Their geometric properties are described by intrinsic length scales arising when analyzing a complete system of equations taking into account molecular effects. Due to the difference of kinetic coefficients (medium discerntion) the variability scales of different variables do not coincide. The synchronism condition (condition of the similarity of scales of different natures) describes the boundaries in the flow regime map. The emerging high gradient surfaces increase the influence of "weak" stratification, stabilize the Eigen-forms of vortices, enlarge the class of forms of motion, lead to strong non-local interaction of different flow elements including waves, density wake and isolated vortices.

Литература

1. Batchelor G.K. An Introduction to Fluid Dynamics. Cambridge University Press, 1970.
2. Debisschop J.R., Sapin S., Delville J., Bonnet J.P. Supersonic mixing layer analysis by laser planogram and hot – wire based POD. — Eddy structure identification in free turbulent shear flows. IUTAM Symposium. Poitiers, 12–14 October 1992.
3. Hussain F. Study of coherent structures via vortex dynamics: core dynamics, helical waves and interaction with fine scale turbulence. Eddy structure identification in free turbulent shear flows. IUTAM Symposium. Poitiers, 12–14 October 1992.
4. Perry A.E., Chong M.S. A description of eddying motions and flow patterns using critical – points concepts. — Annual Review of Fluid Mechanics. 1987, vol.19, pp.125–155.
5. Belyaev V.S., Savinkov A.M., Chashechkin Yu.D. Dynamics of laminar vortex rings in a stratified liquid. — Zhurnal Prikladnoi Mekhaniki i Tekhnicheskoi Fiziki. 1987, 1, pp.37–47.
6. Kistovich A.V., Chashechkin Yu.D. Structure of transient boundary flow on sloping plane in a continuously stratified medium. — Doklady Acad. Sci. 1992, vol.325, 4, pp.833–837.
7. Chashechkin Yu.D. Hydrodynamics of a sphere in a stratified liquid. — Izvestiya AS, Fluid Dynamics, 1989, vol.24, 1, pp.1–7.
8. Makarov S.A., Neklyudov V.I., Chashechkin Yu.D. Spatial Structure of Two – dimensional Monochromatic Internal – Wave Beams in an Exponentially Stratified Liquid. — Izvestiya AS, Atmospheric and Oceanic Physics, 1990, vol.26, 7, pp.548–554.
9. Voeikov I.V., Chashechkin Yu.D. Hydrodynamics of a cylinder in a stratified fluid. Preprint IPM RAS, 1992, 519, p.49 (in Russian).

10. Kistovich A.V., Chashechkin Yu.D. Generation dissipative – gravity waves by heat convection in stratified fluid. — Zhurnal Prikladnoi Mekhaniki i Tekhnicheskoi Fiziki. 1991, 3, pp.49–55.
11. Phillips O.M. On flows induced by diffusion in a stably stratified fluid. — Deep–Sea Research, 1970, vol.17, pp.435–443.
12. Chashechkin Yu.D., Popov V.A. Colour schlieren method. — Doklady AS, 1981, vol.269, 5, pp.1130–1133.
13. Gvozdev A.V., Neklyudov V.I., Chashechkin Yu.D. Comparative analysis of a dynamic characteristics of contact transducers in a continuously stratified fluid. — Izmeritel'naya Tekhnika. 1990, 3, pp.33–35.
14. Chashechkin Yu.D., Makarov S.A. Transient internal waves. Doklady AS. 1984, vol.276, 5, pp.1246–1250.
15. Voeikov I.V., Chashechkin Yu.D. Generation of discontinuities in the wake past a cylinder in flow of continuously stratified liquid. — Izvestiya RAS, Fluid Dynamics. 1993, 1, pp.20–26.
16. Sysoeva E.Ya., Chashechkin Yu.D. Spatial structure of the wake behind a sphere in a stratified fluid. — Zhurnal Prikladnoi Mekhaniki i Tekhnicheskoi Fiziki. 1988, 5, pp.59–65.
17. Sysoeva E.Ya., Chashechkin Yu.D. Vortex system in the stratified wake of a sphere. — Fluid Dynamics. 1991, vol.26, 4, pp.544–551.

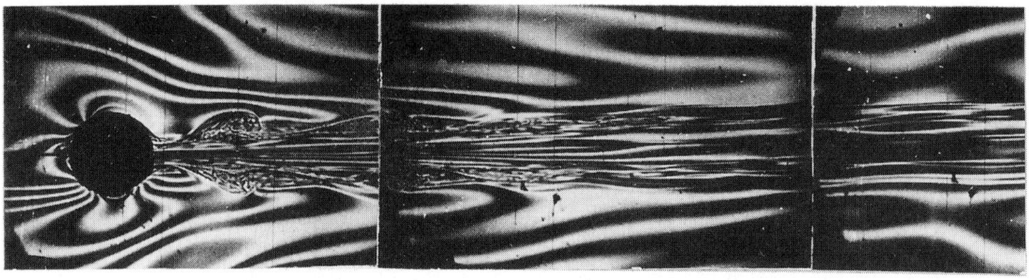

Fig. 1. Black and white replica of the colour schlieren image of the flow around the sphere (T_b=3s; d=4cm; U=2.0cm/s; Fr=0.23; Re=800)

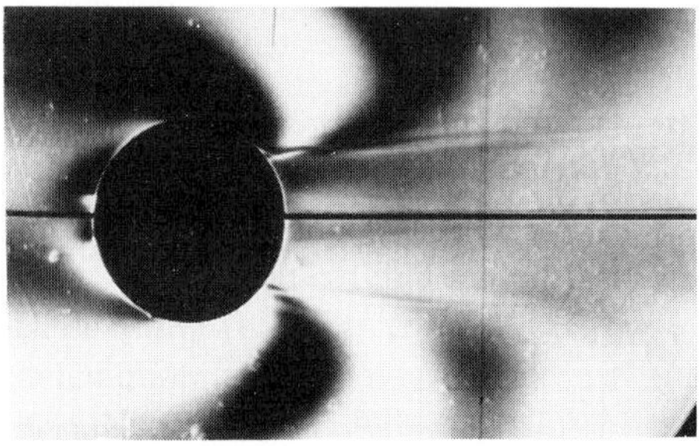

Fig. 2. Schlieren image of the laminar wake and attached internal waves past the horizontally towing sphere (T_b=4.5s; d=5cm; U=0.3cm/s; Fr=0.04; Re=150)

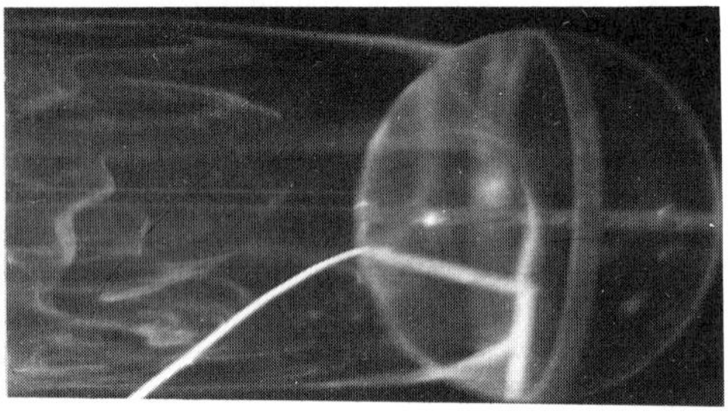

Fig. 3. Prismatic density wake with immersed pair of vortex columns. Electrolytic precipitation. (T_b=4.5s; d=5cm; U=0.7cm/s; Fr=0.1; Re=350)

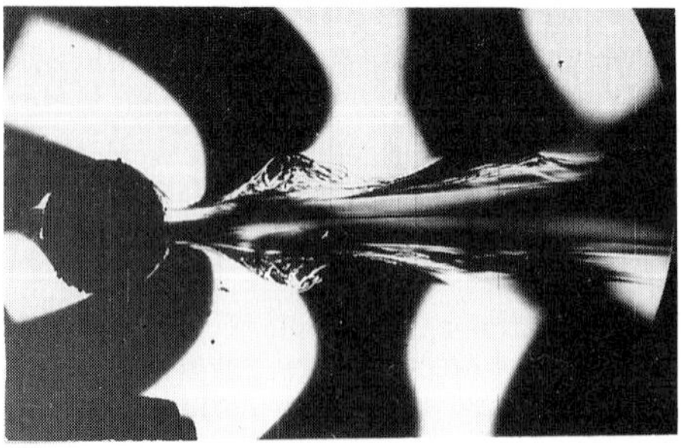

Fig. 4. Vortex billows in the wake. The shape of the wake boundary corresponds to the pattern of attached internal waves. (T_b=4.5s; d=4cm; U=0.95cm/s; Fr=0.16; Re=380)

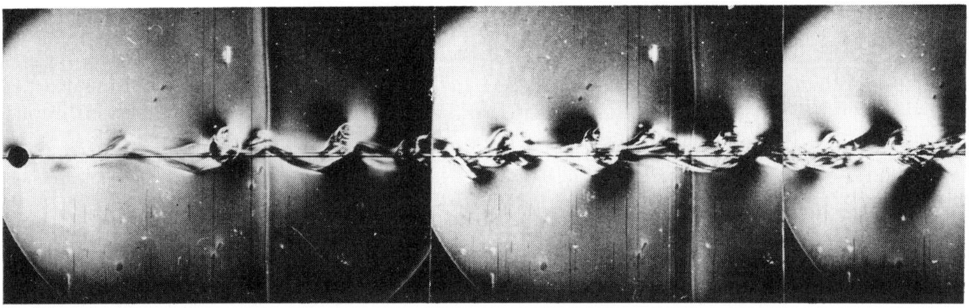

Fig. 5. Formations of eddies due to reconnection of vortex lines in turning domains of the meandering wake past a sphere. (T_b=10s; d=1cm; U=5.7cm/s; Fr=9.1; Re=570; M_ν=d/L_ν=0.29)

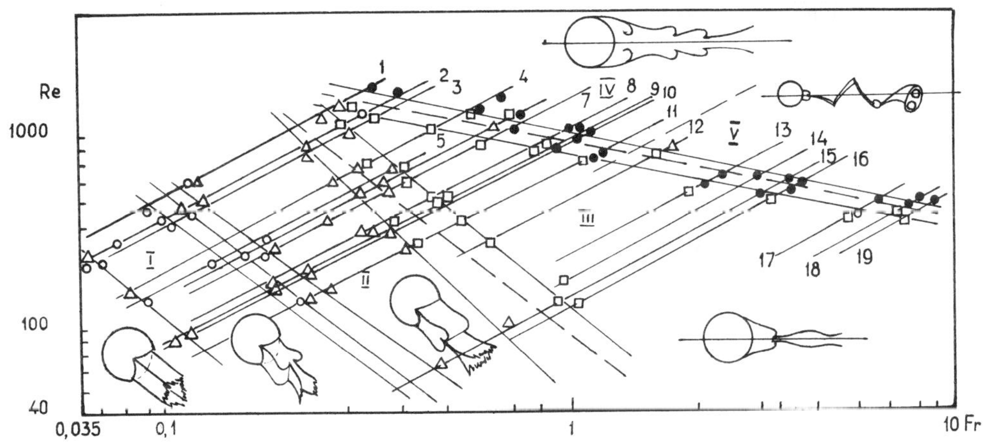

Fig. 6. Flow regime map.
Symbols:
I — Attached pair of vertical vortex columns;
II — Vortex – wave billows in the varicose wake;
III — Accumulation and shedding of vorticity;
IV — Axisymmetric shedding vortex puffs, $M_\nu > 1$;
V — Compact secondary vortices, $M_\nu < 1$

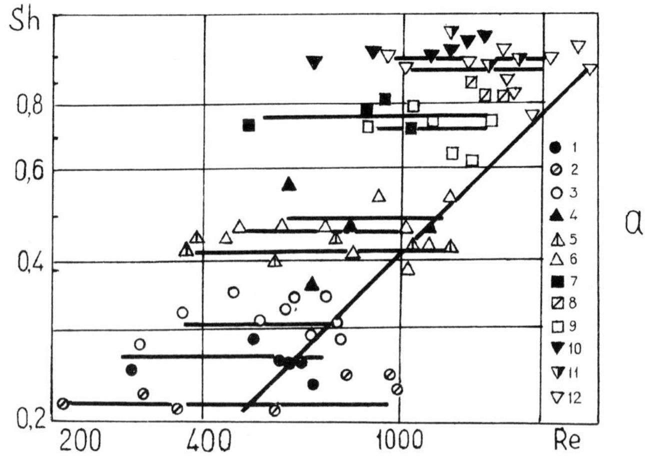

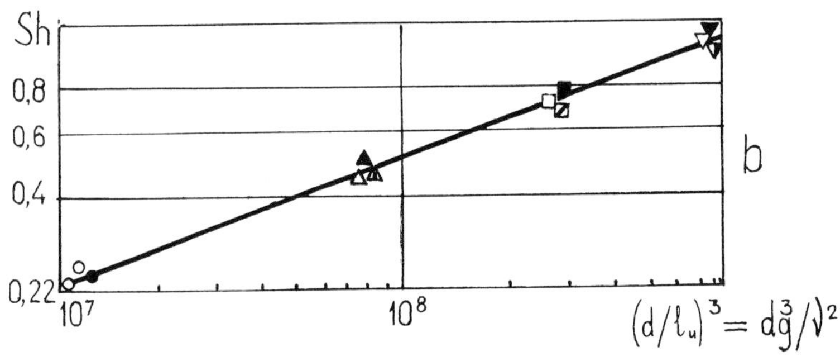

Fig. 7.a. Dependence of the Strouhal number on the Reynolds number

Symbols:

1 — d=1cm; T_b=8s; 2 — d=1cm; T_b=4.5s

3 — d=1cm; T_b=3.5s; 4 — d=2cm; T_b=8s

5 — d=2cm; T_b=4.5s; 6 — d=2cm; T_b=3.5s

7 — d=3cm; T_b=8s; 8 — d=3cm; T_b=4.5s

9 — d=3cm; T_b=3.5s; 10 — d=4cm; T_b=8s

11 — d=4cm; T_b=4.5s; 12 — d=4cm; T_b=3.5s

Fig. 7.b. Strouhal number as a function of the ratio of lengths scales.

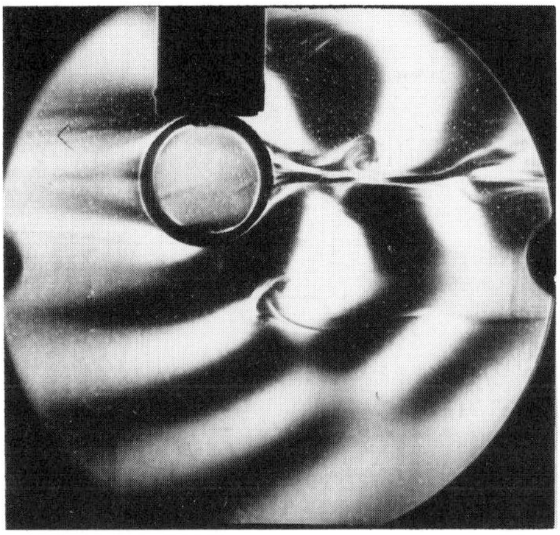

Fig. 8. Isolated suspended vortices past a cylinder. (T_b=30s; d=5cm; U=0.15cm/s; Fr=0.15; Re=79)

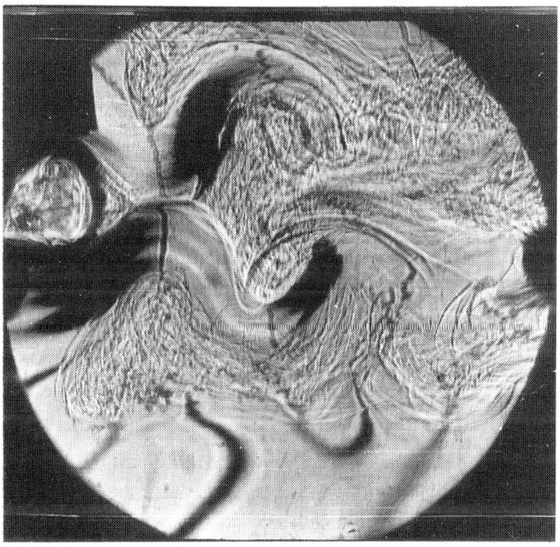

Fig. 9. Sequence of soaring vortices near pulsating wake (T_b=15s; d=5cm; U=0.64cm/s; Fr=0.3; Re=320; C=Λ/d=128; M_ν=1)

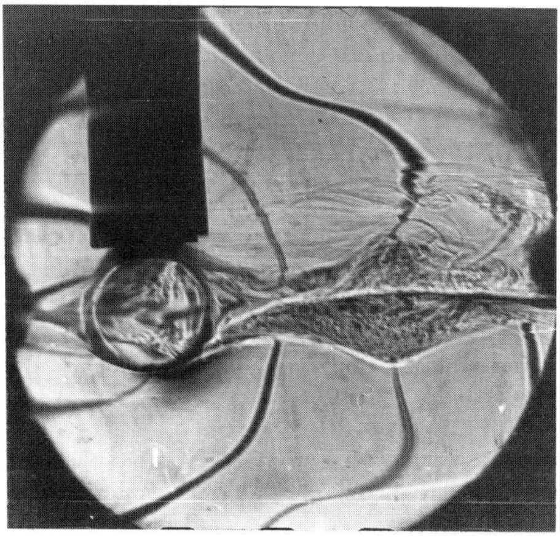

Fig. 10. The wake past a cylinder with laminar vortex bubles. (T_b=14s; d=5cm; U=0.75cm/s; Fr=0.3; Re=375)

Application of Digital Particle Image Velocimetry to a turbulent pipe flow

J.G.Th. van der Hoeven, J. Westerweel, F.T.M. Nieuwstadt
Laboratory for Aero- and Hydrodynamics
Delft University of Technology
Rotterdamseweg 145, 2628 AL Delft, Netherlands

R.J. Adrian
Department of Theoretical and Applied Mechanics
University of Illinois at Urbana-Champaign
104 S. Wright Street, Urbana Il 61801, U.S.A.

Summary
Presented are the results of probably the first successful application of digital particle image velocimetry (DPIV) in a fully developed turbulent flow. The flow under study was a small region close to the wall (ranging from $40 < y^+ < 90$) in a smooth pipe flow, at a relatively low Reynolds number of 5,300, based on the bulk velocity and the diameter of the pipe. The advantage of DPIV in comparison with the traditional implemented PIV method is the enormous reduction in analysis time. The time needed for 1000 interrogations with the conventional PIV analysis is typically 1000 seconds, whereas for DPIV about 60 seconds are needed. This reduction in analysis time is achieved at the cost of a small reduction in accuracy. The present DPIV experiments were carried out with a simple video camera with an effective resolution of 256x512 pixels. Despite this rather poor resolution, the results demonstrate that DPIV is sufficiently accurate to yield statistically significant results. The accuracy is estimated at 0.2-0.4 u^+, which corresponds to a measured particle image displacement of 0.1-0.2 pixel. The turbulent statistics show satisfactory agreement in comparison with data obtained with the conventional PIV method. Visual comparison of vector maps representing instantaneous fluctuating velocity fields obtained from the presented DPIV measurements and conventional PIV experiments previously carried out in the same facility, show that both techniques are able to capture similar flow patterns. A first investigation to detect flow structures gives promising results.

1 Introduction
It is generally accepted that turbulent flow is not random, but consists of spatially and temporal coherent structures. These flow patterns can be observed e.g in flow visualization. With traditional single-point measurement techniques, like hot-wire and laser Doppler anemometer (LDA), it is difficult to observe the spatial character of these structures. Therefore, additional multi-point techniques are required.
Particle Image Velocimetry (PIV) is such a technique. With PIV one can obtain quantitative information on the instantaneous flow pattern in a plane. This yields direct information on the spatial coherence of structures. In addition, the spatial distribution of derived quantities can be computed, such as the out-of-plane vorticity component and the invariants of the 2D deformation tensor. It is believed that these quantities may be relevant for the dynamics of coherent structures.

J. P. Bonnet and M. N. Glauser (eds.), Eddy Structure Identification in Free Turbulent Shear Flows, 405–414.

The technique of PIV starts with a flow which is seeded with small particles that accurately follow the flow. The particles are usually illuminated with a laser sheet. Traditionally, doubly or multiply-exposed images of the particles are recorded on photographic film. The film negative is then analyzed by means of scanning it with an optical/electronic interrogation technique. The principle of this technique is as follows. A small interrogation spot in the negative is illuminated with a laser beam. This results in a Young's fringes interference pattern in the far field of the transmitted light beam. The distance of these fringes is directly proportional to the local fluid velocity. Their orientation is perpendicular to the displacement of the particle-image pairs on the negative (Adrian, 1986). This pattern is subsequently digitized with a typical resolution of 256x256 pixels, for analysis with a computer. The fringe pattern is equivalent to the Fourier transform of the auto-correlation of the interrogation image. Hence, one can alternatively carry out the analysis by direct observation of the interrogation image and by computation of the auto-correlation directly.
This method, based on the analysis of either the fringe pattern or the direct computation of the correlation function, yields accurate local velocity measurements, usually better than 1%. In addition, the spatial resolution is rather high, up to 100x100 interrogation spots per photograph. However, this technique is rather slow, up to several hours per photograph are needed. This rather large analysis time is due, partly to the developing of the film, and partly to the scanning process, which requires the traversing of the camera and laser beam across the negative. Therefore, the conventional PIV technique is only appropriate to treat of a small set of images.

For the study of the dynamics of coherent structures, like e.g. in turbulent boundary layer flows, one is interested in the time evolution of the flow or in the statistics of the observed flow patterns. In this case analysis of large numbers of images is essential. A slow analysis method is then unacceptable. Therefore, we have developed an alternative PIV technique based on a digital analysis of the images instead of the optical (fringe) analysis. (From hereafter we will denote the alternative digital method as DPIV and the conventional method as OPIV.)
First, instead of using photographic film, the complete image of the particles is recorded digitally with a CCD-array, with a typical resolution of 512x512 pixels. Second, the digitized image is then analyzed by interrogating small regions, usually 32x32 pixels or 64x64. In such an interrogation window the distance between particle images is obtained from the correlation function calculated by FFT's. For further details on this alternative analysis for large numbers of images we refer to Westerweel *et al.* (1990).
Using this technique the analysis time per complete image is now reduced to seconds rather than hours. Therefore, sequences of images (100-1000 frames) can be analyzed within reasonable time. However, there is a small reduction in the measuring accuracy, which is now 1-3%, and a somewhat lower spatial resolution, i.e 1,000-4,000 interrogations/image.
Some additional advantages arise due to the digitization of the complete image before analyzing. First, standard image processing techniques can be applied, e.g. to correct for noise or non-uniform background lighting. Second, series of *singly* exposed images can be analyzed using the *cross*-correlation technique on two successive images to obtain the velocity information (Willert and Gharib, 1991). In this way, one gets rid of the velocity direction ambiguity present using the *auto*-correlation technique on *doubly* exposed images. In addition, cross correlation allows a more accurate determination of the displacement of the correlation peak.

Present are the results of the application of DPIV in a fully developed turbulent air

flow. The flow under study was a smooth pipe flow at a relatively low Reynolds number. Recently PIV measurements using the conventional analysis technique were carried out in the same pipe flow facility under the same flow conditions. In that case a total of 33 photographs was taken in a plane through the centreline of the pipe, across the full diameter. For these measurements the negatives of the PIV photographs were analyzed in the interrogation system described by Landreth and Adrian (1990). The results of these experiments were presented together with the results of LDA measurements and DNS calculations for comparison by Westerweel *et al.* (1992). Some experimental conditions and data of the PIV measurements are summarized in Table 1. This will be the set of measurement data, with which we will compare our present DPIV data.
The main reason for doing these experiments was the opportunity to compare the results of the, probably, first application of DPIV measurements in a fully turbulent flow with the results of measurements with the conventional implemented PIV carried out in the same flow facility and under the same flow conditions. To investigate the statistical quantities of flow properties, like vorticity, which are associated with flow structures a larger number of images is required. Therefore, under the same flow conditions, a set of 501 images of doubly exposed particles was taken using a simple video camera. The data was analyzed in terms of statistics, such as mean and rms profiles. Subsequently, conditional sampling was applied.

Table 1 Some experimental conditions and data in the PIV measurements (Westerweel *et al.* (1992)).

Re	5,277	
Re_*	366.3	
U_b	626.6	(mm/s)
u_*	43.5	(mm/s)
c_f	9.56×10^{-3}	
ν/u_*	0.347	(mm)
D	127	(mm)

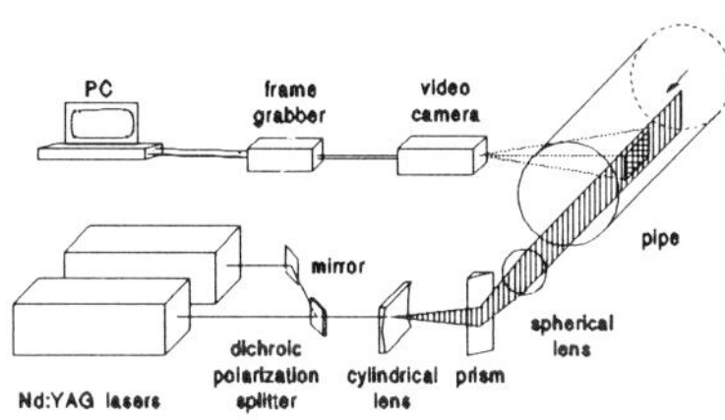

Figure 1. Schematic diagram of the experimental set-up

2 Experimental Set-up

A schematic of the experimental set-up is given in figure 1. Except for the image acquisition, the experimental set-up and flow conditions are the same as described in more detail by Westerweel *et al.* (1992).

Turbulent pipe flow

The images were taken near the end of a 17 m long smooth pipe with an inner diameter of 127 mm. The Reynolds number was 5,277, based on the bulk velocity and the diameter of the pipe. To verify if the air flow was fully developed at the test section LDA measurements were carried out. For the results, which confirmed that the velocity profile was fully developed, and a description of these experiments we refer to the paper by Westerweel *et al.* (1992). The flow was seeded with 1-2 μm diameter oil droplets generated by two atomizers. Injection was about 14 pipe diameters downstream of the flow inlet.
The optical set-up for the pulsed laser light sheet is given in figure 1. Two Nd:YAG pulsed lasers with a repetition rate of 20 Hz were accurately synchronized with a time delay of 0.50000 ms. The orthogonally polarized laser beams were combined into a double pulsed beam by a polarization splitter plate. A prism at the pipe outlet

reflected the beam 90 degrees into the pipe. A spherical lens combined with a cylindrical lens transformed the beam into a thin, vertical light sheet. The thickness of the light sheet was about 1-2 mm.

Image acquisition

The image acquisition system was built up with commercially available components. A simple video camera (Panasonic GP-MF702) with 649x491 (non-square) pixels was used in the interlaced mode, with a low cost Data Translation frame grabber in a PC-AT. The frame rate of the video camera was 30 Hz. The camera was synchronized with the two lasers to obtain doubly exposed images. The memory of the frame grabber could store 4 images. To empty its memory took about 1-2 minutes per 4 frames. Therefore, the 501 taken images are not a 'continuous' time series.
Limited by the number of pixels of the video camera (because the camera was operated in interlaced mode only 256 lines in y-direction are available), the overall image size was chosen rather small, 20 x 15 mm^2. Expressed in viscous units the length and height of the image was 60 x^+ and 50 y^+, respectively. The camera was focused on a small part of the wall boundary layer, 15 to 30 mm distance from the wall, which corresponds to y^+ ranging from 40 to 90. This particular location was chosen, based on the previous OPIV measurements, which revealed it as an interesting area.

3 Results

First, the results of the different steps of the image analysis are presented and discussed. Next, the results of turbulence statistics are presented in comparison with the results of the OPIV experiments in the full pipe diameter. In addition, some results of conditionally sampled statistics are presented, compared with the results from the OPIV data and a direct numerical simulation (DNS) carried out at the same Reynolds number (Eggels *et al.*, 1993).

Preprocessing

The digitally recorded video images can be enhanced relatively easy using standard image processing routines. The camera was used in the interlaced mode. Therefore, as a first step, the black lines were removed and we interpolated the remaining 256 lines to 512 lines. Next, the images were normalized to correct for possible non-uniform background lighting. For these operations we used the software package TCL-image.

Interrogation analysis

In general, PIV data sets contain erroneous vectors. These are due to an insufficient local density of particle-images, or to a non-uniform particle displacement distribution within the interrogation area (Keane and Adrian 1990). After a test analysis of a small number of images it turned out that the particle density was too low for the analysis with 32x32 $pixel^2$ interrogation areas. Therefore, we increased the size of the interrogation window to 64x64 $pixels^2$. This corresponds to a spatial averaging over an area of 7.5 y^+ x 8.2 x^+ units. The overlap of neighbouring interrogation areas was 75%. The combination of the used camera (649x491 px^2) and frame grabber (512x513 px^2) yielded images with 506x480 px^2 exposed pixels. This yielded a set of 501 instantaneous velocity fields of 27x27 nodes each. A vector map of one of the obtained instantaneous velocity fields is shown in figure 2.
The mean displacement of the particle images, as a function of distance from the wall, varied from 7.5 to 8.5 pixels. This means a relative particle displacement compared to the length of the interrogation area of 8/64. This value is in agreement

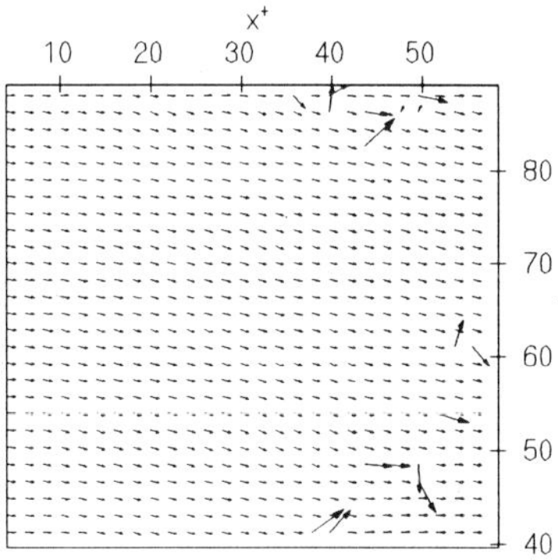

Figure 2. Vector map of the instantaneous velocity field obtained from the interrogation of a DPIV image. The map contains about 2% of spurious vectors.

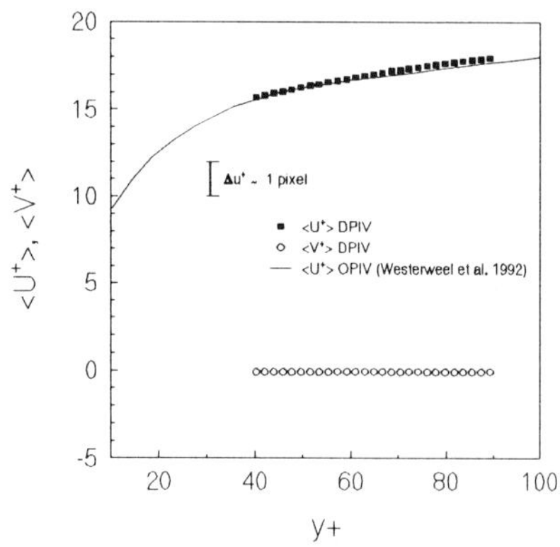

Figure 3. Mean velocity profiles, normalized by friction velocity. The bar denotes the velocity corresponding to a particle displacement of 1 pixel.

with the recommendations by Keane and Adrian (1990).

The analysis of one frame took less than 45 seconds (on our laboratory computer HP9000/720). That is, 60 seconds per 1000 vectors. For comparison only we mention here that for the analysis of an OPIV negative 1000 seconds per 1000 interrogations were needed (disregarding the time needed for the development of the film).

Postprocessing: erroneous vectors

As the example of a raw DPIV vector map shows there are some erroneous vectors (see figure 2). For large numbers of images it would be an enormous laborious task to detect and replace these spurious vectors manually. In addition, visual inspection of PIV measurement data is subjective, and therefore not reproducible and not optimal. Therefore, an algorithm has been written, which rejects vectors as spurious based on a deviation from the local median. An account of the details of this procedure and its efficiency is described by Westerweel (1993). After rejection, vectors are automatically replaced by a new velocity vector which value is computed by means of the interpolation of the eight neighbouring vectors.

Counting the number of detected outliers per image we found that about two third contained less than 1% of spurious vectors (for one third of the images the number of detected outliers was zero). We disregarded all the images which contained more than 5% of spurious vectors. This yielded that 93%, i.e. 464 of the images was used for further analysis.

Turbulence statistics: mean and rms profiles

The turbulent flow statistics were obtained from the remaining 464 instantaneous velocity fields by averaging the data along lines of constant y in each field and then calculating the line averages over the full set. To normalize the flow quantities with the viscous length scale, ν/u_*, and the friction velocity, u_*, we used the results obtained previously with the OPIV measurements (see table 1). The results are

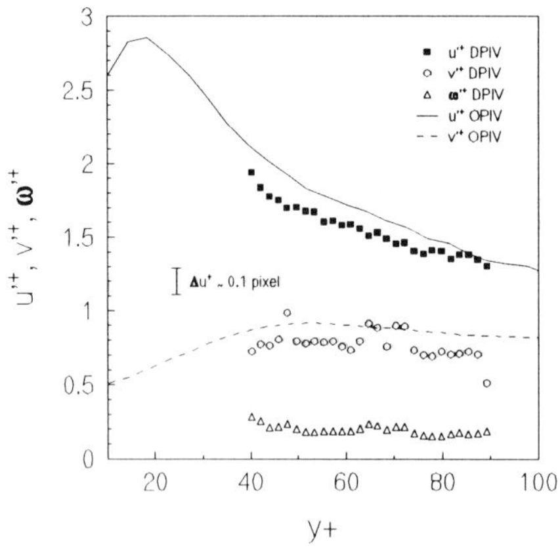

Figure 4. RMS profiles, normalized by friction velocity. The bar denotes the velocity difference corresponding to a measured particle image displacement of 0.1 pixel.

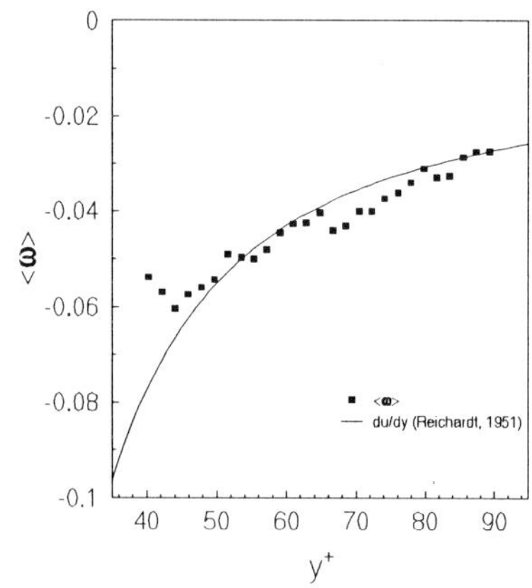

Figure 5. Mean vorticity profile, in viscous units. The line represents the computed profile obtained with Reichardts' formulae for a fully developed turbulent pipe flow.

presented in the figures 3-5. The normalized mean velocity profile (figure 3) and the normalized rms profiles (figure 4) are compared with the results from the OPIV data. As mentioned before, we used the results obtained with OPIV to normalize the DPIV data. Therefore, the agreement between the two mean profiles is rather good. In figure 3 the bar represents the velocity corresponding to a measured particle image displacement of 1 pixel.

In the rms profiles we see a difference of about 10%. This can be explained in terms of the differences in size of the interrogation windows between the two experiments. One should be aware of the fact that because of the finite size of the interrogation window the measured velocity field is a low-pass filtered representation of the observed flow field. The filter length is proportional to the size of the interrogation window. In the OPIV measurement the diameter of the interrogation window was about 4 viscous units, while in the DPIV measurement it was about 8 viscous units. Therefore, the turbulent fluctuations in the DPIV measurements are not fully resolved, which results in a slightly smaller value of rms fluctuating velocities. The bar in figure 4 shows the accuracy in the velocity, corresponding to 0.1 pixel. This indicates that the particle image displacement is estimated within sub-pixel accuracy.

We subsequently computed calculated the out-of-plane component of the vorticity, given by $\omega_z = \partial v/\partial x - \partial u/\partial y$. As an additional check of the accuracy of the measured data we compared the mean out-of-plane vorticity component with the derivative of Reichardts' formulae for the universal profile for fully developed pipe flow (Reichardt, 1951). This is shown in figure 5. Except for the scattering in the measurements the agreement is good. Landreth and Adrian (1990) suggest to apply a careful smoothing prior to estimation of the vorticity, which to suppress the amplification of noise in the vorticity data due to differentiation of the velocity data. Since this would imply a further loss of spatial resolution we computed the vorticity directly from the measured displacement field. The agreement between the experimental and expected

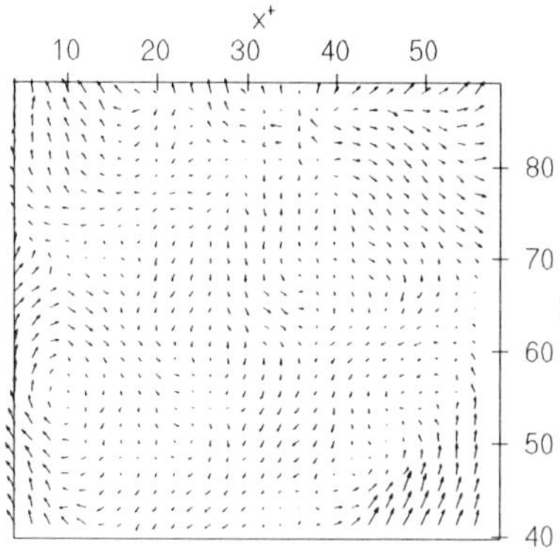

Figure 6. Vector map of the instantaneous fluctuating velocity field as obtained after subtraction of the local mean velocity from the field shown in fig. 2.

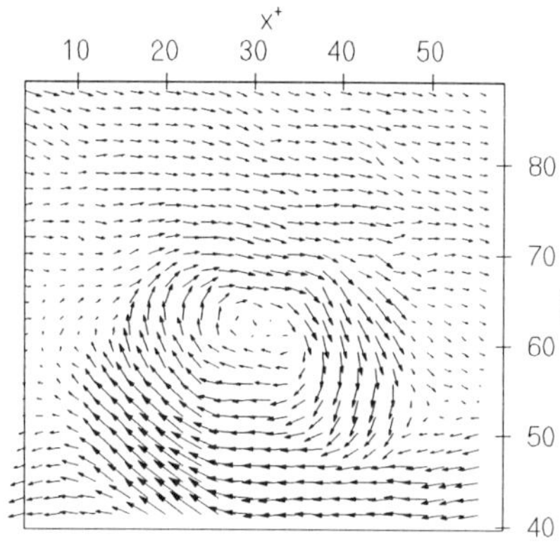

Figure 7. As in figure 6, but obtained from another DPIV image.

result in figure 5 demonstrates that the vorticity - without the additional smoothing - yields significant results.

Turbulence statistics: structures

Visual inspection of the instantaneous fluctuating velocity fields reveals 'structure-like' features. As an illustration two fields are shown in figures 6 and 7. The fluctuating fields are obtained after subtraction of the local mean velocity. For demonstration reasons only, the result was multiplied with a factor of 10. When this procedure is carried out on the vector field shown in figure 2 we obtain the result shown in figure 6. The fluctuating velocity components are rather small, especially if seen in comparison with the instantaneous fluctuating field shown in figure 7. This field reveals a dominant coherent movement in the flow. Figure 8 shows a detail of an instantaneous fluctuating velocity field obtained with OPIV (taken from Westerweel *et al.*, 1992). The similarity between the large patterns is striking! We emphasize that these flow patterns were found in the same pipe flow facility at equal flow conditions, but in different experiments. It should be repeated that the OPIV experiments covered the full pipe diameter instantaneously with sufficient spatial resolution. Whereas for the DPIV experiments we focused on a relatively small region of the pipe flow, due to the relatively low spatial resolution of the image acquisition components. However, visual comparison of the two instantaneous fluctuating fields (figs. 6 and 6) demonstrates that both techniques are able to capture similar structures. Despite the lower spatial resolution and measurement accuracy of DPIV in comparison with OPIV it is possible with DPIV to resolve coherent flow structures in turbulent flows. The obvious advantage of DPIV is that the time needed to record and process the digital image to yield the result shown in figure 7 is only a fraction of the time needed for the analysis of an image with OPIV.

Dealing with large numbers of images, fully automatic algorithms are needed to

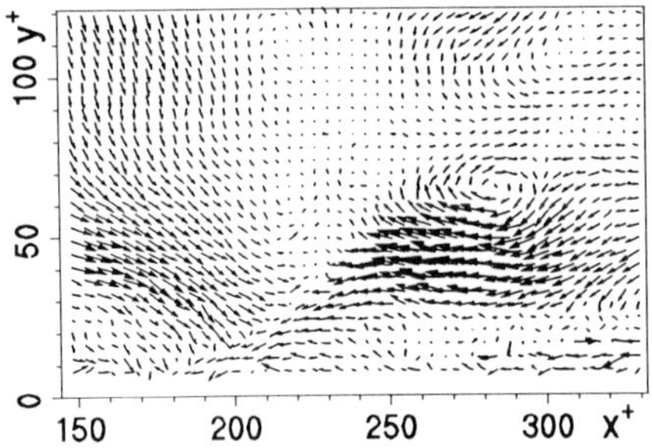

Figure 8. Detail near the wall from the instantaneous fluctuating velocity field as obtained from the interrogation of an OPIV photograph (Taken from Westerweel, et al. (1992)).

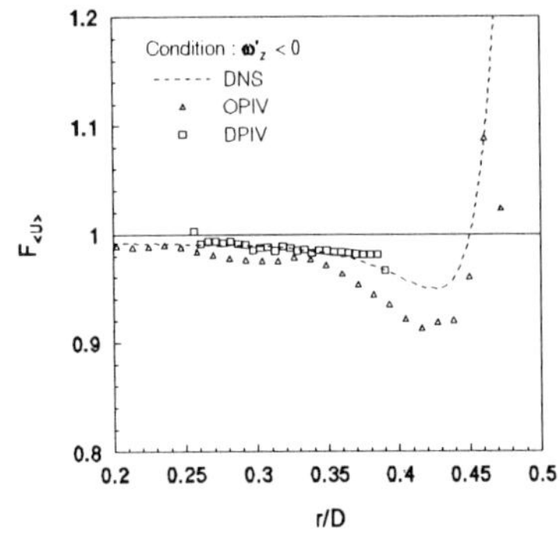

Figure 9. Relative importance of the vortical flow structures to the mean velocity as a function of r/D.

search for flow structures. Recently, Eggels *et al.* (1993) presented some results of a quantitative comparison study of vortical flow structures generated by a direct numerical simulation (DNS) in the same geometry (viz. a smooth pipe) and flow conditions as the OPIV measurements by Westerweel *et al.* (1992) and the DPIV measurements presented here. For obvious reasons of comparison, our data has been processed exactly in the same way.

The assumption underlying their conditional sampling criterion is based on observations of the instantaneous 2-D fluctuating vorticity distributions as obtained with OPIV. It seems that the regions with negative vorticity fluctuations, $\omega'_z < 0$, tend to be more organized in flow structures. Therefore, the regions where $\omega'_z > 0$ were neglected. The former regions are referred to as vortical flow structures. To investigate the relevance of these vortical flow structures to the turbulence flow statistics conditional sampling is used. Therefore, each mean turbulent flow quantity $<\phi>$ is assumed to consist of two contributions: $<\phi>=<\phi>_{vfs}+<\phi>_r$, with the conditionally sampled averages over the vortical flow structures $<\phi>_{vfs}$ and over the remaining flow regions, $<\phi>_r$, respectively,

$$<\phi>_{vfs}=\frac{1}{N}\sum_{i=1}^{N_{vfs}}\phi(i) \qquad <\phi>_r=\frac{1}{N}\sum_{i=1}^{N_r}\phi(i)$$

N equals the total number of data samples, N_{vfs} denotes the number of samples which, according to the conditioning criterion, correspond to vortical flow structures, N_r denotes the number of samples corresponding to all other flow regions. The fraction of samples corresponding to either of the flow types are denoted with n_{vfs} or n_r, with $n_{vfs} + n_r = 1$. To investigate the relevance of the vortical flow structures to the mean turbulence statistics the averaged contribution of both vortical flow structures and all the other flow regions are compared, expressed in the following parameter:

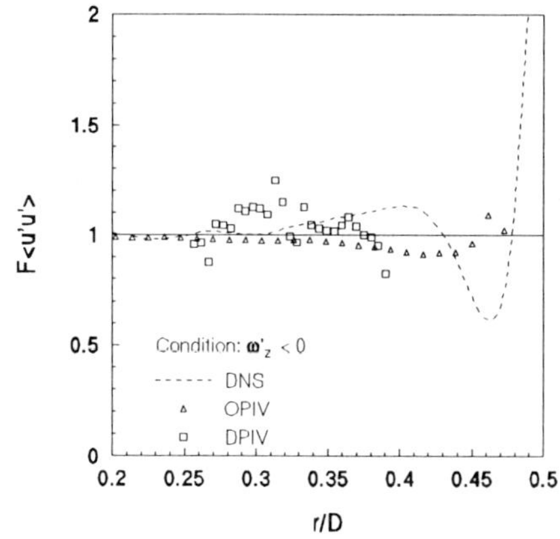

Figure 10. Relative importance of the vertical flow structures to the mean of the streamwise velocity fluctuations, as a function of r/D.

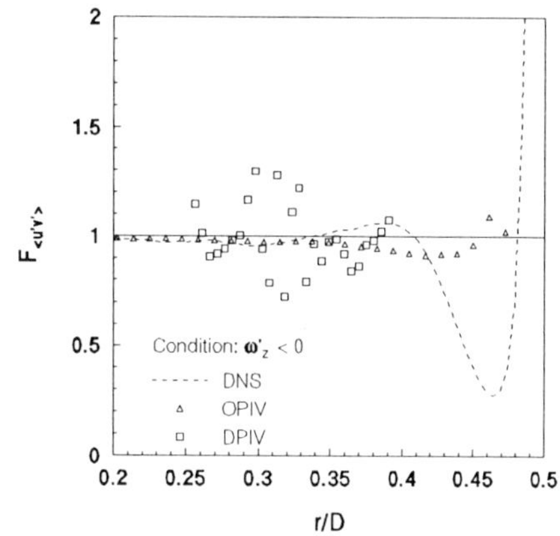

Figure 11. Relative importance of the vortical flow structures to the mean Reynolds shear stress as a function of r/D.

$$F_{<\phi>} = \frac{\dfrac{<\phi>_{vfs}}{n_{vfs}}}{\dfrac{<\phi>_r}{n_r}} = \frac{\dfrac{1}{N_{vfs}}\sum_{i=1}^{N_{vfs}} \phi(i)}{\dfrac{1}{N_r}\sum_{i=1}^{N_r} \phi(i)}$$

If $F_{<\phi>}$ equals one, the averaged contribution of both flow regions is equal. For a more detailed description and for a discussion of the results obtained from the DNS and the OPIV data we refer to the paper by Eggels *et al.* (1993).
The results of this conditional sampling are given in the figures *** in comparison with the results obtained from the OPIV and DNS data sets. In these figures the relative contribution to the mean flow velocity, $<U>$, the mean Reynolds stress, $<u'v'>$, and the turbulence intensity in flow direction, $<u'u'>$ are shown. In the investigated area, $40 < y^+ < 90$, the value of $F_{<\phi>}$ for all three mean quantities is approximately equal to one. $F_{<U>}$ approximately equal to one is indicating, that the vortical flow structures travel with a velocity similar to the local mean flow velocity. The scattering in the DPIV results for $F_{<u'v'>}$ and $F_{<u'u'>}$ is relatively large compared with the results obtained with OPIV.

4 Conclusions

The measurements presented here are probably the first successful application of DPIV in a fully developed turbulent flow. The flow under study was a relatively small area close to the wall in a pipe flow, at a Reynolds number of 5,300. The dimensions of the flow region, ranging from $40 < y+ < 90$, were defined by the resolution of the camera and the frame grabber. A set of 501 directly recorded digital images were analyzed. The analysis time of each image in 730 interrogation positions took less than 45 seconds. This is a considerable improvement in processing speed in comparison with the conventional method applied in OPIV. Hence, we could easily analyze this large amount of images in a very short time. The turbulence statistics of first and second order quantities, calculated from the data are in fair agreement to previously carried out OPIV measurements spanning the full diameter of the same pipe flow. The accuracy is estimated at 0.2-0.4 u+, which corresponds to a measured particle image displacement of 0.1-0.2 pixel. Visual inspection of the instantaneous fluctuating velocity fields shows that both techniques are able to capture similar dominant structures. In addition, some results of a numerical comparison of structures found in the two data sets (DPIV and OPIV) were presented. Summarizing, the results of the application of DPIV in a turbulent flow are encouraging.

References

Adrian, R.J., 1986 *Multi-point optical measurements of simultaneous vectors in unsteady flow - a review.* Int. J. Heat and Fluid Flow 7, 127

Eggels, J.G.M., Westerweel, J., Nieuwstadt, F.T.M. and Adrian, R.J., 1993 *Com parison of vortical flow structures in DNS and PIV studies of turbulent pipe flow.* Proc. of the Int. Conf. on near-wall turbulent flows. (Tempe, AZ) March 15-18, 1993

Keane, R.D. and Adrian, R.J., 1990 *Optimization of particle image velocimeters. Part I: Double pulsed systems.* Meas. Sci. Technol. Vol. 1, pp. 1202-1220

Landreth, C.C. and Adrian, R.J., 1990 *Measurement and refinement of velocity data using high image density analysis in particle image velocimetry.* in: Applications of laser anemometry to fluid mechanics, R.J. Adrian et al (Eds.) Springer, p. 484-497

Reichardt, H., 1951 *Vollständige Darstellung der turbulenten Geschwindigkeitsvertei lung in glatten Leitungen.* Z. angew. Math. Mech. Vol. 31, 208

Westerweel, J., 1993 *Efficient detection of spurious vectors in particle image velocimetry data.* Accepted for publication in Experiments in Fluids.

Westerweel, J., Adrian, R.J., Eggels, J.G.M., and Nieuwstadt, F.T.M., 1992 *Measurements with particle image velocimetry on fully developed turbulent pipe flow at low Reynolds number.* Proc. 6th Int. Symposium on Application of Laser Technology to Fluid Mechanics, Ladoan-Instituto Superior Technico, Lisbon

Westerweel, J., Flor, J.B., and Nieuwstadt, F.T.M., 1990 *Measurement of dynamics of coherent flow structures using particle image velocimetry.* in: Applications of Laser Techniques in Fluid Mechanics (ed. R.J. Adrian *et al.*) Springer, Berlin

Willert, C.W. and Gharib, M.E. 1991 *Digital particle image velocimetry.* Exp. in Fluids, Vol. 10, pp. 181-193

Coherent structures in a simulated turbulent mixing layer

Robert D. Moser and Michael M. Rogers
NASA-Ames Research Center, Moffett Field, CA 94035

A direct numerical simulation of a plane turbulent mixing layer has been performed. The simulation was initialized using two turbulent velocity fields obtained from direct numerical simulation of a turbulent boundary layer at momentum thickness Reynolds number 300 (Spalart 1988). The mixing layer is allowed to evolve long enough for self-similar linear growth to occur, with the visual thickness Reynolds number reaching 14,000.

The simulated flow is examined for evidence of the coherent structures expected in a mixing layer (rollers and rib vortices). Before the onset of self-similar growth, such structures are present, with properties similar to the corresponding laminar or transitional structures. In the self-similar growth regime, however, only the rollers are present with no indication of rib vortices and no indication of conventional pairing. This results in a reduction of mixing and layer growth.

1. Introduction

Since Brown and Roshko's (1974) discovery of large-scale, apparently two-dimensional, structures in turbulent mixing layers, there has been considerable research aimed at determining the origin, universality, and dynamical significance of these structures (see Ho & Huerre, 1984 for a review). The structures appear to be related to the two-dimensional rollers that form due to the instability of a laminar shear layer, and it has been widely assumed that these turbulent coherent structures will behave in the same way as their pre-transition counterparts. However, it has also been suggested that two-dimensional turbulent rollers are not a universal feature of mixing layers (Chandrsuda *et al.*, 1978), implying that the occurrence of such structures depends on the character of the inlet disturbances, which may be facility dependent.

The assumption that the turbulent rollers are dynamically similar to their nonturbulent counterparts is attractive because much is already known about laminar rollers and their instabilities. In particular, it is known that an array of such rollers is unstable to subharmonic disturbances, which lead to pairing (Kelly 1967, Pierrehumbert & Widnall 1982) and indeed

J. P. Bonnet and M. N. Glauser (eds.), Eddy Structure Identification in Free Turbulent Shear Flows, 415–428.

pairings have been reported in turbulent mixing layers. It is also known that the rollers are unstable to three-dimensional disturbances that result in their bending and the formation of the so-called rib vortices in the braid regions between the rollers (Pierrehumbert & Widnall 1982, Corcos & Lin 1984). Recent results (Rogers & Moser 1992) suggest that this three-dimensional instability is governed by two interacting mechanisms, one characteristic of flows dominated by rotation (as in the roller), which leads to the bending of the roller, and the other associated with strain-dominated flows (as in the braid or saddle region), which leads to the formation of the rib vortices. As pointed out by Rogers & Moser (1992), the magnitude of the strain in the braids and rollers of the two-dimensional rolled up mixing layer is about the same at the center line, so the difference in character between the two regions is caused by the reduced levels of spanwise vorticity in the braid region. Thus, if the turbulent rollers lead to braid regions that are substantially depleted of spanwise vorticity, we might expect rib vortices to form in the turbulent flow as well. There is experimental evidence for rib vortices in turbulent mixing layers (*e.g.* Breidenthal 1981, Jimenez 1983, Bernal & Roshko 1986). However, it is not clear under what circumstances the laminar stability ideas discussed above are applicable to turbulent mixing layers.

To address the issues raised above, a direct numerical simulation of a turbulent mixing layer was performed. The simulation was designed to model an experimental situation in which the splitter plate boundary layers are turbulent and no external perturbations are present. The resulting flow fields are examined for evidence of the large scale rollers and other flow features that are expected by analogy with the structure in nonturbulent flows (*i.e.* spanwise rollers and streamwise rib vortices). In what follows, §2 is a description of the simulation, §3 is a discussion of the large scale rollers and their pairing, and §4 contains a description of the three-dimensional structure. Discussion and concluding remarks are given in §5.

2. The Simulation

The numerical simulation discussed in this paper was performed by solving the three-dimensional time-dependent incompressible Navier-Stokes equations. For computational efficiency, a temporally evolving mixing layer was simulated rather than the spatially evolving layer typical of experiments. Comparisons of direct numerical simulations show that the temporally and spatially evolving mixing layers are qualitatively and quantitatively similar (Buell, Moser & Rogers 1992). In this study, the solution domain is periodic in the streamwise (x) and spanwise (z) directions with periods $125\delta_m^0$ and $31.25\delta_m^0$ respectively, where δ_m^0 is the initial momentum thickness of the layer. The domain is infinite in the cross-stream (y) direction. A Galerkin spectral method (Spalart, Moser & Rogers 1991) was used to solve the equations. A passive scalar with Schmidt number 1.0 that goes to zero and one in the free streams, is also simulated.

Initial conditions were generated using two realizations of a turbulent boundary layer with a momentum thickness Reynolds number of 300, as computed by Spalart (1988). One boundary layer was used for each side of the mixing layer, with equal and opposite free stream velocities attained at $y = \pm\infty$. The resulting initial mixing layer profile had a Reynolds number of 800 based on the velocity difference across the layer (ΔU) and the momentum thickness, δ_m, defined as

$$\delta_m = \int_{-\infty}^{\infty} \frac{1}{4} - U^2 \, dy,$$

where U is the mean velocity normalized by ΔU. In the remainder of the paper ΔU and the initial momentum thickness δ_m^0 are used to nondimensionalize all quantities, unless otherwise noted.

The flow was allowed to evolve for a total time of 187.5, at which time the momentum thickness had grown by a factor or three. As can be seen in figure 1, the momentum thickness begins growing linearly, which is characteristic of self-similar evolution, at $t \approx 70$. Also, the integrated (in y) turbulent kinetic energy density q_i^2 grows linearly and the integrated dissipation rate ϵ_i is approximately constant by $t \approx 120$. These are the expected behaviors for a self-similar mixing layer. The self-similar growth rate $d\delta_m/dt$ (normalized by ΔU) in the simulation is 0.014, which is in fairly good agreement with the value of 0.016 found in the experiments of Hussain & Zaman (1985). The discrepancy with the value of 0.02 found by Brown & Roshko (1974) and Rodi(1975) is larger. These observations provide evidence that the simulation represents a self-similar turbulent mixing layer for $t > 120$.

At $t = 120$, when the layer becomes self-similar, the momentum thickness of the simulated layer has only grown by a factor of 2, which is small compared to the growth required to attain self-similarity in many experiments (*e.g.* a factor of about 10 in Hussain & Zaman, 1985 and a factor of about 9 in Mehta & Westphal 1985). There are two likely explanations for this difference. First, the inlet mean velocity profiles are different in the experiments and the computations. The experimental profiles, being asymmetric, may require more development to become self-similar. Second, the experiments may inadvertently include organized disturbances (*e.g.* two-dimensional disturbances resulting from the receptivity of the splitter-plate tip). It may take a long development for the effects of these disturbances to be eliminated. It is also not clear if the effects of such disturbances would extend into the self-similar regime.

In the following sections, the simulated mixing layer is examined for coherent structures at two times, one in the self-similar regime ($t = 187.5$) and one prior to the self-similar regime ($t = 78.5$). The earlier time is late enough for mixing-layer structures to have formed from the initial turbulent boundary layers, but is before the layer has reached self-similarity.

3. Spanwise Rollers and Pairing

Contours of spanwise vorticity (ω_z) in an x–y plane of the mixing layer (figure 2) show that at both times examined ($t = 78.5$ and 187.5) there are amalgamations of spanwise vorticity (rollers) interspersed with thinner regions that have less spanwise vorticity (braids). At the later time, the braid regions still contain many small-scale vorticity fluctuations. However, at the earlier time there are at least some braid regions that are nearly devoid

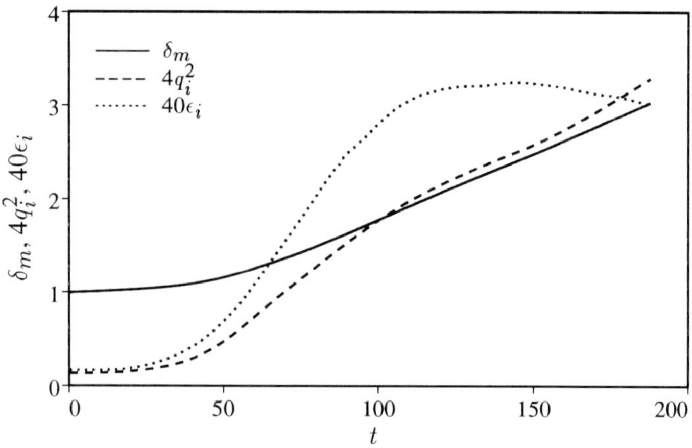

Figure 1. Evolution of momentum thickness (δ_m), integrated (in y) kinetic energy density (q_i^2), and integrated dissipation rate (ϵ_i).

of spanwise vorticity. These "clean" braids are similar to those found in laminar and transitional mixing layers (*e.g.* Rogers & Moser 1992, Lasheras, Cho & Maxworthy, 1986), and to those found in some flow visualizations of turbulent layers (*e.g.* Brown & Roshko 1974).

To study the large-scale structures in more detail, a diagnostic that will locate the rollers and braid regions in x and z is needed. This is similar to the situation in experiments, where a structure must be detected so that it can be included in an ensemble average of similar structures. In this case, however, we wish to map the location and size of the structures in space; no averaging is to be done. The results of Rogers & Moser (1992) suggest that the most important difference between the braids and rollers is the strain dominance of the former and the rotation dominance of the latter. To this end, we consider the two quantities:

$$\Omega = -\frac{1}{2}\int_{-\infty}^{\infty} \omega_z \, dy \quad \text{and} \quad \Sigma_{\mathrm{ij}} = \int_{-\infty}^{\infty} \mathrm{S_{ij}} \, \mathrm{dy},$$

where S_{ij} is the strain-rate tensor, and the cross-stream integral eliminates some of the contributions of small-scale features. When normalized by the local thickness, these quantities are measures of the large-scale vorticity and strain-rate tensor. The layer may be said to be strain dominated where the principal value of Σ_{ij} in the plane normal to the mean vorticity is greater than Ω. This condition reduces to

$$\frac{\partial V}{\partial x} = \frac{\partial}{\partial x}\int_{-\infty}^{\infty} v \, dy > 0.$$

It is expected that $\partial V/\partial x$ will be negative where there are rollers and will be positive where there are braids, and indeed this is the case for the flow considered here. However,

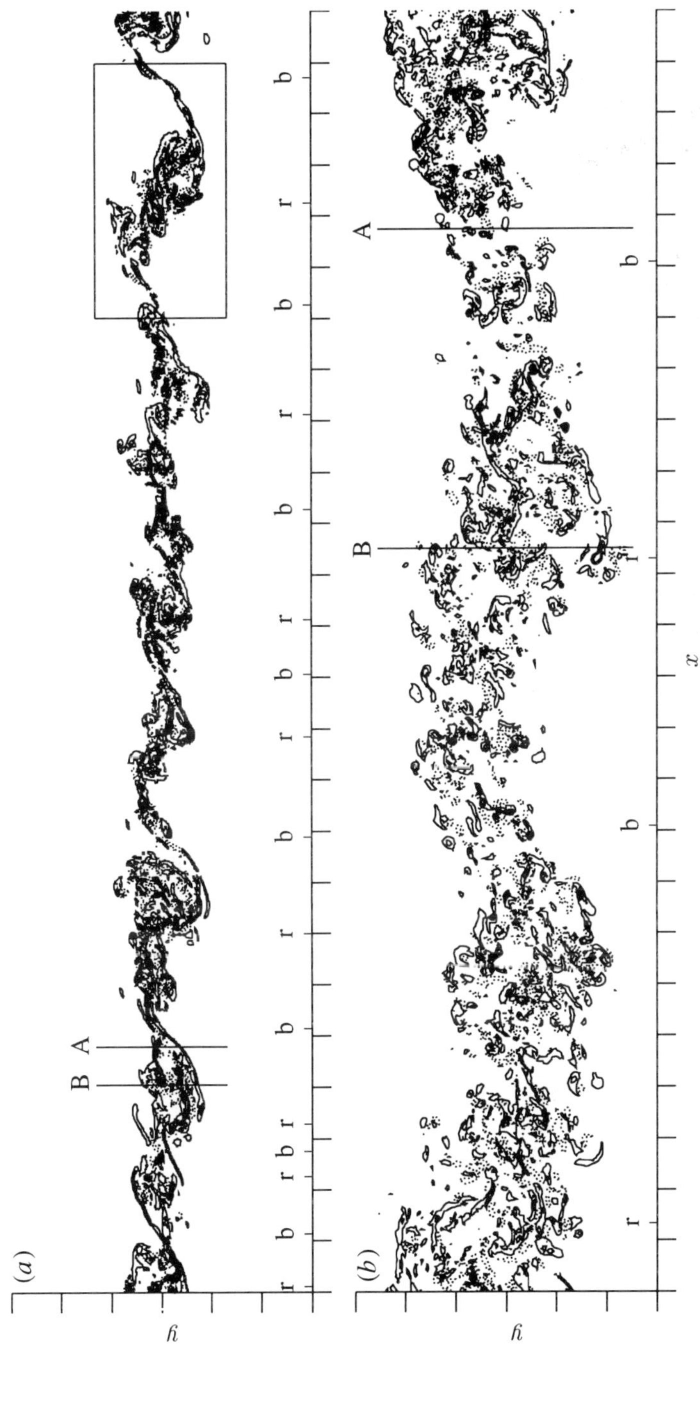

Figure 2. Contours of spanwise vorticity in x–y planes at (a) $z = 18.8$ and $t = 78.5$, (b) $z = 0$ and $t = 187.5$. Contour increments are ± 0.25, positive contours are dotted, and tic marks are at $5\delta_m^0$ increments. Locations marked "b" and "r" are the centers of the braids and rollers respectively as determined from the V diagnostic (see text). The vertical lines mark the locations of the z–y planes depicted in figures 5 and 6. The evolution of the region marked by the box in (a) is shown in figure 4.

Figure 3. Locations in x and z where $V = \int_{-\infty}^{\infty} v \, dy < 0$ (shown in black) at (a) $t = 78.5$ and (b) $t = 187.5$. Braid regions are located at boundaries with black to the left, rollers are located at boundaries with black to the right. Tic marks are at $5\delta_m^0$ intervals and the extra long tic marks on the z axes indicate the z location of the planes shown in figures 2 and 4.

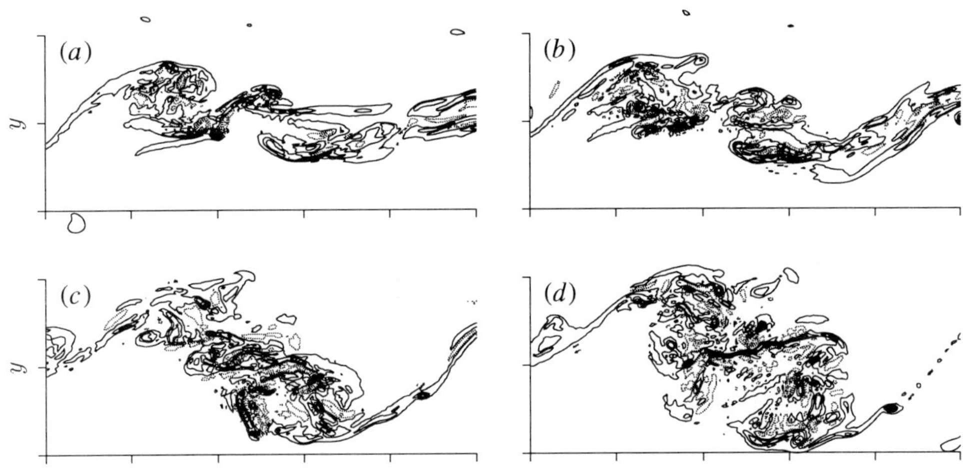

Figure 4. Contours of spanwise vorticity in the $z = 18.8$ x–y plane at (a) $t = 59.4$, (b) $t = 66.3$, (c) $t = 78.5$, and (d) $t = 85.9$. The domain depicted is marked with a box in figure 2(a). The contour increment is ± 0.4, positive contours are dotted, and tic marks are at $5\delta_m^0$ intervals.

variations in $\partial V/\partial x$ are still dominated by small scales, so we consider V instead, which will be decreasing with x in the rollers and increasing with x in the braids. A plot showing where (in x and z) V is positive and negative shows the rollers and braids. In figure 3, braids are seen as boundaries between white and black where black is on the left. A variety of other diagnostics have been tried (*e.g.* integrated in y enstrophy, scalar thickness, integrated strain-rate), but they were all overly sensitive to small-scale features.

The V diagnostic suggests that there are rollers and braid regions in the flow at both times, in agreement with the contour plots in figure 2. The centers of the braids and rollers as determined from V in the planes depicted are marked in figure 2. The V diagnostic also gives an indication of the spanwise coherence of the structures. At $t = 78.5$, the structures have a spanwise extent as large as the streamwise spacing, or larger, but they do not span the entire spanwise domain. There are 8 or 9 rollers in streamwise domain shown. In contrast, the structures at $t = 187.5$ *do* span the domain in z and there are only two structures in x. Since the spanwise domain size is smaller than the streamwise spacing of the structures at $t = 187.5$, it is not possible for the rollers to have the spanwise variations present at the earlier time. Thus the two-dimensionality of the rollers at this time is an artifact of the computation. Remarkably, the number of structures in the streamwise direction has decreased by a factor of about 4 between the times depicted, while the momentum thickness has only increased by a factor of 2. This is consistent with the fact that the rollers appear more elongated at $t = 187.5$ (figure 2(b)).

The roller structures in this flow have clearly increased their streamwise length scale, so one wonders if they did so by the pairing mechanism associated with laminar rolled-up

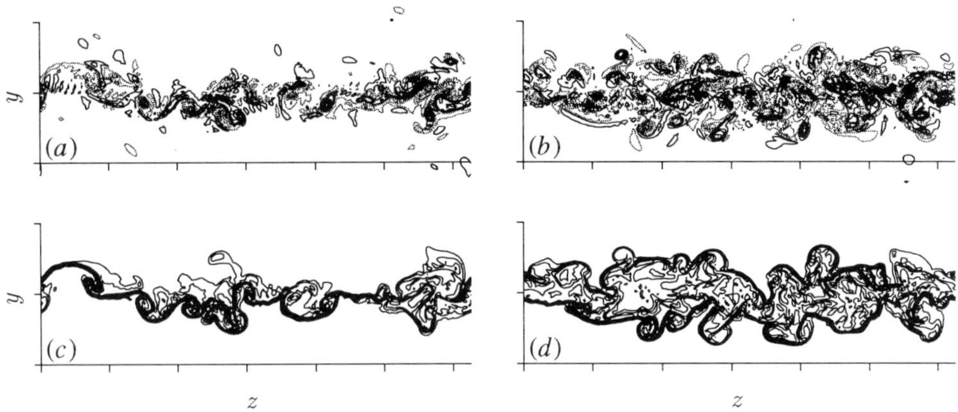

Figure 5. Contours of $(a{,}b)$ streamwise vorticity and $(c{,}d)$ passive scalar in the z–y plane at the line marked $(a{,}c)$ A (braid region) and $(b{,}d)$ B (roller) in figure 2(a) at $t = 78.5$. The contour increment is $(a{,}b)$ 0.3, $(c{,}d)$ 0.1, positive contours are dotted, and tic marks are at $5\delta_m^0$ intervals.

mixing layers. At the earlier time (before self-similarity), pairing apparently does occur. The two rollers in the box in figure 2(a) are rotating one above the other, as can be seen in figure 4. However, after self-similarity no such pairings have been detected. When the self-similar structures change scale, the vorticity from one roller appears to ooze gradually into its neighbors, without the co-rotation characteristic of pairing. This may be the reason why Hussain & Zaman (1985) were unable to detect pairings in their self-similar turbulent mixing layer.

4. Rib Vortices

In laminar and transitional mixing layers, the dynamics of the braid region and those of the roller are different due to the strain dominance of the former and the rotation dominance of the latter (Rogers & Moser 1992). In particular, the braid region is apparently susceptible to a three-dimensional instability, resulting in the formation of streamwise rib vortices. If coherent rib vortices are present in the turbulent flow, they should be relatively easy to detect. In contrast, three-dimensional instabilities of the roller lead to a bending of the roller, which would be difficult to detect in a turbulent flow. We will thus examine the simulated flow at $t = 78.5$ and $t = 187.5$ for evidence of rib vortices.

Contours of streamwise vorticity and the passive scalar in the braid regions (z–y planes marked by the vertical lines labeled "A" in figure 2) and rollers (plane marked "B" in figure 2) are shown for both times in figures 5 and 6. These rollers and braids were selected because they are the most spanwise coherent (see figure 3). At $t = 78.5$, the strong streamwise vorticity in this plane is dominated by compact approximately circular regions of intense vorticity. The passive scalar contours indicate a sharp interface in the

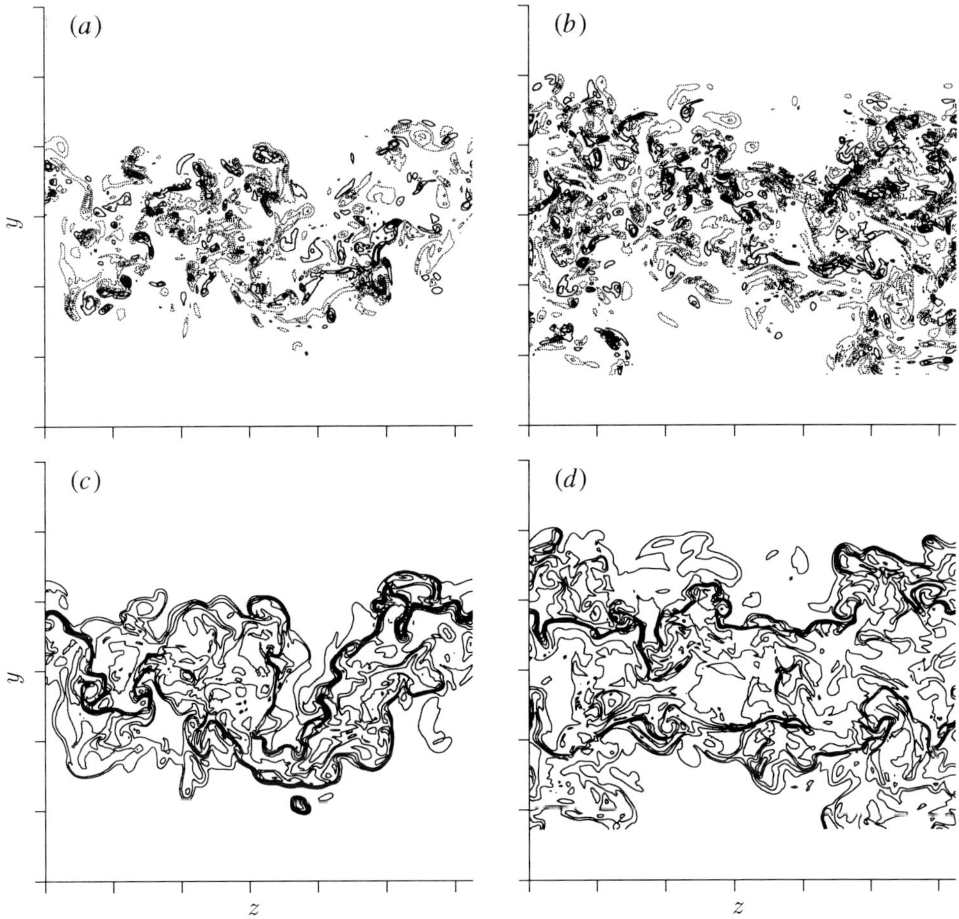

Figure 6. Contours of (a,b) streamwise vorticity and (c,d) passive scalar in the z–y plane at the line marked (a,c) A (braid region) and (b,d) B (roller) in figure 2(b) at $t = 187.5$. The contour increment is (a,b) 0.3, (c,d) 0.1, positive contours are dotted, and tic marks are at $5\delta_m^0$ intervals.

scalar, and this interface is rolled up around the streamwise vortices. These features are characteristic of ribs in the braid regions of nonturbulent mixing layers. In contrast, the streamwise vorticity at $t = 187.5$ (figure 6) is not well organized, although there are a few small circular regions of intense vorticity. The scalar at this time does not have the single sharp interface with discrete rollups seen at the earlier time. Comparing the braid region (figure 6(a)) to the roller core (figure 6(b)) suggests that at $t = 187.5$ the major difference between the braid and roller is that the vortical region in the braid is thinner in the y direction.

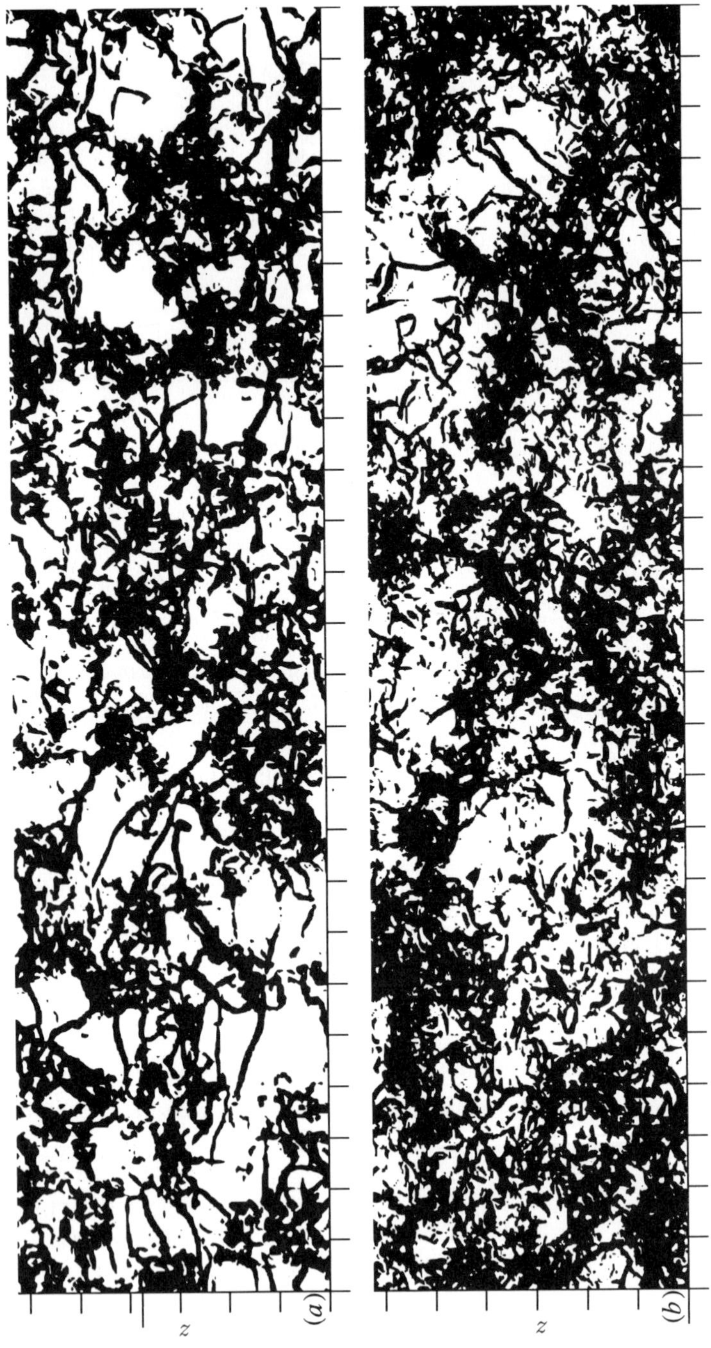

Figure 7. Plan view of high-enstrophy structures (enstrophy greater than (*a*) 2.6, (*b*) 2.0) at (*a*) $t = 78.5$ and (*b*) $t = 187.5$. Tic marks are at $5\delta_m^0$ intervals and the extra long tic marks on the z axes indicate the z location of the planes shown in figures 2 and 4.

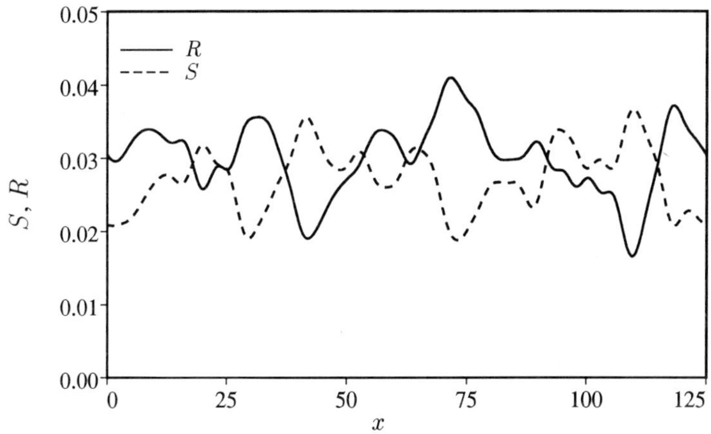

Figure 8. Streamwise (x) variation of the average principal two-dimensional strain rate (S) and average rotation rate ($R = -\omega_z/2$) at $t = 187.5$. The average is taken over the spanwise direction, over the vertical direction in the vortical region ($|\omega_z| > 0.04$), and in the streamwise direction using a Gaussian filter of width 5.

A top view of the high-enstrophy regions of the mixing layer at the two times is shown in figure 7. At $t = 78.5$, many of the braid regions are visible as (white) regions with relatively little vorticity (compare to figure 3). In this figure, the ribs are the long thin roughly streamwise vortices spanning many of the braid regions. At $t = 187.5$, however, the braid regions are not clearly visible, though there is some tendency for the vorticity to be denser in the rollers. Rib vortices are also not so obvious at this time. Although there are some small-scale vortices visible in this figure, none of them extends across an entire braid region. In addition, there does not appear to be any preference for these vortices to occur in the braid region. The small-scale vortices at this time appear to be the "worms" that have been found to be a common feature of turbulent flows (Jimenez, 1992). Jimenez has suggested that turbulent flows have worm vortices with circulation Reynolds numbers (Γ/ν) between 200 and 400. Estimates of the circulations of the strong vortices in figure 6(a) are of this order. However, estimates of rib-vortex circulations in the flow at $t = 78.5$ (figure 5(a)) are as much as 800, suggesting that these vortices represent more than just small-scale turbulent structure.

While the structure of the turbulence in the braids and rollers appears to be similar at $t = 187.5$, the large-scale deformation imposed on the turbulence is different in the two regions. This can be quantified by computing average rotation and strain rates in different regions of the mixing layer. At $t = 187.5$, the rollers are mostly two-dimensional (see figure 3), so a spanwise average can be used. In addition, averages were taken across the vortical region in y (the region where the magnitude of span averaged ω_z is greater than 0.04) and locally in x (using a Gaussian filter with a width of 5). The resulting average two-dimensional principal strain rate and rotation rate ($-\omega_z/2$) are shown as a function of x in figure 8. As expected, the braid regions are strain dominated with strain rates as much as twice the rotation rates, while the rollers are rotation dominated, also by as much as a factor of two. One might expect the effects of this difference to be similar to the effects of such mixed strains on homogeneous turbulence (*i.e.* intermediate between plain strain and shear in the braids and intermediate between pure rotation and shear in the rollers). However, homogeneous turbulence subjected to such strains has not been thoroughly investigated. In any case, the differences between the turbulence in the braids and rollers are clearly too subtle to be detected by the crude visual diagnostics used in figures 6 and 7.

5. Discussion and Conclusions

The observations presented above show that there is a fundamental difference between the large-scale structures in a self-similar turbulent mixing layer and the large-scale structures in the same layer before self-similarity is achieved. Prior to selfsimilarity, the structures correspond closely to the transitional structures that have been studied extensively (Rogers & Moser 1992, Moser & Rogers 1993, and others), with rollers that pair and rib vortices spanning the braid regions. There is no such correspondence in the self-similar regime, which has neither pairings nor rib vortices. It appears that the turbulence in the braid region is not qualitatively different from that in the roller.

Since the "rollers" in this self-similar turbulent mixing layer do not appear to have the dynamical properties expected of such structures, it is not clear what relevance they have to the global development of the mixing layer. For example, if the rollers do not pair, then they will not entrain large amounts of free-stream fluid the way nonturbulent rollers do. Thus, they will not be responsible for a large portion of the mixing layer growth, nor will they dominate the mixing process as was assumed by Broadwell & Breidenthal (1982) in developing their mixing model. The only dynamical significance of the roller structures appears to be that they set the deformation environment in which the turbulence evolves, creating strain-dominated and rotation-dominated regions. The effects of this on the turbulence are yet to be determined.

The lack of ribs and pairings in our self-similar flow is different from several experiments (*e.g.* Brown & Roshko 1974, Konrad 1976, Breidenthal 1980, Jimenez 1983) in which measurements show the layer to be self-similar while flow visualization suggests that ribs and pairings are present. A possible explanation for this difference is the disturbance environment in the experiments, which may include much stronger two-dimensional or quasi-two-dimensional disturbances than are present in the current simulations. Such disturbances might arise due to the receptivity of the splitter-plate tip (Ho & Huerre 1984). The experimental disturbances may also include strong streamwise vortices from imperfections in the facility upstream of the splitter plate tip as in Jimenez (1983). The presence of strong two-dimensional disturbances may also explain the somewhat higher self-similar growth rate in the experiments, because organized pairings of two-dimensional rollers would promote spreading of the layer.

REFERENCES

BERNAL, L. P. & ROSHKO, A. 1986 Streamwise vortex structure in plane mixing layers. *J. Fluid Mech.* **170,** 499–525.

BREIDENTHAL, R. 1981 Structure in turbulent mixing layers and wakes using a chemical reaction. *J. Fluid Mech.* **109,** 1–24.

BROADWELL, E. J. & BREIDENTHAL, R. E. 1982 A simple model of mixing and chemical reaction in a turbulent shear layer. *J. Fluid Mech.* **125,** 397-410.

BROWN, G. L. & ROSHKO, A. 1974 On density effects and large structure in turbulent mixing layers. *J. Fluid Mech.* **64,** 775–816.

BUELL, J. C., MOSER, R. D. & ROGERS, M. M. 1992 A comparison of spatially and temporally developing mixing layers. *in preparation*

CHANDRSUDA, C., MEHTA, R. D., WEIR, A. D. & BRADSHAW, P. 1978 Effect of free-stream turbulence on large structure in turbulent mixing layers. *J. Fluid Mech.* **85,** 693–704.

CORCOS, G. M. & LIN, S. J. 1984 The mixing layer: deterministic models of a turbulent flow. Part 2. The origin of the three-dimensional motion. *J. Fluid Mech.* **139,** 67–95.

HO, C.-M. & HUERRE, P. 1984 Perturbed free shear layers. *Ann. Rev. Fluid Mech.* **16,** 365–424.

HUSSAIN, A. K. M. F. & ZAMAN, K. M. B. Q. 1985 An experimental study of organized motions in the turbulent plane mixing layer. *J. Fluid. Mech.* **159,** 85–104.

JIMENEZ, J. 1983 A spanwise structure in the plane shear layer. *J. Fluid Mech.* **132,** 319–336.

JIMENEZ, J. 1992 Kinematic alignment effects in turbulent flows. *Phys. Fluids A.* **4,** 652-654.

KELLY, R. E. 1967 On the stability of an inviscid shear layer which is periodic in space and time. *J. Fluid Mech.* **27,** 657–689.

KONRAD, J. H. 1976 An experimental investigation of mixing in two-dimensional turbulent shear flows with applications to diffusion-limited chemical reactions. *Intern. Rep.* CIT-8-PU, Calif. Inst. Technol. Pasadena, CA.

LASHERAS, J. C., CHO, J. S. & MAXWORTHY, T. 1986 On the origin and evolution of streamwise vortical structures in a plane, free shear layer. *J. Fluid Mech.* **172,** 231–258.

MEHTA, R. D. & WESTPHAL, R. V. 1985 An experimental study of plane mixing layer development. *NASA TM86698.*

MOSER, R. D. & ROGERS, M. M. 1993 The three-dimensional evolution of a plane mixing layer: pairing and transition to turbulence. *J. Fluid. Mech.* to appear.

PIERREHUMBERT, R. T. & WIDNALL, S. E. 1982 The two- and three-dimensional instabilities of a spatially periodic shear layer. *J. Fluid Mech.* **114,** 59–82.

RODI, W. 1975 A review of experimental data of uniform density free turbulent boundary layers. in *Studies in Convection.* **1,** B. E. Launder, ed., Academic Press, 79-165.

ROGERS, M. M. & MOSER, R. D. 1992 The three-dimensional evolution of a plane mixing layer: The Kelvin–Helmholtz rollup. *J. Fluid. Mech.* **243,** 183-226.

SPALART, P. R. 1988 Direct simulation of a turbulent boundary layer up to $R_\theta = 1410$. *J. Fluid Mech.* **187,** 61-98.

SPALART, P. R., MOSER, R. D. & ROGERS, M. M. 1991 Spectral methods for the Navier–Stokes equations with one infinite and two periodic directions. *J. Comp. Phys.* **96,** 297–324.

Identification of Zones in a Free Evolving Vortex Ring

P. Orlandi and R. Verzicco
Dipartimento di Meccanica e Aeronautica
Università di Roma "La Sapienza" Italy

Summary

A direct numerical simulation of a three-dimensional free evolving vortex ring has been used to understand some of the features of the vortex ring dynamics. In experiments a persistent wake was observed and by this direct simulation the causes of the wake formation have been clarified. By a kinematic alignment study of vorticity with the eigenvectors of the strain rate tensor, together with vorticity visualisations, explanations on the growth of the "Widnall" instabilities have been obtained. The analysis has shown the presence of two zones, the first characterised by intense vorticity and low strain rate and the second with opposite features. In the former there is the ring core that maintains its coherence in the long term, in the latter there are elongated structures contributing to the deformations of the ring.

1 Introduction

Round jets are important flows in several practical applications, for example in combustion chambers where a very efficient combustion is related to a large spreading of the jet and to the mixing of the two reacting fluids. The control of the jet is obtained only if it is well known the vorticity dynamics of the basic structures leading the jet initially wavy and finally turbulent. In a detailed flow visualisation Lasheras *et al.* [1] observed that coaxial jets, at least in their first stage of evolution, can be considered as formed by vortex rings which interact, deform their structure and by pairing and tearing reach the turbulent status. Vortex rings are thus block structures of round jets and then the knowledge of the evolution of a free vortex ring is a basic study for the turbulent jets control.

The vortex ring has been experimentally and theoretically investigated in the past; among the large number of experimental studies to refer to Glezer [2] who produced flow visualisations for laminar and turbulent vortex rings showing the formation of a wake. A further aspect of the vortex ring dynamics is the so called "Widnall instability" associated to the evolution of three-dimensional azimuthal disturbances.

J. P. Bonnet and M. N. Glauser (eds.), Eddy Structure Identification in Free Turbulent Shear Flows, 429–440.

The theoretical analysis by Widnall *et al.* [3] was limited to thin rings, but by flow visualisations they evaluated the development of the number of waves with the size of the ring core.

In this study a numerical simulation of the Navier-Stokes equations, by a finite difference scheme, has been performed to understand, by computer visualisations of vorticity components, the reasons of the wake formation and of the azimuthal deformations. Moreoveor by the instantaneous velocity field, the strain-rate tensor, the eigenvalues and the eigenvectors have been calculated as was done by Ashurst *et al.* [4] in isotropic turbulence. By these quantities the alignment between vorticity vector and eigenvectors of the strain rate-tensor is evaluated. The knowledge of $Q_s = S_{ij}S_{ji}$ and of the determinant R_s of the strain-rate tensor reveals some of the features of the flow. For example the sign of R_s tell us whether two eigenvalues are positive or negatives, in the case of positive R_s the vortices are rodlike while for negative R_s the vortices are sheetlike. On performing these calculations a threshold value of the vorticity magnitude together with a particular region of the flow were selected. Results have shown that where vorticity is high, i.e. in the core, there is high probability to have $R_S > 0$ while the opposite occurs in the region near the axis where the rate of strain is greater than vorticity. This is a condition for a growing instability.

2 The Model

The Navier-Stokes equations for an incompressible viscous flow in primitive variables and in cylindrical coordinates have been solved by a finite difference scheme. The scheme is second order accurate in space and time and it locates the unknowns on a staggered grid with the velocities on the faces of the computational cell, the pressure at the centre and the vorticities at the edges. This choice gives the most compact form of *div grad* and *curl* discrete operators and yields solenoidal vorticity and velocity fields within round-off errors. The solution of the equations has been performed by a fractional step method that evaluates the velocity field in two steps: in the first a non-solenoidal field is obtained then, by the introduction of a scalar Φ, the non-solenoidal field is projected into a solenoidal one. The quantity Φ is calculated by a Poisson equation which does not require explicit boundary conditions. A third-order Runge-Kutta scheme, which compute the viscous terms implicitly and the convective terms explicitly, has been used to advance in time, it consists of three substeps and at each one the scalar Φ is calculated.

More details of the numerical scheme are given by Verzicco and Orlandi [5], here we wish to recall that the equations have been non-dimensionalised by the ring toroidal radius r_0 and its initial circulation Γ, resulting the Reynolds number defined as $Re_\Gamma = \Gamma/\nu$, where ν is the kinematic viscosity. The results here presented have been obtained for $Re_\Gamma = 5500$.

The vortex ring at $t = 0$ had a Gaussian distribution of azimuthal vorticity ω_θ

perturbed with a sinusoidal radial displacement of the toroidal axis with $n = 5$ wavenumber. The maximum amplitude of of the displacement is $\epsilon = 0.02 r_0$.

The expression for ω_θ is

$$\omega_\theta = \frac{\Gamma}{2\pi} e^{-\left(\frac{\rho}{\sigma}\right)^2} \quad \rho^2 \equiv (z - z_0)^2 + (r - r_0(\theta))^2$$

$$r_0(\theta) = r_0(1 + \epsilon \sin(n\theta))$$

The displacement along the radius does not introduce ω_z and the radial vorticity ω_r is evaluated by $\nabla \cdot \omega = 0$.

Calculations in the whole domain have shown the fulfilment of the symmetries in the azimuthal direction. Then a grid of $65 \times 97 \times 129$ has been used to discretize the domain of dimensions $2\pi/5 \times 4r_0 \times 6r_0$ respectively, and periodicity has been imposed in z. The field has been replicated in the azimuthal direction for the graphic visualisation.

From the velocity components, vorticities and the strain-rate symmetric tensor have been evaluated by:

$$S_{11} = \frac{1}{r}\frac{\partial v_\theta}{\partial \theta} + \frac{v_r}{r}, \quad S_{22} = \frac{\partial v_r}{\partial r}, \quad S_{33} = \frac{\partial v_z}{\partial z},$$

$$S_{12} = \frac{1}{2}\left\{r\frac{\partial v_\theta / r}{\partial r} + \frac{1}{r}\frac{\partial v_r}{\partial \theta}\right\}, \quad S_{13} = \frac{1}{2}\left\{\frac{1}{r}\frac{\partial v_z}{\partial \theta} + \frac{\partial v_\theta}{\partial z}\right\}, \quad S_{23} = \frac{1}{2}\left\{\frac{\partial v_r}{\partial z} + \frac{\partial v_z}{\partial r}\right\},$$

$$\omega_z = \frac{1}{r}\frac{\partial r v_\theta}{\partial r} - \frac{1}{r}\frac{\partial v_r}{\partial \theta}, \quad \omega_r = \frac{1}{r}\frac{\partial v_z}{\partial \theta} - \frac{\partial v_\theta}{\partial z}, \quad \omega_\theta = \frac{\partial v_r}{\partial z} - \frac{\partial v_z}{\partial r}$$

The eigenvalues and eigenvectors of the S_{ij} tensor have been calculated by a standard Jacobi method.

The skew-symmetric rotation-rate tensor is related to the vorticity components through:

$$R_{12} = \frac{1}{2}\left(\omega_z - \frac{v_\theta}{r}\right) = -R_{21}, \quad R_{13} = \frac{\omega_r}{2} = -R_{31}, \quad R_{23} = \frac{\omega_\theta}{2} = -R_{32} \quad 1$$

3 Results

At $t = 0$ the ring starts the self-induced translation and it sheds a vorticity wake. The reason of the wake, in a companion paper [6], was proved to be related to the readjustment of the initial Gaussian distribution of ω_θ towards a distribution $\omega_\theta / r = f(\psi)$ which is the solution of the Euler equations. This is strictly valid for an axisymmetric condition, which approximately holds in the first eddy-turnover time ($t \simeq 15$) where the azimuthal disturbances do not grow sufficiently. This wake has a low vorticity content and it is deformed by the strain field, however it is rapidly dissipated and does not affect substantially the ring during its initial translation.

a) b)

Figure 1: Surface contour plots of axial and azimuthal vorticity components a) $t = 30$, b)$t = 57$.

In Fig.1a the surface plots of azimuthal and axial vorticities respectively at a value -1.4 and $\pm.05$ at $t = 30$ are shown and, on making this picture, we have visualised only two opposite structures for a better comprehension of the vortex dynamics. From this picture it is evinced that counter-rotating axial vortices are generated near the core; these, at their turn, induce a hierarchy of opposite sign vortices spanning until the axis, where they become so small to be dissipated by the viscosity. We think that these axial vortices could be responsible of the dye accumulation in the ring centre, depicted in the photograph by Glezer [2]. These vortices and in particular those with a fat structure, encircling the core, cause the radial deformation of the ring that, at this time, is not appreciable. Fig.1b shows that at $t = 57$ the structures of axial vorticity are not visible any longer at the axis of the ring because these have been convected far from the ring and mainly dissipated by viscosity. On the other hand,

a) b)

Figure 2: Surface contour plots of radial and azimuthal vorticity components a) $t = 30$, b)$t = 57$.

the positive axial vorticity within the ring core, represented as the white region on the right side of Fig.1b, has a sufficient strength to substantially deform the ring itself in the radial direction. The amount of ω_z from $t = 30$ has increased for the larger deformation and this intensification causes the birth of opposite sign vorticity localised in a large region encircling the ring core (black structure on the right side).

Similar visualisations were obtained by plotting radial together with azimuthal vorticity. If the same levels, as those of Fig.1, are chosen we observe in Fig.2a that

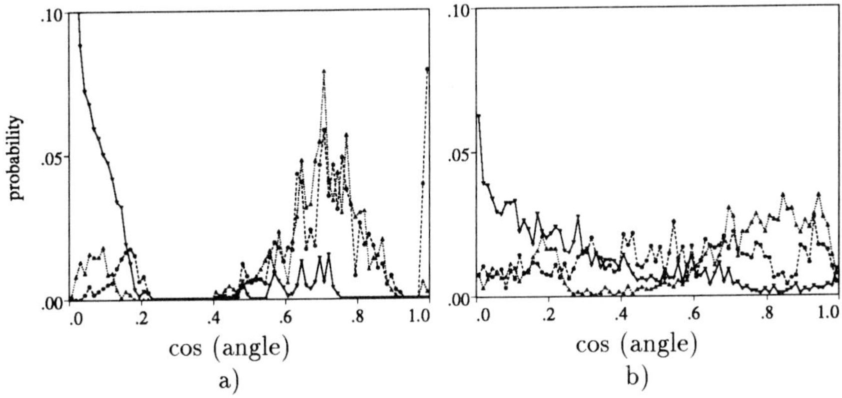

Figure 3: Probability distribution of ,—- $S_{\alpha\theta}$, $---S_{\alpha r}$, $\cdots S_{\alpha z}$, with $\omega_M = 1.5$, a) $t = 30$, b) $t = 57$.

the radial vorticity is localised in smaller regions than those of axial vorticity. Radial vorticity causes deformations in the vertical direction. At $t = 30$, being the maximum value of the vorticity components in the ratio $\omega_\theta : \omega_r : \omega_z = 2.0 : 0.12 : 0.4$, it follows that the deformations of the ring are weakly appreciable, on the contrary at $t = 57.5$ the ratio becomes $2.2 : 0.48 : 0.75$ with the consequence that the deformations are more pronounced and of the same order in the radial and in the vertical direction.

The picture of vorticity components interconnection and their action on the core was, to a grand extent, enlighten by looking at the plots of velocity vectors in horizontal and vertical planes, these pictures are not shown for sake of shortness. Here we wish to present the analysis of the strain field and how it is correlated with the vorticity. This procedure has been used in literature to analyse turbulent fields [4] and only in one case, to our knowledge, it was used to analyse the interaction of well defined vortical structures [7]. In the present case we wish to apply this analysis to the simpler and more studied vortex ring. The eigenvalues α, β and γ of the strain-rate tensor and the eigenvectors $\mathbf{S}_\alpha, \mathbf{S}_\beta$ and $\mathbf{S}_\gamma$ which permit the estimation of the vorticity alignment have been used for the analysis. As usual the convention $\alpha > \beta > \gamma$ is done ($\alpha + \beta + \gamma = 0$ follows from the incompressibility). When $\beta > 0$ ($R_s = \alpha\beta\gamma < 0$) we are in presence of sheetlike regions (ribbons) that are dominant in turbulent flows or when instabilities are growing. The direct simulation is very appropriate for this analysis because it allows the selection of the region of interest and, within this, to consider only the points where the vorticity magnitude is larger than a threshold value ω_M.

The kinematic analysis of the region of the ring core has been performed, by choosing two values for ω_M. At $\omega_M = 1.5$ we focused the attention on the core and we expect to find results congruent with those found by Ref.[4], where regions of

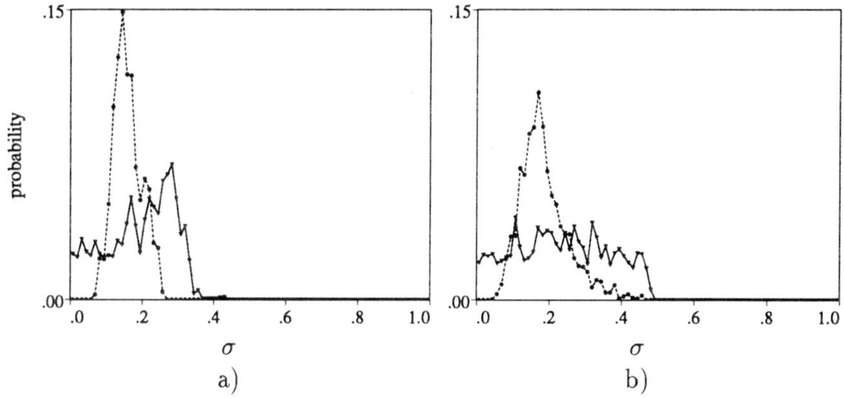

Figure 4: Probability distribution of —— $\sigma = |\beta|/\alpha$ and $- - -\sigma = \alpha/\omega_\beta$ with $\omega_M = 1.5$, a) $t = 30$, b) $t = 57$.

very intense vorticity in isotropic turbulence were considered. In fact Fig.3a shows that at $t = 30$, when the ring is not appreciably deformed, α lies in a $r - z$ plane and in this plane it is of preference oriented around 45^0. We observe also that there is a probability that the largest strain is directed towards the radial direction and zero probability to be directed along the vertical direction. It follows that the ring is deformed more along the radial than along the vertical direction, and this behavior is confirmed by the vorticity visualisations (Fig.1). At $t = 57$, when the deformation has reached a substantial level, Fig.3b shows that the probability that the positive strain α is lying in the $r - z$ plane is reduced, and there is not a preferred orientation at 45^0 any more. Furthermore, being the amount of radial vorticity larger than at $t = 30$, the highest positive strain deforms the core more in the vertical than in the radial direction.

The analysis of the vorticity field has shown that with the constraint $\omega_M > 1.5$, ω_β is the largest component, thus it is aligned as observed in Ref. [4]. We have evaluated the ratio between the α strain and the vorticity ω_β, and Fig.4a shows that, at $t = 30$ this ratio is small ($\sigma = 0.15$) meaning that the core tend to resist to the strain deformation. At $t = 57.5$ the probability is distributed on a broader range (Fig.4b) and the ring core is less resistant to the strain deformation, nevertheless the value of σ at which the peak probability occurs is still low as we would expect from a structure where a preferential orientation of the vorticity is still visible. In Fig.4a the probability distribution of the ratio between β and α has also been reported showing that the shape of the strain tensor is $\alpha : \beta : \gamma = 3 : 1 : -4$ or $4 : 1 : -5$ (probability with peaks at $0.25 < \sigma < 0.30$). These values are in agreement with those found in Ref.[4] or by Jimènez [8] for the Burgers vortex at $Re \approx 100$. The similarity of the strain-rate tensor, despite of the different structures and Reynolds numbers, confirms

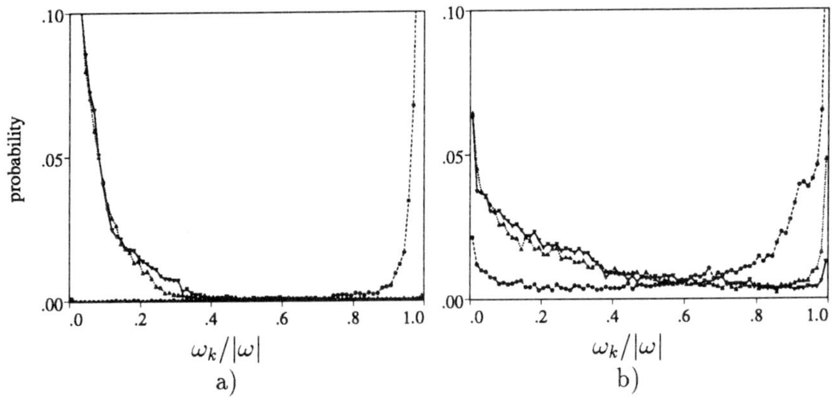

Figure 5: Probability distribution of ,— $k = \alpha$, $---k = \beta$, $\cdots k = \gamma$, with $\omega_M = 0.5$, a) $t = 30$, b) $t = 57$.

that the shape of S_{ij} and its alignment with respect to the vorticity is a kinematic mechanism that is independent of the dynamical process involved in the generation of the structure.

On the other hand in the vortex ring these low ratios, between β and α, must be expected also because the core , at $t = 30$, does not largely differ from an axisymmetric configuration in which the ratio between β and α is very low. We wish to recall that for two dimensional flows β/α is 0 but in the axisymmetric case $\beta \equiv S_{11} = v_r/r$; this strain is small (but not 0) in the core region and it can be very large near the axis of symmetry. This quantity gives information about the vortex stretching in the azimuthal direction that is the responsible for the increase of ω_θ. At $t = 57$ (Fig.4b) there is a substantial probability to have β comparable with α, confirming that large perturbations with respect to the axisymmetric configuration have developed; the probability to have α close to β, is a characteristic of flows without regions of intense concentrated vorticity.

From the comparison between our results and those in isotropic turbulence we think that in isotropic turbulence the intense vortex tubes, the so called "worms", should present deformations similar to those in the first stage of the Widnall instabilities.

The same calculations have been performed by fixing the vorticity threshold at $\omega_M = 0.5$. If we define $\rho_{n/s}$ as the ratio between the number of points where $R_s < 0$ and that where $R_s > 0$, the analysis has shown that, with the high threshold , $\rho_{n/s} = 1.03$ and that it increases up to $\rho_{n/s} = 1.3$ by decreasing the threshold. This outcome shows that, because $R_s < 0$ represents unstable sheetlike vortices, the part outside and that encircling the core contribute to the growth of the azimuthal instabilities.

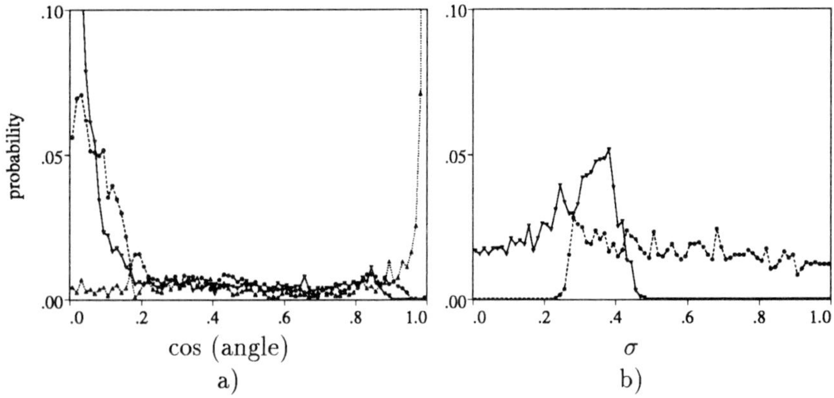

Figure 6: Probability distribution in the axis region at $t = 30$ with $\omega_M = 0.2$, a) of —- $S_{\beta\theta}$, $- - -S_{\beta r}$, $\cdots S_{\beta z}$ and b) of — $\sigma = |\beta|/\alpha$, $- - -\sigma = \alpha/\omega_\beta$.

From this observations we expect that the vorticity alignment holds at the early stage of the evolution (Fig.5a) when the core of the ring is approximately an axisymmetric structure while in the long-term (Fig.5b) it is less pronounced due to the generation of smaller scales. Furthermore when the ring deformation is well developed (Fig.5b) there is also a small probability that the vorticity is aligned with the maximum strain α. This is a very interesting situation because this alignment can lead to the occurrence of a finite time singularity in the 3-D Euler equations.

We were interested also to analyse the region near the axis, dominated by ω_z,as shown by the visualisation in Fig.1a. Because in this region the vorticity level is low ($(\omega_z)_{max} = 0.48$) we have selected as a threshold level $\omega_M = 0.2$. Fig.6a shows that the intermediate strain β is poorly aligned in the radial and azimuthal directions and it has large probability of alignment in the vertical direction. This result is confirmed by the visualisation of Fig.1a where elongated structures of vertical vorticity (aligned with the local intermediate strain) are visible. Although in this region the vorticity level is low, also in this case ω_β is the highest component and ω_α can reach a high level. The ratio α/ω_β in Fig.6b shows that it is almost equally distributed for $\sigma > .25$ while is zero for smaller value, with the consequence that the associated vortices are not persistent and they can easily change their shape, as observed in the vorticity visualisations. Also in this case the shape of the strain tensor approximately is $\alpha :| \beta |= 2.5 : 1$ with a zero probability of almost equal intermediate and positive strain. This is a configuration related to fat non-circular vortices that easily deform.

The analysis has shown that in this region there is a large number of points with the amount of $S_{ij}S_{ji}$ greater than the amount of $R_{ij}R_{ji}$, a further proof of the instability of the inner region of the ring. This is shown in Figs.7a-b, where scatter-plots of $S_{ij}S_{ji}$ versus $R_{ij}R_{ji}$ indicate that inside the core of the ring the vorticities

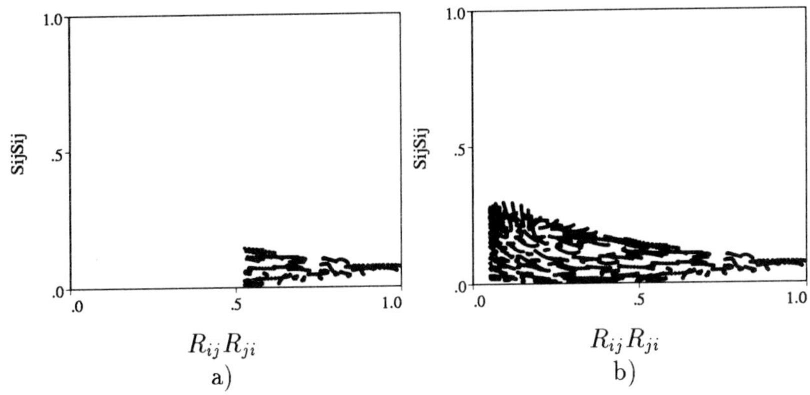

Figure 7: Scatter-plots of rotation-rate *vs* strain-rate at $t = 30$, a) region of the ring core and $\omega_M = 1.5$, b) region around the axis and $\omega_M = 0.5$.

are intense and low levels of strain do not appreciably influence the structure. This does not hold in the region near the axis where the vorticity is low and strain-rates are quite high. In the long-term azimuthal instabilities increase in amplitude and at $t = 57$ we observed that the effect of the strain-rate is sensible also inside the ring core (Figs.8).

Different zones within the flowfield can be characterised by the rate of deformation tensor A_{ij} which can be decoupled in a symmetric S_{ij} and an antisymmetric R_{ij} tensor. According to the definition by Chong *et al.* [9] we are in presence of an eddy region when R_{ij} is the dominant part of A_{ij}, $(-S_{ij}S_{ji}/R_{ij}R_{ji} \lll 1)$. From Eq.[1] it follows that an eddy region is characterised by intense vorticity and low strain. A similar classification was used by Kevlahan *et al.* [10] which divide the field into an eddy region $(-S_{ij}S_{ji}/R_{ij}R_{ji} < \sqrt{3}/3)$, a shear region $(\sqrt{3}/3 < -S_{ij}S_{ji}/R_{ij}R_{ji} < \sqrt{3})$, a convergence region $(-S_{ij}S_{ji}/R_{ij}R_{ji} > \sqrt{3})$ and a stream region where $R_{ij}R_{ji}$ and $S_{ij}S_{ji}$ are small (Fig.9) (as Fig.6 of [10]).

By a procedure similar to that used by Wray & Hunt [11] we have gone through the identification of zones in the ring. In this case instead of considering a single plane of the 3-D field we evaluated the average of $-S_{ij}S_{ji}/R_{ij}R_{ji}$, along θ to have a global picture of the ring. Fig.10 shows $\Omega(r,z) = \overline{-S_{ij}S_{ji}/R_{ij}R_{ji}}$, at $t = 30$ and $t = 57$, together with the average of the velocity vectors in a frame moving with the ring. In both cases an eddy region in the core of the ring is characterized by an almost solid body rotation. This region represents the inner part of the core, where the vorticity is very intense, and it maintains its strength in time. The shear region is a thin transitional zone between the eddy and the convergence zone and it does not play a significant role. On the contrary growth of the convergence zone, characteristic of the instabilities, shows that large strains are induced and these, at their turn, produce

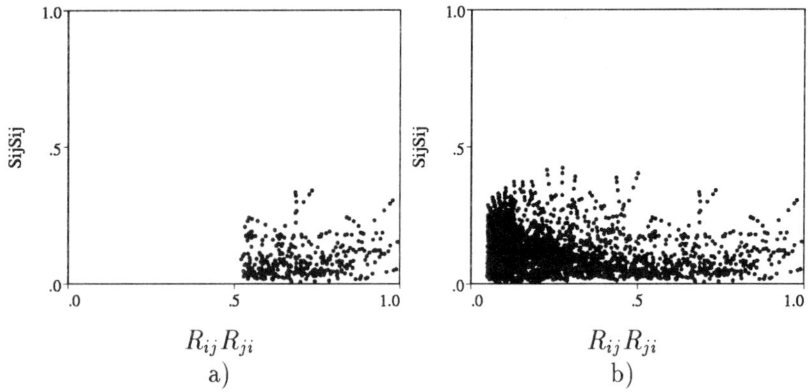

Figure 8: The same as Figs.7 at $t = 57$

further deformations. Fig.10b shows that the deformations in the rear part of the ring, larger than those upstream. The effects of the fat structure, shown in Fig.1b, are represented by the thicker convergence zone in the inner part of the ring.

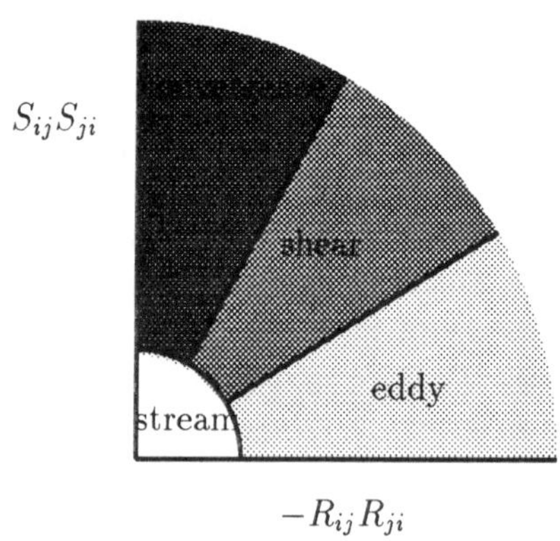

Figure 9: Schematic draft of the regions in the $R_{ij}R_{ji}$-$S_{ij}S_{ji}$ plane: stream, eddy, shear and convergence.

4 Conclusions

In this paper a direct simulation of a free vortex ring has been used to analyse its vorticity and strain field. The vortex ring has been chosen for its coherence and because it can be considered as a basic structure of several flows of practical importance. By

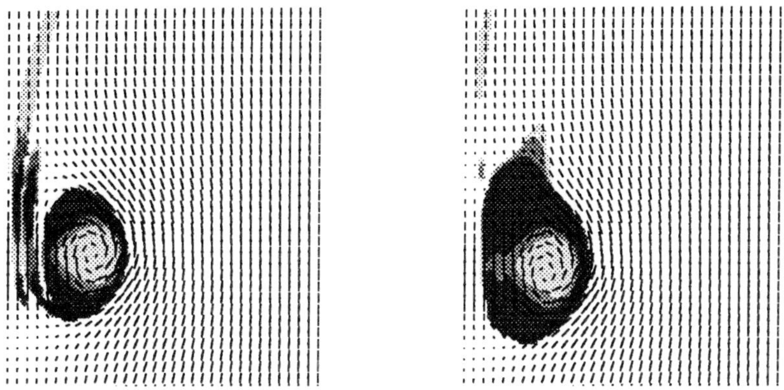

a) b)

Figure 10: Colour-contours of the *r.m.s.* of $\Omega(r, z)$ and velocity vectors, a) $t = 30$, b)$t = 57$. (for colours see Fig.9).

visualisations of the vorticity field, an explanation of the wake formation and of the growth of azimuthal instabilities was obtained. This analysis gives insights impossible to obtain by dye injection in real experiments. The flow structures were also investigated by kinematic considerations on the alignment between the eigenvectors of the strain-rate tensor and the vorticity components. The choice of different zones and different values of the threshold vorticity ω_M, to condition the alignment analysis, has shown that the field can be divided into two regions. The first is constituted by high level coherent vorticity, strongly aligned with the intermediate strain and the structure is mainly axisymmetric. In the second region, with smaller vorticities, there is a reduced probability of alignment between vorticity and intermediate strain and higher probability of negative R_S. It means that regions of low vorticity are more unstable and less resistant to external strain. This result is consistent with the findings of Jimènez [8] in homogeneous turbulence. The similarity between the core of the ring region and the intense vorticity structures in isotropic turbulence, brought us to draw the tentative conclusions that in isotropic turbulence the intense vortex tubes should have instabilities similar to those in vortex rings. Results have also shown that the coherence of regions with high vorticity is decreases in time and this indicates the tendency to generate the smaller structures present in turbulent rings.

The limitation of this work, as usual for the direct simulation, is on the Reynolds number. Before performing simulations at high Reynolds numbers where we expect to have the ring in a turbulent status, we intend to address the study on the interaction between two rings, giving insights to vortex pairing in a round jet, which is known to be the mechanism largely contributing to the spreading of the jet.

Acknowledgements

This work was supported by a grant from the Agenzia Spaziale Italiana.

References

[1] Lasheras,J.C., Lecuona,A. and Rodriguez,P. Three-dimensional structures of the vorticity field in the near region of laminar co-flowing forced jets. The global geometry of turbulence. Jimènez Editor, Plenum Press 1991.

[2] Glezer, A. " The formation of vortex rings ", Phys. Fluids, 31, pp.3532–42, 1988.

[3] Widnall, S.E., Bliss, D.B. & Tsai, C.Y. "The instability of short waves on a vortex ring" J. Fluid Mech, 66, pp.35–47,1974.

[4] Ashurst, W.T., Kerstein, A.R., Kerr, R.M. and Gibson, C.H., " Alignment of vorticity and scalar gradient with strain rate in simulated Navier-Stokes turbulence", Phys. Fluids, 30, pp.2343-2353, 1987.

[5] Verzicco, R. and Orlandi P., "A finite difference scheme for direct simulation in cylindrical coordinates ", Presented at the ICOSAHOM 92, Montpelier, 1992.

[6] Verzicco, R. Orlandi P. and Shariff,K.R., "A numerical simulation of three-dimensional vortex ring instabilities" Internal Report, 1992.

[7] Borotav, O.N. , Pelz, R.B. and Zabusky, N.J. " Reconnection in orthogonally interacting vortex tubes: Direct numerical simulations and quantifications ", Phys. Fluids, A,4, pp.581-605, 1992.

[8] Jimènez, J., " Kinematic alignment effects in turbulent flows", Phys. Fluids, A 4, pp.652-654, 1992.

[9] Chong, M.S., Perry, A.E. and Cantwell B.J., " A general classification of three-dimensional flow fields", Phys. Fluids, A 5, pp.765-777, 1990.

[10] Kevlahan, N., Mahesh, K. and Lee S., " Evolution of the shock front and turbulence structures in the shock/turbulence interaction", Proceedings of the 1992 Summer Program, Centre for Turbulence Research NASA.

[11] Wray, A.A. and Hunt, J.C.R., " Algorithms for Classifications of Turbulent Structures", Topological Fluid Mechanics, Proceedings of the IUTAM Symposium, Moffat & Tsinober Editors, Cambridge 1989.

XI. Compressible Flows

COHERENT STRUCTURES IN SHEAR FLOWS, CHEMICAL TRACERS AND TURBULENT COMBUSTION

J.P. CHOLLET, R.J. GATHMANN
Laboratoire des Ecoulements Géophysiques et Industriels
Université J.Fourier, Institut National Polytechnique de Grenoble, C.N.R.S.

1 Introduction

Coherent structures of turbulent flows are of special interest because of the way they influence flow mixing. Their possibly large length and time scales induce transport of significant amounts of fluid from a location to an other along noticeable distances. Quantities which undergo such a turbulent mixing are momentum, temperature, but also concentrations of species when multi-component fluid is considered. Chemical reactions can act as local sources producing or consuming species concentrations, they can also contribute to the energy balance through heat release.

section 2 begins with some general considerations about coherent structures especially with regard to scalar transport. There is indeed no indisputable definition of coherent structures. Therefore, the concept must be discussed with the various and complementary approaches of experimentalists, numericists, theoricians,.... Coherence whose practical evidence is almost always associated to scalar fields is viewed in terms of evolutions with both time t and space $\vec{x}$. Moreover, instability processes, especially those associated to shear, seem to be a common feature of flows which develop structures. Section 3 recalls the basic frame of equations coupling fluid dynamics to chemistry kinetics and the numerical methods to be used for the purpose of computing solutions with the help of appropriate numerical methods. Real flows are considered in section 3 for two cases of interest when studying high speed aircraft engines : supersonic mixing layers between walls and underexpanded jets.

2 Coherent structures in turbulence

There is no indisputable definition of coherent structures because of the different ways they are approached. The evidence of these structures came first from visualisations of scalar fields which trace the flow and clearly demonstrate that turbulent flow cannot be reduced to mean values with random fluctuations added. An example comes from weather common observation, large eddy evolutions are highlighted by clouds which act as more or less passive flow tracors. Various experiments have also been performed in laboratory, some of them using dying properties of appropriate

J. P. Bonnet and M. N. Glauser (eds.), Eddy Structure Identification in Free Turbulent Shear Flows, 443–451.

chemical tracors such as in [2]. Let us notice that flow structures cannot be visualised as far as they refer to flow variables such as pressure, vorticity or velocity. Nevertheless, in compressible flows ρ density fields can be directly analysed through appropriate optical device. Flow pictures, whatever they are obtained with experimental or numerical tools, must be cautiously interpreted because of bias which depend on the method in use. For instance, experimental visualisations of ρ fields are often characterized by integration along significantly thick fluid layers. Another example is provided by fields computed from numerical simulations which are often displayed under isovalue surfaces of vorticity variables $|\vec{\omega}_0(\vec{x}, t_i)|$ for the threshold value ω_0 at specified times t_i, which does not bring information on time evolution and deals with surface elements which are not attached to a fluid particle. Therefore various methods must be used to probe structures, by combining large scale global visualisations to more quantitative techniques. As structures cannot really be defined in the frame of so many different flow analyse techniques, they remain essentialy associated to remarkable length and time scales whose specificity can be higlighted. The way these space or time scales appear depends on the technique in use : frequency spectra tend to extract characteristic time scales when three dimensional pictures of isovalue surfaces yields length scale distributions. In an attempt to formulate, even vaguely, a condition for the observation of coherent structure in any field $\theta(\vec{x}, t)$, it could be written that a time τ and a length $l \approx u\tau$ can be found, at least inside some time-space interval around $(\vec{x}, t)$, such that :

$$(\theta(\vec{x}, t) - \theta(\vec{x} + \vec{l}, t + \tau))^2 < \epsilon \tag{1}$$

where ϵ is some arbitrary finite real value, and u a typical velocity.

Let us still discuss about structures once they have developed. Whatever quantity is used to trace the structure, it defines some subdomain in space which can be identified as belonging to the structure at a given time. Systems of equations which are considered hereafter (section 3) for non reacting gas flows reduce to Euler equation if diffusivity is neglected. The hyperbolic character of this set of equations comes from the three real characteristic directions : $dx/dt = u$, $dx/dt = u + c$, $dx/dt = u - c$, where c is sound velocity. Because of the non linearity of equations, all these directions are time space dependent with $u(\vec{x}, t)$ and $c(\vec{x}, t)$. For slightly compressible flows (low value of Mach number), the only direction to be taken into account corresponds to the advection through the velocity field $dx/dt = u$. A structure can be viewed as some domain in space whose identity remains preserved when following such a characteristic direction. The quantity which is advected can be either a scalar (species concentration, temperature,...) or some kinematics variables such as the vorticity vector $\vec{\omega}$ which is prefered to the velocity because the Helmoltz equation can be interpreted as a balance between advection, diffusion and production through three dimensional eddy stretching. Nevertheless, there is a major difference between the advection of any scalar and the advection of ω since vorticity is related to the advecting velocity fields (through $\vec{\omega} = \nabla \otimes \vec{u}$) and, at least in three dimensions, vorticity stretching induces vorticity production which means that ω does not remain

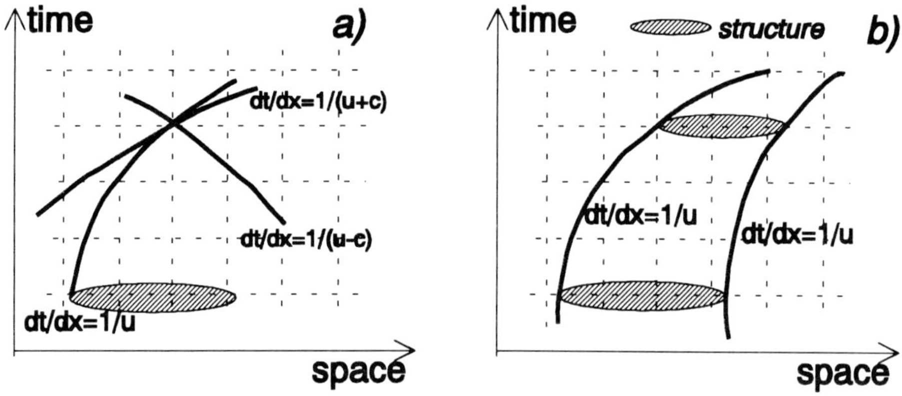

Figure 1: Characteristic directions for flow advection and coherent structures : **a)** compressible (subsonic), **b)** incompressible

invariant along a characteristic direction.

Figures such that Fig.1 almost intuitively suggest conditions to let structures persist or collapse. Although only one direction in space is plotted on these schemes, conclusions hold for time evolving three dimensional flows with characteristic surfaces instead of characteristic lines. Eddy structures should get stronger if the velocity flow $\vec{u}(\vec{x},t)$ is such that characteristic lines tend to be parallel one to the other or slightly convergent. On the opposite, diverging directions will unavoidably make the structure die out. When characteristic lines converge too much, two of them tend to intersect at some $(\vec{x},t)$, which means a singularity. Such an event can presumably be associated to vortex breakdown with the structure collapsing. Unfortunately, this crude analysis (such as fig.1) cannot be extended to more quantitative formulation because of : *(i)* the evolution of velocity $\vec{u}(\vec{x},t)$ along characteristic lines which makes their slope evolve with time and space and differ from a line to the other *(ii)* stretching of vortex lines which makes the vorticity $\vec{\omega}$ advected along the characteristic evolve. Structures as being mainly associated to fluid transport, at least as far as their persistence is concerned, have then much less chance to survive at very high Reynolds numbers because of strong velocity variability for small and even moderate length scales. From this point of view it can be also infered that fully compressible flows behave quite differentely because of the three characteristic directions u, $u \pm c$ (fig.1). There is little chance to develop coherent structures which can be preserved when following three different directions and coherent structures should presumably develop in gaseous flows as far as compressibility effects are weak enough. Typical features of gas flows (shocks, acoustic waves,...) are of course fully different from sructures discussed here.

Diffusivity is a character of real flows but does not play any explicit role in the above-mentioned process, just tending to diffuse the borders of the structure. There is a basic difference between flow with diffusion under usual molecular viscosity-diffusivity and flow with a significant energy at small scales which although inducing eddy diffusivity-like behaviour produces vorticity stretching and can significantly perturbe some local homogeneity of the velocity field. This difference must be taken into consideration, for instance when deriving subgrid scale models.

Up to his point, the discussion has dealt with the evolution of pre-existing structures. The way structures arise from laminar or turbulent flows seems much less clear. Boundary or initial conditions can be sowed with structures as in experiments where inlet flows are not always free from any coherent eddies or in numerical simulations where vortices are explicitely introduced. Structures develop otherwise from processes which make energy concentrate at some given length scale. Hydrodynamics instabilities have the capability of triggering structures whose development depends nevertheless from strongly complex non linear interactions. Kelvin Helmoltz instabilities are presumably the most frequent process to originate structures at both large and small scales since amplification can occur on a wide range of wave numbers as soon as deformation creates inflexion in velocity profiles, which can occur at scales ranging from the mean shear to much more local characteristic lengths.

3 Models and numerical simulations

Coherent structures can be highlighted only through numerical techniques dealing explicitely with both time and space evolution. Turbulent closure models as being built from time averages make coherent structures disappear even if their tracks can be observed on steady recirculations.

The set of unknown variables is $\mathbf{U}(\vec{u}, p, \rho, T, (Y_i, i = 1, n))$ with p as pressure, T temperature and Y_i mass fraction for each of the n species which make up the fluid. Related variables can be prefered, such as vorticity $\vec{\omega}$ or scalars built from linear combinations of some of the Y_i.

The basic set of equations is made of :

- partial derivative equations which can be written for any variable θ under the general form :
$$\partial\theta/\partial t + \vec{u}\nabla\theta = D\nabla^2\theta + S_\theta \tag{2}$$
even if a conservative form is prefered [10] for numerical reasons. D is a diffusivity coefficient and S_θ is a source term (vorticity stretching, heat release, chemical reaction rate,...)

- differential equations for the source terms S_Y in chemically reacting flows, typically : $dY_i/dt = \mathcal{F}_k(\rho, T, Y_i, Y_j)$ with as many equations as elementary steps k of the chemistry scheme

- algebraic equations, for instance ideal gas formulation $p/\rho = RT$ and partial equilibrium equations for some steps of chemistry kinetics.

- formulation of boundary and initial conditions

As coherent structures are associated to distributions in space, they mostly depend on the partial derivative equations, the only ones (possibly also with boundary conditions) to depend on space derivatives.

In order to properly compute coherent structures, numerical methods used to solve the set of equations must compute advective terms without excessive diffusion and dispersion, the later because of the necessity of accurate phase evaluation. A code has been specifically built with : *(i)* a PPM operator splitted in each of the space directions for the hyperbolic contribution, *(ii)* a finite volume diffusive-reactive operator. Details are given in [7, 10, 9].

Although turbulent combustion deals with gas flows, analysis of the impact of flow structures on reactive incompressible flows with flow dynamics not affected by chemistry is a case of interest. [6, 4] show that coherent structures for mixing are well traced by conserved inert scalar (typically $Y_A - Y_B$ for $A + B \rightarrow P$) and that the reaction develops on these structures more or less intermitentely depending on chemical reaction characteristic times. Some preliminary estimation of fractal dimension of the isovalue surfaces of these scalar fields show that they remain sligthly greater than 2 and go to 3 only for higher Schmidt number (that is when diffusivity gets small). Although they depend on the value chosen as the threshold, isovalues of the reaction rate $\dot{w}$ fiels can also be analysed in the same way and provide dimensions higher than for the inert scalar, especially when the reaction gets faster.

Another way of analysing structures goes through lagrangian tracors which give insight to both space and time evolutions, through their trajectory plots. Moreover, tracors can be used to record variable evolutions, including reaction rates $\dot{w}$ [3, 13].

4 High speed jets and mixing layers

Three dimensional simulations have been used to compute round underexpanded jets. At the outlet of scramjet engines, the strong difference between flow and ambiant pressures induces flow configurations which combine strong plane shocks and regions of strong mixing and vorticity. The simulation of [8] includes NO chemistry under a reduced Zeldovich scheme. NO production is actually significantly coupled to structures which develop behind the steady shock. Although flow depends on heat release from chemical reactions, this chemistry actually behaves as being rather passive and recirculating eddies determine a time of residence of reacting species which strongly contributes to NO production. The jet of figure 2 is run at pressure rates weaker thant in conditions [8, 3] and no shock develops. Isovorticity surfaces demonstate that structures build up, first from an instability which creates kinds of λ sharp structures in the mixing layer region at the edge of the jet. Then as proceeding farther in the streamwise directions, interactions build up interacting and more com-

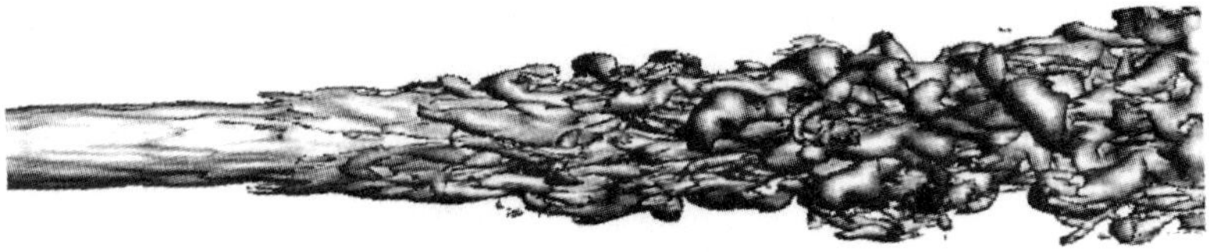

Figure 2: 3D numerical simulation of an underexpanded jet : vorticity modulus at 75 % of the inital vorticity

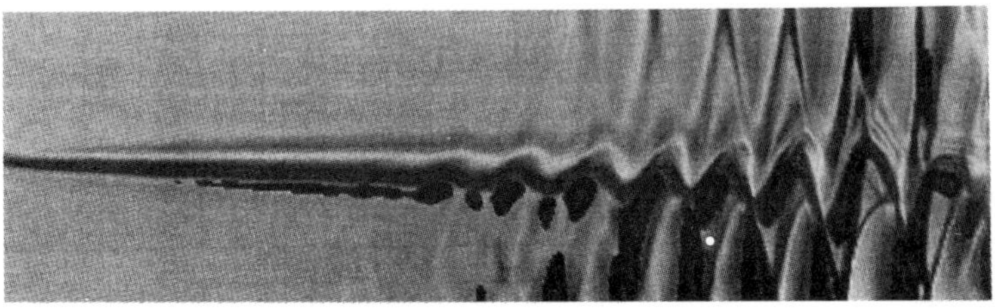

Figure 3: Two dimensional supersonic mixing layer between rigid walls : $\rho(\vec{x}, t)$

plex flow patterns with increasing three dimensionality and less evidence of coherent structures.

Another flow of interest for basic studies of scramjet engines is the supersonic mixing layer between rigid walls [12]. Theoretical studies of linear unstabilities [1] provide results which can be compared with results from numerical simulations. Because of reflective effect of walls (wall friction is neglected), unstable modes can develop provided that flow is supersonic. An extensive set of computations under two dimensional approximation has been run [11] and show good agreement with Tam's results, a significant persistence of amplified modes beyond the transition region and a prevailing effect of modes associated to the slower stream. The sensitivity to flow parameters appears when changing Mach number values, especially under the convective Mach number definition. A typical picture is given in fig.3 with a combination of evolving eddy structures and pressure waves which radiate from the mixing region and are reflected by the walls.

Two dimensions simulations could nevertheless be misleading for these mixing layers because three dimensional simulations clearly show [11] that Kelvin Helmoltz instability modes are more amplified than two dimensional compressible modes. Three dimensionality can be highlighted by pictures at times t_i of vorticity fields with eddy structures very similar to those displayed in figure 2 which show strong oblique structures. A more quantitative analysis is based on Fourier transforms of recorded

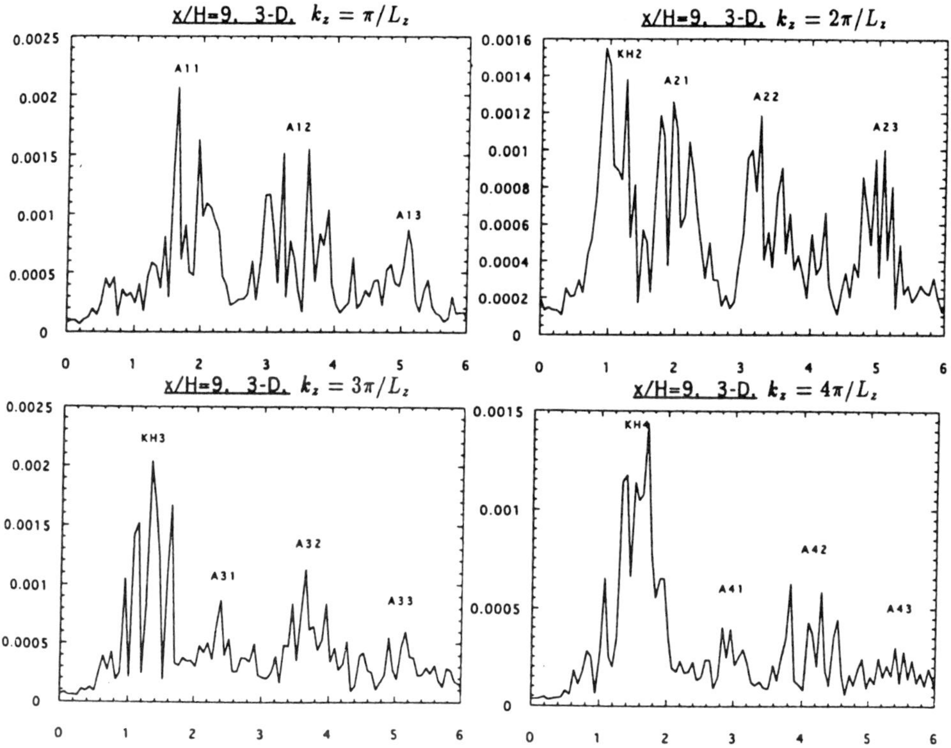

Figure 4: Pressure frequency spectra for different spanwise wavenumbers k_z

pressure values in specified locations. Figure 4 shows that, besides A_{ij} compressible modes, Kelvin Helmoltz oblique modes develop.

Remarkable feature of these layers is that transition to turbulence proceeds through strong shocks in two dimensional approximation when it is mostly driven by Kelvin Helmoltz modes when three dimensionality is fully taken into account. Vortices elongated along streamwise direction appear also to develop noticeably and persist beyond transition section.

Hydrogen-oxygen reduced chemistry scheme is now under consideration in this flow configuration in order to test how located heat release can generate mixing eddy sructures. Moreover as the two steps scheme explicits the radical OH formation, a study of the distribution of these species Y_{OH} or the production consumption rate $\dot{w}_{OH}$ will be feasible.

5 Conclusions and prospects

Coherent structures develop from unstability processes and they evolve as identified fluid patches advected by the flow. Therefore their persistance is improved by a relatively low variability of the velocity field. Instead of being generated from unstabilities, structures can also be introduced through boundary conditions, this is the reason why temporal approximations tend to artificially produce or at least preserve eddy structures.

In compressible flows, structures interact with other features such as pressure waves and shocks which originate density strong gradients or discontinuities which themselves can originate shear instabilities, as observed behind strong shocks which develop under expanded jets.

In order to quantitatively analyse the results from simulations, computed values in specified locations were recorded as they evolve in time, and then Fourier transformed both in time and spanwise z direction. This technique makes possible the analyse of unstable modes which are effectively amplified in flows under consideration and the way these modes trigger transition to developed turbulence. This analysis can also help to trace how structures persist when three dimensionality dominates.

It must also be pointed to the fact that accurate computation of compressible flow needs specific algorithm in order to compute and locate strong gradients as exactly as possible. This condition is of main importance in combustion where heat release reaction can make gradient especially strong. Nevertheless, numerical simulations is always restrictred to those chemical steps whose characteristic time is not too small and is at least of the order of the integration time step Δt.

References

[1] Tam, C.K.W., Hu, F.Q., 1989, The instability of acoustic wave modes of supersonic mixing layers inside a rectangular channel, *J.Fluid Mech.*, **203**, 51-76.

[2] Breidenthal, R., 1981, Structure in turbulent mixing layers and wakes using a chemical reaction, *J.Fluid Mech.*, **109**, 1-24

[3] Chollet, J.P., 1992, Turbulent eddy structures, combustion and chemical reactions, *NATO workshop on vortex flow and related numerical methods*, St Pierre de Chartreuse *to be published*

[4] Chollet, J.P., Vallcorba, M.R., Gathmann, R., 1989, Direct numerical simulations of turbulent combustion in mixing layers, *Turbulence and Coherent Structures*, M. Lesieur and O. Métais, Ed., Kluwer Acad.pub.

[5] Chollet, J.P., Gathmann, R.J., 1991, Reactive mixing layer between pressure reflecting boundaries, *8th symp. on Turb. Sh.Flows*, Münich.

[6] Chollet, J.P., Gathmann, R.J., Vallcorba, M.R., 1991, Reactive mixing layers and turbulent combustion, *Advances in Turbulence 3*, Stockholm, Springer-Verlag.

[7] Gathmann, R.J., 1993, Simulation d'écoulements gazeux supersoniques soumis à réaction chimique, *thèse de l'Université J.Fourier*, Grenoble

[8] Gathmann, R.J., Baudoin C., Chollet J.P, 1992, Direct numerical simulation of nitric oxyde evolution in underexpanded jets, *ASME- paper*, 92-GT-372

[9] Gathmann, R.J., 1991, Analysis of the effect of side-walls and concomitant boundary layers on supersonic 2-D shear-layer instability, *1st europ. symp. on aerothermodynamics for space vehicles ESTEC/ESA*, Noordwijk, the Netherlands.

[10] Gathmann, R.J., Chollet, 1991, Direct numerical simulations of supersonic turbulent combustion, *4th symp. on comp.fluid dynamics*, Davis, California.

[11] Gathmann, R.J., Si Ameur, M., Mathey, F., 1993, Numerical simulations of three dimensional transition in the compressible confined shear layer, *preprint*

[12] Samimy, M., Elliott, G.S., 1990, Effects of compressibility on the characteristics of free shear layers, *AIAA Journal*, **23**, num 3, 439-445.

[13] Vallcorba, M.R., 1992, Ecoulements cisaillés réactifs : simulations numériques de couche de mélange, *thèse de l'Université J.Fourier-Grenoble 1.*

SUPERSONIC MIXING LAYER ANALYSIS BY LASER PLANOGRAM AND HOT-WIRE BASED POD

Debisschop J.R., Sapin S., Delville J. & Bonnet J.P.
C.E.A.T Laboratoire d'Etudes Aérodynamiques
43 route de l'Aérodrome
F-86036 Poitiers cedex

1.Introduction.

An experimental investigation of plane supersonic mixing layers is presented with emphasis on the compressibility effects on the large scale structures. It is well known that, in incompressible mixing layers, 2D large scale structures predominate and play a major role in the development of the flow. In the case of supersonic mixing layers, this organization seems to be strongly affected by the compressibility. In this study, laser-planogram visualizations are associated with hot-wire rake (HWR) measurements to examine the turbulent organization of supersonic mixing layers for various regimes. From the HWR measurements, a scalar Proper Orthogonal decomposition (POD) based on the longitudinal component of the velocity is applied. The Pseudo-Flow-Visualization (PFV) is used to analyze the HWR signals. These results are then compared with equivalent ones obtained in the subsonic regime.

2.Flow configurations.

The compressible mixing layers, formed by the confluence of plane supersonic and subsonic streams, are studied in a high pressure blowdown supersonic wind-tunnel at high Reynolds numbers. Two test-cases are presented here, with convective Mach numbers, M_c, corresponding to 0.525 and 0.64. A third one corresponds to the subsonic regime and is described in details in [5]. The following table gives the principal characteristics of each flow: (subscript $_1$ is for the high speed side and $_2$ for the low speed one).

J. P. Bonnet and M. N. Glauser (eds.), Eddy Structure Identification in Free Turbulent Shear Flows, 453–462.

Test-case	M_1	M_2	R_{e1}	R_{e2}	$r = \frac{U_2}{U_1}$	$s = \frac{\rho_2}{\rho_1}$	M_c
A	2.03	0.41	$\simeq 7.5\ 10^8 m^{-1}$	$\simeq 7.\ 10^7 m^{-1}$	0.27	0.57	0.64
B	1.48	0.28	$\simeq 2.5\ 10^8 m^{-1}$	$\simeq 3.\ 10^7 m^{-1}$	0.225	0.70	0.525
	U_1	U_2					
C	45.2 m/s	25.8 m/s	$\simeq 3\ 10^6 m^{-1}$	$\simeq 2\ 10^6 m^{-1}$	0.59	1.00	0.05

3.Laser planogram visualizations.

The laser planogram visualizations are obtained by means of a Q-switch Ruby laser (typically 20 ns of pulse duration). A conventional optical system allows the formation of a light sheet which, in the present study, is perpendicular to the trailing edge of the splitter plate (Y direction) and aligned with the mean flow direction, X. The seeding of the flow is obtained by injecting ether, by natural aspiration, in the boundary layer of the supersonic side, approximately 50 cm upstream of the trailing edge. The laser light sheet is about 20 cm long and the visualizations along the whole window (40 cm long) are reconstructed from two pictures taken at two different times (figure 1). For $M_c = 0.525$, the observed shapes can be associated to the trace of quasi two-dimensional structures; the structure of the flow seems quite comparable to the one that can be observed in incompressible mixing layers [3]. Conversely, for $M_c = 0.64$ it becomes hardly possible to detect the same characteristics. The limits of the layer become straighter and the internal part of the mixing layer looks more blurred. The increase of the convective Mach number seems to entail a disappearance of the quasi 2-D structure in agreement with conclusions from advanced numerical computations and experimental investigations [6, 4]. The visualizations presented here will not be discussed in details in this study. The main use of these results is here to compare with the Pseudo-flow-Visualizations obtained from the HWR.

4.Hot-wires rake measurements.

A specially designed hot-wire rake allows to measure the instantaneous signals from 12 probes placed across the whole mixing layer (Y direction). The rake is typically longer than twice the local vorticity thickness of each mixing layer. The probes are 1×1 mm thick with a 2.5 μm wire diameter. The measurements are carried out in the asymptotic region. In the compressible cases, the high Reynolds numbers based on the wire diameter (typically more than 70) and the high overheat coefficient (0.8) limit some Mach number effects in the transonic domain as demonstrated by [7]. The signals are simultaneously sampled at 100 kHz and low-pass filtered at 50 kHz. An *a-posteriori* calibration technique has been used. This technique is based on the application of the *Very Strong Reynolds Analogy* and uses the values of the mean and rms velocities carried out from previous LDA measurements [2]. The anemometers outputs are then assumed to be directly proportional to the velocity fluctuations. (In the incompressible case, the conventional King's law has been applied.)

5.Scalar POD.

It is now well known that the Proper Orthogonal Decomposition initially introduced by Lumley [8] is a tool for studying the organization of turbulent flow. This technique has been up to now essentially applied to subsonic configurations. For supersonic flows, only a few attempts have been made. Kirby et al. [9] have applied the POD to a simulated supersonic shear layer. These authors used the snapshot POD essentially in order to reduce the size of the information obtained by the computation. In this study, we apply the POD on the longitudinal fluctuating velocity component, the time dependency being removed by a temporal Fourier transform. The eigen-value problem which is solved is then:

$$\int \Psi_{uu}(Y,Y';f)\Phi_u^{*(n)}(Y';f)\,dY' = \lambda^{(n)}(f)\Phi_u^{(n)}(Y;f)\ ,$$

where Ψ is the Fourier transform of the two point correlation tensor, $R_{uu}(Y,Y';\tau)$. In order to apply the POD, the correlation tensor R_{uu} has to be known for each (Y,Y') combination. This is the main use of HWR which allows, specially in supersonic configuration and in intermittent wind tunnels, to access to this kind of information.

Temporal Scales.

In compressible, supersonic flows, the choice of relevant variables can be of first importance. We choose here to use the temporal Reynolds decomposition of the velocity field, leading to consider the conventional velocity fluctuations. The alternative approach based on the Favre mass-weighted decomposition is not used in the present study. Its implications on the POD results will not been discussed here, but remain an interesting field of discussion. For the following discussion, we use as non-dimensional frequency the Strouhal number $St_\omega = f\delta_\omega/U_c$ based on the vorticity thickness and a convection velocity. U_c is $\frac{1}{2}(U_1+U_2)$ for the subsonic flow. The choice of the convection velocity for the supersonic case is not straightforward [10]. We choose here the conventional definition: $\dfrac{a_1U_2+a_2U_1}{a_1+a_2}$, where a_1 and a_2 are the external speeds of sound.

Energy reconstruction.

The following table gives the relative energy of the first modes for the three test-cases.

Test-case	M_c	Relative energy in first mode	Relative energy in first 3 modes
A	0.64	0.48	0.815
B	0.525	0.42	0.76
C	0.05	0.44	0.78

The quick convergence of POD is found for the compressible cases as well as in the subsonic one. As in other studies, in each cases considered here, the energy contained in the first mode of POD reaches about 40% of the global energy.

Eigen-spectra.

Comparable typical shapes of the eigen-spectra are obtained for each configuration as shown on figure 2. Here the eigen-spectra are normalized by the global energy in the flow. It can be noticed that with the dimensionless frequency used, the same typical Strouhal range (0.15-0.25) corresponding to local maxima of energy can be defined whatever the flow regime is. However, a discrepancy appears between the subsonic and compressible cases: the first mode of the subsonic case has its maximum located near $St_\omega = 0.15$, while maximum of the second mode lies near 0.25. On the contrary, for the compressible cases, both modes are located at the same frequency which is typically of the order of 0.20. These average values of the characteristic Strouhal number found from the POD is very close to the ones which can be obtained from one-point conventional spectra [2]. In the compressible regime the Strouhal value is then of the same order than in the subsonic one. Another important feature is that the relative energy contained in the vicinity of the maximum and the sharpness of the peak decreases, when the Mach number increases. At the same time, the higher non-dimensional frequencies are dumped.

Space time correlations (STC).

From the eigen-values and eigen-vectors of the POD, the conventional Space-Time-Correlations can be retrieved easily:

$$R_{uu}(Y,Y';\tau) = \sum_n R_{uu}^{(n)}(Y,Y';\tau)$$

where $R_{uu}^{(n)}$ is the contribution of the individual nth mode [5]. This approach is used here only for the analysis of the contribution of the first mode of the POD: $Ruu^{(1)}$. Two Y locations of the reference probe are considered here: in the vicinity of the mixing layer axis and near its low velocity side edge.

• **On the mixing layers axis.** When the reference probe is near the mixing layer axis, the STC, $< u(Y \sim 0, Y';t).u(Y \sim 0, Y';t+\tau) >$, are presented on Fig. 3-1, 3-2 and 3-3). For the three flow regimes, the integral temporal scale corresponding to the distances between the negative-correlation areas at zero-space separation remains of a comparable order: $\tau U_c/\delta_\omega \sim 4$. This result can be interpreted as a confirmation of the correct choice of the temporal reference scale. The typical shapes of the iso-contours found in the compressible case remains very similar to the ones obtained in the subsonic one. However, the highest levels of correlation are more concentrated for small space and time separations in the compressible cases.

In the three configurations, the first mode of the POD contributes quite significantly to the correlations. In addition, the global characteristics of the shape of the iso-correlations are well predicted by this mode. This feature is shown on figures 3-4, 3-5 and 3-6. The same levels of correlations are used for all the plots. This result is in fact a direct consequence of the POD which is an optimal decomposition for the energy and which is then well representative in the areas where the energy is maximum (ie in the vicinity of the mixing layers axis).

• **At the edge of the mixing layers.** The analysis of the STC for a reference probe located at $Y \sim -\delta_\omega/2$ (ie in the low velocity side edge) shows that the the structure of the supersonic flow differs from the incompressible regime. In the subsonic regime, it is well known that the phase opposition between the fluctuations of the velocity in the high and low velocity sides can generally be associated with the trace of the passage of the large scales structures. An antisymmetry relative to the mixing layer axis is observed (figure 3-7). However, in the compressible regimes, this antisymmetry disappears and a quite different behavior is then noticeable (figures 3-8 and 3-9). For these supersonic configurations, the lower side of the mixing layer is uncorrelated with the upper side for zero time delays; negative and positive correlations between the two sides of the flow are then found only for negative and positive, non-zero time delays respectively.

The first mode contribution to the STC also exhibits a quite different behavior in this location. As already mentioned by Delville et al. [5], in the subsonic mixing layer the first mode of the scalar-u POD is unable to predict the phase opposition between the two sides of the flow (figure 3-10) and in this case, higher modes are necessary in order to enhance the reconstruction. On the contrary, in the compressible regime (figures 3-11 and 3-12), the first mode of the POD is more suited and the overall shapes of the STC are then quite well predicted. Moreover, for the higher convective Mach number, even the correlations reconstructed from the first mode are closer to the measured ones (figure 3-12).

Instantaneous reconstruction of the velocity field.

The use of rakes of hot-wires allows one to know the velocity simultaneously over the whole Y extent of the mixing layer. We can then estimate the contribution $u^{(n)}$ of each mode (n) of the POD to this instantaneous values by using [5]:

$$u^{(n)}(Y,t) = \mathcal{F}^{-1}\left(a^{(n)}(f).\Phi^{(n)}(Y,f)\right), \quad \text{where} \quad a^{(n)}(f) = \int \mathcal{F}\left(u(Y,t)\right)\Phi^{*(n)}(Y,f)\,dY$$

and where ($\mathcal{F}$ is the temporal Fourier transform).

The instantaneous velocities fluctuations measured and the corresponding contributions from the POD are visualized with the PFV technique. By using the coloration rule based on the histogram of the velocity distribution described on figure 4 [1], one can access simply to the main "structure" of the flow. In these PFV, the red dominant hues correspond to positive fluctuations and the blue ones to negative fluctuations. Figure 4 compares the three flow regimes, the PFV of the three flow regimes obtained from direct measurement are plotted together with the ones that can be extracted from each of the first two modes of the POD. For each case, the temporal extent (horizontal) corresponds to 4-5 times the typical temporal scale obtained from the characteristic Strouhal number.

• **Direct measurement.** In the subsonic configuration, the quasi periodic, side to side phase oppositions are observed on fig. 4-1. This can be linked to the passage of the large scales structures. In the case of compressible regimes, (figures 4-2 and 4-3) the general organization disappears and the phase oppositions are no more distinguishable. Near the mixing layer axis, the succession of positive and negative pockets is still more or less present; however they are less continuously distributed across the layer. In addition, the high velocity side exhibits a straight edge which can neither be found for the low velocity side in the compressible case nor for both sides in the subsonic regime. This last feature seems to increase with M_c and is more noticeable for $M_c = 0.64$ (fig. 4-3) than for $M_c = 0.525$ (fig. 4-2).

• **First two modes of the POD.**

For the subsonic regime, the PFVs exhibit the general feature already mentioned [5]: the first mode tends to give instantaneously a single partition in the Y direction. The picture (fig. 4-4) is then quasi symmetrical relative to the mixing layer axis, whilst the second mode (fig. 4-7) gives a dual spatial partition, corresponding to an antisymmetric behavior. The side to side phase opposition shown by the first mode does not match to the direct measurement PFV, the second mode being closer to that characteristic. As found from the STC, higher modes are necessary to get a satisfactory representation.

In the supersonic mixing layers the results are presented on fig. 4-4-5-7-8. Some similarities are evidenced for the behavior of the two first modes in the subsonic side of the subsonic and supersonic mixing layers. In the upper parts of both compressible test-cases, the PFVs of the modes exhibit a quite different feature. Alternate bands can be seen, starting from the high velocity edge. These bands seem to be not correlated with what happens in the opposite side. This phenomena can be interpreted as the loss of side to side instantaneous correlation already observed from the STC. An other striking feature is that these "bands" are tilted upstream in the supersonic cases.

6.Discussion.

The hot-wires rakes measurements confirm the evolution of the structure announced by laser planogram visualizations. For $M_c = 0.525$, the HWR measurements show a modification of the structure but some quasi two-dimensional structures are still observable by visualization. For $M_c = 0.64$ these modifications are increased and the large scale structure is no more observable by visualization. The informations taken from these two techniques seem then to be very complementary. The Strouhal number values are quite similar for supersonic and subsonic mixing layers. This effect confirms that some quasi periodic structure still persists at M_c greater than 0.6. The temporal scale based on the conventional convection velocity seems to be appropriate to the study of compressible mixing layers at least in the present convective Mach number range. The Proper-Orthogonal-Decomposition works quite well for supersonic speeds and at high Reynolds number. The POD also gives some new information on the structure of the supersonic mixing layers. The most important compressibility effect that can be noted is the loss of the side to side phase opposition. It seems that the aspects of the structure change when the Mach number increases. It is noticeable that the the information given by the POD's first mode is more representative in the compressible cases than for the subsonic flow.

Aknowledgments This study is supported by DRET/DGA through grant number 90-068.

References

[1] S. Bellin, J. Delville, J. Vincendeau, J.H. Garem & JP Bonnet, "Large scale structure characterization in a 2D mixing layer by Pseudo-Flow-Visualization and Delocalized Conditional Sampling. *IUTAM Symposium on Eddy Structure Identification, Poitiers*, Oct. 1992.

[2] J.P. Bonnet & J.R. Debisschop, " Analyse Expérimentale de Couches de Mélange Supersoniques", 29eme *Colloque d'Aérodynamique Appliquée AAAF Biscarrosse*, 21-23 sept. 1992

[3] G. Brown & A. Roshko "On Density Effects and Large Scale Structure in Turbulent Mixing Layers."
Journal of Fluid Mechanics, Vol. 64, 1974, pp.775-781.

[4] J.H. Chen, "The Effects of Compressibility on Conserved Scalar Entrainment in a Plane Free Shear Layer" *TSF 8*, Munich, Sept. 1991.

[5] J. Delville, S. Bellin & J.P. Bonnet " Use of the Proper Orthogonal Decomposition in a Plane Turbulent Mixing Layer." *in Turbulence and Coherent structures, Kluwer Acad. Publ., 1990.*

[6] Y. Fouillet, Thesis, 1991, IMG Grenoble.

[7] C.C. Horstman & W.C. Rose, " Hot-Wire Anemometry in Transsonic Flow.", *AIAA Journal, Vol. 15, No. 3, 1977, pp. 395-401.*

[8] J. Lumley "The structure of inhomogeneous turbulent shear flows", *Atmosph. turbulence and radio wave propagation*, edit. Yaglom and Takarsky, Moscow, 1967.

[9] M. KIRBY, J.P. BORIS & L. SIROVICH "A Proper Orthogonal Decomposition of a Simulated Supersonic Shear Layer", *Int. Jour. for numerical methods in fluids, vol. 10, pp 411-428, 1990*

[10] D. Papamoschou, " Structure of the Compressible Turbulent Shear Layer." *AIAA Paper 89-0126, January 1989.*

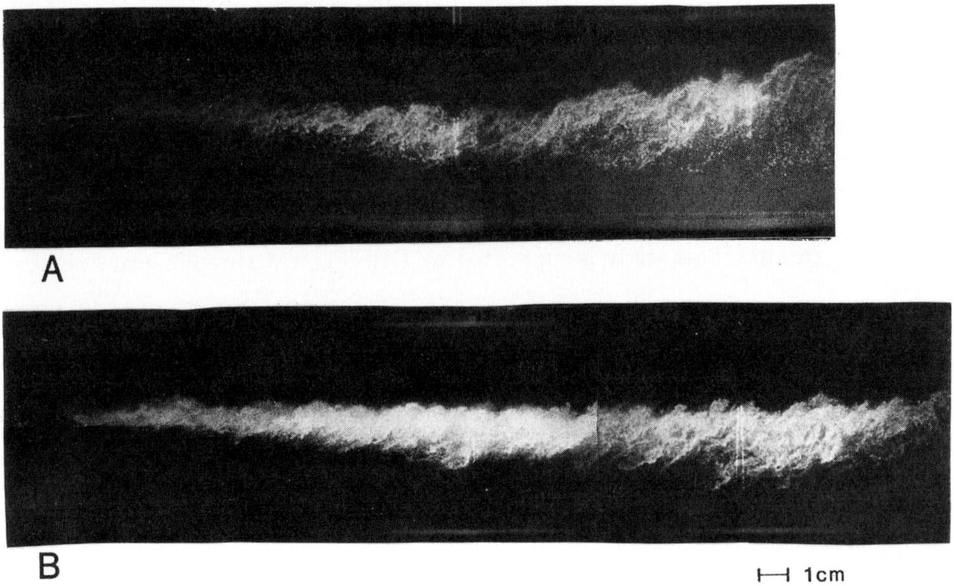

Figure 1: Laser planogram visualizations of supersonic mixing layers for: $A) M_c = 0.525$, $B) M_c = 0.640$

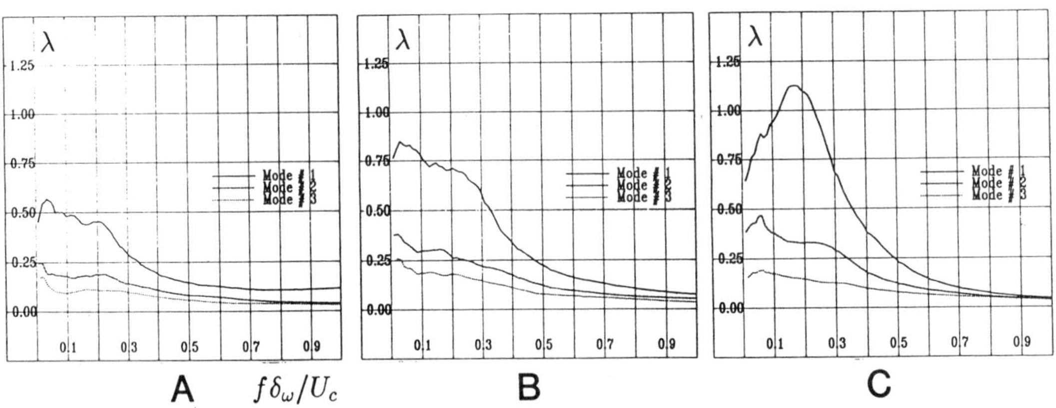

Figure 2: 3 first eigen-spectra of POD $A) M_c = 0.64$, $B) M_c = 0.525$, $C) subsonic$.

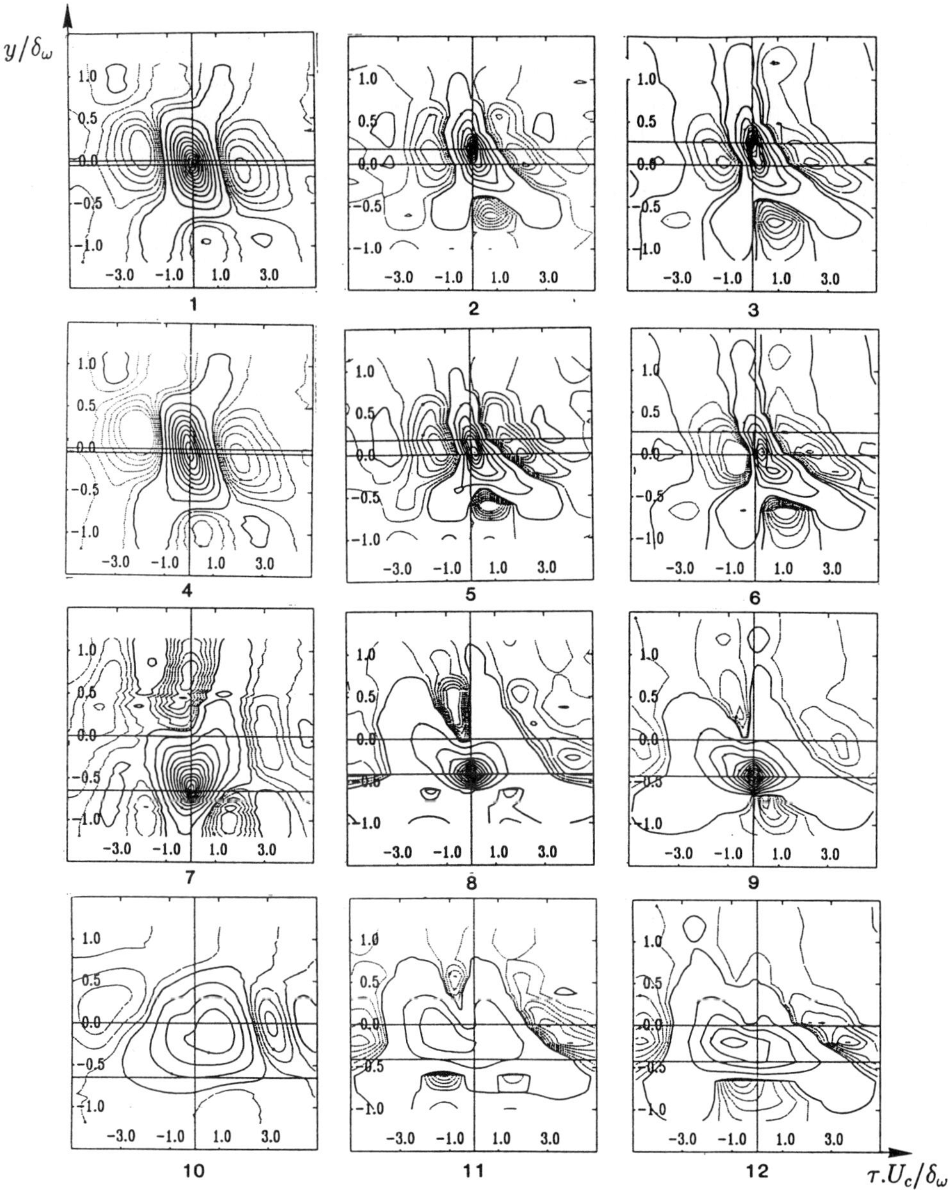

Figure 3: Iso-contours plots of space-time correlations and first mode participation. ——— positive correlations, ... negative correlations. Iso contours level are determined from the measured STC.

	measured		1st mode	
M_c	$Y/\delta\omega \sim 0$	$Y/\delta\omega \sim -0.5$	$Y/\delta\omega \sim 0$	$Y/\delta\omega \sim -0.5$
0.05	1	7	4	10
0.525	2	8	5	11
0.640	3	9	6	12

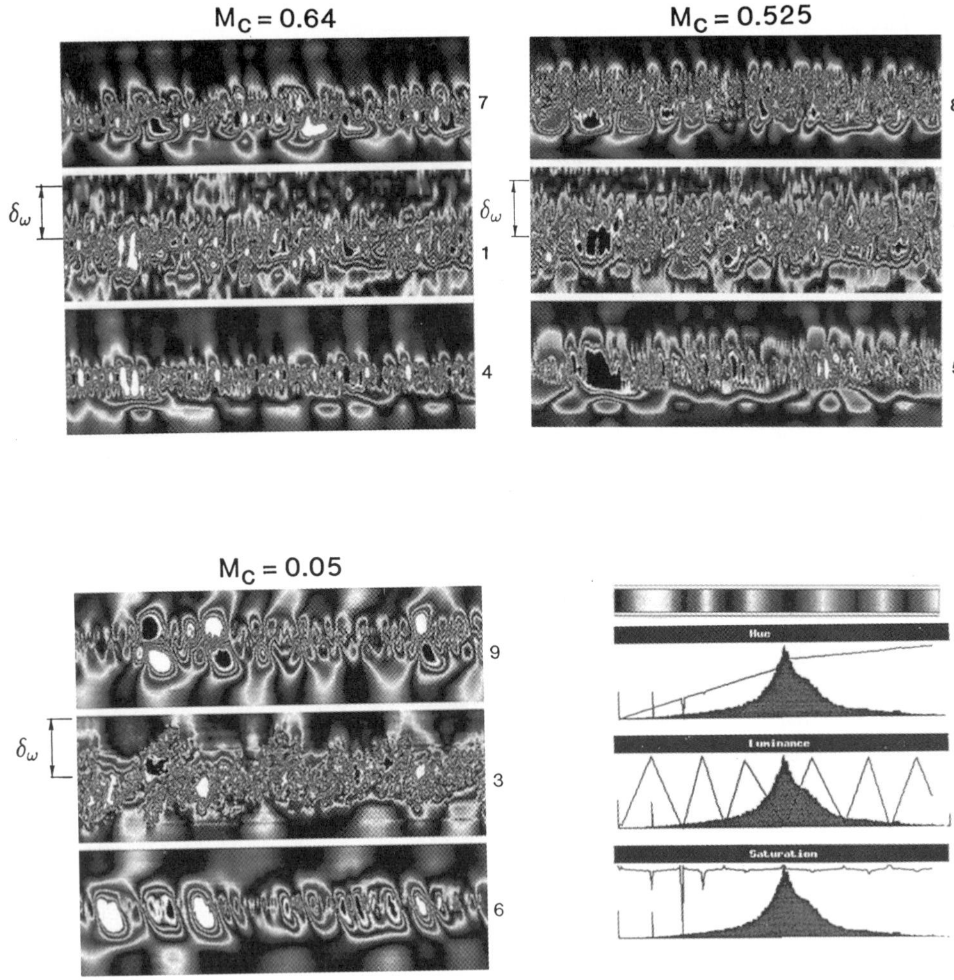

Figure 4: Pseudo Flow Visualizations of fluctuating instantaneous velocity:

M_c	0.64	0.525	0.05
direct meas.	1	2	3
first mode	4	5	6
second mode	7	8	9

Colour version in Colour Plates section, p. xxxi.

Flow Visualization in Compressible Turbulent Shear Layers

H. Rosemann*, P. Dimotakis and J.L. Hall**

Graduate Aeronautical Laboratories, California Institute of Technology, Pasadena, CA 91125

Abstract: Cross-sectional Rayleigh scatter images have been obtained from a two-dimensional, compressible turbulent shear layer with isentropic convective Mach numbers $M_{c1}^{(i)}$ ranging from 0.09 to 0.92. With both streams being subsonic, the same two-dimensional large-scale Brown-Roshko structures have been found as in Schlieren pictures. With a supersonic high speed stream, but at still low $M_{c1}^{(i)}$, these structures disappear from the Schlieren pictures but remain visible in the Rayleigh scatter images. At higher compressibility levels smaller structures form a very irregular interface between the two streams in the Rayleigh scatter images. No strong shocks have been found inside the shear layer. The probability density function of the Rayleigh scatter intensity for the low compressibility case shows a dominating peak representing the mixed fluid in the vortex. In the high compressibility case a broader distribution corresponding to smaller flow areas with different, but within the area constant intensity, was found.

1. Introduction

Two-dimensional compressible turbulent shear layers usually do not show the large-scale, two-dimensional Brown-Roshko vortices as was found in recent experiments by various researchers. Side-view Schlieren pictures from experiments conducted in the GALCIT Supersonic Shear Layer Facility (Hall 1991, Hall *et al.* 1991a) show large-scale vortical structures only for those cases clearly, when the high speed stream is subsonic. In all cases with a supersonic high speed stream, the shear layer appeares as a wedge with very little large-scale structure visible inside.

In the highest compressibility case a complex system of travelling waves on the low speed side was detected. It seemed to be generated by some structure inside the shear layer not visible in the Schlieren pictures. Measurement of the convection velocity yielded $2.1 < M_{c2} < 2.6$, a strong deviation from the value of $M_{c2}^{(i)} = 0.99$

* Institute of Experimental Fluid Dynamics, DLR, D-3400 Göttingen, FR Germany

** Cryofuel Systems Group, University of Victoria, Victoria, BC, Canada V8W 2Y2

J. P. Bonnet and M. N. Glauser (eds.), Eddy Structure Identification in Free Turbulent Shear Flows, 463–474.

predicted by the isentropic theory which is based on total pressure matching at interstitial saddle points between the large-scale vortices in the convective frame of reference, *cf.* Dimotakis (1991), Papamoschou & Roshko (1988).

This discrepancy was already found by Papamouschou (1989) and led him and Dimotakis (1991) to propose the existence of shocks inside the shear layer, which would provide the total pressure loss necessary to allow pressure matching in these cases. Although the shocks would have to be strong to let the model agree with the measured convective Mach numbers, they have not yet been detected in Schlieren pictures.

Therefore new visualization experiments were designed and performed in the GALCIT Supersonic Shear Layer Facility using a light sheet technique with Rayleigh scattering. Because of the stronger three-dimensionality of compressible shear layers, this technique should be better suited to provide further insight into the structure inside the shear layer than the spanwise integrating Schlieren method.

2. Experimental Facility and Instrumentation

The Supersonic Shear Layer Combustion Facility is a two-stream blow-down wind tunnel which was designed to produce chemically reacting flow with H_2, F_2 and NO as reactants. The visualization experiments described here, however, were performed with non-reacting flow only. Both streams are supplied independently by separate delivery networks. Inside the test section the gases pass through diffuser cones, screens, honeycombs, and interchangable nozzle blocks to form the shear layer at the end of a splitter plate. On the low speed side (lower side), always a subsonic nozzle with an exit height of 50.8 mm was used whereas on the high speed side (upper side), one nozzle for a nominal Mach number of $M_1 = 1.5$ with an exit height of 31.8 mm and another one for subsonic velocities with an exit height of 50.8 mm were used. The channel width is 152.4 mm.

The flow channel is bounded by 406 mm long horizontal guide walls, which are equipped with flexible hinges and pressure tabs to control the streamwise pressure gradient by adjusting the guidewall angles. Windows in the sidewalls provide optical access to the flow. Static pressure in the test section was about atmospheric and a typical run time was 3 s. More details on the design of the facility, the control system and the data aquisition system can be found in Hall (1991).

The lightsheet was set up vertically in the streamwise direction using a frequency-doubled Nd:YAG laser with 7 ns pulse width and 0.3 J energy per pulse. The pictures were taken by a cryogenically cooled CCD-camera with a 1024×1024 pixel array. Low signal level required special measures to obtain satisfactory results. A 2.5 mm wide and 76.2 mm long slits was cut into each guidewall on the centerline to send the lightsheet through since too much straylight would have been created at the air/glas interfaces of windows. No disturbing effects from the slits onto the shear layer could be observed. The thickness of the lightsheet varied between 0.5 *mm* and about 1 *mm* in the observation area, which was located about 150 *mm* or three low speed nozzle exit heights downstream of the splitterplate.

For good contrast, species of very different Rayleigh scatter cross sections have been chosen for the low and high speed test gas, namely helium, nitrogen, and ethylene, of which the ratio of the cross sections is 0.015 : 1 : 6. The total amount of scattered light is very low, for nitrogen only a maximum of about 4300 photons can be theoretically received by each pixel of the CCD-camera under the conditions of the experiment and the assumption of no losses in the windows and optics.

The raw images were normalized with a reference light sheet and the intensities scaled so that black always represents pure low speed fluid and white pure high speed fluid. For the purpose of qualitative comparisons and as a first approximation one can view this scaled intensity as the high speed fluid concentration, since the difference of the Rayleigh scatter cross sections between the test gases is fairly high. Also, a geometric transformation was performed to achieve a uniform scale and orientation of all pictures.

3. Results and Discussion

Five different cases with isentropic convective Mach numbers ranging from 0.09 to 0.92 have been investigated. Table 1 shows the main flow parameters. The Reynolds number Re_δ is based on the velocity difference, the local thickness of the shear layer, and an averaged value for the viscosities.

Case	1	2	3	4	5
Gas 1	N_2	N_2	N_2	N_2	He
Gas 2	He	C_2H_4	He	C_2H_4	C_2H_4
M_1	0.65	0.65	1.48	1.48	1.50
M_2	0.10	0.30	0.20	0.30	0.30
$r = U_2/U_1$	0.46	0.46	0.46	0.23	0.09
$s = \rho_2/\rho_1$	0.13	0.92	0.10	0.70	4.0
Re_δ	$2.6 \cdot 10^4$	$1.7 \cdot 10^5$	$3.4 \cdot 10^4$	$7.2 \cdot 10^5$	$6.6 \cdot 10^5$
$M_{c1}^{(i)}$	0.09	0.17	0.19	0.52	0.92
$(\delta/x)_{vis}$	0.097	0.101	0.062	0.114	0.108

TAB. 1 Run parameters

All Schlieren photographs shown here have been taken from earlier experiments conducted in the same facility by Hall *et al.* (1991a) and Hall (1991). The slight differences in the run parameters to the current experiments because of the use of ethylene instead of nitrogen or density matched helium/argon mixtures have been

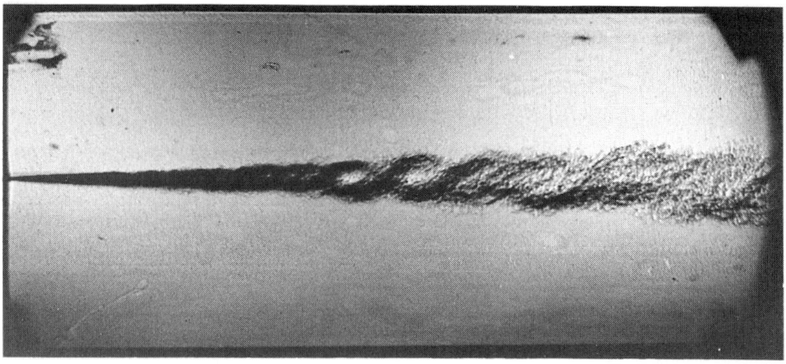

FIG. 1 Schlieren picture of the case 2 flow
$(N_2/(33\%H_2 + 66\%Ar),\ M_1 = 0.59,\ M_2 = 0.27,\ M_{c1}^{(i)} = 0.14)$

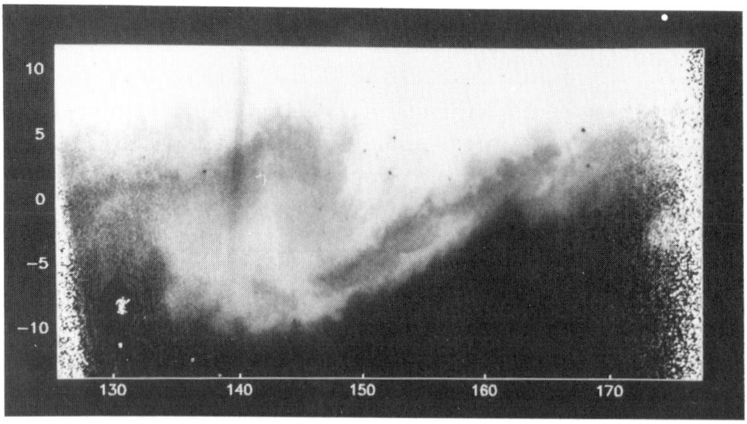

FIG. 2 Rayleigh scatter image of the case 2 flow
$(N_2/C_2H_4,\ M_1 = 0.65,\ M_2 = 0.30,\ M_{c1}^{(i)} = 0.17)$

neglected since the purpose of the Rayleigh scatter experiments was mainly to get more qualitative informations about the topological structure of the shear layer.

Figure 1 shows the Schlieren photograph of the low compressibility case 2 flow and Fig. 2 the corresponding Rayleigh scatter image. The axis labels in the Rayleigh scatter images represent the distance from the tip of the splitter plate in mm. To scale the Rayleigh scatter images to the Schlieren pictures, remember that the exit height of the low speed nozzle is $50.8\,mm$. The high speed stream is always on top.

Figure 1 exhibits the classical behaviour of an incompressible shear layer, as would be expected at $M_{c1}^{(i)} = 0.17$. The Rayleigh scatter image Fig. 2, shows a large spanwise vortex. In the braid region one can see a lengthy dark region. Since this region is be surrounded by an area with a higher concentration of high speed fluid, it has to be a streamwise vortex. Therefore, even in this low compressibility case one can not speak of a purely two-dimensional structure. Along the interface to the low speed side one can see small vortices roll up.

Althoug the quality of the Rayleigh scatter images of the case 1 flow (not shown here) is much lower because of the lower Rayleigh scatter cross section of the test gases, the same basic flow features can be made out as in the case 2 flow.

Figure 3 shows three profiles of the relative intensity distribution across the shear layer derived from Fig. 2 at $x = 138\ mm$, $x = 142\ mm$, and $x = 155\ mm$. The first curve for $x = 138\ mm$ represents a cut through the left side of the vortex. Clearly visible is the steep rise from the low speed side to intensities of about 0.65 to 0.70 inside the vortex. Towards the high speed side follows a dip with $e_{px} = 0.47$ and again a steep rise to pure high speed fluid. The intensity in the center of the vortex at $x = 142\ mm$ reaches values up to 0.8, small areas with $e_{px} \approx 0.5$ and $e_{px} \approx 0.55$ are visible on both sides of the core. At $x = 155\ mm$ lies a cut through the streamwise vortex, which is clearly recognizable by the dip at $y \approx -2\ mm$ with maybe two intensity levels of 0.51 and 0.48. The intensity in the outer regions of the vortex at both sides of that dip reaches again values between 0.60 and 0.65.

The probability density function (PDF) of the relative intensity is presented in Fig. 4. Only data from the shear layer with as little as possible from the free stream regions was used for the evaluation. The shape of the distribution is similar to those measured by Konrad (1976) in an incompressible shear layer (see also Dimotakis 1989). The measurements were performed by Konrad using an aspirating probe at various fixed positions across the shear layer and computing the PDF of the time trace. To compare with the spatial PDF of the present experiment one has to take the average of the temporal PDFs obtained at the different crosswise positions.

The two delta functions in Fig. 4 at $e_{px} = 0$ and $e_{px} = 1$ represent pure low speed and pure high speed flow, respectively. The central region with the peak at about $e_{px} = 0.65$ shows the distribution of the mixed fluid (mixed within the limits of the spatial resolution of the measurement). The average high speed mole fraction and the entrainment ratio can in principle be derived from the PDF, but the limited resolution and accuracy of the data did not allow the calculation of reliable values.

About the same convective Mach number as in case 2 was realized with $M_{c1}^{(i)} = 0.19$ in case 3 with a supersonic high speed side. The Schlieren picture, Fig. 5, shows weak waves on the high speed side coming from the tip of the splitter plate and small disturbances in the upper wall which do not disturb the shear layer. But the most outstanding feature is the missing of large-scale structures.

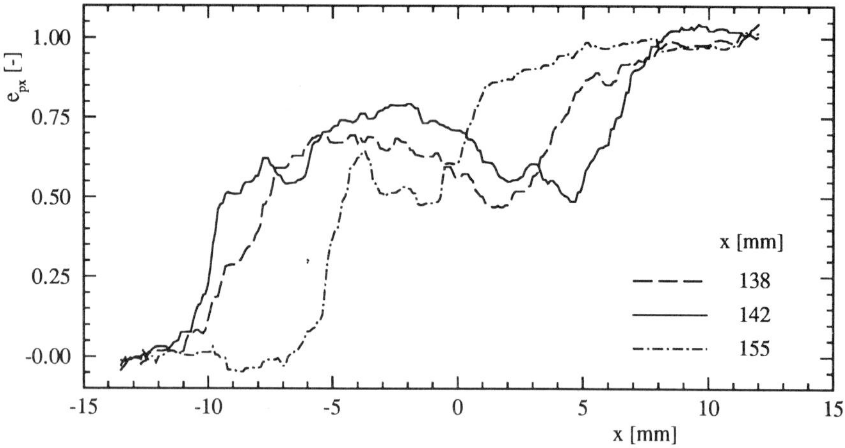

FIG. 3 Intensity profiles for the case 2 flow
(N_2/C_2H_4, $M_1 = 0.65$, $M_2 = 0.30$, $M_{c1}^{(i)} = 0.17$)

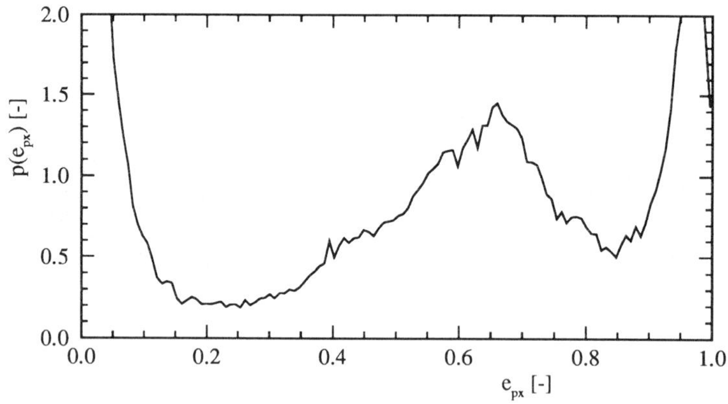

FIG. 4 Probability density function of the intensity for the case 2 flow
(N_2/C_2H_4, $M_1 = 0.65$, $M_2 = 0.30$, $M_{c1}^{(i)} = 0.17$)

The Rayleigh scatter image in Fig. 6, however, shows spanwise vortices, perhaps not as regular as in the previous case. These different results of the two techniques can be explained by the integration across the span performed by the Schlieren method. If the vortices are inclined in the streamwise direction, or if the spanwise coherence is reduced sufficiently, the structures can be averaged out by spanwise integration. A comparison of the diameter of the vortex in Fig. 6 with the thickness of the shear layer in Fig. 5 shows, that, as in the case 2 flow, the vortices still fill out the shear layer completely. Streamwise vortices were not found under these conditions.

FIG. 5 Schlieren picture of the case 3 flow
(N_2/He, $M_1 = 1.48$, $M_2 = 0.20$, $M_{c1}^{(i)} = 0.19$)

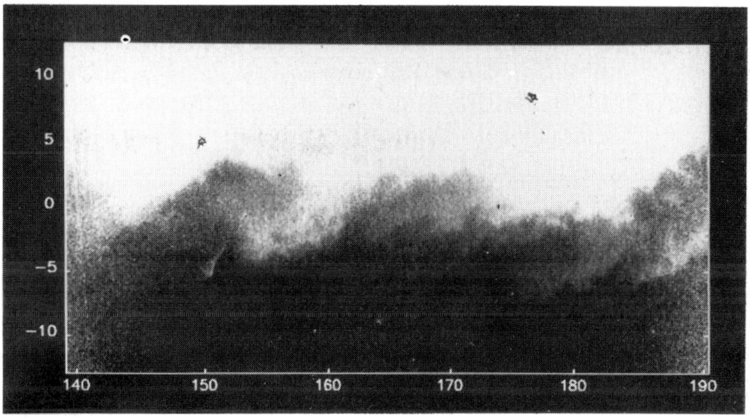

FIG. 6 Rayleigh scatter image of the case 3 flow
(N_2/He, $M_1 = 1.48$, $M_2 = 0.20$, $M_{c1}^{(i)} = 0.19$)

It has to be mentioned, that this flow case was also unusual in other respects. The growth rate was found to be unexpectedly low and Schlieren pictures taken further downstream show a development to larger instabilities again, *cf.* Hall *et al.* (1991a) and Hall (1991). It was presumed, that the combination of a low density ratio with a supersonic high speed stream is responsible for these effects.

The most interesting case in the recent investigations at this facility was of course case 5 from Tab. 1, the Schlieren picture is shown in Fig. 7. Again, no large-scale vortices can be recognized, only in the downstream part of the picture an inclined orientation of the turbulent structure can be made out. The complex

FIG. 7 Schlieren picture of the case 5 flow
$(He/N_2\,,\ M_1 = 1.50\,,\ M_2 = 0.30\,,\ M_{c1}^{(i)} = 0.91)$

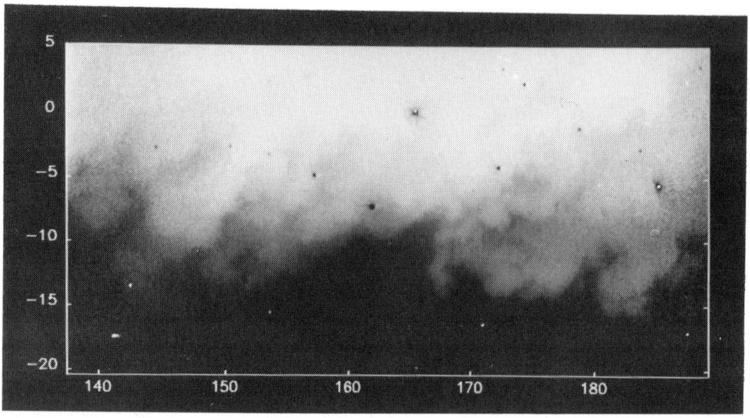

FIG. 8 Rayleigh scatter image of the case 5 flow
$(He/C_2H_4\,,\ M_1 = 1.50\,,\ M_2 = 0.30\,,\ M_{c1}^{(i)} = 0.92)$

wave system in the subsonic low speed flow led to the assumption of structures travelling at supersonic speed with respect to the low speed stream.

Now also in the Rayleigh scatter image, Fig. 8, vortices can not be recognized any more. Instead, there is a very irregular, sharp interface between the shear layer and the free streams, particularly visible on the low speed side. The inclined orientation of the structures in the Schlieren picture is found here again. Pure high and especially pure low speed fluid intrudes from both sides deeply into the shear layer so that in some places, as in the middle of Fig. 8, the concentration gradient is very steep.

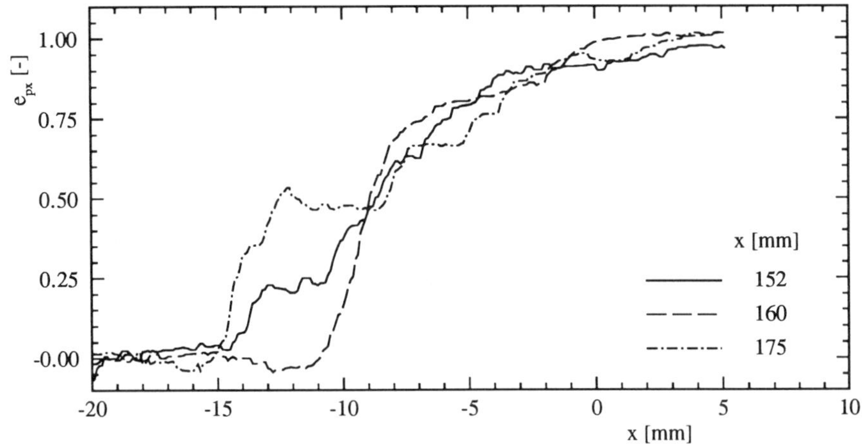

FIG. 9 Intensity profiles for the case 5 flow
$(He/C_2H_4\,,\ M_1 = 1.50\,,\ M_2 = 0.30\,,\ M_{c1}^{(i)} = 0.92)$

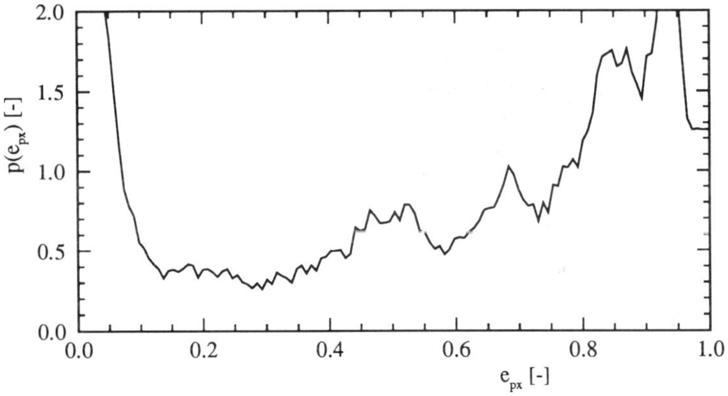

FIG. 10 Probability density function of the intensity for the case 5 flow
$(He/C_2H_4\,,\ M_1 = 1.50\,,\ M_2 = 0.30\,,\ M_{c1}^{(i)} = 0.92)$

The comparison of the two images, Fig. 7 and Fig. 8, shows that the shape of the shear layer must be highly irregular also in the crosswise direction. Only the spanwise integration of such a shape can make the shear layer appear like a nearly perfect wedge in the Schlieren picture.

The intensity profiles of this case look different than those of the incompressible shear layer, Fig. 9. The smaller structures in Fig. 8 correspond to plateaus in the intensity profiles at various levels. In between them are steep gradients, see *e.g.* at $x = 175\,mm$. At $x = 160\ mm$ there is a very sharp step from the low speed side up to an intensity value of 0.7 with a smoother transition to pure high speed fluid.

The PDF for this case is presented in Fig. 10. Compared to the low compressibility case (Fig. 4), the concentration distribution is shifted towards more high speed fluid. This is consistent with the fact that Fig. 8 shows mostly flow regions of relatively high intensity. No single central peak can be found, but there are peaks in the PDF which correspond partially to the plateaus in the intensity profiles in Fig. 9.

The waves system from Fig. 7 or any effects on the structure can not be observed in Fig. 8. Some faint streaks with about the right angle could be detected in other runs when the contrast in the region of the low speed stream was enhanced. Since the Schlieren technique is sensitive to density gradients, a wave showing up clearly in a Schlieren picture can still mean a relatively small difference in the absolut value of the density, which is recorded in a Rayleigh scatter image. Still, it is not quite satisfactory that the wave system could not be detected more safely by the Rayleigh scatter technique.

The shocks proposed to adapt the isentropic theory to compressible shear layers have not been found in Fig. 8. Dimotakis (1991) calculated a pre-shock Mach number of 4.3 meaning a density ratio of 4.7 across a normal shock for the case 5 flow conditions. It is hard to imagine this would not show up in the Rayleigh scatter image. It must therefore be concluded, that shocks of this strength are not present in the flow and that the questions about the nature of the convecting structures, the convection velocity, and the cause for the wave system on the low speed side could not be answered completely by the present experiments.

For completeness, it should be mentioned that the Schlieren picture of a case 4 flow looks similar to Fig. 7 except for the missing waves on the low speed side. The corresponding Rayleigh scatter image shows a smoother transition from low to high speed fluid across the shear layer than Fig. 8, but is also void of large scale vortices.

Even with the limitations with respect to spatial resolution and noise level of the concentration profiles and PDFs, together with the Rayleigh scatter images some important conclusions can be drawn about the development of the shear layer structure and the mixing process with increasing compressibility.

In incompressible shear layers mixing is provided by the large-scale Brown-Roshko structure. They entrain large chunks of fluid from both free stream which are then "digested", mixed down to the molecular level by turbulence and molecular diffusion inside the vortices. This mixing mechanism is very effective. The composition of the mixed fluid is determined by the rate the free stream fluids are entrained with and consequently on finds a pronounced peak in the PDF centered around that value, *cf.* Konrad (1976) and Fig. 4.

As compressibility is increased, the large scale vortices break up into smaller, less coherent ones resulting in reduced entrainment and thereby also in a lower growth rate.

One can speculate, that this broader range of sizes of structures with a higher degree of three-dimensionality involved in the mixing process leads to the broader distribution of high speed mole fractions in Fig. 10. The smaller structures do not span the whole thickness of the shear layer and are therefore not able to entrain only pure fluid from the free streams. They entrain at least partially already premixed fluid, so that mixing becomes a multi-stage process.

The existence of regions of constant concentration with strong gradients in between them in the compressible shear layers seems to indicate, that, within the scale of these smaller structures, the mixing rate is still limited by entrainment, and not by the mixing on the molecular scale.

4. Conclusions

Rayleigh scatter imaging has proven to be a feasible method to obtain cross-sectional images of a flow where mixing of species of different Rayleigh scatter cross section is to be investigated.

The flow visualizations in the convective Mach number range of $0.09 \leq M_{c1}^{(i)} \leq 0.92$ show, in general agreement with the results of Clemens & Mungal (1990), a decrease in two-dimensionality and coherence of the large scale structures with increasing convective Mach number.

Contrary to our case 3 experiment, Clemens & Mungal also detected Brown-Roshko vortices in the Schlieren photographs at these flow conditions with a supersonic high speed stream. The very low density ratio in our experiment, which was found to lead to an irregular growth rate (in combination with the supersonic high speed stream) also, might be responsible for this difference.

The strong shocks, proposed to adapt the isentropic model of convecting structures to the experimental findings for the convection velocities at higher compressibility levels, have not been found in the Rayleigh scatter images.

The development of the results for the concentration distribution with increasing convective Mach numbers can be interpreted as a transition from the one-stage mixing model in the incompressible shear layer, where the large-scale vortex entraines only unmixed free stream fluid and mixes it subsequently, to a multi-stage process where smaller structures entrain unmixed (if they are close to the edges of the shear layer) and already premixed fluid. The fact, that regions of constant concentration with strong gradients in between them exist in the flow, shows that the mixing rate is still limited by entrainment, and not by mixing on the molecular scale.

Acknowledgements

The authors would like to thank Earl Dahl for his help in running the experiments reported here. The work was sponsored by AFOSR, Air Breathing Propulsion, whose support is gratefully acknowledged.

References

DIMOTAKIS, P. E. [1989] "Turbulent Free Shear Layer Mixing", *AIAA 27th Aerospace Sciences Meeting*, 9–12 January 1989 (Reno, Nevada), AIAA-89-0262.

DIMOTAKIS, P. E. [1991] "On the Convective Velocity of Turbulent Structures in Supersonic Shear Layers", *AIAA 22nd Fluid Dynamics, Plasma Dynamics and Laser Conference*, 24–26 June 1991 (Honululu), AIAA-91-1724.

HALL, J. L. [1991] "An Experimental Investigation of Structure, Mixing and Combustion in Compressible Turbulent Shear Layers", *Ph.D. thesis, California Institute of Technology.*

HALL, J. L., DIMOTAKIS, P. E. and ROSEMANN, H. [1991a] "Experiments in non-reacting Compressible Shear Layers", *AIAA 29th Aerospace Sciences Meeting*, 7–10 January 1991 (Reno, Nevada), AIAA-91-0629.

HALL, J. L., DIMOTAKIS, P. E. and ROSEMANN, H. [1991b] "Some Measurements of Molecular Mixing in Compressible Shear Layers", *AIAA 22nd Fluid Dynamics, Plasma Dynamics & Laser Conference*, 24–26 June 1991 (Honolulu, Hawaii), AIAA-91-1719.

KONRAD, J. H. [1976] "An experimental investigation of mixing in two-dimensional turbulent shear flows with applications to diffusion-limited chemical reactions", *Ph.D. thesis, California Institute of Technology.*

PAPAMOSCHOU, D. [1989] "Structure of the Compressible Turbulent Shear Layer", *AIAA 27th Aerospace Sciences Meeting*, 9–12 January 1989 (Reno, Nevada), AIAA-89-0126.

PAPAMOSCHOU, D. and ROSHKO, A. [1988] "The Compressible Turbulent Shear Layer: An Experimental Study", *J. Fluid Mech.* **197**, 453–477.

IDENTIFICATION OF LARGE STRUCTURES IN COMPRESSIBLE MIXING LAYERS USING FILTERED RAYLEIGH SCATTERING

M. Samimy, G.S. Elliott, and S.A. Arnette
Department of Mechanical Engineering
The Ohio State University
Columbus, Ohio 43210, USA

ABSTRACT

High Reynolds number compressible planar free shear layers were investigated using a visualization technique based on filtered Rayleigh/Mie scattering. Two convective Mach numbers, $M_c=0.51$ and 0.86, were studied in the developing and fully developed regions. Scattering was collected from condensed water particles; both product formation and passive scalar methods were utilized. A molecular filter was used to eliminate the background scattering from walls and windows. The large structures in the $M_c=0.51$ case appear as nominally two-dimensional spanwise rollers possessing relatively well defined core and braid regions, and experience amalgamation and tearing processes similar to those observed in incompressible free shear layers. The $M_c=0.51$ plan and spanwise views indicate the presence of organized streamwise vortices in both the developing and fully developed regions. The structures for the $M_c=0.86$ case are much less organized than the $M_c=0.51$ case and highly three dimensional. There is no clear evidence of streamwise vortices for this case. However, the large structures seem to typically be spanwise oblique in the fully developed region.

INTRODUCTION

For the past few years, we have been exploring the effects of compressibility in high Reynolds number turbulent free shear layers. Various experimental techniques such as LDV, two-point space-time correlations, conventional laser sheet lighting, and planar filtered Rayleigh/Mie scattering have been used to study these flows [Samimy and Elliott 1990, Elliott and Samimy 1990, Samimy et al. 1992, Elliott et al. 1992, and Glawe and Samimy 1993]. Mixing layers with convective Mach numbers (M_c) of 0.51, 0.64, and 0.86 have been investigated. While $M_c=0.51$ marks the onset of the compressibility effect on the mixing layer growth rate, the vorticity thickness growth rate of the $M_c=0.86$ case is approximately 50% of that of an incompressible mixing layer with the same velocity and density ratios. The statistical results, which are in accord with the findings of other researchers, indicate that as the compressibility level is increased 1) the mixing layer's growth rate and ability to entrain and disperse solid particles are substantially decreased, 2) spatial correlation levels are drastically reduced, 3) all three measured components of the Reynolds stress ($\sigma_u/\Delta U$, $\sigma_v/\Delta U$, and $\overline{u\,v}/\Delta U^2$) are decreased (the anisotropy ratio, σ_u/σ_v, remains relatively constant but the correlation coefficient, $R_{uv}=-\overline{u\,v}/\sigma_u\sigma_v$, is significantly decreased, especially toward the edges of the mixing layer), 4) the extent of the turbulent mixing region is significantly decreased, and 5) the large scale structures become more 3-D, less organized, and predominately oblique.

J. P. Bonnet and M. N. Glauser (eds.), Eddy Structure Identification in Free Turbulent Shear Flows, 475–486.

In order to further explore the nature of the structures in compressible mixing layers and the effects of compressibility on these structures, we have been developing a qualitative/quantitative planar imaging technique, which is discussed briefly below.

EXPERIMENTAL FACILITY AND INSTRUMENTATION

The high Reynolds number dual-stream wind tunnel at the Aeronautical and Astronautical Research Laboratory at The Ohio State University has a 152.4 x 152.4 mm test section. A 3.18 mm thick steel splitter plate, with a machined angle on the subsonic side of approximately 1° over a 125 mm length and a flat profile on the supersonic side, separates the two flows. The trailing edge of the splitter plate has a thickness of about 0.5 mm. The mass flow of the two streams on either side of the splitter plate can be independently controlled. The streams merge downstream of the splitter plate to form a planar free shear layer. Optical access is provided by windows in the side walls, with a possible side viewing area of 80 mm high and 450 mm long, and top and bottom viewing areas 30 mm wide and 300 mm long.

The convective Mach numbers were changed by using different supersonic nozzles for the top stream. The two convective Mach numbers considered here are 0.51 and 0.86 with flow parameters given in Table 1, where δ and Θ refer to the high speed side 99% incoming boundary layer and momentum thicknesses, respectively.

TABLE I Incoming Flow Conditions

	M_1	M_2	M_c	δ [mm]	Θ [mm]	Re_Θ
Case 1	1.8	0.51	0.51	8.0	0.62	27700
Case 2	3.0	0.45	0.86	9.2	0.37	24700

The light source is a Quanta Ray GCR-4 frequency-doubled (532 nm) Nd:YAG laser which is injection seeded to provide a narrow linewidth and approximately 50 GHz tuning capability. The Nd:YAG laser has a pulse width of 9 ns, which effectively freezes the flow field. A combination of prisms and spherical and cylindrical lenses were used to construct a sheet of light a fraction of a mm thick in the test section. The light sheet was rotated so that streamwise (X-Y), plan (X-Z), and spanwise (Y-Z) planes were visualized.

Two systems are used to capture and store the images. The first system is an ITT 8-bit double intensified CCD camera with video output which is stored on S-VHS. The other system is a Princeton Instruments intensified CCD camera with 14-bit output stored on a 486 PC.

FILTERED RAYLEIGH/MIE SCATTERING

In optical flow diagnostic techniques which utilize the scattering of laser light, there are two main sources of scattering. The first source is the particles or molecules in the flow field. Because of the flow velocity, this scattering possesses a Doppler shift whose magnitude is given by

$$f_D = \frac{1}{\lambda}\left(\underline{k}_s - \underline{k}_0\right) \cdot \underline{V} \tag{1}$$

where $\underline{k}_s$ and $\underline{k}_0$ are the observed and incident unit light wave vectors, respectively, $\underline{V}$ is the velocity vector, and λ is the wavelength of the illuminating light. The second source of collected scattering is unwanted scattering from windows and walls bounding the test section, which experiences no Doppler shift. Figure 1a shows these two components of collected scattering in the frequency domain for a situation in which particles are the scattering medium. The linewidth of the Doppler shifted scattering from the particles in the flow, which are insensitive to molecular motions, is the same as that of the illuminating laser. Similarly, the linewidth of the unshifted scattering is the same as that of the illuminating laser.

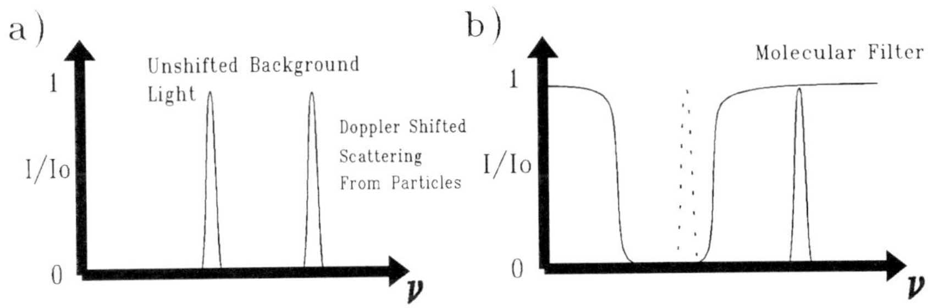

Fig. 1 Filtered Rayleigh/Mie scattering technique using particles as the scattering medium

Miles et al.[1992] have used filtered Rayleigh scattering in flow diagnostics. In this technique, using a properly tuned injection seeded frequency-doubled Nd:YAG laser results in a narrow enough linewidth for the hyperfine absorption bands of iodine to absorb the unshifted background light while allowing the Doppler shifted scattering from particles in the flow to pass. This situation is illustrated in Fig. 1b, where a hypothetical temperature broadened absorption profile with sharp cut-off slopes has been superposed onto Fig. 1a. Inherent to the situation depicted in Fig. 1b is the assumption that the given flow field/optical arrangement provides sufficient Doppler shift for the scattering from the particles in the flow to fall completely outside of the absorption region. Komine et al.[1991] used an iodine molecular filter whose absorption profile exhibited more gradual slopes on both sides of the absorption profile (relative to the sharp cut-off slopes of the temperature broadened profiles considered above) to obtain velocity measurements in a seeded flow. The role of the filter was not to suppress the background noise, but rather to attenuate the collected intensity as a function of frequency, so the amount of Doppler shift present in the signal could be obtained. The seed particles provided scattering in the Mie regime.

The present study focuses on the visualization of large scale structures in compressible mixing layers utilizing filtered Rayleigh/Mie scattering. The filter used in the present visualizations is temperature-broadened optically thick. A representative filter profile is shown in Fig. 1 (b). The filter is a glass cylinder 8.9 cm in diameter and 25.4 cm long, operated at a pressure and temperature of 0.5 torr and 298 K, respectively. These conditions were used to reduce, if not eliminate, the background scattering from the walls and windows. For quantitative planar measurements, we have investigated both temperature and pressure broadened filter profiles in detail and have used the technique to obtain planar velocity measurements in a supersonic jet [Elliott et al. 1993a].

In the present experiments, the scattering particles for the visualizations were produced with two different techniques. Based on the nature of introduction, Clemens and Mungal [1991] referred to them as the product formation and passive scalar methods. In the product formation method (Fig. 2a), moisture is introduced to the warmer subsonic flow which, upon mixing with the colder but dry supersonic flow, condenses and forms small particles. These particles mark the mixed fluid. In the passive scalar method (Fig. 2b) the particles are formed upstream in the supersonic nozzle and mark the colder supersonic flow, which is diluted when mixed with the warmer subsonic flow in the shear layer. Although other investigators have used ethanol vapor [Clemens and Mungal 1992 and Messersmith et al. 1991], in the results presented here the existing water vapor in the air was used to provide condensation and subsequent formation of particles. For the product formation results, water vapor was present in the ambient air introduced in the subsonic side of the flow. The passive scalar method, however, used the small amount of moisture still present in the compressed air after drying. Scalar transport particles formed in the supersonic flow for Case 2 ($M_1=3.0$) but not for Case 1 ($M_1=1.83$).

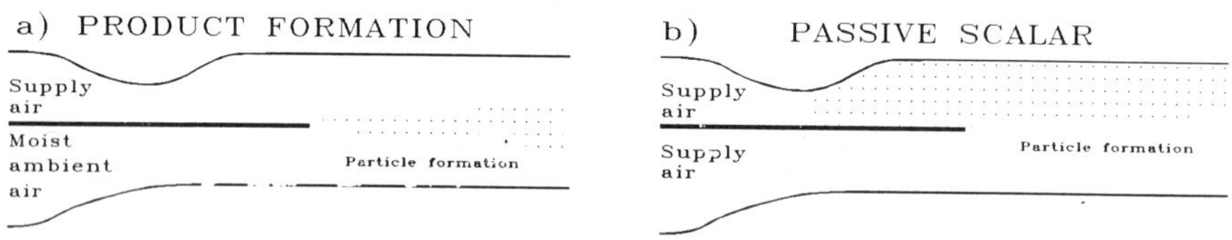

Fig. 2 Schematic of a) product formation, and b) passive scalar methods of particle formation

RESULTS AND DISCUSSION

Earlier LDV measurements have shown that for both convective Mach number cases examined here, the fully developed region (which is characterized not only by linear growth of mixing layer, but also by the collapse of normalized mean and turbulence profiles) begins approximately 150 mm from the splitter plate [Elliott and Samimy 1990]. In this section, results from both the developing and fully developed regions are presented. Although only a few are presented here, hundreds of images have been collected and reviewed for each case. The $M_c=0.86$ images consist of both scalar transport and product formation, all of the $M_c=0.51$ images are from the product formation technique. The flow direction in all streamwise and plan view images is from right to left. As mentioned earlier, the pulse width of the laser is 9 ns, effectively freezing the flow for each image. However, the time lapse between two consecutive images is 0.1 s, making them totally uncorrelated.

Figure 3 shows streamwise (X-Y plane) views of the developing region for M_c=0.51 at four different instances in time. A substantial number of the images for this lower compressibility case (Fig. 3a-3c) show large scale structures, some of them (Fig. 3a and 3b) "roller- type" structures with distinct braid and core regions similar to those observed by Brown and Roshko [1974] in incompressible shear layers. At other instances in time (Fig. 3d), there is no discernable large scale structure.

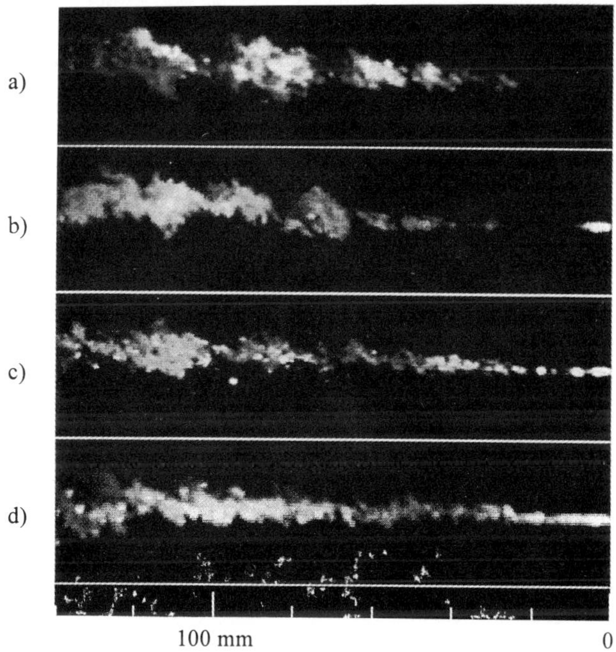

Fig. 3 Streamwise(X-Y) images in the developing region for M_c=0.51 case

The streamwise images for the fully developed region of M_c=0.51 exhibit many of the same characteristics as the upstream images. Figure 4 shows streamwise (X-Y plane) views at four time instances. Again the flow field is dominated by rollers with distinct core and braid regions of the Brown and Roshko type (Fig. 4a and 4b), but their presence is not as distinct in all images (Fig. 4d). Similar structures were also observed by Clemens and Mungal [1992] for M_c=0.28 to 0.50. In some cases, a large region of mixed fluid (Fig. 4c) with little similarity to the well defined rollers of Fig. 4a appears. This region of fluid is in fact two large structures in the process of amalgamation. Collected double-pulse images with time delay from 20 to 100 μs at this convective Mach number indicate that large structures are sometimes stretched until they are torn apart, and sometimes two adjacent structures amalgamate [Elliott et al. 1993b].

Figure 5 shows plan (X-Z plane) view images in the developing and fully developed regions of Case 1. The laser sheet passes through the middle of the mixing layer. The most striking feature of Fig. 5a & 5b is that the product formation seems to begin almost immediately after the end of the splitter plate with streamwise streaks present in the first 30 mm of the flow. These streaks do not seem to exhibit a stationary spanwise location and become less evident further downstream. The streaks resemble streamwise vortices observed in incompressible shear layers by Bernal and Roshko [1986] and Lasheras et al. [1986], although Bernal and Roshko attributed their streaks to upstream conditions that are amplified in the shear layer and remain spanwise stationary. For the images in the fully developed region (Fig. 5c & 5d), cuts made through the middle of the shear layer tend to partially mask the two dimensionality of the structures. However, towards the top edge of the shear layer, the two dimensionality of the large scale structures is much more evident. Note also small scale structures imbedded within the large two-dimensional structures.

a)

b)

c)

d)

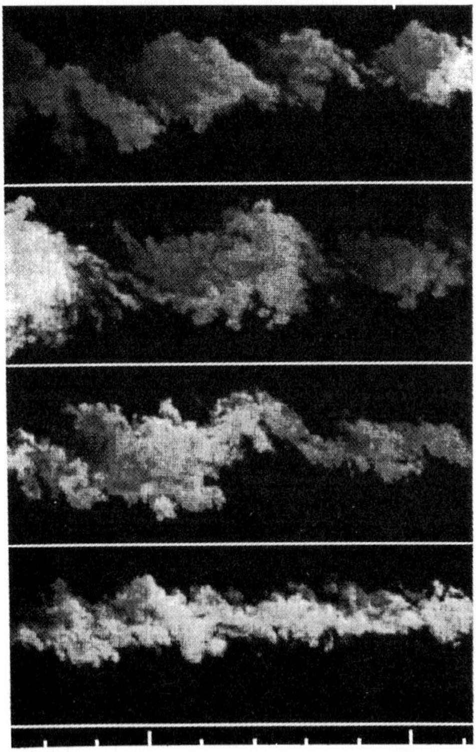

200 mm 150 mm

Fig. 4 Streamwise(X-Y) images in the fully developed region for M_c=0.51 case

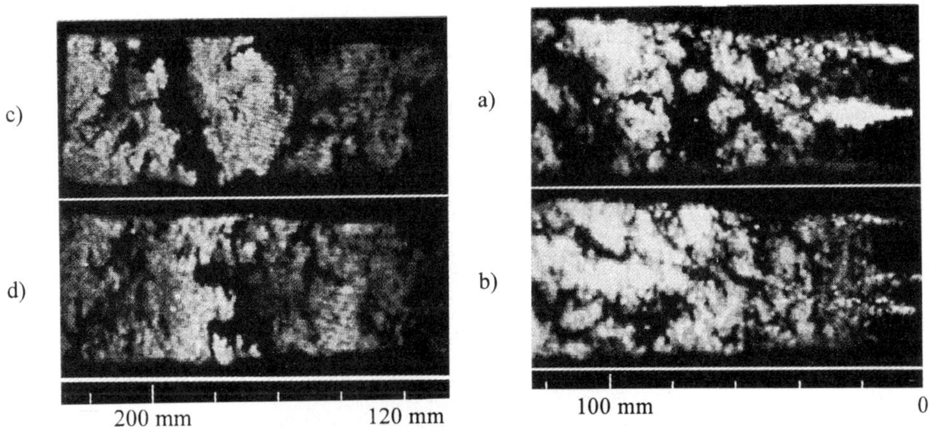

Fig. 5 Plan(X-Z) images in the developing (a & b) and fully developed (c & d) regions for M_c=0.51 case

Figure 6 presents spanwise (Y-Z plane) views in the developing and fully developed regions of the shear layer for different streamwise locations for Case 1. Streamwise streaks in the plan view (Fig. 5) now appear as dots initially (not shown) that grow larger as they convect downstream. Between X= 20 & 50 mm (Fig. 6a & 6b) the regions of stronger intensity have partially dissipated and the shear layer is almost uniform, but wavy across the span. Figure 6c & 6d show the spanwise view cutting through braid regions in the fully developed region of the mixing layer (X=170, 270 mm, respectively), and Fig. 6e shows a spanwise cut through a core region at X=270 mm. When a core is positioned in the laser sheet the mixing layer thickness marked by product formation, is greater than when a braid region is positioned in the laser sheet. Even at X=270 mm, over 500 incoming boundary layer momentum thicknesses downstream of the splitter plate, significant undulations are present in the braid region, possibly indicating the existence of streamwise structures imbedded in the spanwise structures. The persistence of these streamwise undulations is far beyond the downstream locations where Bernal and Roshko observed the existence of streamwise structures in incompressible mixing layers.

For the M_c=0.86 case, the thermodynamic properties of the supersonic stream are such that the small amount of residual moisture in the flow condenses. Thus there is always a passive scalar signal present. Product formation is introduced by supplying moist ambient air to the subsonic stream, with condensation resulting upon mixing of the subsonic air with the cold supersonic air. As a result, when the low speed side is supplied with ambient air rather than dry compressed air, both signals are present. Figure 7 presents streamwise (X-Y plane) views in the developing and fully developed regions of the M_c=0.86 case. The mixing region marked by the product formation is separated from the passive scalar signal in the top freestream in Fig. 7a, 7d, & 7e. This is an

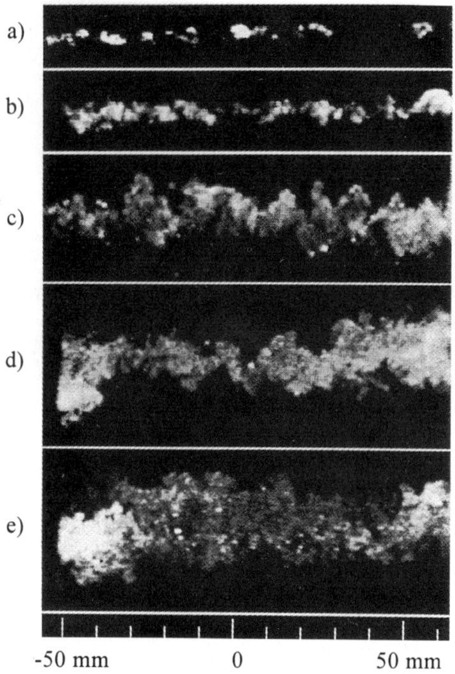

Fig. 6 Spanwise(Y-Z) images in the developing (a & b) and fully developed (c-e) regions for $M_c=0.51$ case

effect of the higher temperature in the boundary layer of the high speed side formed on the splitter plate, which separates the product formation signal in the mixing region from the passive scalar signal in the supersonic stream. Neither the product formation nor the scalar transport in this case gives any indication of large structures with core and braid regions as seen in the $M_c=0.51$ case. However, some large scale structures can be seen in the fully developed region (Fig. 7b & 7d). Unlike the structures in the $M_c=0.51$ case, the three-dimensional and less organized structures in this case can not be accurately followed using double pulse images for even a distance of less than the local vorticity thickness [Elliott et al. 1993b]. One interesting point is that even though there are no well organized structures in this case, the lower edge of the mixing layer is much more corrugated than the upper edge. This means that there is much larger interfacial area (which translates into more entrainment and mixing) on the low speed edge than on the high speed edge. This finding is in accord with our earlier LDV results where significant reduction in the Reynolds stresses and higher order moments of fluctuations was observed on the high speed edge for the higher compressibility case relative to those in the low compressibility case.

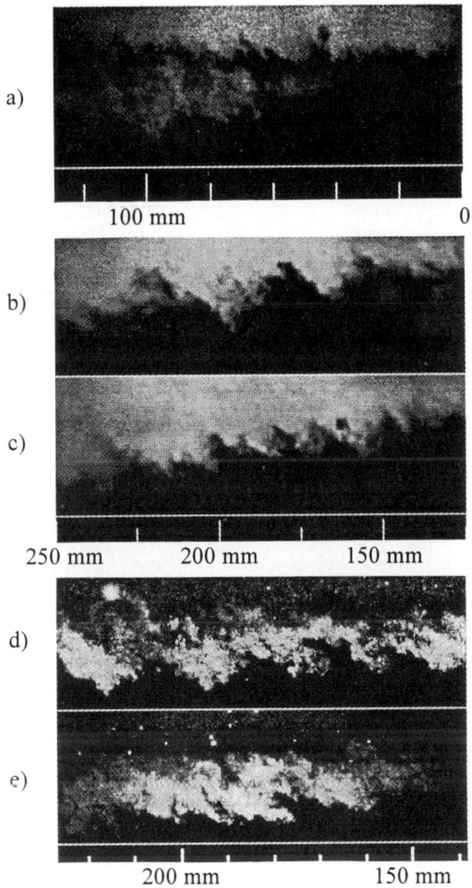

Fig. 7 Streamwise(X-Y) images in the developing (a) and fully developed (b-e) regions for $M_c=0.86$ case

Figure 8 presents three passive scalar plan views (X-Z plane) in the fully developed region of $M_c=0.86$ case. Figure 8a is a cut through the upper edge and Fig. 8b & 8c are cuts through the center of mixíng layer. Although oblique structures are not as well defined as the two dimensional ones found at $M_c=0.51$, two types of spanwise oblique structures seem to typically occur. First there are images, such as Fig. 8a & 8c, which show oblique branching from a streamwise core. Another type of obliquity is seen in Fig. 8b, where an entire large scale structure appears with oblique orientation in the shear layer. Sandham and Reynolds [1991] predicted the presence of oblique structures in their stability analysis and numerical simulations for $M_c=0.80$. Their linear stability analysis showed that as the convective Mach number increases, the spanwise angle of the most amplified waves increases. Also, in low Reynolds number numerical simulations using a random initial disturbance, Sandham and Reynolds [1991] showed that as the compressibility increases from M_c=0.3 to 1.05, the spanwise structures go from spanwise two-dimensional orientation to spanwise oblique orientation. At M_c=0.80, the structures appeared to be in transition with an angle of approximately 45°. This suggests that for Case 2 (M_c=0.86), this state of transition exists, and at higher convective Mach numbers the oblique structures seen experimentally might be more well defined. By separating two fast response pressure transducers in the spanwise direction, Samimy et al. [1992] showed that the peak of the space-time correlation shifted from zero for M_c =0.51 to a delay, depending on the spanwise location of the probes, for the higher convective Mach number case of M_c=0.86. This time delay indicated that the large structures were spanwise oblique for the higher convective Mach number, which is in agreement with the obliquity seen in the plan view images.

a)
b)
c)

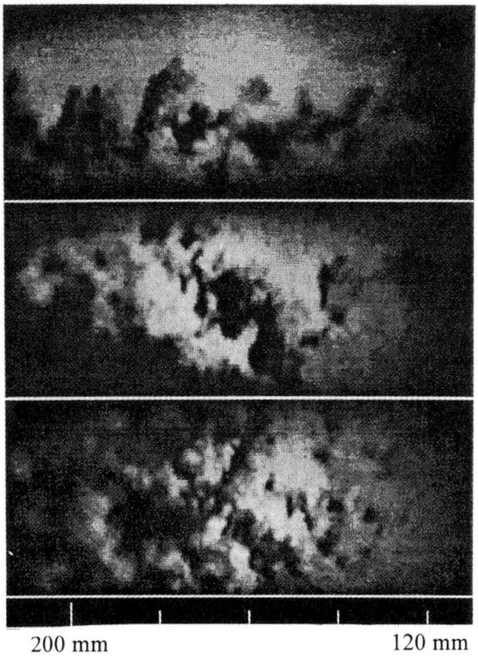

Fig. 8 Plan(X-Z) images in the fully developed region for M_c=0.86 case

Spanwise (Y-Z plane) images of the passive scalar method are shown in Fig. 9 in the developing (Fig. 9a) and the fully developed (Fig. 9b) regions for the M_c=0.86 case. The spanwise views indicate the flow is three dimensional, but the undulations are not as strong as in the M_c=0.51 case. Only a limited number of images show large excursions from the shear layer into the free stream. The streamwise structures seem to grow as they convect downstream, but their significance remains minimal relative to those found in the $M_c=0.51$ case. A recent work by Bonnet et al. [1993] showed the existence of streamwise structures in a flow with a convective Mach number of 0.64. These structures were not as strong as those of the $M_c=0.51$ case, but much more organized than those of the $M_c=0.86$ case. Interestingly, the spanwise scale of these structures did not seem to change significantly with X.

a)

b)

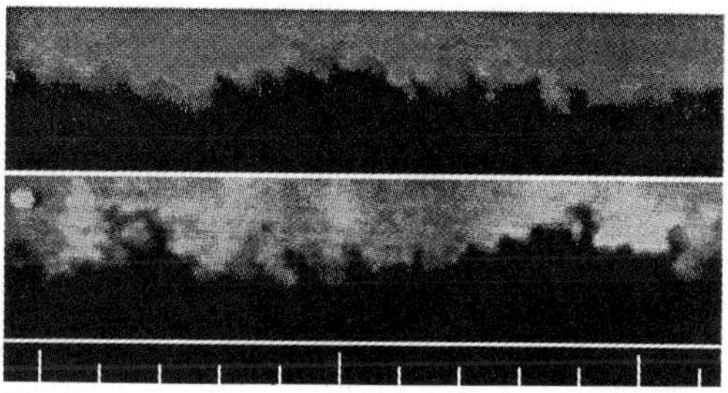

Fig. 9 Spanwise(Y-Z) images in the developing (a) and fully developed (b) regions for $M_c=0.86$ case

CONCLUSIONS

Flow visualizations have been acquired with filtered Rayleigh/Mie scattering in compressible free shear layers. Visualizations for the lower convective Mach number, $M_c=0.51$, showed organized large scale structures which were largely two dimensional, possessing core and braid regions similar to those seen in incompressible shear layers. For the low compressibility case, plan views showed streamwise streaks in the developing region of the mixing layer resembling streamwise vortices observed in incompressible shear layers. The spanwise views for this case suggested that strong streamwise vortices might be present even in the fully developed region. For the higher convective Mach number, $M_c=0.86$, the large scale structures were apparent, but much less organized and random in occurrence relative to the $M_c=0.51$ case. In addition, the structures are strongly three-dimensional, with the visualizations providing ample evidence of spanwise obliquity.

ACKNOWLEDGMENTS

The support of NASA Lewis (NAG 3-764), National Science Foundation (CTS-9006879) and ONR (N00014-90-J-1730) is greatly appreciated.

REFERENCES

Bernal, L.P. and Roshko, A., "Streamwise Vortex Structure in Plane Mixing Layers," Journal of Fluid Mechanics, Vol. 170, pp. 499-525, 1986.

Bonnet, J.P., Debisschop, J.R., and Chambres, O., "Experimental Studies of the Turbulent Structure of Supersonic Mixing Layers," AIAA-93-0217, 1993.

Brown, G.L. and Roshko, A., "On Density Effects and Large Scale Structure in Turbulent Mixing Layers," Journal of Fluid Mechanics, Vol. 64 pp. 775-816, 1974.

Clemens, N.T. and Mungal, M.G., "A Planar Mie Scattering Technique for Visualizing Supersonic Mixing Flows," Experiments in Fluids, Vol. 11, pp. 175-185, 1991.

Clemens, N.T. and Mungal M.G., "Two- and Three- Dimensional Effects in the Supersonic Mixing Layer," AIAA Journal, Vol. 30, pp. 973-981, 1992.

Elliott, G.S. and Samimy, M., "Compressibility Effects in Free Shear Layers," Physics of Fluids A, Vol.2, pp. 1231-1240, 1990.

Elliott, G.S., Samimy, M., and Arnette, S.A., "A Study of Compressible Mixing Layers Using Filtered Rayleigh Scattering Based Visualizations," AIAA Journal, Vol. 30, No. 10, pp. 2567-2569, 1992.

Elliott, G.S., Samimy, M., and Arnette, S.A., "Molecular Filtered-Based Diagnostics in High Speed Flows," AIAA-93-0512, January, 1993a.

Elliott, G.S., Samimy, M., and Arnette, S.A., "The Evolution of Large Scale Structures in Compressible Mixing Layers, 9th Turbulent Shear Flows, Kyoto, Japan, 1993b.

Glawe, D. and Samimy, M., "Experimental Study of Solid Particle Dispersion in Supersonic Mixing Layers," Journal of Propulsion and Power, Vol. 9, No. 1, pp. 83-89, 1993.

Komine, H., Brosnan, S.J., Litton, A.B. and Stappaerts, E.A., "Real-Time Doppler Global Velocimetry," AIAA-91-0337, 1991.

Lasheras, J.C., Cho, J.S., and Maxworthy, T., "On the Origin and Evolution of Streamwise Vortical Structures in a Plane, Free Shear Layer," Journal of Fluid Mechanics, Vol. 172, pp. 231-258, 1986.

Messersmith, N.L., Dutton, J.C., and Krier, H., "Experimental Investigation of Large Scale Structures in Compressible Mixing Layers," AIAA Paper 91-0244, 1991.

Miles, R.B., Forkey, J.N., and Lempert, W.R., "Filtered Rayleigh Scattering Measurements in Supersonic/Hypersonic Facilities," AIAA-92-3894, 1992.

Samimy, M. and Elliott, G.S., "Effects of Compressibility on the Characteristics of Free Shear Layers," AIAA Journal Vol. 28, pp. 957-966, 1990.

Samimy, M., Reeder, M.F., and Elliott, G.S., "Compressibility Effects on Large Scale Structures in Mixing Layers," Physics of Fluids A, Vol. 4, pp. 1251-1258, 1992.

Sandham, N.D. and Reynolds, W.C., "Three-dimensional Simulations of Large Eddies in the Compressible Mixing Layer," Journal of Fluid Mechanics, Vol. 224, pp. 133- 158, 1991.

LIST OF PARTICIPANTS

ADRIAN R.J.
University Illinois 216 Talbot Lab
1045 SO. Wright Street
URBANA Illinois 61801 USA
Tel.: (1) 2173331793
Fax : (1) 2172445707

AHD M.
L.E.D
ENSMA
F-86000 Poitiers
Tel.: (33) 49537090
Fax : (33) 49605000

ANTONIA R.A.
Univ. Newcastle,
Dept. of Mech. Eng.
Rankin Drive
NEWCASTLE NSW 2308, Australia
Tel.: (08) 6149216186
Fax : (08) 6149216946

ASTRUC D.
CERFACS
42, rue Gustave Coriolis
F-31057 TOULOUSE
Tel.: (33) 61193041
Fax : (33) 61193000

ATTIACH J.-C.
TSI France
68 rue de Paris
F-93804 EPINAY Cedex
Tel.: 16 1 48232131
Fax : 16 1 48276492

AUBRY N.
Levich Institute
City College of New York
140th Street & Convent Av.
10031 New York USA
Tel.: (212) 6506844
Fax : (212) 6506835

AZAD R.S.
University of Manitoba,
Dept. of Mechanical & Industrial Eng.
R3T 2N2 Winnipeg, MANITOBA
CANADA
Tel.: (204) 4749608
Fax : (204) 2757507

BACRY E.
Centre de Recherche Paul Pascal
CNRS
Avenue Schweitzer
F-33600 PESSAC
Tel.: (33) 56845665
Fax : (33) 56845600

BARRE S.
IMST IM2
12 avenue du Général Leclerc
F-13003 MARSEILLE
Tel.: (33) 91505439
Fax : (33) 91081637

BELLIN S.
HORWELL
53, Avenue Gabriel Péri
F-92500 RUEIL MALMAISON
Tel.: (33) 1 47086145
Fax : (33) 1 47324448

BELORGEY M.
Lab. de Mécanique des Fluides
Quai Frissard B.P. 265
F-76055 LE HAVRE Cedex
Tel.: (33) 35536951
Fax : (33) 35539703

BENAISSA A.
I.M.S.T.
12, Avenue du Général Leclerc
F-13003 MARSEILLE
Tel.: (33) 91505439
Fax : (33) 91081637

BERKOOZ G.
Cornell Univ.
245 Upson Hall
ITHACA, NY 14853-7501
USA
Tel.: (1) 6072551222

BONNET J.P.
CEAT/LEA
43 rue de l'Aérodrome
F-86036 POITIERS
Tel.: (33) 49537000
Fax : (33) 49537001

BOREE J.
IMFT
2, Avenue du Prof. Camille Soula
F-31400 TOULOUSE
Tel.: (33) 61285891
Fax : (33) 61285899

BRIZZI L.E.
L.E.A
40 Avenue du Recteur Pineau
F-86022 POITIERS
Tel.: (33) 49453656
Fax : (33) 49453663

BROSSARD J.
Lab. de Mécanique des Fluides
Quai Frissard B.P. 265
F-76055 LE HAVRE CEDEX
Tel.: (33) 35536938
Fax : (33) 35539703

BULL M.
Dept. of Mechanical Engineering
University of Adelaide
ADELAIDE, S.A. 5000, Australie
Tel.: (08) 2285460
Fax : (08) 2240464

CENEDESE A.
Rome University
Via Eudossiana 18
00184 ROME, Italy
Tel.: (39) 44585218
Fax : (39) 4817245

CHAPMAN K.
Dept. of Mech. and Aeron. Engg.
Clarkson University
POTSDAM NY 13699, USA
Tel.: (1) 3152686683
Fax : (1) 3152686438

CHASCHECHKIN J.D.
The Inst. for Problems in Mech.
The Russian Academy of Sciences
Prospect Vernadskogo 101
117526 MOSCOW, RUSSIA
Tel.: (7) 0954344341
Fax : (7) 0959382048

CHAUVE M.P.
IMST IM2
12 avenue du Général Leclerc
F-13003 MARSEILLE
Tel.: (33) 91505439
Fax : (33) 91081637

CHOLLET J.-P.
LEGI IMG
BP 53X
F-38041 GRENOBLE CEDEX
Tel.: (33) 76825075
Fax : (33) 76825085

CHRISTESEN
DANTEC
22 Avenue de la Baltique
Villebon
F-91953 LES ULIS
Tel.: (33) 64463029
Fax : (33) 64469314

CLARK D.G.
Queen Mary & Westfield College
Dept. of Aeronautical Engineering
Mile End Road
E1 4NS LONDON, England
Tel.: (44) 0719755500

COLE D.
Dept. of Mech. and Aeron. Engg.
Clarkson University
POTSDAM NY 13699
USA
Tel.: (1) 3152686683
Fax : (1) 3152686438

COUDER Y.
Lab. Phys. Stat.
25 rue Lhomond
F-75231 PARIS Cedex 05
Tel.: (33) 43291225
Fax : (33) 45873489

DEBISSCHOP J.R.
CEAT/LEA
43 rue de l'Aérodrome
F-86036 Poitiers cedex
Tel.: (33) 49537014
Fax : (33) 49537001

DELVILLE J.
CEAT/LEA
43 Rue de l'Aérodrome
F-86036 Poitiers cedex
Tel.: (33) 49537057
Fax : (33) 49537001

DENG G.
D.M.N., L.M.F.
Ecole Centrale de Nantes
1 rue de la Noe
F-44072 NANTES
Tel.: (33)(33) 40371600
Fax : (33) 40474706

DEUTSCH E.
EDF/DER
6 quai Watier
F-78400 CHATOU
Tel.: (33) 30878330
Fax : (33) 30877253

DOHMANN J.
Universitat Dortmund
Fachb CT/EPT
Postfach 500500 W-4600 DORTMUND
Tel.: (49) 2317552314
Fax : (49) 2317553209

DUCOTET C.
L.T.S.I.
23 rue de Dr. Paul Michelon
F-42100 Saint Etienne
Tel.: (33) 77421500
Fax : (33) 77421820

DUSSAUGE J.P.
IMST IM2
12 avenue du Général Leclerc
F-13003 MARSEILLE
Tel.: (33) 91505439
Fax : (33) 91081637

FARGE M.
Lab. de Météorologie Dynamique
Ecole Normale Supérieure
24 rue Lhomond
F-75231 PARIS Cedex 05
Tel.: (33) 43317273
Fax : (33) 43368392

FEDORCHENKO A.
Expert-consultant
INTAR JV-AVANTEC INC
MOSCOW USSR
Tel.: (095) 9211004

FERRE J.A.
Dept. Eng. Quimica i Biochimica
Placa Imperial Tarraco n°1
43005 TARRAGONA Spain
Tel.: (34) 77219101
Fax : (34) 77220006

FIEDLER H.E.
Hermann-Fottinger-Institut.
TU-Berlin Strasse des 17 Juni 135
1000 BERLIN, Germany
Tel.: (49) 30/31423359
Fax : (49) 30/31421101

FUJITA M.
Univ. of Electro-Communications
1-5-1 Chofugaoka Chofu
182 TOKYO, Japan
Fax : (81) 424843327

GATSKI T.B.
Theoretical Flow Physics Branch
NASA Langley RC, MS 156
HAMPTON VA 23681, USA
Tel.: (1) 8048645552
Fax : (1) 8048646134

GAUDET L.
University of Surrey
Dept. of Civil Engg. Surrey
GU2 5XH GUILDFORD, U.K.
Tel.: (44) 483300800
Fax : (44) 483300803

GERMANO M.
Politecnico di Torino
C. so Duca degli Abruzzi 24
10129 TORINO, Italie
Tel.: (39) 1156468
Fax : (39) 15646899

GIESEKE T.
Ohio State University
1079 Rob. Lab.
206 W. 18th Avenue
43210 COLOMBUS, OHIO, USA
Tel.: (1) 4142923163

GLAUSER M.N.
Dept. of Mech. and Aeron. Engg.
Clarkson University
POTSDAM NY 13699, USA
Tel.: (1) 3152686683
Fax : (1) 3152686438

GRAU F.X.
Univers. Tarragona
Pl. Imperial TERRACO 1
43005 SPAIN
Tel.: (34) 77225254
Fax : (34) 77220006

GUEZENNEC Y.
Ohio State University
Mechanical Engineering Dept.
206 W 18th Avenue
43210 COLOMBUS, OHIO, USA
Tel.: (1) 6142921910
Fax : (1) 6142923163

HAAS J.F.
C.E.A.
Centre d'Etudes de Vaujours
F-77181 COURTRY
Fax : 16 1 49367501

HA MINH H.
I.M.F. de Toulouse
Av. du Prof. Camille Soula
F-31400 TOULOUSE
Tel.: 61285834
Fax : 61079613

HASEGAWA T.
Toyohashi University of Technology
Dept. of Energy Eng.
1-1, Tempaku-cho
TOYOHASHI, 441 Japan
Tel.: (81) 532470111
Fax : (81) 532463213

HAYAKAWA M.
Hokkaido University
Dept. of Mechanical Engineering II
Hokkaido University, N13 W8
060 SAPPORO, Japan
Tel.: (81) 0117162111
Fax : (81) 0117460194

HILBERG D.
TU-Berlin HFI Sekr. HF1
Strasse des 17 Juni 135
D-W-1000 BERLIN 12
Germany
Tel.: (49) 3031425614
Fax : (49) 3031421101

Van der HOEVEN J.
Lab. for Aero. Hydrodynamics
Rotterdamseweg 145 2628
AL DELFT Netherlands
Tel.: (31) 15782963
Fax : (31) 15782947

HOLLINGWORTH N.
KLUWER
Spuiboulevard 50, P.O. box 17
3300 AA DORDRECHT, Netherlands
Tel.: (31) 78334306
Fax : (31) 78334254

HOLLOWAY G.
University of New Brunswick
Fredericton, N.B.
E3B 5A3 Canada
Tel.: (506) 4534513
Fax : (506) 4535025

HUNT J.C.R.
Meteorological Office
London Road
BRACKNELL RG12 2SZ
United Kingdom
Fax : (44) 344856909

HUSSAIN F.
University of Houston
Dept. of Mechanical Engineering
4800 Calhoun Road
77204-4792 HOUSTON, Texas USA
Tel.: (1) 7137434545
Fax : (1) 7137434503

INOUE O.
IFS Tohoku University
211 Katahiri, Aoba-Ku
980 SENDAI, Japan
Tel.: (81) 222276200
Fax : (81) 222232748

JEONG J.
Univ. of Houston,
Dept. of Mech. Eng.
University Park, 4800 Calhoun rd
77004-4792 HOUSTON, USA
Tel.: (1) 7434544
Fax : (1) 7434503

KACHANOV Y.S.
Inst. of Theoretical & Applied Mech.
NOVOSIBIRSK, 630090
Russia
Tel.: 7 (3832) 356921
Fax : 7 (3832) 352268

KASTRINAKIS E.
Univ. of Thessaloniki
Dept. of Chemical Eng., Box No 453
54006 THESSALONIKI, Greece
Tel.: (031) 991532
Fax : (031) 206138

KEVLAHAN N.
DAMTP
University of Cambridge
CB3 4EW CAMBRIDGE, U.K.
Tel.: (44) 223337900
Fax : (44) 223337918

KIEHN R.M.
Physics Department
University of Houston
HOUSTON Texas 77004
U.S.A.

LARCHEVEQUE M.
Lab. de Modélisation en Méc.
Université Pierre et Marie Curie
4 Place Jussieu
F-75252 PARIS Cedex 05
Tel.: (33) 1 44275472
Fax : (33) 1 44275259

LAZARIAN A.
D.A.M.P.T.
University of Cambridge
Silver Street
GB-CAMBRIDGE CB3 9EW
Fax : (0223) 337918

LEBRET Y.
LMFN CORIA URA CNRS 230
INSA de Rouen B.P. 8
F-76131 Mont-Saint-Aignan
Tel.: (33) 35146412
Fax : (33) 35146034

LEGENDRE S.
DRET SOR
26 Bd Victor
F-00460 ARMEES
Tel.: (33) 45524526
Fax : (33) 45526520

LESIEUR M.
LEGI
I.M.G., B.P. 53X
F-38041 GRENOBLE Cedex
Tel.: (33) 76825064
Fax : (33) 76825271

LEUCHTER O.
ONERA
8,Rue des Vertugadins
F-92190 MEUDON
Tel.: (33) 1 46235100
Fax : (33) 1 46235151

LO S.H.
QMW Turbulence Unit
Aeronautical Eng. Dept.
Mile End Road
E1 4NS LONDON, U.K.
Tel.: (44) 0719755196
Fax : (44) 0719755500

LOCQUET V.
L.C.P.C.
Domaine du Deffend
F-86800 MIGNALOUX
Tel.: (33) 49468019
Fax : (33) 49468382

LUMLEY J.L.
Cornell Univ.
254 Upson Hall
ITHACA, NY 14853-75101, USA
Fax : (1) 6072551222

LUXTON R.E.
University of Adelaide GPO Box 498
5001 ADELAIDE, S.A.
Australie
Tel.: (61) 2285459
Fax : (61) 82322096

MAEKAWA H.
Univ. of Electro-Communications
1-5-1 Chofugaoka Chofu
182 TOKYO, Japan
Fax : (81) 424843327

MAKITA H.
Toyohashi University of Technology
Dept. of Energy Eng.
1-1, Tempaku-cho, TOYOHASHI, 441
Japan
Tel.: (81) 532470111
Fax : (81) 532463213

MALIK N.
Swiss Federal Institute, Zurich
I.H.W., ETH Hoenggerberg
CH-8093 ZURICH, Switzerland
Tel.: (01) 3773604
Fax : (01) 3712283

MANHART M.
Univ. der Bundeswehr Munchen
LRT/WE7
8014 NEUBIBERG, Germany
Tel.: (089) 60042131
Fax : (089) 60044092

MICHAUX- LEBLOND N.
Lab. de Mécanique des Fluides
Quai Frissard
B.P. 265
F-76055 LE HAVRE CEDEX
Tel.: (33) 35536939
Fax : (33) 35539703

MINIER J.-P.
EDF/DER
6 Quai Watier
F-78400 CHATOU
Tel.: (33) 1 30877140
Fax : (33) 1 30877253

MOCANU F.
The Inst. of Applied Math.
of Romanian Academy
STR. Dreptatii Nr 8, Bl 02,
BUCAREST, Roumanie

MOSER R.
NASA Ames Research
Turbulence Physics Section
CFD Branch, Code RFT
MOFFETT FIELD CA 94035, USA
Fax : (1) 415 6044733

NAKAHMI H.
L.C.P.C.
Domaine du Deffend
F-86800 MIGNALOUX
Tel.: (33) 49468019
Fax : (33) 49468382

NATHAN G.
University of Adelaide
Mech. Engg. Dept., GPO Box 498
5001 ADELAIDE, Australie
Tel.: (61) 82285460
Fax : (61) 82322096

NIESSEN R.
Institut de Saint-Louis ISL
B.P. 34
F-68301 SAINT-LOUIS Cedex
Tel.: (33) 89695000
Fax : (33) 89695002

NOMURA K. K.
University of California,
Dept. of Mech. and Aerospace Engg.
IRVINE, California 92717, USA
Tel.: (1) 7148566131
Fax : (1) 7148568585

NYCHAS S.
Univ. of Thessaloniki
Dept. of Chem. Eng., Box No 453
54006 THESSALONIKI, Greece
Tel.: (031) 991532
Fax : (031) 206138

ORLANDI P.
Dip. di Meccanica e Aeronautica
Via Eudossiana 16
00184 ROMA Italy
Tel.: (39) 644585878
Fax : (39) 6484854

PASCHEREIT C.O.
Hermann Foöttinger Inst. Sekr HF1
Müller-Breslau Str8
D-1000 BERLIN 12
Tel.: (49) 30314214857
Fax : (49) 3031421101

PENOT F.
L.E.T.
40 avenue du Recteur Pineau
F-86022 Poitiers
Tel.: (33) 49453810
Fax : (33) 49453663

PERRY A.
Mech. & Manuf. Eng. Dept.
University of Melbourne
Parkville, Victoria 3052,
Australia

PETROSYAN A.
Space Research Inst.
Profsouznaya 84/32
MOSCOW USSR 117810
Tel.: 0953335356
Fax : 0953107023

PHILLIPS W.R.C
Clarkson University
W228 Old Main Potsdam, N. Y.
13699 POTSDAM, USA
Tel.: (1) 3152683829
Fax : (1) 3152686438

POLLARD A.
Dept. Mechanical Engg.
Queen's University
KINGSTOM K7L 3N6
Canada
Tel.: (1) 6135452569
Fax : (1) 6135456489

REDONDO J.M.
Dept. Fisica Aplicada
Univ. Politecnica de Catalunya
Jordi girona 31
08034 BARCELONA Spain.
Tel.: (34) 37057
Fax : (34) 34016801

ROSEMANN H.
DLR
Bunsenstrasse 10,
D-3400 GÖTTINGEN, Germany
Tel.: (49) 5517092260
Fax : (49) 5517092811

SAMIMY M.
Ohio State University
206 West 18th Avenue
43210 COLOMBUS USA
Tel.: (1) 6142926988
Fax : (1) 6142923163

SCHAEFER H. J.
Institut de Saint-Louis ISL
B.P. 34
F-68301 SAINT-LOUIS Cedex
Tel.: (33) 89695054
Fax : (33) 89695002

SCHON J.P.
C.T.S.I.
23 rue du Dr Paul Michelon
F-42100 Saint-Etienne
Tel.: (33) 77421500
Fax : (33) 77421820

SCHMIED L.
Lab. d'Optoélectronique
IFREMER Université
Zone Portuaire de Bregaillon
F-83507 La Seyne sur Mer
Tel.: (33) 94304900
Fax : (33) 94301372

SITT B.
C.E.A.
Centre d'Etudes de Vaujours
B.P.7
F-77181 COURTRY
Tel.: (33) 1 49368913
Fax : (33) 1 49367501

SORIA J.
Dept. Mech. Engg.
Monash University
CLAYTON VIC 3168
Australia

STAQUET C.
Lab. de Physique
Ecole Normale Supérieure de Lyon
46 allée d'Italie
F-69364 LYON Cedex 07
Tel.: (33) 72728193
Fax : (33) 72728080

STREMLER E.
DeltaLab/Aerometrics
F-38340 VOREPPE
Tel.: (33) 169412214
Fax : (33) 169412217

SULLIVAN P.
Dept. of Mech. engg.
Queens'University
KINGSTON K7L 3N6
Canada
Tel.: (1) 6135452575
Fax : (1) 6135456489

SYSOEVA E.
Inst. for Problems in Mech.
The Russian Academy of Sciences
Prospect Vernadskogo 101
117526 MOSCOW, Russia

TARDU S.
LEGI-IMG
B.P. 53X
F-38041 St MARTIN D'HERES
Tel.: (33) 76825049
Fax : (33) 76825001

TENAUD C.
LIMSI CNRS BP133
F-91403 ORSAY
Tel.: (33) 1 69858076
Fax : (33) 1 69858088

TOQUE N.
CEA
Centre d'Etudes de Vaujours
F-77181 COURTRY
Tel.: (33) 1 49368000
Fax : (33) 1 49367523

TOWNSEND A.A.
Emmanuel College
CB23AP CAMBRIDGE - UK
Tel.: (44) 023333200
Fax : (44) 0233334426

TOY N.
University of Surrey
Dept. of Civil Engg., Surrey
GU2 5XH GUILDFORD UK
Tel.: (44) 483300800
Fax : (44) 483300803

TOYODA K.
Hokkaido Inst. Tech.
Maeda 7-15-4-10 Teine-Ku,
SAPPORO 006, Japan
Tel.: 0116812161
Fax : 0116851030

TSINOBER A.
University of Tel-Aviv
Faculty of Engg.
TEL-AVIV, 69978, Israel
Tel.: 97236408509
Fax : 97236429540

UKEILEY L.
Dept. of Mech. and Aeron. Engg.
Clarkson University
POTSDAM NY 13699, USA
Tel.: (1) 3152686683
Fax : (1) 3152686438

VANDROMME D.
LMFN CORIA
URA CNRS 230, INSA de Rouen B.P. 8
76131 Mont-Saint-Aignan
Tel.: (33) 35146034
Fax : (33) 35146140

VASSILICOS J.C.
University of Cambridge
Dept. of Applied Math & Theor. Physics
Silver Street
CB3 9EW CAMBRIDGE, UK
Tel.: (0223) 337859
Fax : (0223) 337918

VINCENDEAU E.
CEAT/LEA
43 rue de l'Aérodrome
F-86036 POITIERS
Tel.: (33) 49537000
Fax : (33) 49537001

VOKE P.R.
University of Surrey Guilford
GU2 5XH GUILFORD, UK
Tel.: (44) 483300800
Fax : (44) 483306039

WALLACE J.M.
Dept. of Mechanical Engineering
The Univeristy of Maryland
College Park, MD 20742 - USA
Tel.: (301) 4055271
Fax : (301) 3149477

WENGLE H.
Univ. der Bundeswehr Munchen
LRT/WE7
8014 NEUBIBERG, Germany
Tel.: (089) 60042131
Fax : (089) 60044092

YERU R.
CEAT/LEA
43 rue de l'Aérodrome
F-86036 POITIERS
Tel.: (33) 49537000
Fax : (33) 49537001

Authors Index

Mechanics

FLUID MECHANICS AND ITS APPLICATIONS

Series Editor: R. Moreau

Aims and Scope of the Series

The purpose of this series is to focus on subjects in which fluid mechanics plays a fundamental role. As well as the more traditional applications of aeronautics, hydraulics, heat and mass transfer etc., books will be published dealing with topics which are currently in a state of rapid development, such as turbulence, suspensions and multiphase fluids, super and hypersonic flows and numerical modelling techniques. It is a widely held view that it is the interdisciplinary subjects that will receive intense scientific attention, bringing them to the forefront of technological advancement. Fluids have the ability to transport matter and its properties as well as transmit force, therefore fluid mechanics is a subject that is particularly open to cross fertilisation with other sciences and disciplines of engineering. The subject of fluid mechanics will be highly relevant in domains such as chemical, metallurgical, biological and ecological engineering. This series is particularly open to such new multidisciplinary domains.

1. M. Lesieur: *Turbulence in Fluids.* 2nd rev. ed., 1990 ISBN 0-7923-0645-7
2. O. Métais and M. Lesieur (eds.): *Turbulence and Coherent Structures.* 1991 ISBN 0-7923-0646-5
3. R. Moreau: *Magnetohydrodynamics.* 1990 ISBN 0-7923-0937-5
4. E. Coustols (ed.): *Turbulence Control by Passive Means.* 1990 ISBN 0-7923-1020-9
5. A.A. Borissov (ed.): *Dynamic Structure of Detonation in Gaseous and Dispersed Media.* 1991 ISBN 0-7923-1340-2
6. K.-S. Choi (ed.): *Recent Developments in Turbulence Management.* 1991 ISBN 0-7923-1477-8
7. E.P. Evans and B. Coulbeck (eds.): *Pipeline Systems.* 1992 ISBN 0-7923-1668-1
8. B. Nau (ed.): *Fluid Sealing.* 1992 ISBN 0-7923-1669-X
9. T.K.S. Murthy (ed.): *Computational Methods in Hypersonic Aerodynamics.* 1992 ISBN 0-7923-1673-8
10. R. King (ed.): *Fluid Mechanics of Mixing.* Modelling, Operations and Experimental Techniques. 1992 ISBN 0-7923-1720-3
11. Z. Han and X. Yin: *Shock Dynamics.* 1993 ISBN 0-7923-1746-7
12. L. Svarovsky and M.T. Thew (eds.): *Hydroclones.* Analysis and Applications. 1992 ISBN 0-7923-1876-5
13. A. Lichtarowicz (ed.): *Jet Cutting Technology.* 1992 ISBN 0-7923-1979-6
14. F.T.M. Nieuwstadt (ed.): *Flow Visualization and Image Analysis.* 1993 ISBN 0-7923-1994-X
15. A.J. Saul (ed.): *Floods and Flood Management.* 1992 ISBN 0-7923-2078-6
16. D.E. Ashpis, T.B. Gatski and R. Hirsh (eds.): *Instabilities and Turbulence in Engineering Flows.* 1993 ISBN 0-7923-2161-8
17. R.S. Azad: *The Atmospheric Boundary Layer for Engineers.* 1993 ISBN 0-7923-2187-1
18. F.T.M. Nieuwstadt (ed.): *Advances in Turbulence IV.* 1993 ISBN 0-7923-2282-7
19. K.K. Prasad (ed.): *Further Developments in Turbulence Management.* 1993 ISBN 0-7923-2291-6
20. Y.A. Tatarchenko: *Shaped Crystal Growth.* 1993 ISBN 0-7923-2419-6

Kluwer Academic Publishers – Dordrecht / Boston / London

Mechanics

FLUID MECHANICS AND ITS APPLICATIONS

Series Editor: R. Moreau

21. J.P. Bonnet and M.N. Glauser (eds.): *Eddy Structure Identification is Free Turbulent Shear Flows.* 1993 ISBN 0-7923-2449-8

Kluwer Academic Publishers – Dordrecht / Boston / London